INTERPRETING

Engineering Drawings

INTERPRETING

Seventh Edition

CECIL H. JENSEN
JAY D. HELSEL

Engineering Drawings

THOMSON

DELMAR LEARNING

Australia   Canada   Mexico   Singapore   Spain   United Kingdom   United States

THOMSON

DELMAR LEARNING

Interpreting Engineering Drawings, Seventh Edition
Cecil H. Jensen, Jay D. Helsel

Vice President, Technology and Trades ABU:
David Garza

Director of Learning Solutions:
Sandy Clark

Managing Editor:
Larry Main

Senior Acquisitions Editor:
James DeVoe

Senior Production Manager:
John Fisher

Marketing Director:
Deborah S. Yarnell

Marketing Specialist:
Mark Pierro

Director of Production:
Patty Stephan

Content Project Manager:
Jennifer Hanley

Technology Project Manager:
Linda Verde

Editorial Assistant:
Tom Best

Cover Image:
Comstock

Library of Congress Cataloging-in-Publication Data

Jensen, Cecil Howard
   Interpreting engineering drawings / Cecil H. Jensen — 7th ed.
      p. cm.
   Includes index.
   ISBN 1-4180-5573-5
   1. Engineering drawings. I. Title.
   T379.J45 2006
   604.2'5—dc22

                                      2006014643

NOTICE TO THE READER

CONTENTS

Appendix   465

Index   505

PREFACE

We are proud to present the seventh edition of *Interpreting Engineering Drawings.* It is clearly the most comprehensive and up-to-date text of its kind. The authors have worked diligently to provide a text that will best prepare students to enter twenty-first-century technology-intensive industries. It is also useful to those individuals working in technology-based industries who feel the need to enhance their understanding of key aspects of twenty-first-century technology. To that end, the text offers the flexibility needed to provide instruction in as narrow or as broad a customized program of studies as is required or desired. Clearly, it provides the theory and practical application for individuals to develop the intellectual skills needed to communicate technical concepts used throughout the international marketplace.

Flexibility is the key to developing a program of studies designed to meet the needs of every student. *Interpreting Engineering Drawings,* Seventh Edition, is designed to allow instructors and students to pick and choose specific units of instruction based on individual needs and interests. Though students should cover everything offered in the core material in the text (Units 1 through 20), advanced topics are offered throughout the remaining 34 units to provide opportunities for students to become highly skilled in understanding only selected advanced subjects or a broad range of subjects that spread over nearly all aspects of modern industry. Additionally, ancillary materials offered on the Instructor's CD, as well as the Internet Resources listed at the end of each unit, provide for a more in-depth understanding of the material covered. Through the use of these ancillary materials, the depth of understanding achieved is limited only by the student's time constraints and the desire to master the material provided.

It is important to know that the entire text is developed around the latest drafting standards accepted throughout industry. This includes both decimal-inch and metric (millimeter) sizes and related concepts. Both systems are introduced early in the text and are reinforced in both theory and practical application through the broad range of assignments at the end of each unit. These concepts are further reinforced as students are encouraged to use the Appendix at the end of the text. Tables in the Appendix are given in both systems of measure.

Features that made *Interpreting Engineering Drawings* highly successful in previous editions continue to be used in the seventh edition. For example, as always, the text carefully examines the very basic concepts needed to understand technical drawings and meticulously and methodically takes the student through progressively more complex issues. Plenty of carefully developed illustrations, reinforced by the use of a second color, provide a clear understanding of material covered in the written text. Assignments provided at the end of each unit are designed to measure the student's understanding of the material covered as well as reinforce the theoretical concepts. Further, only after the student develops a clear understanding of basic concepts is he or she introduced to more advanced units such as modern engineering tolerancing (geometric tolerancing and true positioning), manufacturing materials and processes, welding drawings, piping, and other similar advanced topics.

Though *Interpreting Engineering Drawings* has always used sketching practices as a means of reinforcing the student's understanding of technical information, the seventh edition greatly expands this important technique. Not only does sketching enhance the student's understanding of technical concepts, it also enhances his or her ability to communicate technical concepts more effectively.

In keeping with the dynamic changes in the field of engineering graphics, various new features have been added to this seventh edition.

New Features of the Seventh Edition

- *Additional assignments.* Significantly more assignments designed to broaden and reinforce basic concepts have been added to Units 1 through 20 (the core of all print-reading concepts for mechanical programs). They also provide increased flexibility in the selection of assignments.

- *Internet resources.* Considerably more information is provided by listing important Internet resources at the end of each unit. This more than doubles the amount of useful information made available to the student.

- *ExamView.* A bank of test questions and answers (test generator) has been provided for each unit in the text for use in evaluating the progress of each student.

- *Instructor's e.resource CD.* The complete Instructor's Manual is now packaged on an e.resource CD for faster and more convenient access to review tests and solutions, assignment solutions, ExamView testbank, PowerPoint® presentations, image library, and other useful teaching materials (ISBN 1418055743).

- *Standards update.* All drawings in the text have been updated to conform to the latest ASME drawing standards.

- *Expanded sketching practice.* The use of sketching techniques as a means of better understanding and communicating technical information has been greatly expanded throughout this text. This becomes increasingly more important as computer-aided drafting continues to rapidly replace board drafting.

- *Co-author.* A co-author has been added to offer his expertise and experience to this edition.

The authors and the publisher hope you find the seventh edition of *Interpreting Engineering Drawings* to be as practical and useful as you have the previous editions. Please feel free to contact us through the publisher if you have questions or comments about the book.

ABOUT THE AUTHORS

Cecil H. Jensen took an early retirement from teaching to devote his full time to technical writing. He held the position of Technical Director at the McLaughlin Collegiate and Vocational Institute, Oshawa, Ontario, Canada, and has more than 27 years of teaching experience in mechanical drafting. He was an active member of the Canadian Standards Association (CSA) Committee on Technical Drawings. Mr. Jensen has represented Canada at international (ISO) conferences on engineering drawing standards, which took place in Oslo, Norway, and Paris, France. He also represented Canada on the ANSI Y14.5M Committee on Dimensioning and Tolerancing. He was the successful author of numerous texts, including *Engineering Drawing and Design, Fundamentals of Engineering Drawing, Geometric Dimensioning and Tolerancing for Engineering and Manufacturing Technology, Computer-Aided Engineering Drawing,* and *Home Planning and Design.* Before he began teaching, Mr. Jensen spent several years in industrial design. He also supervised the evening courses in Oshawa and was responsible for teaching selected courses for General Motors Corporation apprentices.

Jay D. Helsel has worked in education for more than 35 years, having served as a professor of applied engineering and technology courses, chairperson of the Department of Applied Engineering and Technology, and Vice President for Administration and Finance at California University of Pennsylvania. Dr. Helsel has had extensive experience teaching mechanical drafting at both the secondary and postsecondary levels and has worked in industry as well. He holds an undergraduate degree from California University of Pennsylvania, a master's degree from the Pennsylvania State University, and a doctoral degree from the University of Pittsburgh. Dr. Helsel is a full-time writer and has authored publications such as *Engineering Drawing and Design, Fundamentals of Engineering Drawing, Programmed Blueprint Reading, Computer-Aided Engineering Drawing,* and *Mechanical Drawing: Board and CAD Techniques,* as well as various workbooks and other ancillary products associated with these publications.

ACKNOWLEDGMENTS

The author would like to thank and acknowledge the many professionals who reviewed the manuscript, measure helping us publish this textbook. The technical edit was performed by Connie Dotson, Newport Corporation.

A special acknowledgment is due to the following instructors who reviewed the chapters in detail:

Gerald Cavanaugh, Schoolcraft College, Livonia, MI

Michael Hennessey, University of St. Thomas, St. Paul, MN

Dale Howser, Milwaukee Area Technical College, Milwaukee, WI

Rick Jerz, St. Ambrose University, Davenport, IA

Jim Johnson, Lorain County Community College, Elyria, OH

Jeffrey Szymanski, Milwaukee Area Technical College, Milwaukee, WI

BASES FOR INTERPRETING DRAWINGS

Commonly Used Descriptive Terms

When looking at objects, we normally see them as three-dimensional—as having width, depth, and height; or length, width, and height. The choice of terms used depends on the shape and proportions of the object.

Spherical shapes, such as a basketball, would be described as having a certain *diameter* (one term).

Cylindrical shapes, such as a baseball bat, would have *diameter* and *length.* A hockey puck would have *diameter* and *thickness* (two terms).

Objects that are not spherical or cylindrical require *three* terms to describe their overall shape. The terms used for a car would probably be *length, width,* and *height;* for a filing cabinet—*width, height,* and *depth,* even though the longest measurement (length) could be the width, height, or depth; for a sheet of drawing paper—*length, width,* and *thickness.* The terms used are interchangeable according to the *proportions* of the object being described, and the *position* it is in when being viewed. For example, a telephone pole lying on the ground would be described as having *diameter* and *length,* but when placed in a vertical position, its dimensions would be *diameter* and *height.*

In order to avoid confusion, distances from left to right are referred to as *width,* distances from front to back as *depth,* and vertical distances (except when very small in proportion to the others) as *height.*

The Need for Standardization

Engineering drawings are more complicated and require a set of rules, terms, and symbols that everyone can understand and use. A drawing showing a part may be drawn in New York, the part made in California, and then sent to Michigan for assembly. If this is to be successfully accomplished, the drawing must have only one interpretation.

Most countries set up standards committees to accomplish this feat. These committees must decide on factors such as the best methods of representation, dimensioning and tolerancing, and the adopting of drawing symbols. Different styles of lines must be established to represent visible or hidden lines, or to indicate the center of a feature. If only one interpretation of a drawing is to be met, then the rules must be followed and interpreted correctly.

In the United States, drawing standards are established by the *American Society of Mechanical Engineers* (ASME) and in Canada, by the *Canadian Standards Association* (CSA). Members of these committees are part of the worldwide committee on standardization, known as the *International Organization for Standardization* (ISO).

The drawings and information shown throughout this text are based on the ASME-Y14 Series of Drawing Standard Practices. In some areas of drawing practice, such as in simplified drafting, national standards have not yet been established. The authors have, in such cases, adopted the practices used by leading industries in the United States.

Engineering or *technical drawings* furnish a description of the shape and size of an object. Other information necessary for the construction of the object is given in a way that renders it readily recognizable to anyone familiar with engineering drawings.

Pictorial drawings are similar to photographs, because they show objects as they would appear to the eye of the observer, Figure 1–1. Such drawings, however, are not often used for technical designs because interior features and complicated details are easier to understand and dimension on orthographic drawings. The drawings used in industry must clearly show the exact shape of objects. This usually cannot be accomplished in just one pictorial view, because many details of the object may be hidden or not clearly shown when the object is viewed from only one side.

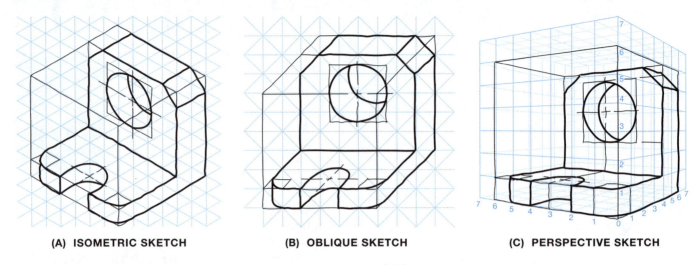

(A) ISOMETRIC SKETCH   **(B) OBLIQUE SKETCH**   **(C) PERSPECTIVE SKETCH**

FIGURE 1–1 ■ *Pictorial sketches.*

For this reason, the drafter must show a number of views of the object as seen from different directions. These views, referred to as front view, top view, right-side view, and so forth, are systematically arranged on the drawing sheet and projected from one another, Figure 1–2. This type of projection is called *orthographic projection* and is explained in Unit 4. The ability to understand and visualize an object from these views is essential in the interpretation of engineering drawings.

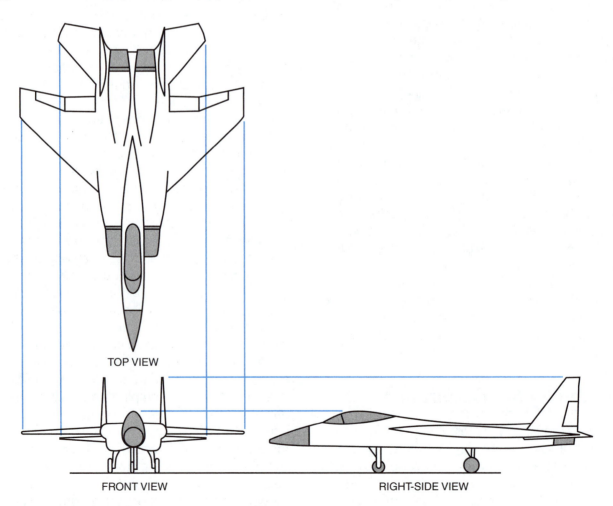

TOP VIEW

FRONT VIEW   RIGHT-SIDE VIEW

FIGURE 1–2 ■ *Systematic arrangement of views.*

ENGINEERING DRAWINGS

Throughout the history of engineering drawings, many drawing conventions, terms, abbreviations, and practices have come into common use. It is essential that all drafters, designers, and engineers use the same practices if drafting and sketching are to serve as a reliable means of communicating technical theory and applications.

An engineering drawing consists of a variety of line styles, symbols, and lettering. When positioned correctly on the drawing paper, they convey precise information to the reader.

LINE STYLES AND LETTERING

Line Styles

A line is the fundamental, and perhaps the most important, single entity on an engineering drawing. Lines are used to illustrate and describe the shape and size of objects that will later become real parts. The various lines used on engineering drawings form the alphabet of the drafting language. Like letters of the alphabet, they are different in their appearance. Two widths of lines are used.

VISIBLE LINES

Thick, continuous lines are used to indicate all visible edges of an object. They are known as *visible* or *object* lines. They should stand out clearly in contrast to other lines, so that the shape of an object is quickly apparent to the eye. Other types of lines are normally drawn as thin lines.

CONSTRUCTION LINES

When first laying out a sketch, light, thin, solid lines are used to develop the shape and location of features. These lines are called *construction* lines, and being very thin and light, are normally left on the sketch. The applications of the other types of lines are explained in detail throughout this text.

Lettering

The most important requirements for lettering used on engineering sketches are legibility and reproducibility. These requirements are best met by the style of lettering known as standard upper-

A B C D E F G H I J K L M N O P
Q R S T U V W X Y Z
1 2 3 4 5 6 7 8 9 0

FIGURE 1–3 ■ *Recommended lettering for use on engineering drawings.*

case Gothic, as shown in Figure 1–3. Suitable lettering size for notes and dimensions is .10 inch (in.) for decimal-inch drawings, and 3 millimeter (mm) for metric drawings. Larger characters are used for drawing titles and numbers, and where it may be necessary to bring some part of the drawing to the attention of the reader.

Symbols and Abbreviations

Symbols and abbreviations are extensively used on engineering drawings. They reduce drawing time and save valuable drawing space. The symbols are truly a universal language, as their meanings are understood in all countries. The first abbreviations and symbols that you will see on the drawings in this text are:

IN., meaning *inch*
mm, meaning *millimeter*
FT, meaning *foot*
Ø, meaning *diameter*
R, meaning *radius*

SKETCHING

Sketching is the simplest form of drawing. It is one of the quickest ways to express ideas. The drafter, technician, or engineer may use sketches to help simplify and explain (communicate) thoughts and concepts to other people. Sketching, therefore, is an important and effective method of communication.

Sketching is also a part of drafting and design because the drafter frequently sketches ideas and designs prior to making the final drawing using computer-aided drafting (CAD). Practice in sketching helps develop a good sense of proportion and accuracy of observation. It is also effective in resolving problems in the early stages of the design process.

CAD is replacing board drafting because of its speed, versatility, and economy. Sketching, like drafting, is also changing, and cost-saving methods are being used to produce a sketch. For example, grid-type sketching paper is used to reduce sketching time and to produce a neater and more accurate sketch. This is because grid-type sketching paper has a built-in ruler for measuring distance and lines act as a straightedge when lines are drawn.

Not all of the drawing needs to be drawn freehand, if faster methods can be used. For example, long lines can be drawn faster and more accurately when a straightedge is used. Large circles and arcs may be drawn or positioned by using a compass. Small circles and arcs may be drawn with the aid of a circle template.

Materials for Sketching

Sketching has two main advantages over formal drawing. First, only a few materials and instruments are required to produce a sketch. Second, you can produce a sketch anywhere. If many sketches are to be made, such as when working from this text, the sketching materials described next should be considered.

SKETCHING PAPER

This type of paper has light, thin lines, and the sketch is made directly on the paper. Various grid sizes (spacings) and formats are available to suit most drawing requirements. The two basic types of sketching paper are two-dimensional and three-dimensional sketching paper.

Two-Dimensional Sketching Paper. This type of sketching paper is primarily used for drawing one-view sketches and orthographic views, which are covered in this unit and in Unit 4. The paper has uniformly spaced horizontal and vertical lines that form squares. These are available in a variety of grid sizes, Figure 1–4. The most commonly used spaces or grids are the decimal-inch, fractional-inch, and centimeter. These spaces are further subdivided into smaller spaces, such as eighths or tenths of one inch or 1 mm. Because the units of measure are not shown on these sheets, the spaces can represent any desired unit of length.

Three-Dimensional Sketching Paper. Three-dimensional sketching paper is designed for sketching pictorial drawings. There are three basic types: isometric, oblique, and perspective, Figure 1–5.

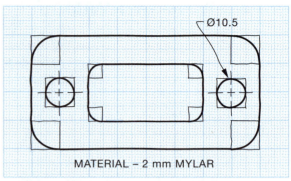

MATERIAL – 2 mm MYLAR

(A) ONE-VIEW SKETCH ON DECIMAL-INCH (.01 INCH DIVISIONS) SKETCHING PAPER

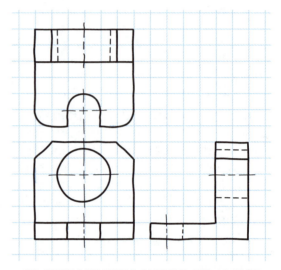

(B) ORTHOGRAPHIC SKETCH ON .25 INCH DIVISION SKETCHING PAPER

FIGURE 1–4 ■ *Two-dimensional sketching paper.*

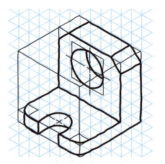 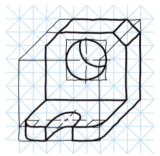

(A) ISOMETRIC SKETCHING PAPER   **(B) OBLIQUE SKETCHING PAPER**

FIGURE 1–5 ■ *Three-dimensional sketching paper.*

Isometric sketching paper has evenly spaced lines running in three directions. Isometric sketching is covered in Unit 7.

Oblique sketching paper is similar to two-dimensional sketching paper except that 45°

lines that pass through the intersecting horizontal and vertical lines are added in one or both directions. Oblique sketching is covered in Unit 7.

One-, two-, and three-point perspective sketching papers are designed with worm's- and bird's-eye views. The spaces on the receding axes are proportionately shortened to create a perspective illusion. The sketches made on this type of paper provide a more realistic view than the sketches made on the isometric and oblique sketching papers.

PENCILS AND ERASERS

Soft lead pencils (grades F, H, or HB), properly sharpened, are the best for sketching. Erasers that are good for soft leads, such as a plastic eraser or a kneaded-rubber eraser, are most commonly used.

TRIGONOMETRY SET

This small, compact math set includes a compass, plastic ruler, and triangles. These drawing tools are very useful for sketching.

TEMPLATES

A circle template will improve the quality of your sketches by making circles and arcs neat and uniform. It will also reduce sketching time. Elliptical circle templates, which are used for pictorial sketching, are normally made available in the drafting classroom for use by students.

Sketching Techniques

With reference to Figure 1–6, the following sketching techniques were used:

- A 1-inch grid subdivided into tenths was selected for the part to be sketched. It required decimal-inch dimensioning. The part was sketched to half scale (half size). This type of sketching paper simplified the measuring of sizes and spacing and ensured accuracy when parallel and vertical lines were drawn. The grid lines also acted as guidelines for the lettering of notes and helped produce neat, legible lettering.

- A straightedge was used for drawing long lines. This method of drawing lines was faster and more accurate than if the lines were drawn freehand.

- A circle template was used for drawing the circular holes. Freehand sketching of round holes is time consuming and is not accurate or pleasing to the eye.

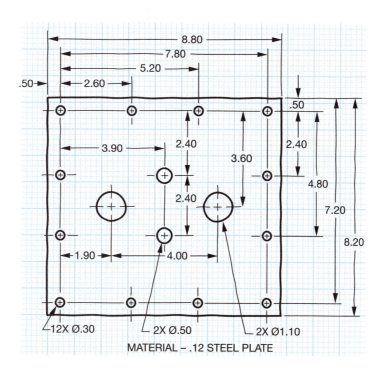

MATERIAL – .12 STEEL PLATE

FIGURE 1–6 ■ *Sketch of a cover plate.*

INFORMATION SHOWN ON ASSIGNMENT DRAWINGS

Assignment problems are either in inch units of measurement or in millimeters (metric). Metric assignments are distinguishable by the letter M shown after the assignment number located at the bottom right-hand corner of the assignment sheet. Circled numbers and letters shown in color are used only to identify lines, distances, and surfaces so that questions may be asked about these features, as shown on Assignment A-14. For purposes of clarity, the actual working drawing is shown in black. The information shown in color is for instructional purposes only and would not appear on working drawings found in industry.

REFERENCES

ASME Y14.2M-1992 (R2003) Line Conventions and Lettering

ASME Y14.38-1999 Abbreviations and Acronyms

INTERNET RESOURCES

Incompetech. For information on grid sheets, see: http://www.Incompetech.com

IDS Development—Nebraska Education. For information on the various line types used on engineering drawings see: http://idsdev.mccneb.edu/djackson/lineintro.htm

Wikipedia, the Free Encyclopedia. For information on engineering drawings and various line types, see: http://en.wikipedia.org/wiki/Engineering_drawing

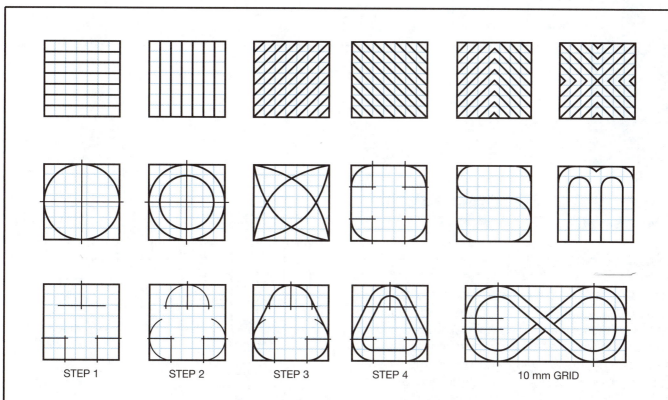

STEP 1 STEP 2 STEP 3 STEP 4 10 mm GRID

ASSIGNMENT: ON A CENTIMETER GRID SHEET (1 mm SQUARES),
SKETCH THE SHAPES SHOWN ABOVE. ALLOW 5 mm
BETWEEN BLOCKS. THICK OBJECT LINES ARE TO
BE USED FOR THE SQUARES AND LINE FEATURES.
THIN LINES ARE TO BE USED FOR THE CONSTRUCTION
LINES IN STEPS 1 THROUGH 4.

**SKETCHING LINES,
CIRCLES, AND ARCS** | **A–1M**

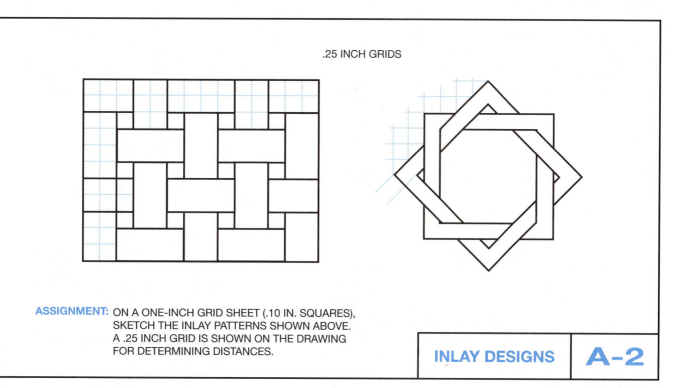

.25 INCH GRIDS

ASSIGNMENT: ON A ONE-INCH GRID SHEET (.10 IN. SQUARES),
SKETCH THE INLAY PATTERNS SHOWN ABOVE.
A .25 INCH GRID IS SHOWN ON THE DRAWING
FOR DETERMINING DISTANCES.

INLAY DESIGNS | **A–2**

LINES USED TO DESCRIBE THE SHAPE OF A PART

Visible Lines

Most objects drawn in engineering offices are complicated and contain many surfaces and edges. In Unit 1, thick, solid lines, called *visible lines,* were introduced. Visible lines are thick, bold lines that clearly stand out on the drawing and define the exterior shape of the object. However, many interior features, such as lines and holes, cannot be seen when viewed from the outside of the part. In order to show these hidden features, a special style of line, called a *hidden line,* is used. Hidden lines, along with visible lines, are used on engineering drawings to fully describe both the exterior and interior of the part.

Hidden Lines

Hidden lines are used to describe features that cannot be seen. They are positioned on the view in the same manner as visible lines. These lines consist of short, evenly spaced thin dashes and spaces. The dashes are three to four times as long as the spaces. These lines should begin and end with a dash in contact with the line in which they start and end, except when such a dash would form a continuation of a visible line. Dashes should join at corners. Figure 2–1 shows examples of hidden line applications.

Break Lines

Break lines serve many purposes. For example, they are used to shorten the view of long uniform sections, which saves valuable drawing space, Figure 2–2(A). They also are used to remove a segment of a part that serves no useful purpose on the drawing, thus saving valuable drawing or sketching time, Figure 2–2(B). The break line shown in this Figure is one of several break

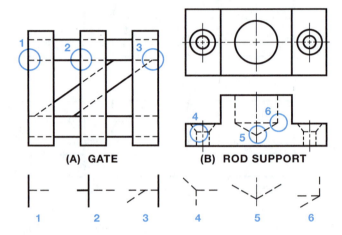

(A) GATE   (B) ROD SUPPORT

FIGURE 2–1 ■ *Hidden lines.*

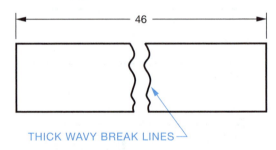

THICK WAVY BREAK LINES

(A) TO SHORTEN LENGTH

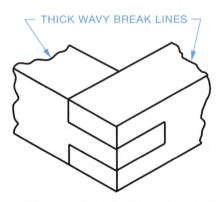

THICK WAVY BREAK LINES

(B) NOT SHOWING UNNECESSARY DETAIL

FIGURE 2–2 ■ *The use of break lines.*

line styles used on engineering drawings. Other break lines are explained in Unit 8. This particular type of break line is shown as a thick, solid line because it forms part of the outline of the object being drawn. It is the third line style used to show the outline of a part.

TITLE BLOCKS AND TITLE STRIPS

Title blocks vary greatly and are usually preprinted. They are located in the lower right-hand corner of the drawing media. The arrangement and size of the title block are optional, but the following four items must be shown:

- Drawing number
- Name of firm or organization
- Title or description
- Scale

Provision may be made within the title block for other pertinent information, such as the date of issue, signatures, approvals, and tolerance notes. A typical title block is shown in Figure 2–3.

In classrooms, where smaller sheet sizes are used, a title strip is commonly used. A typical title strip is shown in Figure 2–4(A). Unless otherwise designated by your instructor, the title strip shown in Figure 2–4(B) will be used on your sketching assignments.

NORDALE MACHINE COMPANY		
PITTSBURGH, PENNSYLVANIA		
PHONE 1-800-564-7832		EMAIL NORDALE@att.net
COVER PLATE		
MATL- SAE 1020 STL	NO. REQD- 4	
SCALE- 1 : 5	DN BY D Scott	C2694
DATE- 04/07/04	CH BY R Jenson	

FIGURE 2–3 ■ *A typical title block.*

DRAWING TO SCALE

When objects are drawn at their actual size, the drawing is called *full scale* or *scale 1:1.* Many objects, however, including buildings, ships, and airplanes, are too large to be drawn full scale. Therefore, they must be drawn to a *reduced scale.* An example would be the drawing of a house to a scale of 1:48 (1/4" = 1 foot) in the inch-foot scale.

Frequently, small objects, such as watch parts, are drawn larger than their actual size in order to clearly define their shapes. This is called drawing to an *enlarged scale.* The minute hand of a wrist watch, for example, could be drawn to scale 5:1 or 10:1.

Many mechanical parts are drawn to half scale, 1:2, and fifth scale, 1:5. Notice that the scale of the drawing is expressed in the form of a ratio. The left side of the ratio represents a unit of measurement

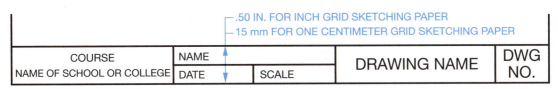

(A) TYPICAL TITLE STRIP LAYOUT

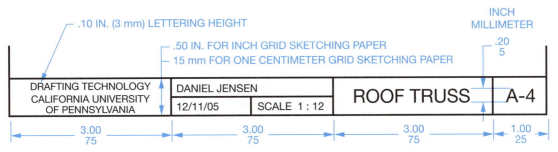

(B) RECOMMENDED LAYOUT AND LETTERING SIZES FOR SKETCHING PAPER

FIGURE 2–4 ■ *Title strips.*

of the size drawn. The right side represents the measurement of the actual object. Thus, 1 unit of measurement on the drawing equals 5 units of measurement on the actual object.

REFERENCE

ASME Y14.2M-1992 (R2003) Line Conventions and Lettering

INTERNET RESOURCES

IDS Development—Nebraska Education. For information on the various line types used on engineering drawings, see: http://idsdev. mccneb.edu/djackson/lineintro.htm

Metrication.com. For information on metrication in drafting and engineering, see: http://www. metrication.com

Integrated Publishing. For information on title blocks, see: http://www.tpub.com/engbas/3-15htm

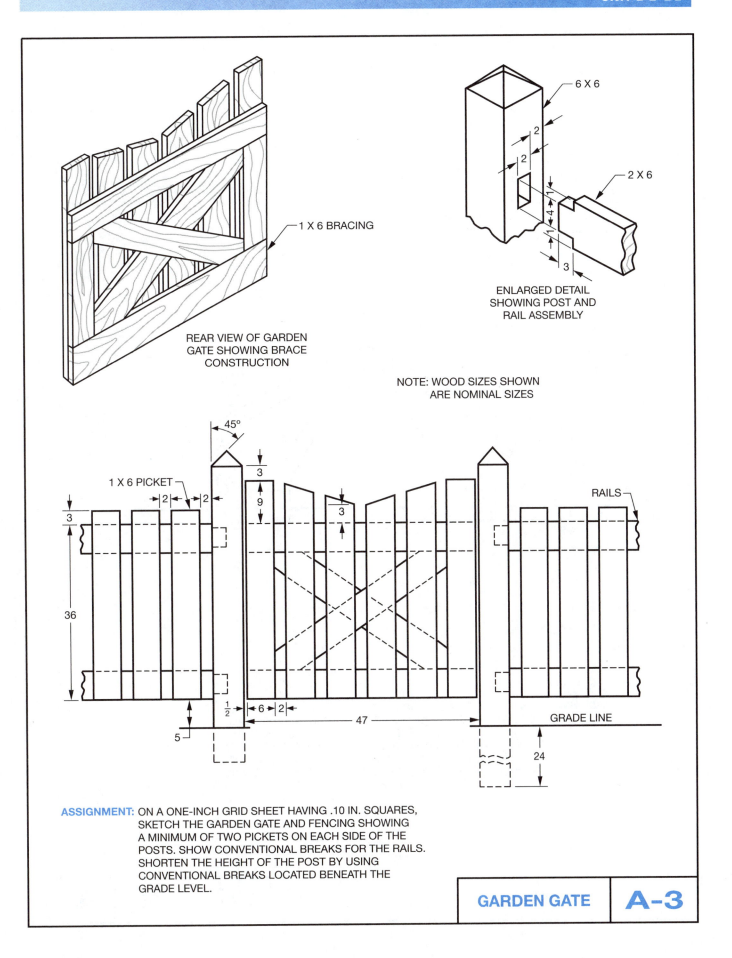

1 X 6 BRACING

REAR VIEW OF GARDEN
GATE SHOWING BRACE
CONSTRUCTION

6 X 6

2 X 6

ENLARGED DETAIL
SHOWING POST AND
RAIL ASSEMBLY

NOTE: WOOD SIZES SHOWN
ARE NOMINAL SIZES

45°

1 X 6 PICKET

RAILS

GRADE LINE

ASSIGNMENT: ON A ONE-INCH GRID SHEET HAVING .10 IN. SQUARES, SKETCH THE GARDEN GATE AND FENCING SHOWING A MINIMUM OF TWO PICKETS ON EACH SIDE OF THE POSTS. SHOW CONVENTIONAL BREAKS FOR THE RAILS. SHORTEN THE HEIGHT OF THE POST BY USING CONVENTIONAL BREAKS LOCATED BENEATH THE GRADE LEVEL.

GARDEN GATE   **A-3**

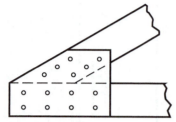

ENLARGED VIEW SHOWING NAILING
ARRANGEMENT OF .50 IN. GUSSETS

NOTE: LUMBER SIZES SHOWN
ARE NOMINAL INCH SIZES

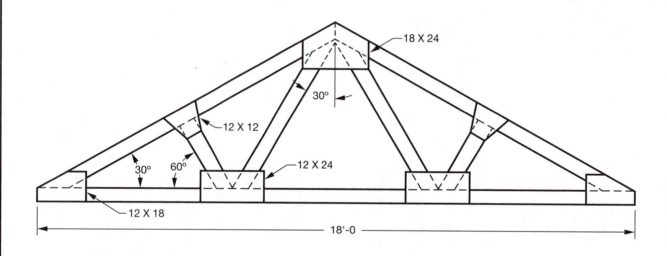

ASSIGNMENT: ON A DECIMAL-INCH GRID SHEET HAVING .10 IN. DIVISIONS, SKETCH THE LEFT HALF OF THE ROOF TRUSS TO THE SCALE OF 1 IN. = 1 FT. EXTEND THE TRUSS A SHORT DISTANCE BEYOND THE CENTER OF THE TRUSS AND USE CONVENTIONAL BREAKS ON THE TRUSS MEMBERS. INCLUDE AN ENLARGED VIEW (2 IN. = 1 FT) OF THE END GUSSET ASSEMBLY SHOWING THE NAILING REQUIREMENTS.

ROOF TRUSS | **A–4**

CIRCULAR FEATURES

Circular features consist of full circles and arcs (parts of circles). Typical drawings with circular features are illustrated in Figure 3–1. Example 1 simply consists of center lines and two circles having the same center point (*concentric circles*). In Example 2, notice that there are four small circles, two half circles, and four quarter circles (rounded corners). The half and quarter circles are called *arcs*. A point where a straight line joins a curved line is called a *point of tangency,* as shown in Example 3.

CENTER LINES

Due to tooling and manufacturing requirements, circular, cylindrical, and symmetrical parts, including holes, must have their centers located. A special line, referred to as a *center line,* is used to locate these features.

A center line is drawn as a thin, broken line of long and short dashes, spaced alternately, as shown in Figure 3–2. The long and short dashes may vary in length, depending on the size of the drawing. Center lines may be used to indicate center points, axes (singular, *axis*) of cylindrical parts, and axes of symmetrically shaped surfaces or parts. Solid center lines are often used on small holes (Figure 3–1, Example 2), but the broken line is preferred. Center lines should project for a short distance beyond the outline of the part or feature to which they refer. They may be lengthened (extended) for use as extension lines for dimensioning purposes. In this case, the extended portion is not broken, as shown in Figure 3–2, Example 1.

SKETCHING CIRCLES AND ARCS

Circular features include both full circles and parts of circles called arcs. These features may be drawn with a circle template, a compass, or freehand. Because speed and accuracy of detail are important in the process of preparing sketches useful in communicating technical ideas, basic drafting instruments such as a circle template or compass are commonly used.

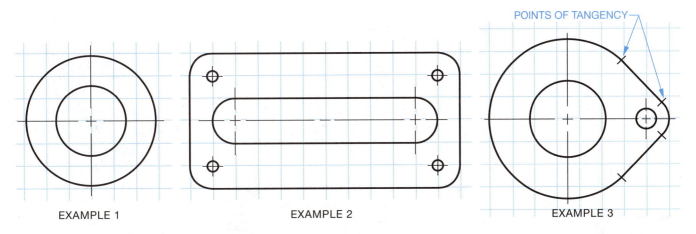

EXAMPLE 1   EXAMPLE 2   POINTS OF TANGENCY   EXAMPLE 3

FIGURE 3–1 ■ *Illustrations of simple objects having circular features.*

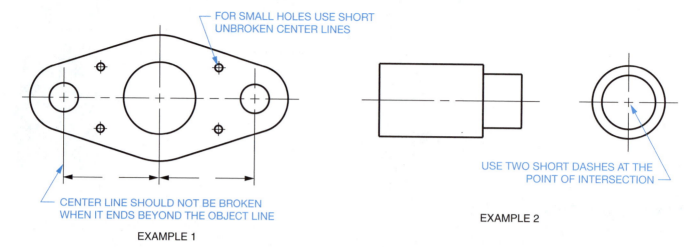

FOR SMALL HOLES USE SHORT UNBROKEN CENTER LINES

CENTER LINE SHOULD NOT BE BROKEN WHEN IT ENDS BEYOND THE OBJECT LINE

EXAMPLE 1

USE TWO SHORT DASHES AT THE POINT OF INTERSECTION

EXAMPLE 2

FIGURE 3–2 ■ *Center line application.*

There are several ways to sketch circles and arcs and no single method is considered best. The method chosen is influenced by what instruments are available, and by personal preference.

Using a Circle Template

Circle templates are often used to draw circles and arcs on sketches to improve quality and speed up the process. Circle templates are made of thin plastic sheets with multiple holes having a range of diameters up to 1.50 inches (approximately 38 mm). The holes are labeled with their respective sizes in decimal-inches or millimeters and each hole has register marks for quick and accurate alignment with vertical and horizontal center lines, as shown in Figure 3–3.

To construct a circle using a circle template, proceed as follows:

- Locate the center of the circle or arc by drawing its center lines, Figure 3–3(A).

- Using the appropriate hole size, place the circle template over the center lines and align the register marks with the center lines, Figure 3–3(B).

- Using a pencil, trace around the hole in the template to draw the circle or arc, Figure 3–3(C).

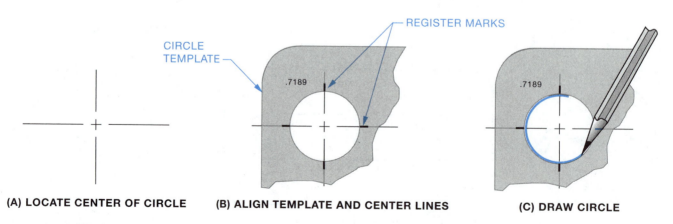

REGISTER MARKS

CIRCLE TEMPLATE

.7189

.7189

(A) LOCATE CENTER OF CIRCLE   **(B) ALIGN TEMPLATE AND CENTER LINES**   **(C) DRAW CIRCLE**

FIGURE 3–3 ■ *Drawing a circle using a circle template.*

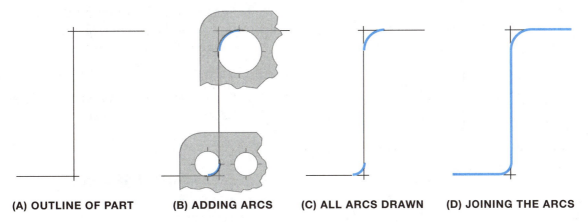

| (A) OUTLINE OF PART | (B) ADDING ARCS | (C) ALL ARCS DRAWN | (D) JOINING THE ARCS |

FIGURE 3–4 ■ *Constructing arcs using a circle template.*

To construct an arc (rounds or fillets) using a circle template requires a different technique, Figure 3–4.

- Sketch construction lines, outlining the part that requires arcs, Figure 3–4(A).

- Using the appropriate hole size, place the circle template over the sketching area and align the circumference of the circle with the two constructions lines and draw the arc, Figure 3–4(B). The arc line should be a thick, solid line.

- Repeat this procedure for the remaining arcs to be drawn, Figure 3–4(C).

- Join these arcs with straight object lines, Figure 3–4(D).

Using a Compass

Though a circle template is recommended for sketching circles up to its largest hole size (generally 1.50 inches in diameter), a *compass* may be used for larger circles and arcs. The compass is a drafting tool that is often used to improve quality and efficiency in the sketching process. When used for sketching, most any size and type of compass is adequate. The compass found in the instrument set described in Unit 1 generally holds a common pencil and is sharpened using a standard classroom pencil sharpener.

The following procedure for laying out and drawing circles with a compass is illustrated in Figure 3–5.

- Locate the center of the circle by drawing center lines, Figure 3–5(A).

- Estimate the length of the radius and mark it off on the center lines, Figure 3–5(B).

- Set the compass point on the intersection of the center lines and adjust the compass lead to the radius mark.

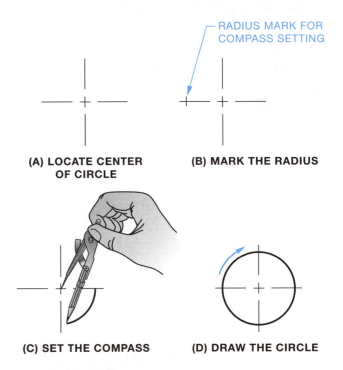

| (A) LOCATE CENTER OF CIRCLE | (B) MARK THE RADIUS |
| (C) SET THE COMPASS | (D) DRAW THE CIRCLE |

RADIUS MARK FOR COMPASS SETTING

FIGURE 3–5 ■ *Drawing a circle using a compass.*

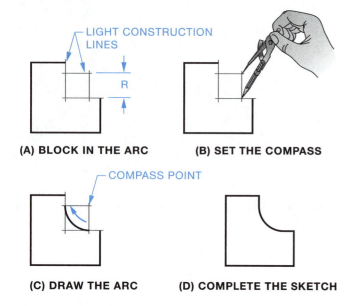

(A) BLOCK IN THE ARC   **(B) SET THE COMPASS**

(C) DRAW THE ARC   **(D) COMPLETE THE SKETCH**

FIGURE 3–6 ■ *Drawing an arc using a compass.*

- Set the compass point on the intersection of the center lines and adjust the compass lead to the radius mark, Figure 3–6(B).

- Draw the arc as shown in Figure 3–6(C).

- Sketch tangent lines and other details as necessary, Figure 3–6(D).

Using Freehand Sketching Techniques

Though the use of a circle template or compass is preferred for drawing circles and arcs, at times the instruments may not be available and circles and arcs will need to be sketched freehand. One common method, shown in Figure 3–7 is as follows:

- Sketch vertical and horizontal construction lines to locate the circle or arc, as shown in Figure 3–7(A). Estimate the length of the radius (plural, *radii*) and mark it off on the center lines.

- With the radius marks as guides, sketch a square using construction lines into which you will then sketch the circle or arc, Figure 3–7(B).

It is generally good practice to first sketch the circle or arc using construction lines and then darken the line when you are satisfied with the size and shape. Making the sketch on a grid sheet adds to the efficiency of using this and other methods for drawing circles and arcs.

- Proceed to draw the circle by starting the arc in the lower right quadrant, Figure 3–5(C).

- Complete the circle by rotating the compass in a clockwise direction. Left-handed individuals may find it easier to reverse the direction of compass rotation, Figure 3–5(D).

The following procedure for laying out and drawing arcs is illustrated in Figure 3–6.

- Use construction lines to locate and block in the extent of the arc. Notice that the radius of the arc is used to locate its center, Figure 3–6(A).

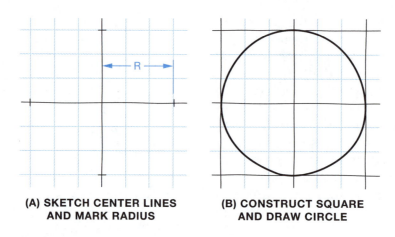

**(A) SKETCH CENTER LINES
AND MARK RADIUS**   **(B) CONSTRUCT SQUARE
AND DRAW CIRCLE**

FIGURE 3–7 ■ *Sketching a circle within a square.*

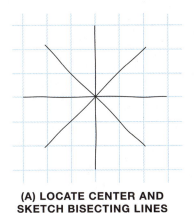

(A) LOCATE CENTER AND SKETCH BISECTING LINES

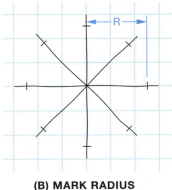

(B) MARK RADIUS

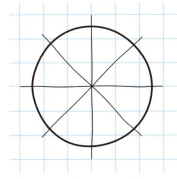

(C) SKETCH CIRCLE THROUGH RADIUS MARKS

FIGURE 3–8 ■ *Alternate method for sketching a circle.*

Another common method, shown in Figure 3–8, is as follows:

- Begin by locating the center and constructing vertical and horizontal center lines, as shown in Figure 3–8(A). Next, sketch bisecting construction lines through the center as shown.

- Estimate the length of the radius and mark off this distance on all the lines, Figure 3–8(B).

- Sketch the circle or arc by connecting the radius marks. You may find it easier to sketch the bottom of the curve. If so, sketch it first and then turn the paper so that another portion of the circle is on the bottom, and then sketch it. Continue in this manner until the circle or arc is complete, as shown in Figure 3–8(C).

Sketching a Complete View Containing Circles and Arcs

The following procedure for laying out and sketching a complete view containing straight lines, circles, and arcs is illustrated in Figure 3–9.

- Lay out center lines and radius marks for all circles and arcs, Figure 3–9(A).

- Use a circle template, compass, or freehand sketching technique to draw circles and arcs, Figure 3–9(B).

- Sketch construction lines to lay out straight tangent lines that do not follow grid lines, Figure 3–9(C).

- Darken all lines as appropriate, Figure 3–9(D).

REFERENCE

ASME Y14.2M-1992 (R2003) Line Conventions and Lettering

INTERNET RESOURCES

Goldengrovehs. For additional information on sketching circular features, see: http://www.goldengrovehs.sa.edu.au/home/tech/yr8drawassignments/Drawingassignments.htm

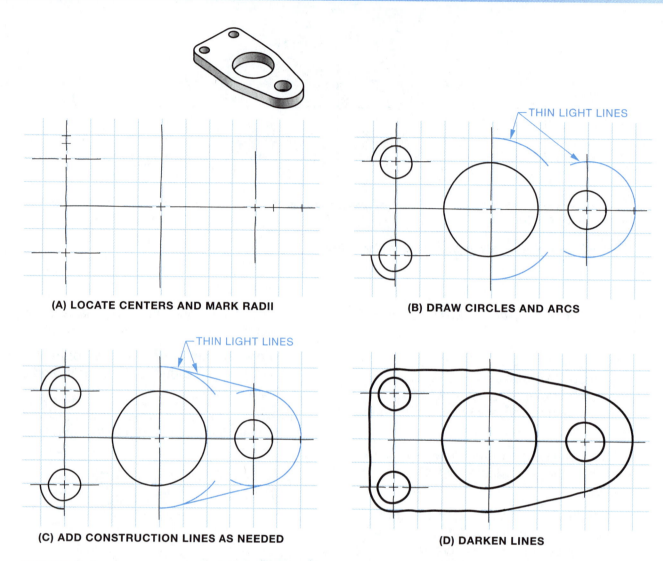

(A) LOCATE CENTERS AND MARK RADII

(B) DRAW CIRCLES AND ARCS

THIN LIGHT LINES

(C) ADD CONSTRUCTION LINES AS NEEDED

THIN LIGHT LINES

(D) DARKEN LINES

FIGURE 3–9 ■ *Sketching a complete view containing straight lines, circles, and arcs.*

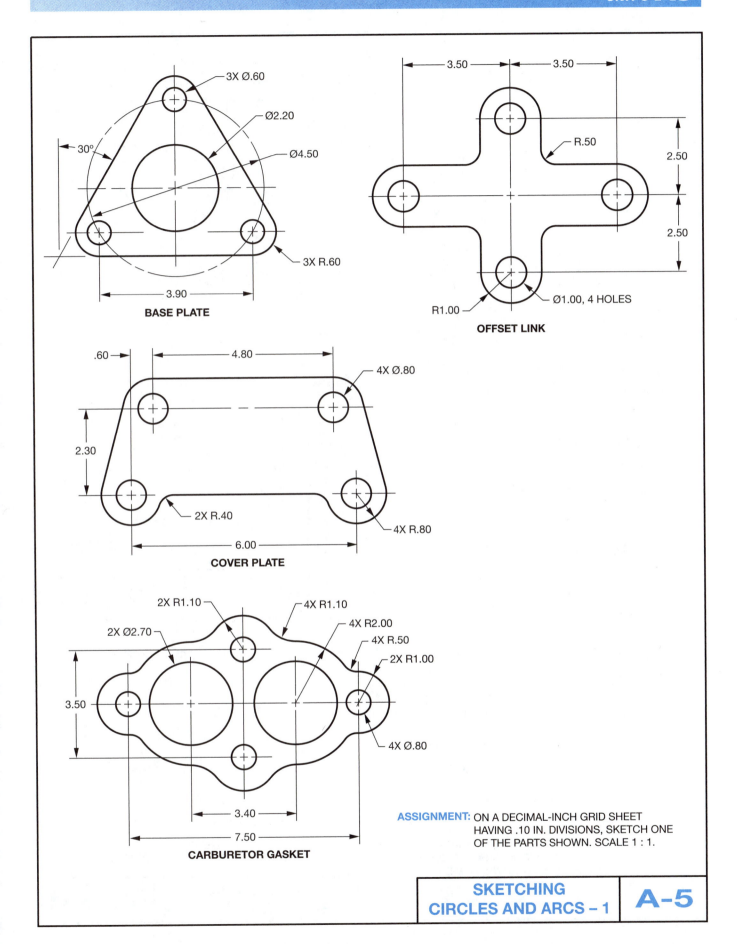

BASE PLATE

3X Ø.60
Ø2.20
Ø4.50
30°
3X R.60
3.90

OFFSET LINK

3.50
3.50
R.50
2.50
2.50
Ø1.00, 4 HOLES
R1.00

COVER PLATE

.60
4.80
4X Ø.80
2.30
2X R.40
4X R.80
6.00

CARBURETOR GASKET

2X R1.10
4X R1.10
2X Ø2.70
4X R2.00
4X R.50
2X R1.00
3.50
4X Ø.80
3.40
7.50

ASSIGNMENT: ON A DECIMAL-INCH GRID SHEET HAVING .10 IN. DIVISIONS, SKETCH ONE OF THE PARTS SHOWN. SCALE 1 : 1.

SKETCHING CIRCLES AND ARCS – 1

A-5

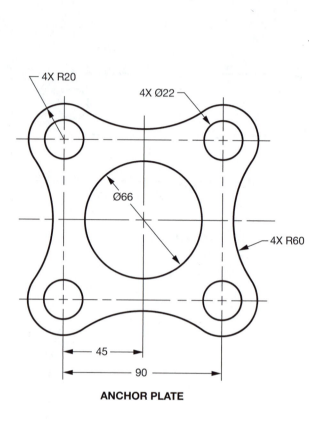

4X R20

4X Ø22

Ø66

4X R60

45

90

ANCHOR PLATE

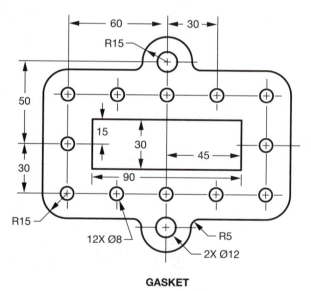

20

45

12 WIDE, 2 SLOTS

R20

℄

R

55

Ø8
6 HOLES EQUALLY
SPACED ON Ø60

Ø40

R40

SHAFT SUPPORT

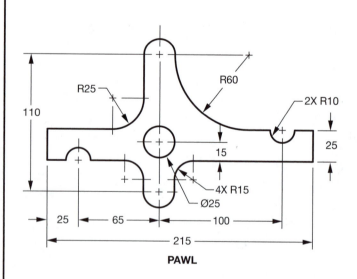

R25

R60

2X R10

110

15

25

4X R15

Ø25

25

65

100

215

PAWL

60

30

R15

50

15

30

45

30

90

R15

R5

12X Ø8

2X Ø12

GASKET

METRIC
DIMENSIONS IN MILLIMETERS

**SKETCHING
CIRCLES AND ARCS – 2**

A-6M

WORKING DRAWINGS

A *working drawing* is a drawing that supplies information and instructions for the manufacture or construction of machines or structures. Generally, working drawings are classified into two groups: *detail drawings* (Figure 4–1), which provide the necessary information for the manufacture of the parts for a specific product or structure, and *assembly drawings* (Figure 4–2), which supply information necessary for their assembly.

Because working drawings may be sent to another plant, another company, or even to another country to manufacture, construct, or assemble the final product, the drawing should conform to the drawing standards of that company. As a result, most companies follow the drawing standards of their country. For example, drawing standards approved and adopted by the American Society of Mechanical Engineers (ASME) have been adopted by most industries throughout the United States. Similarly, the Canadian Standards Association sets the drawing

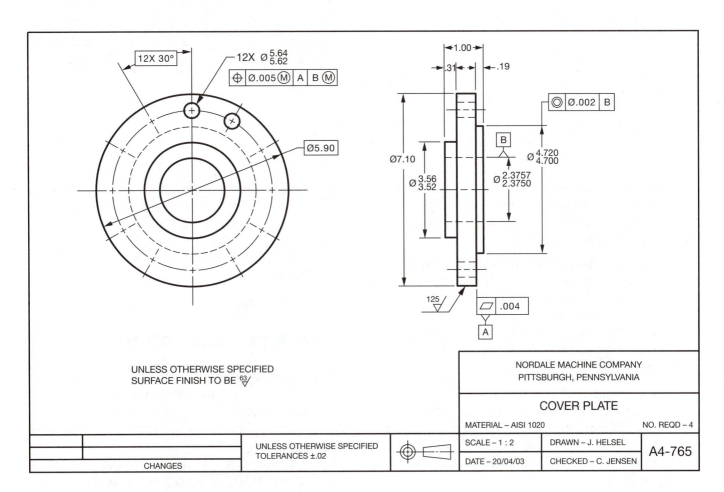

FIGURE 4–1 ■ *A simple detail drawing.*

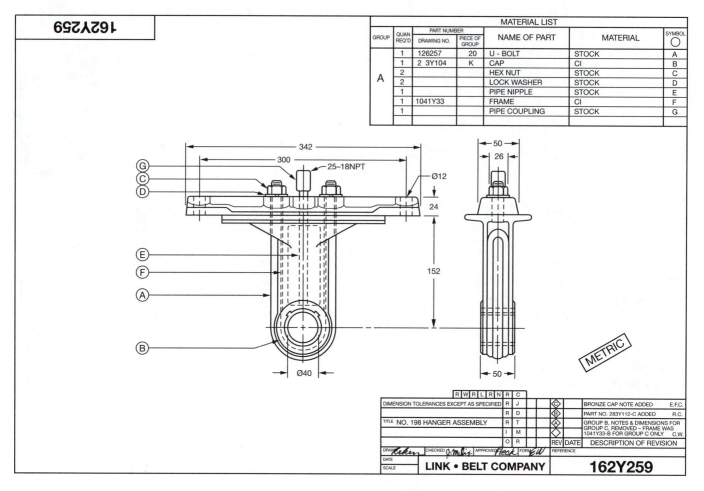

MATERIAL LIST							
GROUP	QUAN REQ'D	PART NUMBER		NAME OF PART	MATERIAL		SYMBOL ◯
		DRAWING NO.	PIECE OF GROUP				
A	1	126257	20	U - BOLT	STOCK		A
	1	2 3Y104	K	CAP	CI		B
	2			HEX NUT	STOCK		C
	2			LOCK WASHER	STOCK		D
	1			PIPE NIPPLE	STOCK		E
	1	1041Y33		FRAME	CI		F
	1			PIPE COUPLING	STOCK		G

FIGURE 4–2 ∎ *An assembly drawing.*

standards for industries throughout Canada. Fortunately, these two sets of standards are similar.

The information found on working drawings may be classified under three headings:

- **Shape** or **shape description.** This refers to the selection and number of views and other details used to show or describe the shape of the part. Though multiview drawings are generally used as working drawings, pictorial drawings are also sometimes used.

- **Dimensions** or **size description.** Approved dimensioning methods for engineering drawings are explained throughout this text starting in Unit 5. The units of measurement recommended are the decimal-inch and the millimeter.

- **Specifications.** Additional information such as general notes, type of material, heat treatment, surface texture finish, and other similar data needed to manufacture the part are included on the drawing or in the title block.

ARRANGEMENT OF VIEWS

Because several views of a part are normally required to describe its shape, the manner in which the views are positioned on the drawing must be clearly understood and have only one interpretation. Two systems of arranging or positioning of views are used on engineering drawings. These systems are known as *first-angle* and *third-angle orthographic projection.*

Third-angle orthographic projection is used by many countries, including the United States and Canada, and thus in this text. Most European and Asiatic countries have adopted first-angle projection. The shapes and sizes of views are identical in both systems; only the positioning of views differs.

ISO PROJECTION SYMBOL

Because these two types of arrangements or views are used on engineering drawings, it is necessary to be able to identify the type of projection used. The International Organization for Standardization (ISO) has recommended that one of the symbols shown in Figure 4–3 be shown on all engineering drawings to indicate the type of projection used. Its preferred location is in the lower right-hand corner of the drawing, adjacent to the title block, Figure 4–4.

THIRD-ANGLE PROJECTION

The third-angle system of projection is used almost exclusively on mechanical engineering drawings in North America because it permits each feature of the object to be drawn in true proportion and without distortion along all dimensions.

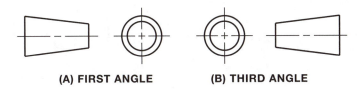

(A) FIRST ANGLE   **(B) THIRD ANGLE**

FIGURE 4–3 ■ *ISO projection symbols.*

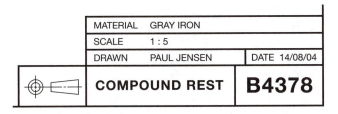

MATERIAL	GRAY IRON	
SCALE	1 : 5	
DRAWN	PAUL JENSEN	DATE 14/08/04
COMPOUND REST		B4378

FIGURE 4–4 ■ *The ISO symbol is located adjacent to the title block on the drawing.*

Three views are usually sufficient to describe the shape of an object. The views most commonly used are the front, top, and right side, Figure 4–5(A). In third-angle projection the object may be assumed to be enclosed in a glass box, Figure 4–5(B). A view of the object drawn on each side of the box represents that which is seen when looking perpendicularly at each face of the box. If the box were unfolded as if hinged around the front face, the desired orthographic projection would result, Figures 4–5(C) and 4–5(D). These views are identified by names as shown. With reference to the front view:

- The top view is placed above.
- The bottom view is placed underneath.
- The left view is placed on the left.
- The right view is placed on the right.
- The rear view is placed at the extreme left or right, whichever is convenient.

Before a drawing is made, the drafter must decide on the number of views necessary to adequately show the part, and which of the six sides of the part would make the best principal (front) view. Factors such as the most informative view and the avoidance of hidden lines help influence the decision making. The front view of the drawing need not be the "front" of the finished part.

The front view on the drawing shows the width and height of the object. As previously mentioned, the term *length* should be avoided when describing the views in orthographic projection, because it normally refers to the longest dimension. When describing the width, height, or depth of a part, any one of these may be the longest measurement.

Once the front view is selected, the next step is to decide what other views are required to adequately show the shape and features of the part. Seldom are more than three views necessary to completely describe the part. Therefore, the simple object shown in Figure 4–6 can be used to illustrate the position of these principal dimensions.

In Figure 4–6 the object is shown in (A) pictorial form, and (B) orthographic projection. The orthographic drawing uses each view to represent the exact shape and size of the object and the relationship of the three views to one another. This principle of projection is used in all mechanical drawings. The isometric drawing shows the relationship of the front, top, and right-side surfaces in

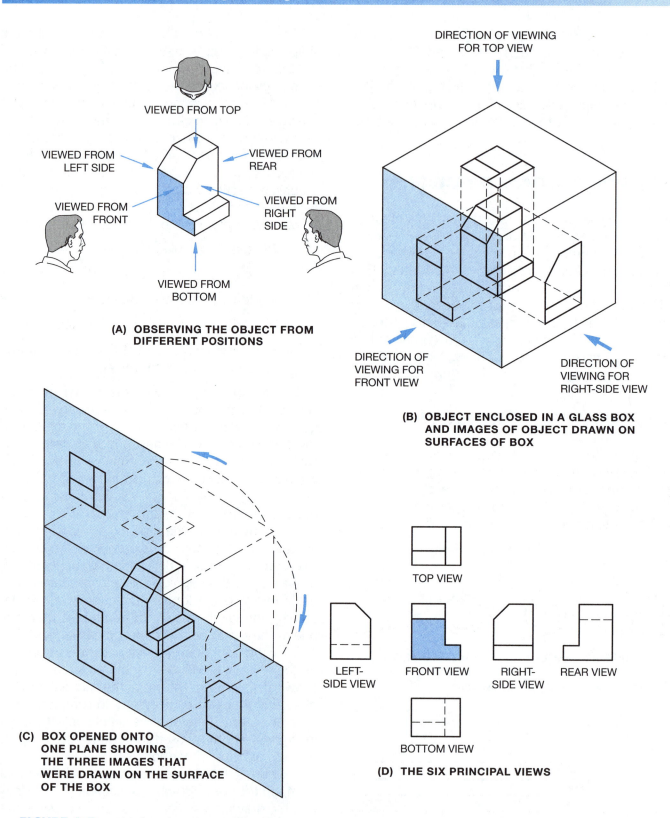

(A) OBSERVING THE OBJECT FROM DIFFERENT POSITIONS

DIRECTION OF VIEWING FOR TOP VIEW

VIEWED FROM TOP

VIEWED FROM LEFT SIDE

VIEWED FROM REAR

VIEWED FROM FRONT

VIEWED FROM RIGHT SIDE

VIEWED FROM BOTTOM

DIRECTION OF VIEWING FOR FRONT VIEW

DIRECTION OF VIEWING FOR RIGHT-SIDE VIEW

(B) OBJECT ENCLOSED IN A GLASS BOX AND IMAGES OF OBJECT DRAWN ON SURFACES OF BOX

(C) BOX OPENED ONTO ONE PLANE SHOWING THE THREE IMAGES THAT WERE DRAWN ON THE SURFACE OF THE BOX

TOP VIEW

LEFT-SIDE VIEW

FRONT VIEW

RIGHT-SIDE VIEW

REAR VIEW

BOTTOM VIEW

(D) THE SIX PRINCIPAL VIEWS

FIGURE 4–5 ■ *Third-angle orthographic projection.*

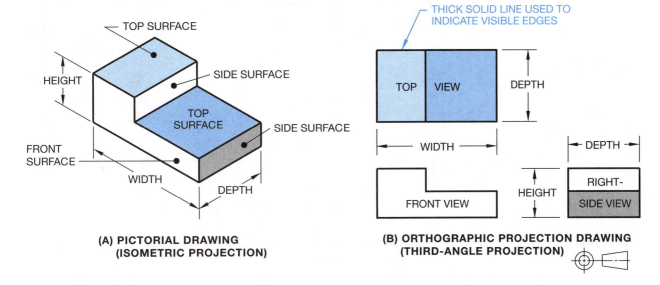

FIGURE 4–6 ■ *A simple object shown in (A) pictorial form, and (B) orthographic projection.*

a single view. Typical parts with flat surfaces are shown in both pictorial and third-angle projection in Figure 4–7.

Objects with Circular Features

Typical parts with circular features are illustrated in Figure 4–8. Note that the circular feature appears circular in one view only and that no line is used to indicate where a curved surface joins a flat surface. Hidden circles, like hidden edges of flat surfaces, are represented on drawings by a hidden line. Objects having circular features are shown in both pictorial and third-angle projection in Figure 4–8. Often only two views are required to show the shape of the part, as shown in Figures 4–8(E) and 4–8(F).

SKETCHING VIEWS IN THIRD-ANGLE PROJECTION

Objects drawn using *orthographic projection* are represented with one or more views, depending on the shape and complexity of the part. For example, a part made from a flat sheet of material having uniform thickness, such as a gasket, can easily be represented with one view and a note describing the material and its thickness. More complex objects may require two, three, or more views for complete shape description.

Figure 4–9 is a pictorial drawing of an object (latch) that would best be described using front,

top, and right-side views. The arrow points to the view that best describes the shape of the object and therefore will become the front view.

Step 1 in Figure 4–10 shows a front view of the latch and a top view that has been projected from it. In some cases, features are projected back-and-forth from view to view as a means of efficiently arriving at finished details in all views. For example, the outline of the front view is first drawn and projected upward to develop the top view. The edges of the small holes are then projected back down to the front view so that the hidden edges can be accurately located. Adequate space must be provided between views for dimensions. You will learn about dimensioning practices in Unit 5.

The next step is to develop the third view. The use of a miter line provides a fast and accurate method of constructing the third view once two views have been developed.

Using a Miter Line to Construct the Right-Side View

- Given the top and front views (Step 1, Figure 4–10), project horizontal lines to the right of the top view (Step 2). The projection lines should be thin, light lines that later can be easily erased or simply ignored.

- Decide how far from the front view the side view is to be drawn (distance D).

- Construct a 45° miter line as shown.

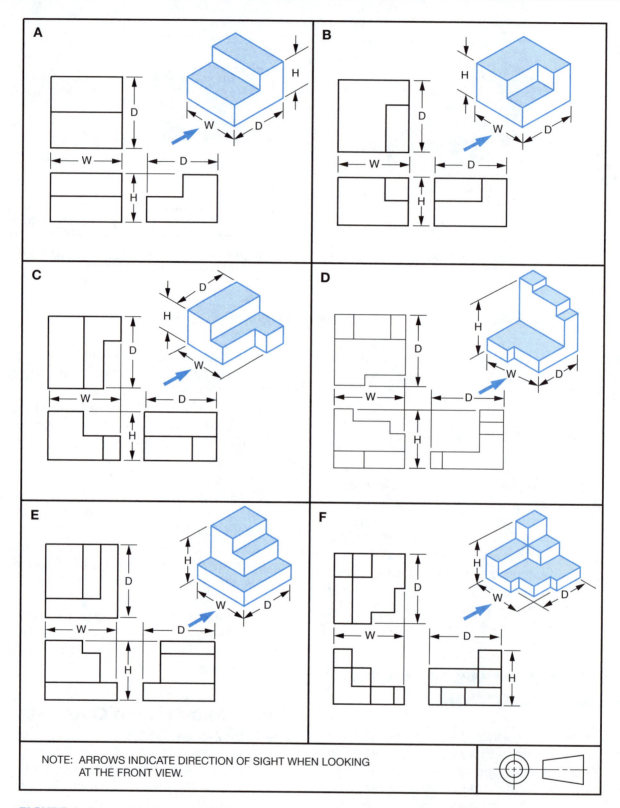

FIGURE 4–7 ■ *Illustrations of simple objects with flat surfaces drawn in third-angle projection.*

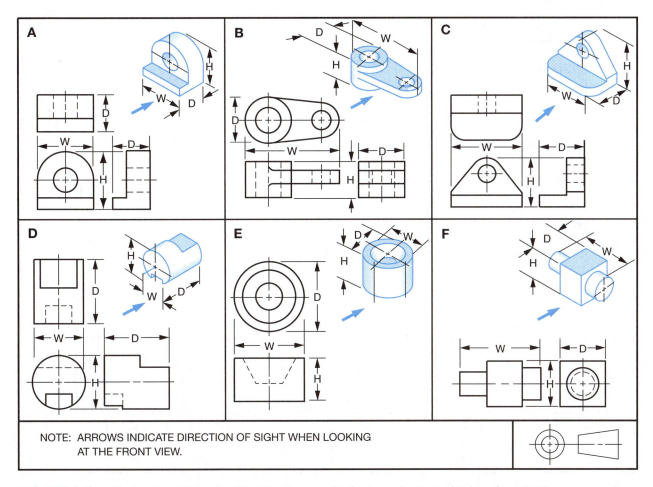

NOTE: ARROWS INDICATE DIRECTION OF SIGHT WHEN LOOKING AT THE FRONT VIEW.

FIGURE 4–8 ■ *Illustrations of simple objects having circular features drawn in third-angle projection.*

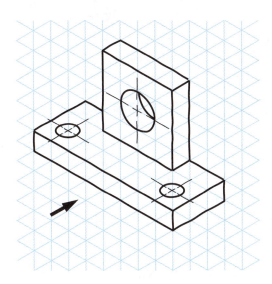

FIGURE 4–9 ■ *Pictorial drawing of a latch.*

- Drop vertical projection lines from points where the horizontal projection lines intersect the miter line.

- Run horizontal projection lines to the right from the front view. The intersections of the vertical and horizontal projection lines are used to locate points on which the right-side view is developed (Step 3).

Using a Miter Line to Construct the Top View

- Given the front and right-side views (Step 1, Figure 4–11), project vertical lines up from the right-side view (Step 2). Decide how far from the front view the top view is to be drawn (distance D).

- Construct a 45° miter line as shown.

- Run horizontal projection lines from points where the vertical projection lines intersect the miter line.

- Run vertical projection lines up from the front view. The intersections of the vertical and horizontal projection lines are used to locate points on which the top view is developed (Step 3).

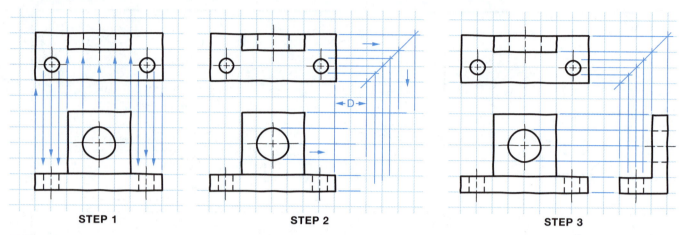

FIGURE 4–10 ■ *Using a miter line to construct the right-side view.*

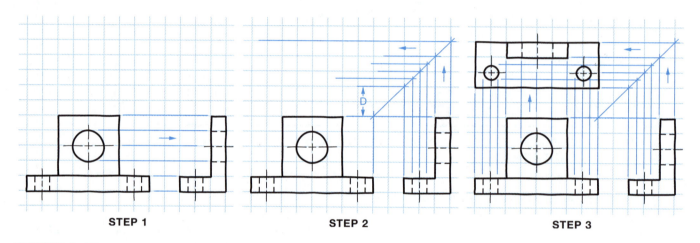

FIGURE 4–11 ■ *Using a miter line to construct the top view.*

After some practice using the miter-line method, you should be able to construct the various views simply by projecting details visually from one view to another by following the grid lines. With practice, your sketching efficiency and accuracy will improve rapidly.

REFERENCES

ASME Y14.3M-1994 (R1999) Multi- and Sectional-View Drawings

INTERNET RESOURCES

Animated Worksheets. For information on third-angle projection and miter lines, see: http://www.animatedworksheets.co.uk

Drafting Zone. For information on dimensioning angles, see: http://www.draftingzone.com

technologystudent.com. For information on third-angle projection and related subjects, see: http://www.technologystudent.com/designpro/ortho2.htm

Wikipedia, the Free Encyclopedia. For information on third-angle projection, see: http://en.wikipedia.org/wiki/Orthographic_projection

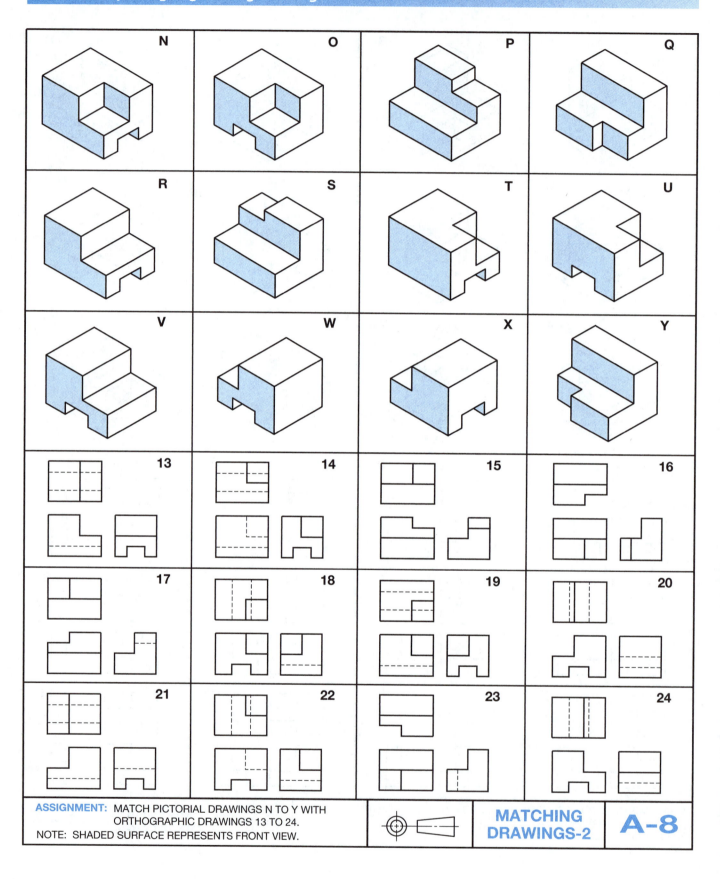

ASSIGNMENT: MATCH PICTORIAL DRAWINGS N TO Y WITH ORTHOGRAPHIC DRAWINGS 13 TO 24.
NOTE: SHADED SURFACE REPRESENTS FRONT VIEW.

MATCHING DRAWINGS-2

A-8

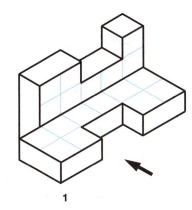

1

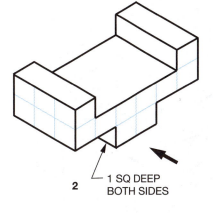

2

1 SQ DEEP
BOTH SIDES

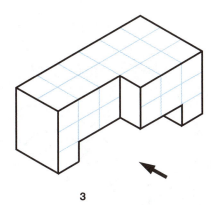

3

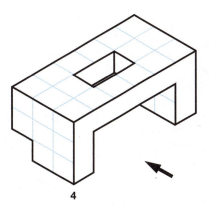

4

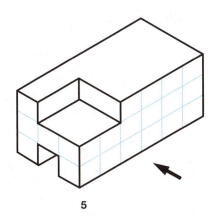

5

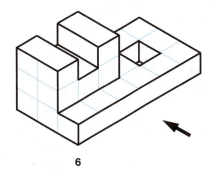

6

ASSIGNMENT: ON A .25 INCH GRID SHEET SKETCH THE TOP, FRONT, AND RIGHT SIDE VIEWS OF THE OBJECTS SHOWN USING THIRD-ANGLE ORTHOGRAPHIC PROJECTION. ONE SQUARE ON THE OBJECT REPRESENTS ONE SQUARE ON THE SKETCHING PAPER. ALLOW ONE SQUARE BETWEEN VIEWS AND A MINIMUM OF TWO SQUARES BETWEEN OBJECTS.

NOTE: ARROW INDICATES DIRECTION OF FRONT VIEW.

**ORTHOGRAPHIC SKETCHING
VISIBLE AND HIDDEN LINES**

A-9

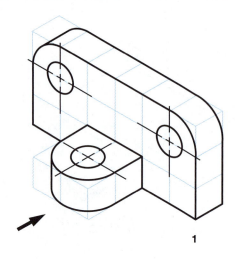

1

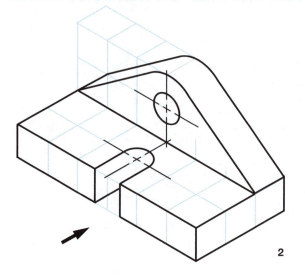

2

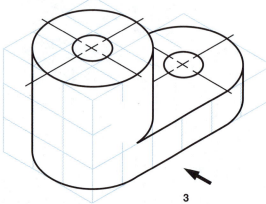

3

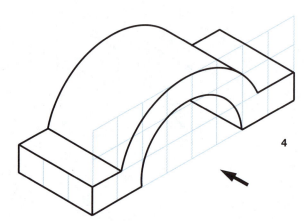

4

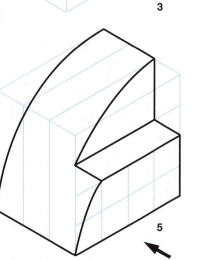

5

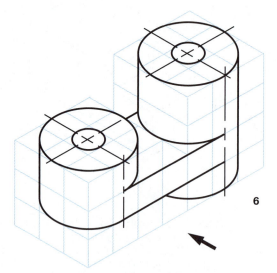

6

ASSIGNMENT:
ON A .25 INCH GRID SHEET SKETCH THE TOP, FRONT, AND RIGHT SIDE VIEWS OF THE OBJECTS SHOWN USING THIRD-ANGLE ORTHOGRAPHIC PROJECTION. ONE SQUARE ON THE OBJECT REPRESENTS ONE SQUARE ON THE SKETCHING PAPER. ALLOW ONE SQUARE BETWEEN VIEWS AND A MINIMUM OF TWO SQUARES BETWEEN OBJECTS.

NOTE: ARROW INDICATES DIRECTION OF FRONT VIEW.

ORTHOGRAPHIC SKETCHING OF PARTS HAVING CIRCULAR FEATURES

A-10

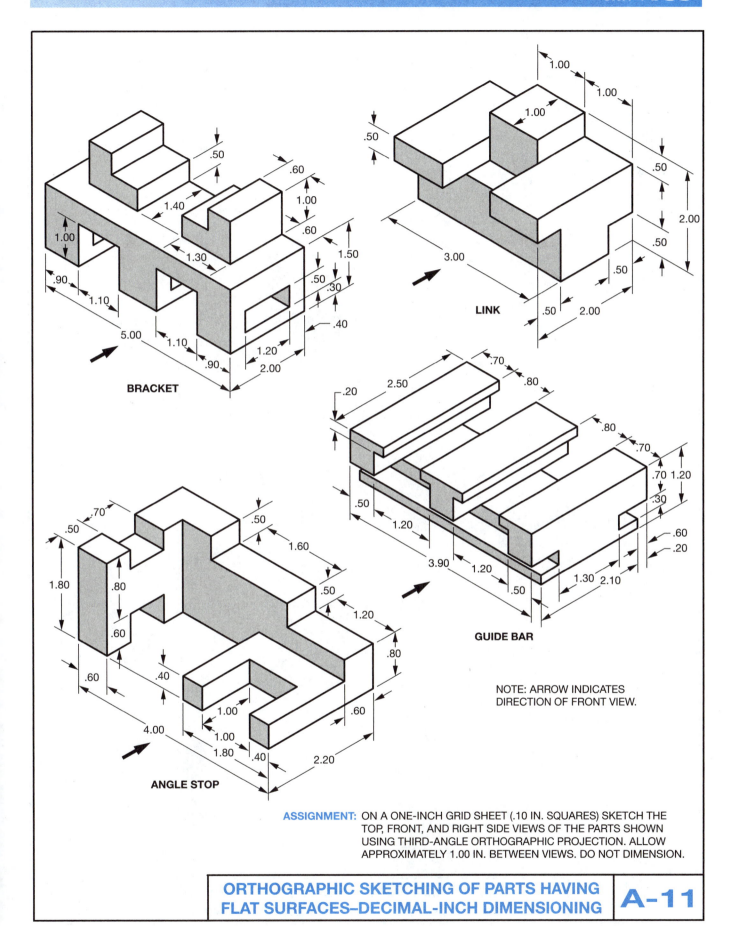

BRACKET

LINK

GUIDE BAR

ANGLE STOP

NOTE: ARROW INDICATES
DIRECTION OF FRONT VIEW.

ASSIGNMENT: ON A ONE-INCH GRID SHEET (.10 IN. SQUARES) SKETCH THE
TOP, FRONT, AND RIGHT SIDE VIEWS OF THE PARTS SHOWN
USING THIRD-ANGLE ORTHOGRAPHIC PROJECTION. ALLOW
APPROXIMATELY 1.00 IN. BETWEEN VIEWS. DO NOT DIMENSION.

**ORTHOGRAPHIC SKETCHING OF PARTS HAVING
FLAT SURFACES–DECIMAL-INCH DIMENSIONING**

A-11

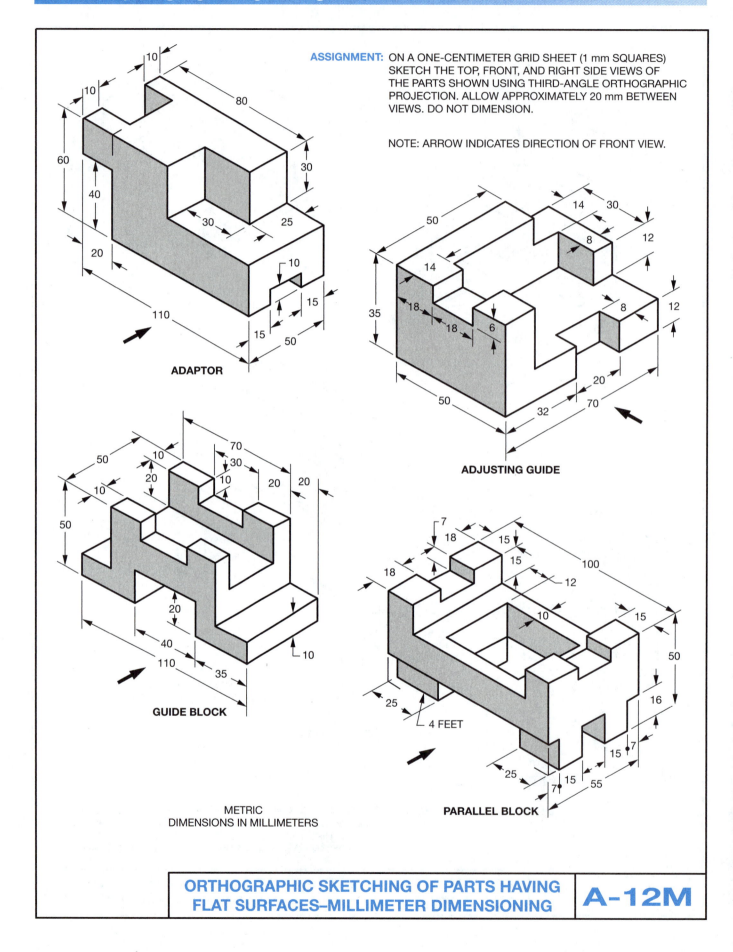

ASSIGNMENT: ON A ONE-CENTIMETER GRID SHEET (1 mm SQUARES) SKETCH THE TOP, FRONT, AND RIGHT SIDE VIEWS OF THE PARTS SHOWN USING THIRD-ANGLE ORTHOGRAPHIC PROJECTION. ALLOW APPROXIMATELY 20 mm BETWEEN VIEWS. DO NOT DIMENSION.

NOTE: ARROW INDICATES DIRECTION OF FRONT VIEW.

ADAPTOR

ADJUSTING GUIDE

GUIDE BLOCK

PARALLEL BLOCK

METRIC
DIMENSIONS IN MILLIMETERS

**ORTHOGRAPHIC SKETCHING OF PARTS HAVING
FLAT SURFACES–MILLIMETER DIMENSIONING**

A–12M

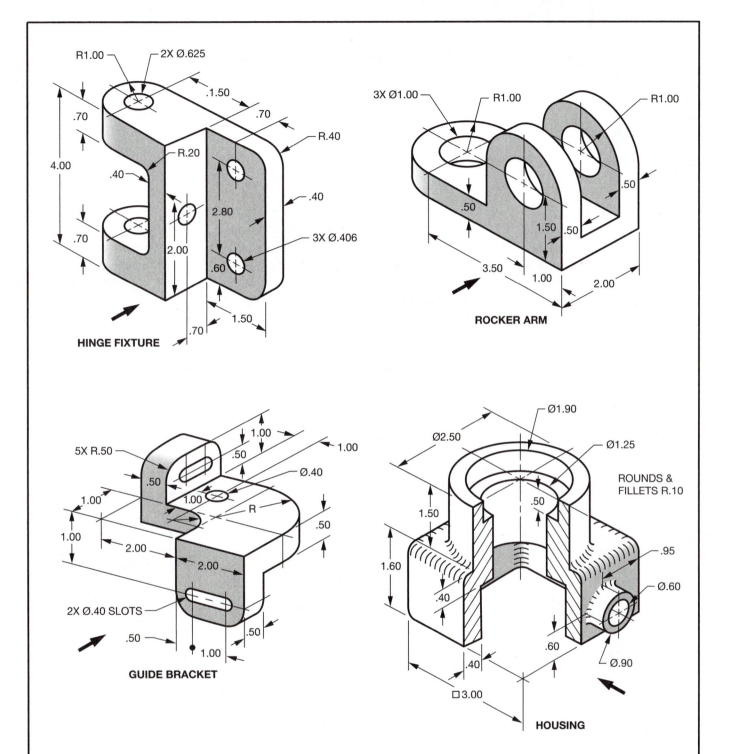

R1.00 — 2X Ø.625

.1.50

.70

R.40

.70

R.20

.40

R.20

4.00

2.80

.40

.70

2.00

.60

3X Ø.406

.70

1.50

HINGE FIXTURE

3X Ø1.00 — R1.00 — R1.00

.50

.50

1.50

.50

3.50

1.00

2.00

ROCKER ARM

5X R.50

1.00

.50

.50

Ø.40

1.00

1.00

R

1.00

2.00

2.00

.50

.50

2X Ø.40 SLOTS

.50

1.00

GUIDE BRACKET

Ø1.90

Ø2.50

Ø1.25

ROUNDS &
FILLETS R.10

.50

1.50

.95

1.60

.40

Ø.60

.60

Ø.90

.40

□3.00

HOUSING

ASSIGNMENT: ON A ONE-INCH GRID SHEET (.10 IN. SQUARES) SKETCH THE TOP,
FRONT, AND RIGHT SIDE VIEWS OF ONE OF THE PARTS SHOWN USING
THIRD-ANGLE ORTHOGRAPHIC PROJECTION. ALLOW APPROXIMATELY
1.00 IN. BETWEEN VIEWS. DO NOT DIMENSION.

NOTE: ARROW INDICATES DIRECTION OF FRONT VIEW.

**ORTHOGRAPHIC SKETCHING OF PARTS HAVING
CIRCULAR FEATURES–DECIMAL-INCH DIMENSIONING** | **A-13**

DIMENSIONING

Dimensions are shown on drawings by extension lines, dimension lines, leaders, arrowheads, figures, notes, and symbols. These lines and dimensions define such geometrical characteristics as distances, diameters, angles, and locations, Figure 5–1. The lines used in dimensioning are thin in contrast to the outline of the object. The dimension must be clear and concise, permitting only one interpretation. In general, each surface, line, or point is located by only one set of dimensions. Exceptions to these rules are the two types of rectangular coordinate dimensioning discussed in Unit 18.

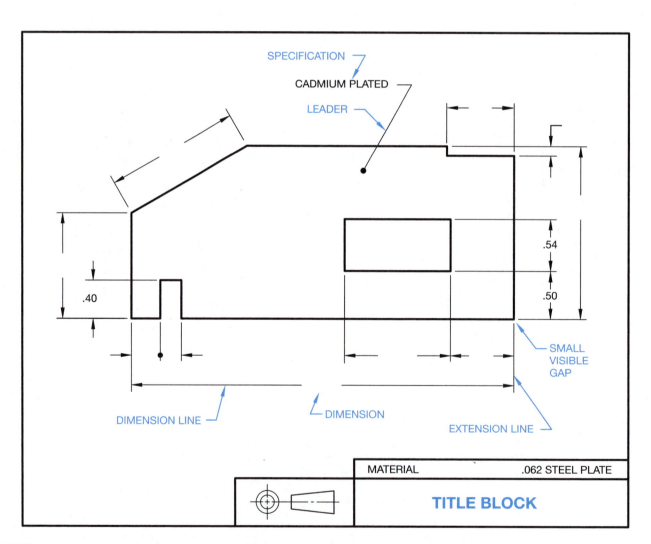

FIGURE 5–1 ■ *Basic dimensioning elements.*

READING DIRECTION

Dimensions and notes on engineering drawings should be placed to read from the bottom of the drawing, Figure 5–1.

DIMENSIONING FLAT SURFACES

Dimension Lines

Dimension lines denote particular sections of the object. They should be drawn parallel to the section they define. Dimension lines terminate in arrowheads, which touch an extension line and are broken in order to allow the insertion of the dimension, Figure 5–1. Where space does not permit the insertion of the dimension line and the dimension between the extension lines, the dimension line may be placed outside the extension line. The dimension can also be placed outside the extension line if the space between the extension lines is limited. In restricted areas, a common industrial practice is to replace the two arrowheads with a circular dot. These methods are shown in Figure 5–2.

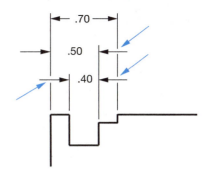

FIGURE 5–3 ■ *Breaks in extension lines.*

Extension Lines

Extension lines denote the points or surfaces between which a dimension applies. They extend from object lines and are drawn perpendicular to the dimension lines, as shown in Figures 5–1 and 5–2. A small gap (.03 to .06 in.) is left between the extension line and the outline to which it refers.

Where extension lines cross arrowheads, a break in the extension line is preferred, Figure 5–3.

Leaders

Leaders are used to direct dimensions or notes to the surface or points to which they apply, Figure 5–4. A leader consists of a line with or without a short horizontal bar adjacent to the note or dimension, and an inclined portion that terminates with an arrowhead touching the line or point to which it applies. A leader may terminate with a dot when it refers to a surface within the outline of a part.

NOTE: If by chance a dimension is omitted on a drawing, contact the drafting department. *Never scale a drawing for a missing dimension.*

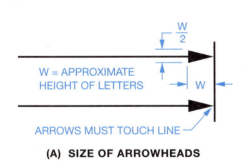

(A) SIZE OF ARROWHEADS

W = APPROXIMATE HEIGHT OF LETTERS

ARROWS MUST TOUCH LINE

.08 — 1.25 — .30

.30

OR

.08 — 1.25 — .30 .30

A CIRCULAR DOT REPLACES THE TWO ARROWHEADS AND DIMENSION LINES

(B) APPLICATION OF ARROWHEADS AND DIMENSIONS

FIGURE 5–2 ■ *Arrowheads.*

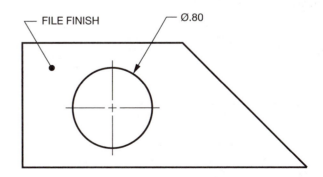

FILE FINISH   Ø.80

FIGURE 5–4 ■ *Using leaders for dimensions and notes.*

Linear Units of Measurement

Although the metric system of dimensioning is expected to become the official standard of measurement for engineering drawings, most drawings in existence in North America today are dimensioned in decimal inches. For this reason, drafters and persons involved in the reading of engineering drawings should be familiar with all the dimensioning systems they may encounter.

The dimensions used in this book are primarily decimal inch. However, metric dimensions are used very frequently. When metric units of measurement are used, the drawing prominently displays the word METRIC and a note stating the dimensions are in millimeters.

INCH UNITS OF MEASUREMENT

The Decimal-Inch System (U.S. Customary). In the decimal-inch system, parts are designed in basic decimal increments, preferably .02 inch, and are expressed as two-place decimal numbers. Using the .02 module, the second decimal place (hundredths) is an even number or zero, Figure 5–5. Sizes other than these, such as .25, are used when they are essential to meet design requirements. When greater accuracy is required, sizes are expressed as three- or four-place decimal numbers such as 1.875 or 4.5625.

The Fractional-Inch System. This system was replaced by the decimal-inch system of dimensioning engineering drawings over 50 years ago. Due to existing tools (drills, reamers, etc.) and pipe sizes being made to fractional-inch sizes, the reader should be aware of this system of dimensioning. In this system, sizes are expressed in common fractions, the smallest divisions being 64ths, Figure 5–6. Sizes other than common fractions are expressed as decimals.

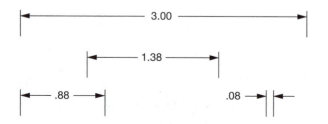

FIGURE 5–5 ■ *Decimal-inch dimensioning.*

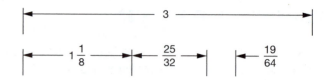

FIGURE 5–6 ■ *Fractional-inch dimensioning.*

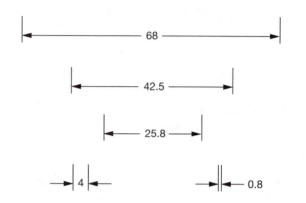

FIGURE 5–7 ■ *Metric (millimeter) dimensioning.*

SI (METRIC) UNITS OF MEASUREMENT

The standard metric units on engineering drawings are the millimeter for linear measure and the micrometer for surface roughness, Figure 5–7.

In metric dimensioning, as in decimal-inch dimensioning, numerals to the right of the decimal point indicate the degree of precision.

Whole dimensions do not require a zero to the right of the decimal point.

2 not 2.0
10 not 10.0

A millimeter value of less than 1 is shown with a zero to the left of the decimal point.

0.2 not .2 or .20
0.26 not .26

Commas should not be used to separate groups of three numbers in metric values. A space should be used in place of the comma.

32 541 not 32,541
2.562 826 6 not 2.5628266

Identification. A metric drawing should include a general note, such as UNLESS OTHERWISE SPECIFIED, DIMENSIONS ARE IN MILLIMETERS. In addition, a metric drawing should be identified by the word METRIC prominently displayed near the title block.

Choice of Dimensions

The choice of the most suitable dimensions and dimensioning methods depends, to some extent, on whether the drawings are intended for unit production or mass production.

Unit production refers to applications in which each part is to be made separately, using general-purpose tools and machines. Details on custom-built machines, jigs, fixtures, and gages required for the manufacture of production parts are made in this way. Frequently, only one of each part is required.

Mass production refers to parts produced in quantity, for which special tooling is usually provided. Most part drawings for manufactured products are considered to be for mass-produced parts.

Functional dimensioning should be expressed directly on the drawing, especially for mass-produced parts. This will result in the selection of datum features on the basis of function and assembly. For unit-produced parts, it is generally preferable to select datum features on the basis of manufacture and machining, Figure 5–8.

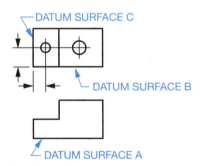

(A) THE PART WITH DATUM SURFACES SELECTED

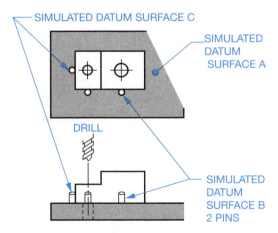

(B) POSITIONING PART FOR DRILLING HOLES

FIGURE 5–8 ■ *Selection of datum surfaces for dimensioning.*

Basic Rules for Dimensioning

- Place dimensions between the views when possible, Figure 5–9(A).

- Place the dimension line for the shortest width, height, and depth nearest the outline of the object, Figure 5–9(B). Parallel dimension lines are placed in order of their

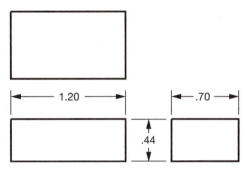

(A) PLACE DIMENSIONS BETWEEN VIEWS

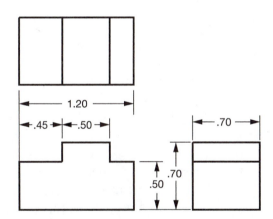

(B) PLACE SMALLEST DIMENSION NEAREST THE VIEW BEING DIMENSIONED

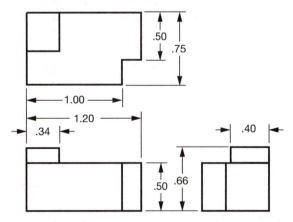

(C) DIMENSION THE VIEW THAT BEST SHOWS THE SHAPE

FIGURE 5–9 ■ *Basic dimensioning rules.*

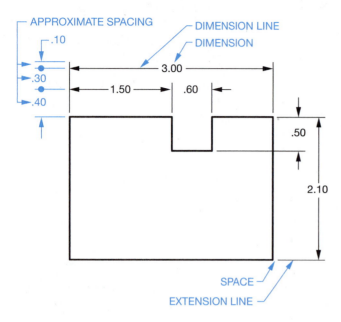

FIGURE 5–10 ■ *Placement of dimensions.*

size, making the longest dimension line the outermost line.

- Place dimensions near the view that best shows the characteristic contour or shape of the object, Figure 5–9(C). In following this rule, dimensions will not always be between views.

- When several dimension lines are directly above or next to one another, it is good practice to stagger the dimensions in order to improve the clarity of the drawing. The spacing suitable for most drawings between parallel dimension lines is .30 in. (8 mm), and the spacing between the outline of the object and the nearest dimension line should be approximately .40 in. (10 mm), Figure 5–10. Other rules for the placement of dimensions will appear as you proceed through later units and assignments throughout the book.

- On large drawings, dimensions can be placed on the view to improve clarity.

Dimensioning Cylindrical Features

Features shown as circles are normally dimensioned by one of the methods shown in Figure 5–11. Where the diameters of a number of concentric cylinders are to be given, it may be more convenient to show them on the side view. The diameter symbol Ø should always precede the diametral dimension.

The radius of the arc is used in dimensioning a circular arc. The letter R is shown before the radius dimension to indicate that it is a radius. Approved methods for dimensioning arcs are shown in Figure 5–12.

Dimensioning Cylindrical Holes

Specification of the diameter with a leader, as shown in Figure 5–13, is the preferred method for designating the size of small holes. For larger diameters, use one of the methods illustrated in Figure 5–11. When the leader is used, the symbol Ø precedes the size of the hole. The note end of the leader terminates in a short horizontal bar that is adjacent to the beginning or the end of the note. When two or more holes of the same size are required, the number of holes is specified. If a blind hole is required, the depth of the hole is included in the dimensioning note; otherwise, it is assumed that all holes shown are through holes.

DRILLING, REAMING, AND BORING

Drilling is the process of using a drill to cut a hole through a solid, or to enlarge an existing hole. For some types of work, holes must be drilled smooth and straight, and to an exact size. In other work, accuracy of location and size of the hole are not as important.

When accurate holes of uniform diameter are required, they are first drilled slightly undersize and then reamed. *Reaming* is the process of sizing a hole to a given diameter with a reamer in order to produce a hole that is round, smooth, straight, and accurate.

Boring is one of the more dependable methods of producing holes that are round and concentric. The term *boring* refers to the enlarging of a hole by means of a boring tool. The use of reaming is limited to the sizes of available reamers. However, holes may be bored to any size desired.

The degree of accuracy to which a hole is to be machined is specified on the drawing.

The use of operational names, such as *turn, bore, grind, ream, tap,* and *thread,* with dimensions should be avoided. Although the drafter should be aware of the methods by which a part can be produced, the method of manufacture is better left to the shop. If the part is adequately dimensioned, and

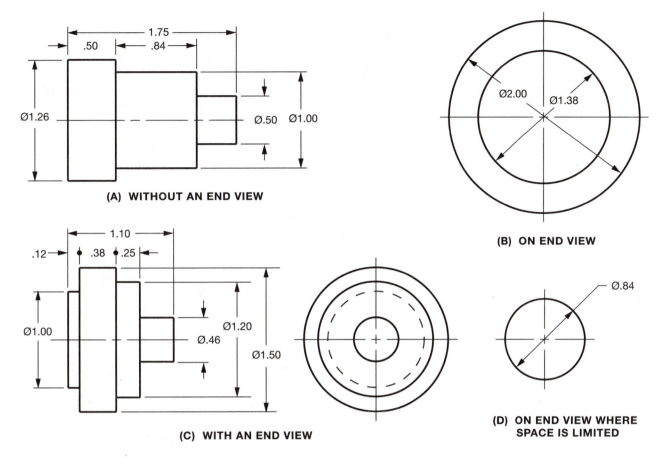

FIGURE 5–11 ■ *Dimensioning diameters.*

(A) WITHOUT AN END VIEW

(B) ON END VIEW

(C) WITH AN END VIEW

(D) ON END VIEW WHERE SPACE IS LIMITED

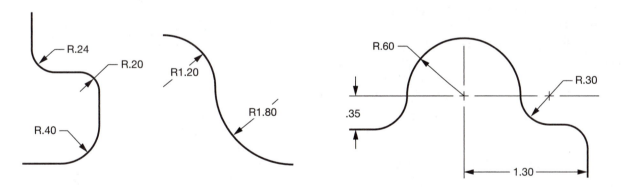

FIGURE 5–12 ■ *Dimensioning radii.*

has the surface texture symbols showing the finish quality desired, it remains a shop problem to meet the drawing specifications.

Dimensioning Rounds and Fillets

A round, or radius, or chamfer is put on the outside of a piece to improve its appearance and to avoid forming a sharp edge that might chip off under a sharp blow or cause interference. It is also a safety feature. A fillet is additional metal allowed in the inner intersection of two surfaces, Figure 5–14. This increases the strength of the object. A general note, such as ROUNDS AND FILLETS R10 or ROUNDS AND FILLETS R10 UNLESS OTHERWISE SHOWN, is normally used on the drawing instead of individual dimensions.

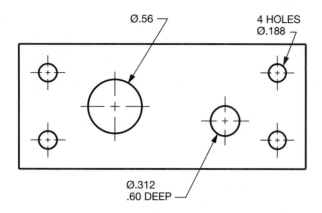

FIGURE 5–13 ■ *Dimensioning circular holes.*

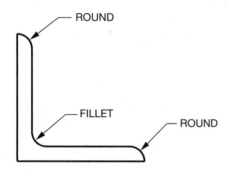

FIGURE 5–14 ■ *Fillets and rounds.*

Dimensioning Repetitive Features

Repetitive features and dimensions may be specified on a drawing by the use of an X in conjunction with the numeral to indicate the "number of times" or "places" they are required. A space is inserted between the X and the dimension, as shown in Figure 5–15.

Identifying Similarly Sized Features

Where many similarly sized holes or features appear on a part, some form of identification may be desirable in order to ensure the legibility of the drawing, Figure 5–16.

REFERENCES

ASME Y14.5M-1994 (R1999) Dimensioning and Tolerancing

ASME Y14.2M-1992 (R2003) Line Conventions and Lettering

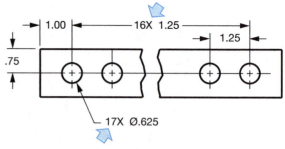

EXAMPLE 1

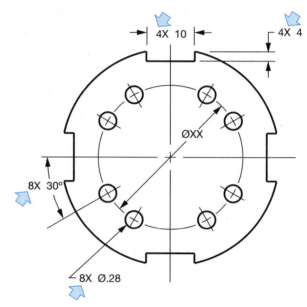

EXAMPLE 2

FIGURE 5–15 ■ *Dimensioning repetitive features.*

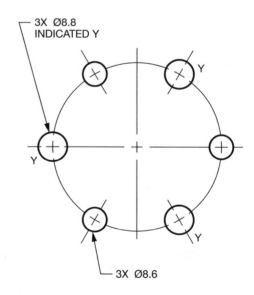

FIGURE 5–16 ■ *Identifying similarly sized holes.*

INTERNET RESOURCES

Drafting Zone. For information on dimensioning circular features, see: http://www.draftingzone.com

Integrated Publishing. For information on dimensioning, see: http://www.tpub.com/engbas/3-201htm

IDS Development—Nebraska Education. For information on the various line types used on engineering drawings, see: http://idsdev.mccneb.edu/djackson/lineintro.htm

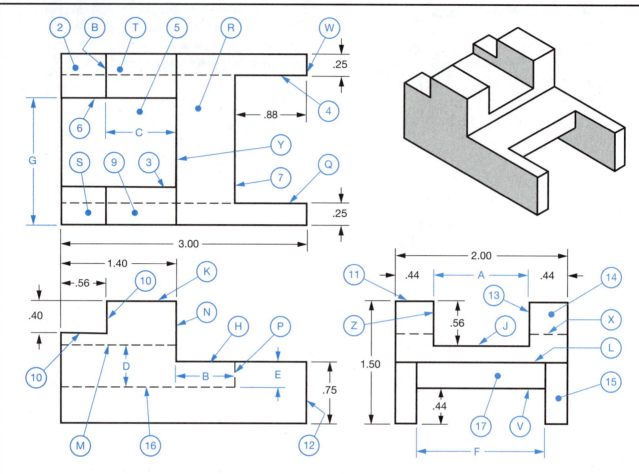

QUESTIONS:

1. What is the name of the object?
2. What is the drawing number?
3. How many castings are required?
4. What material is the part made of?
5. What is the overall width?
6. What is the overall height?
7. What is the overall depth?
8. Calculate distances A through G.
9. Which line in the top view represents surface (P)?
10. Which line in the side view represents surface (5)?
11. Which line in the side view represents surface (R)?
12. Which surface in the top view does line (K) in the front view represent?
13. Which surface in the top view does line (M) in the front view represent?
14. Which line in the side view represents the same surface representing the front view?
15. What kind, or type of line, is line (M)?
16. Which front view line does line (X) in the side view represent?

17. Which front view line does line (Y) in the top view represent?
18. Which line in the front view does the surface (15) in the side view represent?
19. Which front view line represents surface (R) in the top view?
20. Which surface in the side view represents line (N) of the front view?
21. Which line in the side view represents surface (2)?
22. Which surface in the side view does line (P) represent?
23. Which surface in the top view does line (11) represent?
24. Which line in the side view does line (3) in the top view represent?
25. Which line in the side view does line (16) in the front view represent?
26. Which surface in the side view does line (W) represent?

QUANTITY	875	
MATERIAL	MALLEABLE IRON	
SCALE	NOT TO SCALE	
DRAWN		DATE

FEED HOPPER **A-14**

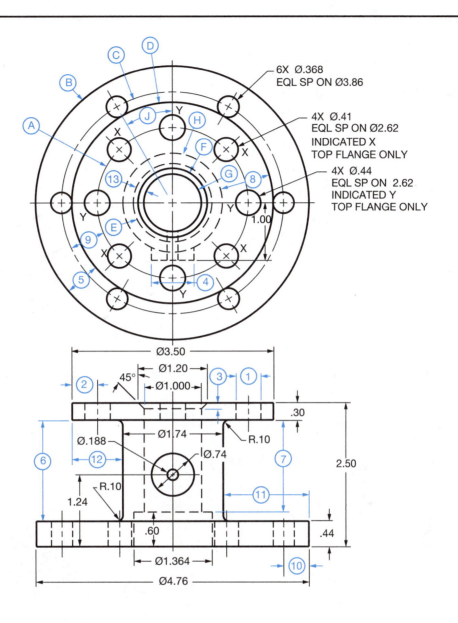

6X Ø.368
EQL SP ON Ø3.86

4X Ø.41
EQL SP ON Ø2.62
INDICATED X
TOP FLANGE ONLY

4X Ø.44
EQL SP ON 2.62
INDICATED Y
TOP FLANGE ONLY

QUESTIONS:

1. What are the diameters of circles (A) to (H)?

2. How many holes are in the bottom surface?

3. How many holes are in the top surface?

4. How deep is the Ø1.000 hole from the top of the coupling?

5. What is angle (J)?

6. How thick is the largest flange?

7. What size bolts would be used for the Y holes located on the top flange?
   Allow approximately .06 total clearance. (Refer to the bolt sizes in the appendix.)

8. What size bolts would be used for the bottom flange? Allow approximately
   .06 total clearance. (Refer to the bolt sizes in the appendix.)

9. Calculate distances (1) to (13).

MATERIAL	GRAY IRON	
SCALE	NOT TO SCALE	
DRAWN		DATE

COUPLING **A-15**

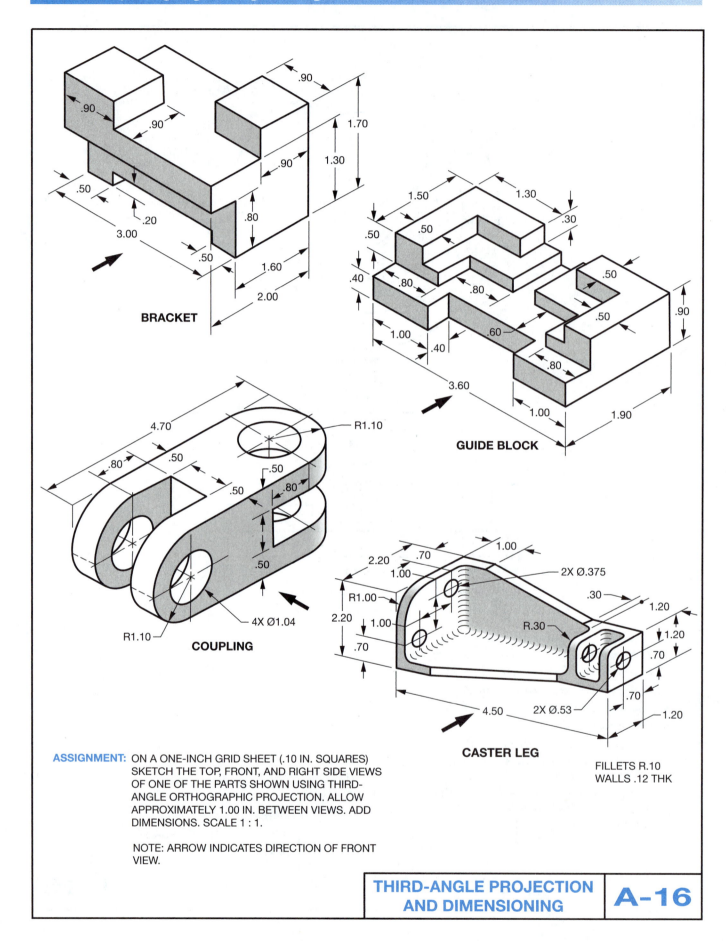

.90

.90

.90

.90

1.70

1.30

.50

.20

.80

3.00

.50

1.60

2.00

BRACKET

1.50

1.30

.30

.50

.50

.40

.80

.80

.50

1.00

.40

.60

.50

.80

3.60

1.00

1.90

.90

GUIDE BLOCK

4.70

R1.10

.80

.50

.50

.50

.80

.50

R1.10

4X Ø1.04

COUPLING

2.20

.70

1.00

1.00

2X Ø.375

R1.00

.30

1.20

2.20

1.00

R.30

1.20

.70

.70

.70

1.20

4.50

2X Ø.53

CASTER LEG

FILLETS R.10
WALLS .12 THK

ASSIGNMENT: ON A ONE-INCH GRID SHEET (.10 IN. SQUARES) SKETCH THE TOP, FRONT, AND RIGHT SIDE VIEWS OF ONE OF THE PARTS SHOWN USING THIRD-ANGLE ORTHOGRAPHIC PROJECTION. ALLOW APPROXIMATELY 1.00 IN. BETWEEN VIEWS. ADD DIMENSIONS. SCALE 1 : 1.

NOTE: ARROW INDICATES DIRECTION OF FRONT VIEW.

THIRD-ANGLE PROJECTION AND DIMENSIONING

A-16

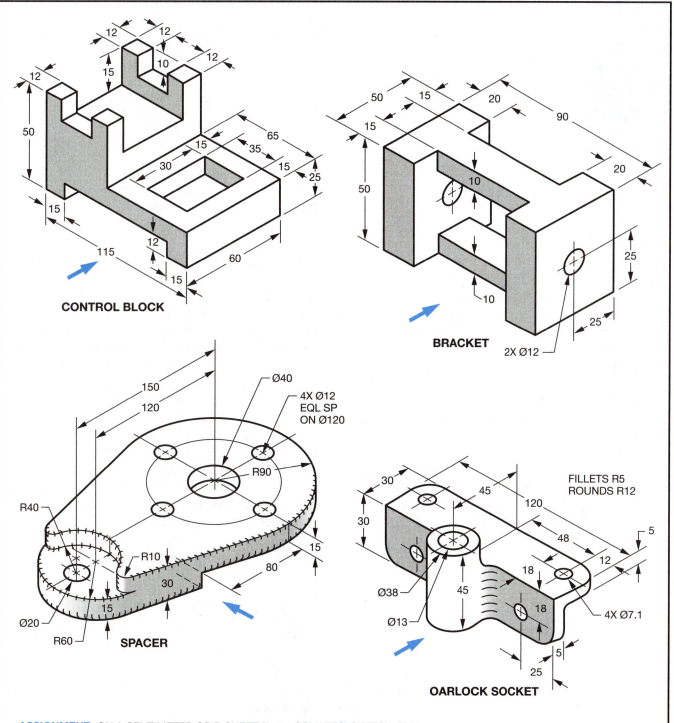

CONTROL BLOCK

BRACKET

2X Ø12

Ø40
4X Ø12
EQL SP
ON Ø120

150
120
R90
R40
R10
R60
R30
Ø20
R15
80
15
30
15

SPACER

FILLETS R5
ROUNDS R12
30
45
120
30
48
5
18
12
Ø38
45
18
Ø13
4X Ø7.1
5
25

OARLOCK SOCKET

ASSIGNMENT: ON A CENTIMETER GRID SHEET (1 mm SQUARES) SKETCH ONE
OF THE PARTS SHOWN USING THIRD-ANGLE ORTHOGRAPHIC
PROJECTION. ALLOW APPROXIMATELY 20 mm BETWEEN
VIEWS AND ADD DIMENSIONS. FOR THE SPACER, SKETCH ONLY
THE TOP AND FRONT VIEWS AND USE THE SCALE 1 : 2.

NOTE: ARROW INDICATES DIRECTION OF FRONT VIEW.

METRIC
DIMENSIONS IN MILLIMETERS

**THIRD-ANGLE PROJECTION
AND DIMENSIONING**

A–17M

INCLINED SURFACES

If the surfaces of an object lie in either a horizontal or vertical position, the surfaces appear in their true shape in one of the three views, and these surfaces appear as a line in the other two views.

When a surface is sloped or inclined in only one direction, that surface is not seen in its true shape in the top, front, or side views. It is, however, seen in two views as a distorted surface. On the third view it appears as a line.

The true length of surfaces A and B in Figure 6–1 is seen in the front view only. In the top and side views, only the width of surfaces A and B appears in its true size. The length of these surfaces is foreshortened.

Where an inclined surface has important features that must be shown clearly and without distortion, an auxiliary or helper view must be used. These views are discussed in detail later in the book.

Illustrations of simple objects having inclined surfaces appear in Figure 6–2.

MEASUREMENT OF ANGLES

Some objects do not have all their features positioned in such a way that all surfaces can be in the horizontal and vertical planes at the same time. The design of the part may require that some of the lines in the drawing be shown in a direction other than horizontal or vertical. This will require that some lines be drawn at an angle.

The amount of this divergence, or obliqueness, of lines may be indicated by either an offset dimension or an angle dimension, as shown in Figure 6–3.

Angle dimensions may be expressed in degrees and decimal parts of a degree. They may also be expressed in degrees, minutes, and seconds. The former method is now preferred.

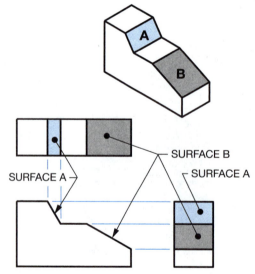

SURFACE B

SURFACE A →

SURFACE A

NOTE: THE TRUE SHAPES OF SURFACES A AND B DO NOT APPEAR ON THE TOP OR SIDE VIEWS.

FIGURE 6–1 ■ *Inclined surfaces.*

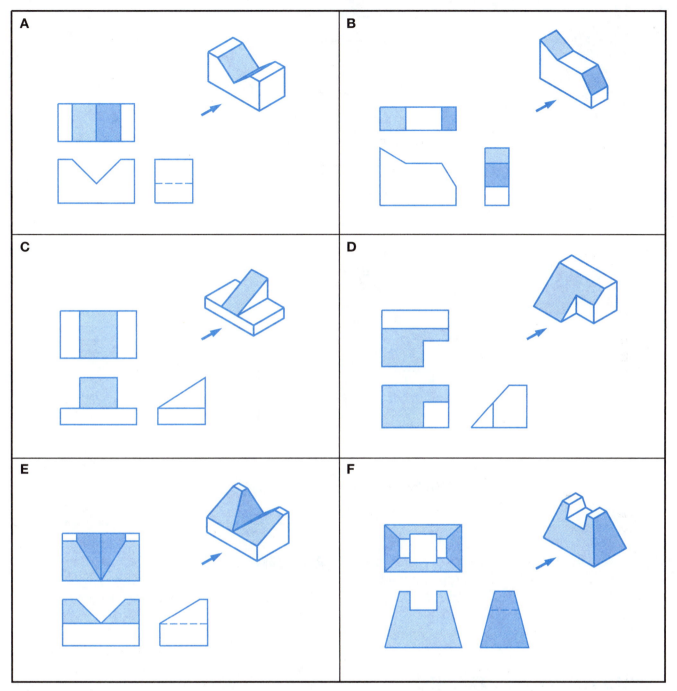

NOTE: ARROWS INDICATE DIRECTION OF SIGHT WHEN LOOKING AT FRONT VIEW

FIGURE 6–2 ■ *Illustrations of simple objects having inclined surfaces.*

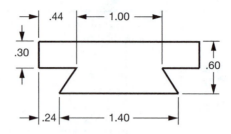

(A) LINEAR MEASUREMENTS

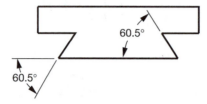

(B) ANGLE MEASUREMENTS USING DECIMAL DEGREES

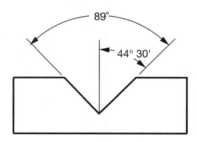

(C) ANGLE MEASUREMENTS USING DEGREES AND MINUTES

FIGURE 6–3 ■ *Dimensioning angles.*

The symbols for degrees (°), minutes ('), and seconds (") are included with the appropriate values. For example, 2°; 30°; 28°10'; 0°15'; 27°13'15"; 0°0'30"; 0.25°; 30°0'0"; ±0°2'30"; and 2°±0.5° are all correct forms.

SYMMETRICAL OUTLINES

Symmetrical outlines or features may be indicated on a drawing by means of the symmetry symbol shown in Figure 6–4. Two thick parallel lines are placed on the center lines above and below the feature. The use of this symbol means the part or feature is symmetrical about the center line or feature.

MACHINE SLOTS

Slots are used chiefly in machines to hold parts together. The two principal types are *T slots* and *dovetails,* Figure 6–4.

A dovetail is a groove or slide whose sides are cut on an angle. This forms an interlocking joint between two pieces, enabling the slot to resist pulling apart in any direction other than along the lines of the dovetail slide itself.

The dovetail is commonly used in the design of slides, including lathe cross slides, milling machine table slides, and other sliding parts.

The two parts of a dovetail slide are shown in Figure 6–4.

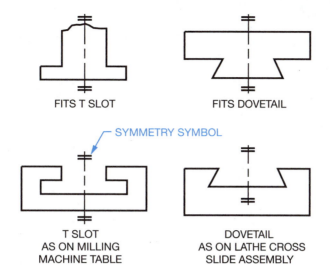

FIGURE 6–4 ■ *Using the symmetry symbol to indicate symmetry on machining slots.*

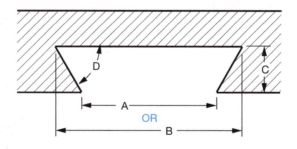

FIGURE 6–5 ■ *Dimensions for dovetails.*

When dovetail parts are to be machined to a given width, they may be gauged by using accurately sized cylindrical rods or wires.

Dovetails are usually dimensioned as in Figure 6–5. The dimensions limit the boundaries within which the machinist works.

The edges of a dovetail are usually broken to remove the sharp corners. On large dovetails the external and internal corners are often machined as shown in Figure 6–6 as at A or B.

REFERENCE DIMENSIONS

When a reference dimension is shown on a drawing for information only and is not used for the manufacture of the part, it must be clearly labeled. The approved method for identifying reference dimensions on a drawing is the enclosure of the dimensions inside parentheses, Figure 6–7. Previously, reference dimensions were identified by placing the abbreviation REF after or below the dimension.

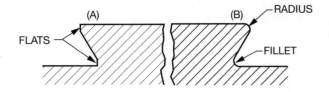

FIGURE 6–6 ■ *Corners used on large dovetails.*

REFERENCES

ASME Y14.5M-1994 (R1999) Dimensioning and Tolerancing

INTERNET RESOURCES

Drafting Zone. For information on drafting symbols and reference dimensioning, see: http://www.draftingzone.com

Metrication.com. For metrication in drafting and engineering, see http://www.metrication.com

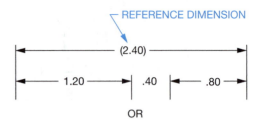

OR

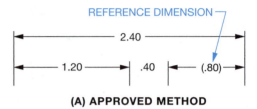

(A) APPROVED METHOD

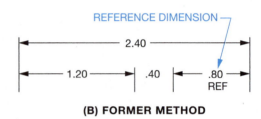

(B) FORMER METHOD

FIGURE 6–7 ■ *Reference dimensions.*

QUESTIONS:

1. Calculate distances A to G.
2. At what angle is line ⑥ to the vertical?
3. At what angle is line ⑦ to the horizontal?
4. Locate surface ⑥ in the side view.
5. Locate surface ① in the side view.
6. Locate surface ⑥ in the top view.
7. Which lines in the side view are represented by line ② in the front view?
8. Locate ⑱ in the top view.
9. Locate surface ⑨ in the side view.
10. Locate surface ⑫ in the front view.
11. Locate surface ③ in the top view.
12. Which lines in the side view are represented by point ④ in the front view?
13. Which line in the side view is line ⑯ in the top view?
14. Locate surface ⑩ in the side view.
15. Locate surface ⑩ in the front view.
16. Locate surface ⑫ in the side view.
17. Which line does point ④ represent in the top view?
18. Locate line ㉔ in the top view.
19. Locate line ㉘ in the top view.
20. Locate line ㉕ in the top view.
21. Which line in the front view is surface ⑨ in the top view?

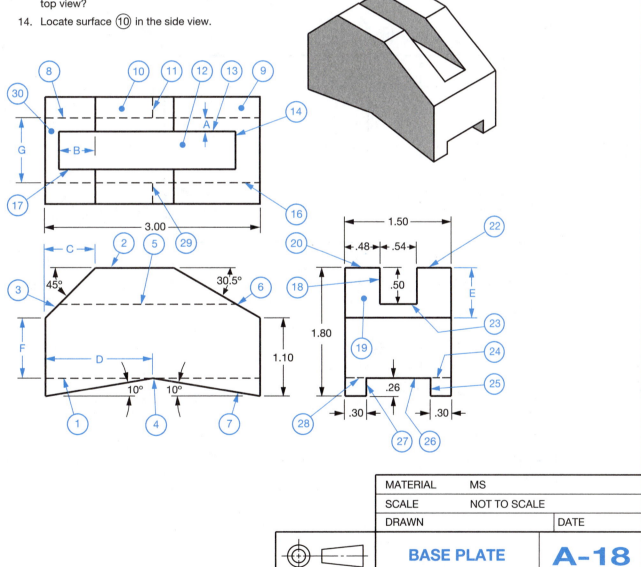

MATERIAL	MS	
SCALE	NOT TO SCALE	
DRAWN		DATE

BASE PLATE   **A-18**

QUESTIONS:

1. In which view is the shape of the dovetail shown?
2. In which view is the shape of the T slot shown?
3. How many rounds are shown in the top view?
4. In which view is a fillet shown?
5. Which line in the top view represents surface (R) of the side view?
6. Which line in the front view represents surface (R)?
7. Which line of the top view represents surface (L) of the side view?
8. Which line in the front view represents surface (L)?
9. Which line in the side view represents surface (A) on the top view?
10. Which dimension in the front view represents the width of surface (A)?
11. What type of lines are (B), (J), and (K)?
12. How far apart are the two hidden edge lines on the side view?
13. What dimension indicates how far line (J) is from the base of the slide?
14. How wide is the opening in the dovetail?
15. Which two lines in the top view indicate the opening of the dovetail?
16. At what angle to the horizontal is the dovetail cut?
17. In the side view, how far is the lower left edge of the dovetail from the left side of the piece?
18. What are the lengths of dimensions Y, V and X?
19. What is the height of the dovetail?

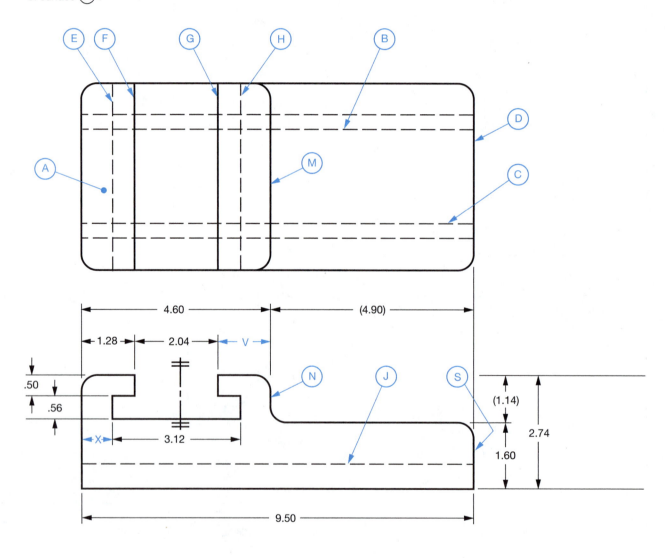

20. How much material remains between the surface represented by line Ⓠ and the top of the dovetail after the cut has been taken?

21. What is the vertical distance from the surface represented by line Ⓠ to that represented by line Ⓣ?

22. Which dimension represents the distance betwen lines Ⓕ and Ⓖ?

23. What is the overall height of the T slot?

24. What is the distance between the bottom of the T slot and the top of the dovetail?

25. What is the width of the bottom of the T slot?

26. What is the height of the opening of the bottom of the T slot?

27. What is the horizontal distance from line Ⓝ to line Ⓢ?

28. What is the unit of measurement for the angles shown?

29. How many reference dimensions are shown on the drawing?

30. What is the size of the largest reference dimension?

ROUNDS AND FILLETS R.38

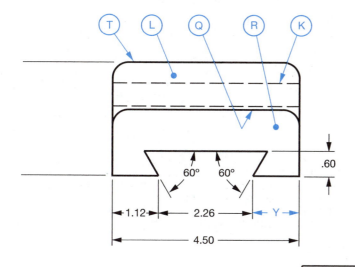

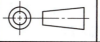

MATERIAL	GRAY IRON	
SCALE	NOT TO SCALE	
DRAWN		DATE

COMPOUND REST SLIDE   **A-19**

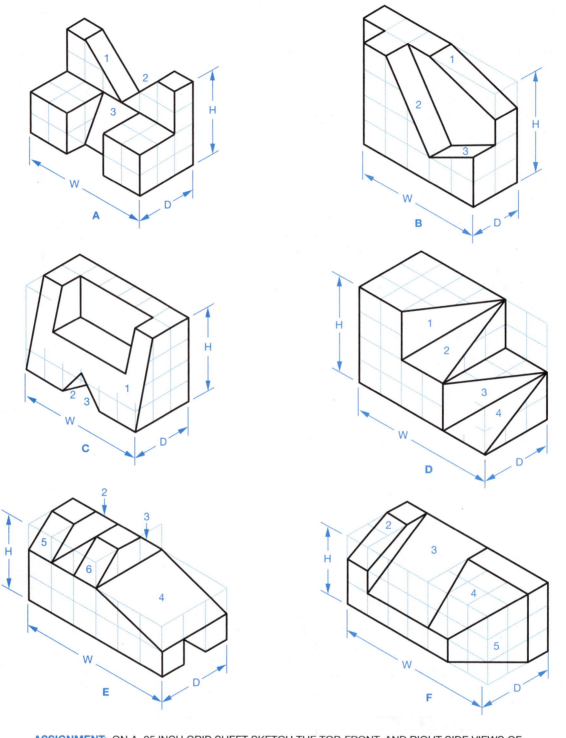

ASSIGNMENT: ON A .25 INCH GRID SHEET SKETCH THE TOP, FRONT, AND RIGHT SIDE VIEWS OF THE SIX PARTS SHOWN. EACH SQUARE SHOWN ON THE OBJECT REPRESENTS ONE SQUARE ON THE GRID SHEET. ALLOW ONE GRID SPACE BETWEEN VIEWS AND A MINIMUM OF TWO GRID SPACES BETWEEN THE OBJECTS. IDENTIFY THE SLOPED SURFACES ON THE THREE VIEWS WITH THE CORRESPONDING NUMBERS SHOWN ON THE PICTORIAL DRAWING.

ORTHOGRAPHIC SKETCHING OF OBJECTS HAVING SLOPED SURFACES USING GRID LINES | **A-20**

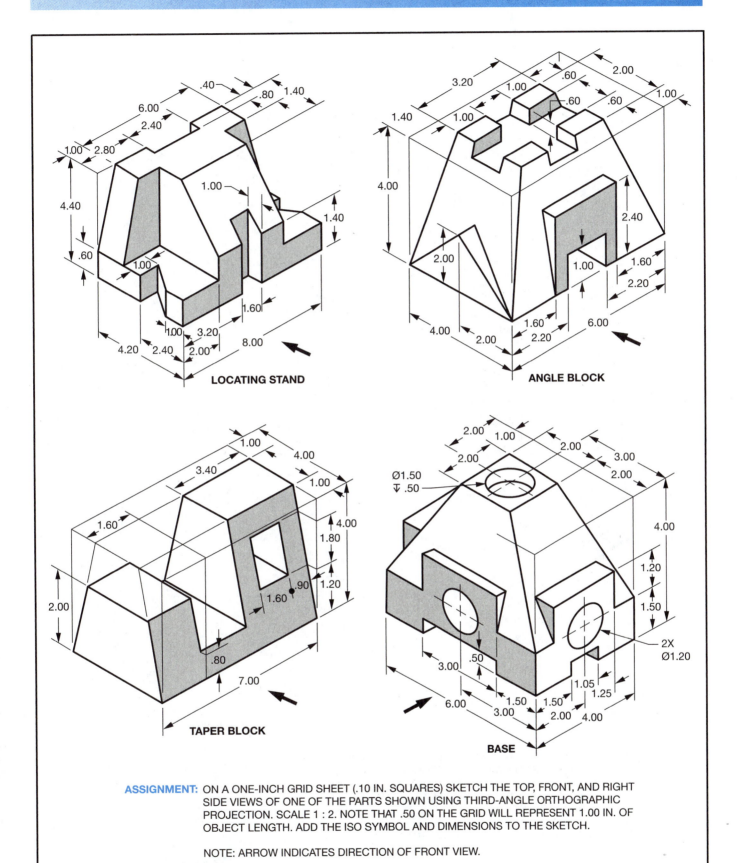

LOCATING STAND

ANGLE BLOCK

TAPER BLOCK

BASE

ASSIGNMENT: ON A ONE-INCH GRID SHEET (.10 IN. SQUARES) SKETCH THE TOP, FRONT, AND RIGHT SIDE VIEWS OF ONE OF THE PARTS SHOWN USING THIRD-ANGLE ORTHOGRAPHIC PROJECTION. SCALE 1 : 2. NOTE THAT .50 ON THE GRID WILL REPRESENT 1.00 IN. OF OBJECT LENGTH. ADD THE ISO SYMBOL AND DIMENSIONS TO THE SKETCH.

NOTE: ARROW INDICATES DIRECTION OF FRONT VIEW.

ORTHOGRAPHIC SKETCHING OF PARTS HAVING SLOPED SURFACES USING DECIMAL-INCH DIMENSIONING   **A-21**

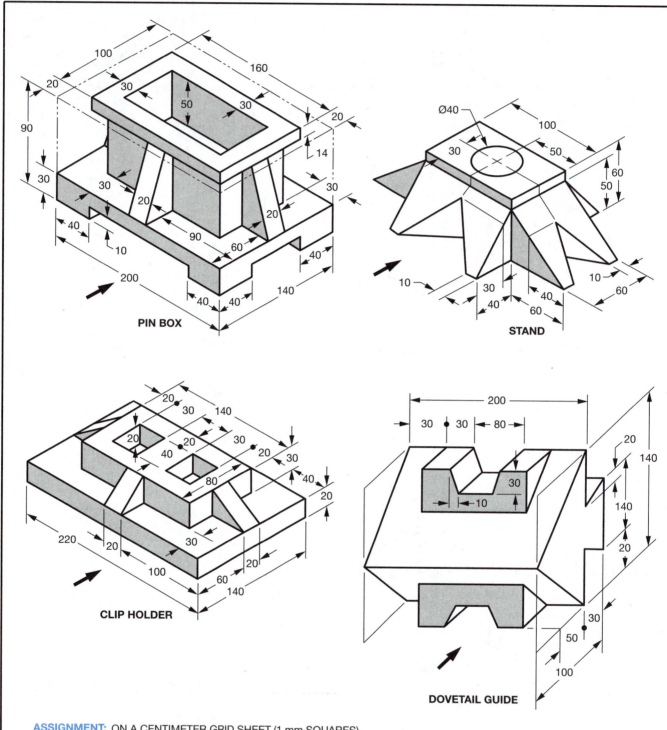

PIN BOX

STAND

CLIP HOLDER

DOVETAIL GUIDE

ASSIGNMENT: ON A CENTIMETER GRID SHEET (1 mm SQUARES)
SKETCH THE TOP, FRONT, AND RIGHT SIDE VIEWS OF ONE
OF THE PARTS SHOWN USING THIRD-ANGLE ORTHOGRAPHIC
PROJECTION. ADD THE ISO SYMBOL AND DIMENSIONS TO
THE SKETCH. SCALE 1 : 2.

NOTE: ARROW INDICATES DIRECTION OF FRONT VIEW.

METRIC
DIMENSIONS IN MILLIMETERS

**ORTHOGRAPHIC SKETCHING OF PARTS HAVING
SLOPED SURFACES USING MILLIMETER DIMENSIONING**

A-22M

PICTORIAL SKETCHING

Pictorial sketching is widely used in industry because this type of sketching is easy to read and understand, Figure 7–1. It is also a quick and easy means of communicating technical ideas. Isometric sketching, one of several types of pictorial drawing, is the most frequently used. With the use of pictorial grid sheets and ellipse templates, pictorial drawings can be sketched quickly and accurately.

Viewing Direction

The pictorial sketch may be drawn so the part is viewed from above (bird's eye view), or from below (worm's eye view), Figure 7–2. The part features you wish to show normally govern the viewing direction selected.

ISOMETRIC SKETCHING

All isometric sketches are started by constructing the isometric axes, which includes a vertical line for height and isometric lines to the left and right, at an

angle of 30° from the horizon, for width and depth. The three faces seen in the isometric view are the same faces that would be seen in the normal orthographic views: top, front, and side, as shown in Figure 7–3(A). Figure 7–3(B) shows the selection of

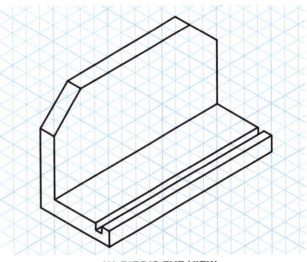

(A) BIRD'S EYE VIEW

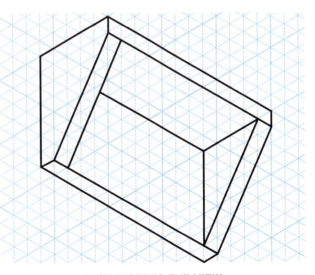

(B) WORM'S EYE VIEW

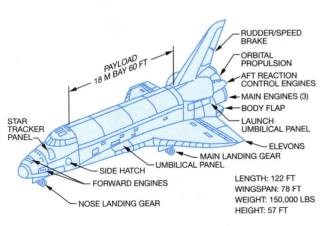

PAYLOAD
18 M BAY 60 FT

RUDDER/SPEED
BRAKE

ORBITAL
PROPULSION

AFT REACTION
CONTROL ENGINES

MAIN ENGINES (3)

BODY FLAP

LAUNCH
UMBILICAL PANEL

ELEVONS

MAIN LANDING GEAR

UMBILICAL PANEL

STAR
TRACKER
PANEL

SIDE HATCH

FORWARD ENGINES

NOSE LANDING GEAR

LENGTH: 122 FT
WINGSPAN: 78 FT
WEIGHT: 150,000 LBS
HEIGHT: 57 FT

SPACE SHUTTLE ORBITER

FIGURE 7–1 ■ *Application of a pictorial sketch.*

FIGURE 7–2 ■ *Viewing direction.*

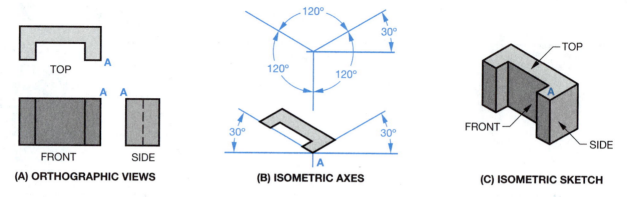

(A) ORTHOGRAPHIC VIEWS   **(B) ISOMETRIC AXES**   **(C) ISOMETRIC SKETCH**

FIGURE 7–3 ■ *Isometric axes and projection.*

the front corner "A" and the construction of the isometric axes. Figure 7–3(C) shows the completed isometric view. All lines are drawn to their true length, measured along the isometric axes, and hidden lines are usually omitted.

Isometric Grid Sheets

This type of isometric sketching paper has evenly spaced lines running in three directions. Two sets of lines are sloped in the direction of the isometric axes. The third set of lines is vertical and passes through the intersection of the sloping lines, as in Figure 7–2. The most commonly used grids are the inch, which is further subdivided into either 4 or 10 equal grids, and the centimeter, which is further subdivided into 10 equal grids of 1 mm. No units of measure are shown on these sheets; therefore the spaces could represent any convenient unit of size.

Inclined Surfaces

Many objects have inclined surfaces that are represented by sloping lines in orthographic views. On isometric drawings, sloping surfaces appear as *nonisometric lines.* To create them, their endpoints, which are found on the ends of isometric lines, are joined with a straight line. Figure 7–4 shows how to construct nonisometric lines.

Circles and Arcs

A circle on the three faces of an object drawn in isometric has the shape of an ellipse, as shown in Figure 7–5. Practically all circles and arcs shown on isometric sketches are made with the use of an isometric ellipse template. The template shown in

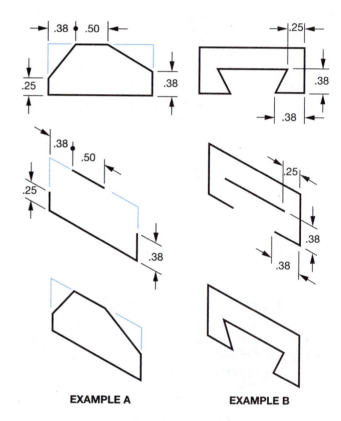

EXAMPLE A   **EXAMPLE B**

FIGURE 7–4 ■ *Construction of nonisometric lines.*

Figure 7–5 combines ellipses, scales, and angles. Markings on the ellipse coincide with the center lines of the holes, speeding up the drawing of circles and arcs.

Basic Steps to Follow for Isometric Sketching

To save time and to make a more accurate and neater-looking sketch, use an isometric ellipse template for drawing arcs and circles and a straightedge for drawing long lines.

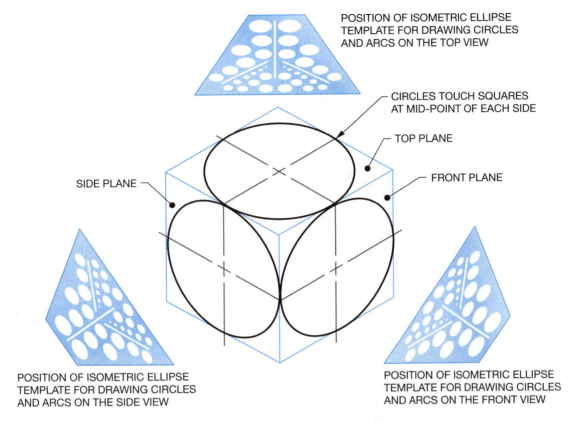

POSITION OF ISOMETRIC ELLIPSE TEMPLATE FOR DRAWING CIRCLES AND ARCS ON THE TOP VIEW

CIRCLES TOUCH SQUARES AT MID-POINT OF EACH SIDE

TOP PLANE

FRONT PLANE

SIDE PLANE

POSITION OF ISOMETRIC ELLIPSE TEMPLATE FOR DRAWING CIRCLES AND ARCS ON THE SIDE VIEW

POSITION OF ISOMETRIC ELLIPSE TEMPLATE FOR DRAWING CIRCLES AND ARCS ON THE FRONT VIEW

FIGURE 7–5 ■ *Using the isometric ellipse template for drawing circles and arcs.*

A commonly used technique for sketching is to sketch a box having the maximum height, width, and depth of the object, and then the parts of the box, which are not part of the object, are removed, leaving the parts that form the total object, Figure 7–6.

Step 1. Build a Frame. The frame (or box) is the overall size of the part to be drawn. It is drawn with construction lines.

Step 2. Block in the Overall Sizes for Each Detail. These subblocks or frames enclose each detail. They are drawn with construction lines.

Step 3. Add the Details. Lightly sketch the shapes of the details using construction lines. For circles, draw squares equal to the size of the diameter. Also sketch in the lines to represent the center lines of the circle.

Step 4. Darken the Lines. Using a soft lead pencil, darken in the visible object lines.

OBLIQUE SKETCHING

This method of pictorial drawing is based on the procedure of placing the object with one face parallel to the frontal plane and placing the other two faces on oblique (or receding) planes, to left or right, top or bottom, at any convenient angle. The three axes of projection are vertical, horizontal, and receding. Figure 7–7 illustrates a cube drawn in typical positions with the receding axes at 60°, 45°, and 30°. This form of projection has the advantage of showing one face of the object without distortion. The face with the greatest irregularity of outline or contour, the face with the greatest number of circular features, or the face with the longest dimension faces the front, Figure 7–8.

Two types of oblique projection are used extensively. In *cavalier oblique,* all lines are made to their true length, measured on the axes of the projection. In *cabinet oblique,* the lines on the receding axis are shortened by one-half their true length to compensate for distortion and to approximate more closely what the human eye would see. For this reason, and because of the simplicity of projection, cabinet oblique is a commonly used form of pictorial representation, especially when circles and arcs are to be drawn. Figure 7–9 shows a comparison of cavalier and cabinet oblique. Note that hidden lines are omitted unless required for clarity. Most of the drawing techniques for isometric projection apply to oblique projection.

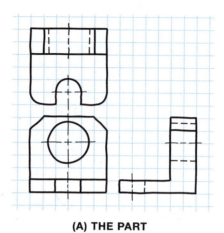

(A) THE PART

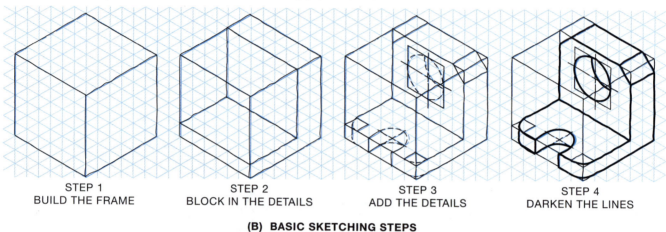

STEP 1
BUILD THE FRAME

STEP 2
BLOCK IN THE DETAILS

STEP 3
ADD THE DETAILS

STEP 4
DARKEN THE LINES

(B) BASIC SKETCHING STEPS

FIGURE 7–6 ∎ *Basic steps to follow for isometric sketching.*

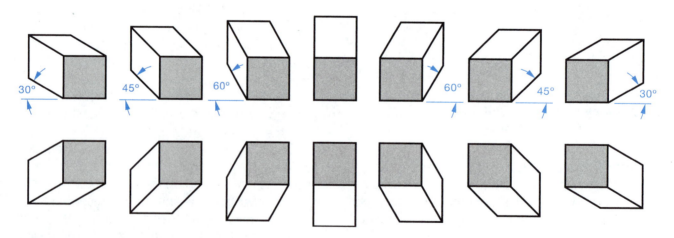

FIGURE 7–7 ∎ *Typical positions of receding axes for oblique projection.*

Oblique Grid Sheets

This type of sketching paper is similar to the two-dimensional sketching paper except that 45° lines, which pass through the intersecting horizontal and vertical lines, are added in either one or both directions. The most commonly used grids are the inch, which is subdivided into smaller evenly spaced grids, and the centimeter. As there are no units of measurements shown on these sheets, the spaces can represent any convenient unit of length, Figure 7–10.

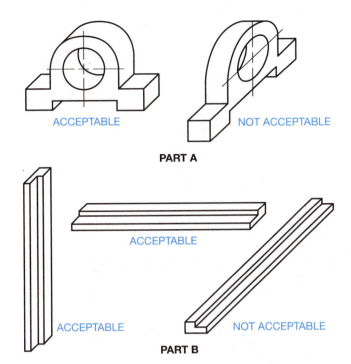

ACCEPTABLE NOT ACCEPTABLE

PART A

ACCEPTABLE

ACCEPTABLE NOT ACCEPTABLE

PART B

FIGURE 7–8 ■ *Two general rules for oblique projection.*

Inclined Surfaces

Angles that are parallel to the picture plane are drawn as their true size. Other angles can be laid off by locating the ends of the inclined line.

A part with notched corners is shown in Figure 7–11(A). An oblique drawing with the angles parallel to the picture plane is shown in Figure 7–11(B). In Figure 7–11(C) the angles are parallel to the profile plane. In each case, the angle is laid off by measurement parallel to the oblique axes, as shown by the construction lines. Because the part, in each case, is drawn in cabinet oblique, the receding lines are shortened by one-half their true length.

Circles and Arcs

Whenever possible, the face of the object having circles or arcs should be selected as the front face, so that such circles or arcs can be easily drawn in their true shape, Figure 7–12.

When circles or arcs must be drawn on one of the oblique faces, the following method is recommended. With reference to Figure 7–12(B):

- Block off an oblique square with center lines equal to the diameter of the circle required. Blocking in the circle first also helps to get the proper size and shape of the ellipse. If an ellipse template is available, select an ellipse that fits within the square and touches the sides of the square at its midpoints. Using thick, dark lines (object lines), draw the oblique circle (ellipse), Figure 7–12(C).

- If an ellipse template is not available, lightly sketch an ellipse within this square with the circumference of the ellipse making contact with the square at its midpoints, Figure 7–12(B).

- Using object lines, darken the oblique circle, Figure 7–12(C).

Basic Steps to Follow for Oblique Sketching (Figure 7–13)

Step 1. Build a Frame. The frame or box is the overall size of the part to be drawn. It is drawn with light, thin lines.

Step 2. Block in the Overall Size of Each Detail. These subblocks or frames enclose each detail. For circles, draw squares equal to the diameter size.

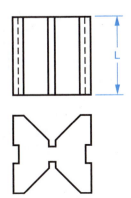 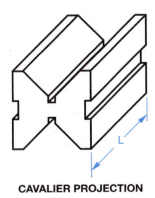 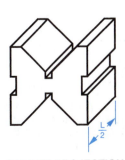

CAVALIER PROJECTION   **CABINET PROJECTION**

FIGURE 7–9 ■ *Types of oblique projection.*

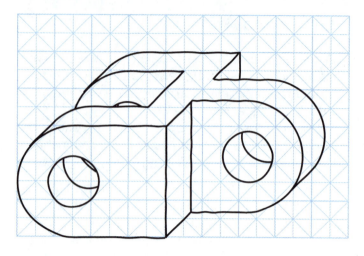

FIGURE 7–10 ■ *Oblique sketching paper.*

Also sketch the center lines. They are drawn using light, thin lines.

Step 3. Add the Details. Lightly sketch the shape of the details in each of their frames. These details are drawn using light, thin lines. If an oblique circle (ellipse) template is available, the arcs and circles are drawn using thick, dark (visible object) lines.

Step 4. Darken the Lines. Use a soft lead pencil to darken the lines.

REFERENCES

ASME Y14.4M-1989 (R1999) Pictorial Drawing

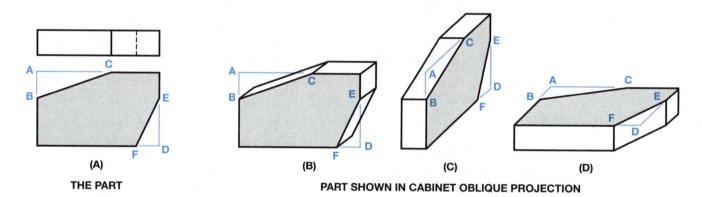

(A)	(B)	(C)	(D)
THE PART	PART SHOWN IN CABINET OBLIQUE PROJECTION		

FIGURE 7–11 ■ *Drawing inclined surfaces.*

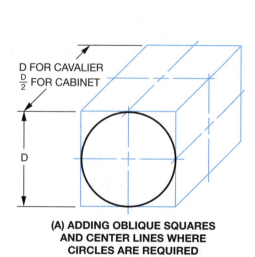

(A) ADDING OBLIQUE SQUARES AND CENTER LINES WHERE CIRCLES ARE REQUIRED

D FOR CAVALIER
$\frac{D}{2}$ FOR CABINET

D

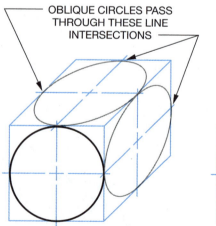

OBLIQUE CIRCLES PASS THROUGH THESE LINE INTERSECTIONS

(B) LIGHTLY SKETCHING IN THE SIZE AND LOCATION OF OBLIQUE CIRCLES

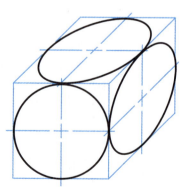

(C) COMPLETING THE OBLIQUE CIRCLES

FIGURE 7–12 ■ *Sketching oblique circles.*

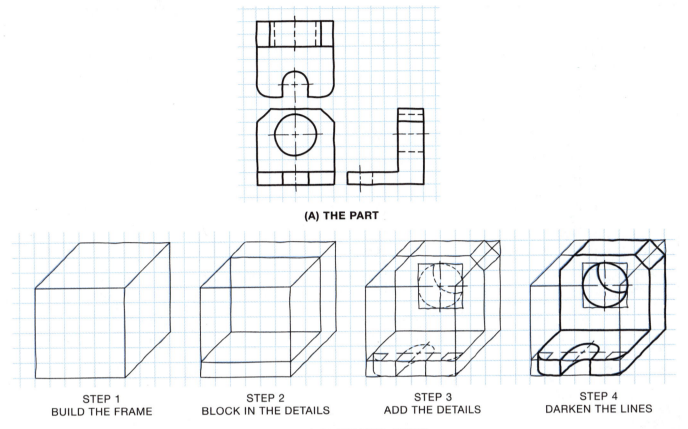

(A) THE PART

STEP 1
BUILD THE FRAME

STEP 2
BLOCK IN THE DETAILS

STEP 3
ADD THE DETAILS

STEP 4
DARKEN THE LINES

(B) BASIC SKETCHING STEPS

FIGURE 7–13 ■ *Basic steps to follow for oblique sketching.*

INTERNET RESOURCES

. Animated Worksheets. For information on
isometric and perspective drawings, see:
http://www.animatedworksheets.co.uk

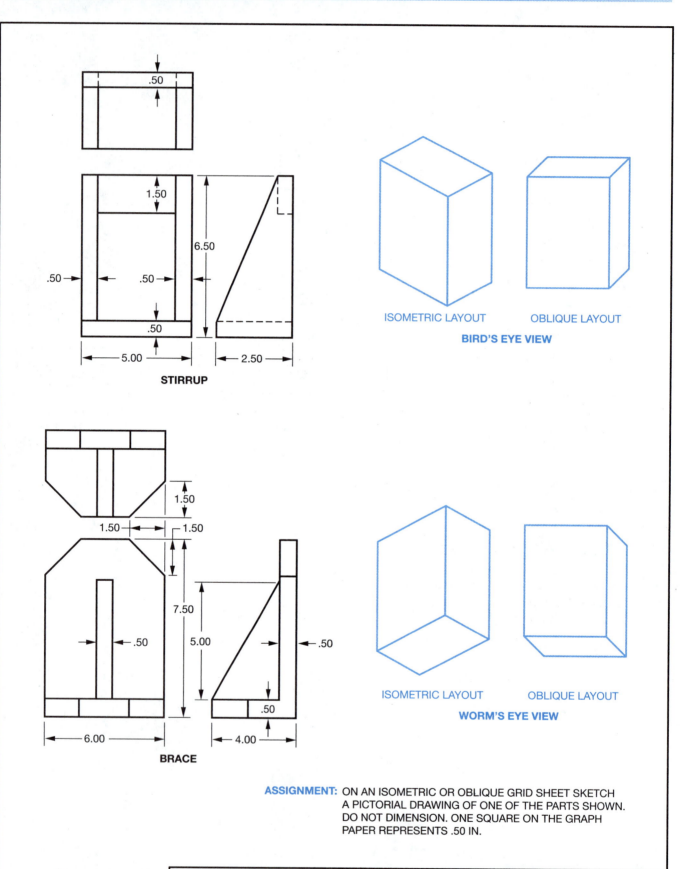

.50

1.50

6.50

.50 — — .50 — — — .50

.50

— 5.00 — — 2.50 —

STIRRUP

ISOMETRIC LAYOUT   OBLIQUE LAYOUT

BIRD'S EYE VIEW

1.50

1.50 — — 1.50

7.50

.50

5.00   — .50

.50

— 6.00 —   — 4.00 —

BRACE

ISOMETRIC LAYOUT   OBLIQUE LAYOUT

WORM'S EYE VIEW

ASSIGNMENT: ON AN ISOMETRIC OR OBLIQUE GRID SHEET SKETCH A PICTORIAL DRAWING OF ONE OF THE PARTS SHOWN. DO NOT DIMENSION. ONE SQUARE ON THE GRAPH PAPER REPRESENTS .50 IN.

PICTORIAL SKETCHING OF PARTS HAVING FLAT SURFACES USING DECIMAL-INCH DIMENSIONING   **A-23**

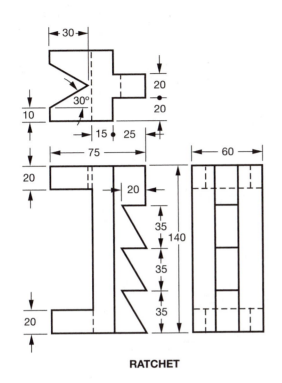

RATCHET

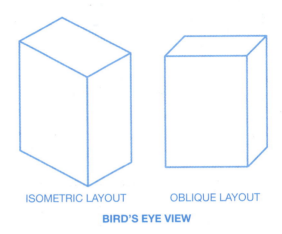

ISOMETRIC LAYOUT   OBLIQUE LAYOUT

BIRD'S EYE VIEW

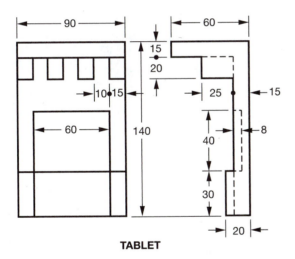

TABLET

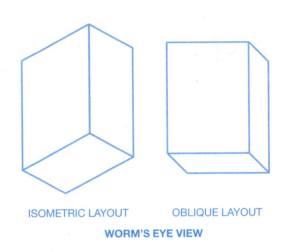

ISOMETRIC LAYOUT   OBLIQUE LAYOUT

WORM'S EYE VIEW

ASSIGNMENT: ON AN ISOMETRIC OR OBLIQUE GRID SHEET SKETCH
A PICTORIAL DRAWING OF ONE OF THE PARTS SHOWN.
DO NOT DIMENSION. ONE SQUARE ON THE GRAPH
PAPER REPRESENTS 10 mm.

METRIC
DIMENSIONS IN MILLIMETERS

**PICTORIAL SKETCHING OF PARTS HAVING FLAT
SURFACES USING MILLIMETER DIMENSIONING**

A-24M

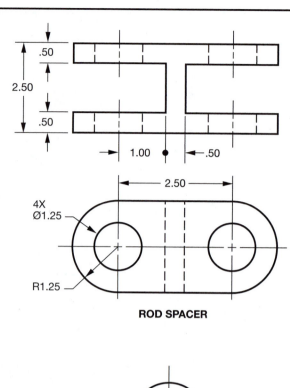

4X
Ø1.25

R1.25

.50

2.50

.50

1.00

.50

2.50

ROD SPACER

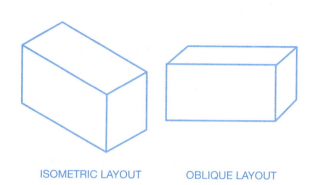

ISOMETRIC LAYOUT   OBLIQUE LAYOUT

BIRD'S EYE VIEW

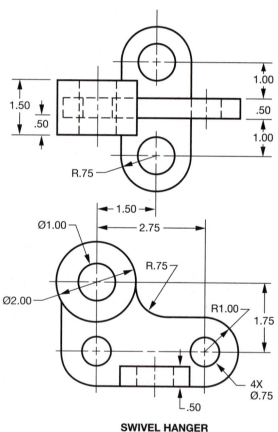

1.50

.50

R.75

1.00

.50

1.00

1.50

2.75

Ø1.00

R.75

Ø2.00

R1.00

1.75

4X
Ø.75

.50

SWIVEL HANGER

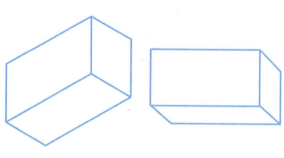

ISOMETRIC LAYOUT   OBLIQUE LAYOUT

WORM'S EYE VIEW

ASSIGNMENT:
ON AN ISOMETRIC OR OBLIQUE GRID SHEET SKETCH
A PICTORIAL DRAWING OF ONE OF THE PARTS SHOWN.
DO NOT DIMENSION. ONE SQUARE ON THE GRAPH
PAPER REPRESENTS .25 IN.

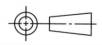

**PICTORIAL SKETCHING OF PARTS HAVING CIRCULAR
FEATURES USING DECIMAL-INCH DIMENSIONING**

A-25

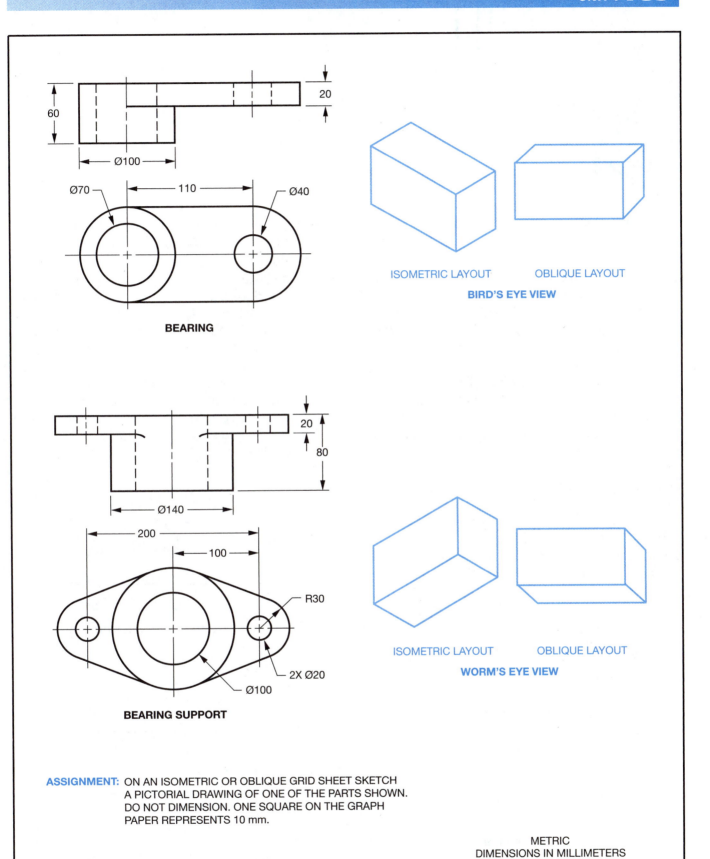

60

20

Ø100

Ø70   110   Ø40

BEARING

ISOMETRIC LAYOUT   OBLIQUE LAYOUT

BIRD'S EYE VIEW

20

80

Ø140

200

100

R30

2X Ø20

Ø100

BEARING SUPPORT

ISOMETRIC LAYOUT   OBLIQUE LAYOUT

WORM'S EYE VIEW

ASSIGNMENT: ON AN ISOMETRIC OR OBLIQUE GRID SHEET SKETCH
A PICTORIAL DRAWING OF ONE OF THE PARTS SHOWN.
DO NOT DIMENSION. ONE SQUARE ON THE GRAPH
PAPER REPRESENTS 10 mm.

METRIC
DIMENSIONS IN MILLIMETERS

**PICTORIAL SKETCHING OF PARTS HAVING
CIRCULAR FEATURES USING METRIC DIMENSIONING**

A-26M

MACHINING SYMBOLS

When preparing working drawings of parts to be cast or forged, the drafter must identify part surfaces that require machining or finishing. This is done by adding a machining allowance symbol to the surface or surfaces that must be finished by the removal of material, Figure 8–1. Figure 8–2 shows the current machining symbol and those which were formerly used on drawings. This information is essential in order to alert the patternmaker and diemaker to provide extra metal on the casting or forging to allow for the finishing process. Depending on the material to be cast or forged, between .04 and .10 inch is usually allowed on small castings and forgings for each surface that requires finishing, Figure 8–3.

Like dimensions, machining symbols are not duplicated on the drawing. They should be used on the same view as the dimensions that give the size or location of the surfaces. The symbol is placed on the line representing the surface or on a leader or an extension line locating the surface. The symbol and the inscription should be oriented so they may be read from the bottom or right-hand side of the drawing.

Where all the surfaces are to be machined, a general note, such as FINISH ALL OVER (FAO), may be used and the symbols on the drawings omitted.

The outdated machining symbols shown in Figure 8–2(B) are found on many older drawings still in use. When called upon to make changes or revisions on a drawing already in existence, a drafter must adhere to the drawing conventions shown on that drawing.

The machining symbol does not indicate surface-finish quality. The surface texture symbol, which is

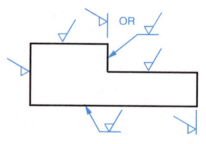

(A) RECOMMENDED SYMBOL

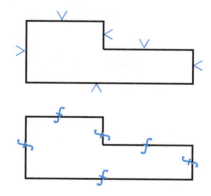

(B) FORMER MACHINING SYMBOLS

FIGURE 8–2 ■ *Application of machining symbol.*

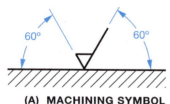

(A) MACHINING SYMBOL

FINISHED SURFACE

ORIGINAL SURFACE

EXTRA MATERIAL PROVIDED TO PRODUCE A DESIRED SURFACE FINISH

(B) MEANING

FIGURE 8–1 ■ *Machining symbol.*

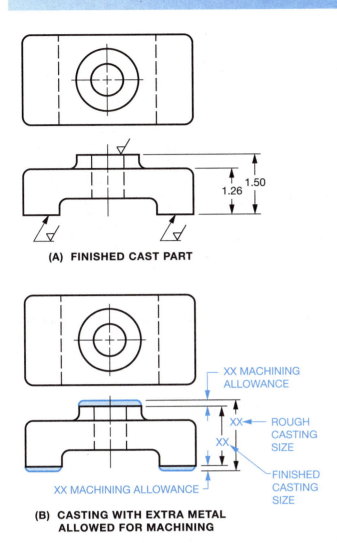

(A) FINISHED CAST PART

(B) CASTING WITH EXTRA METAL ALLOWED FOR MACHINING

FIGURE 8–3 ■ *Allowance for machining.*

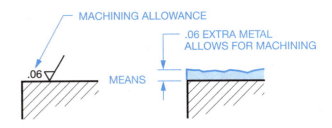

FIGURE 8–4 ■ *Indicating the value of the machining allowance.*

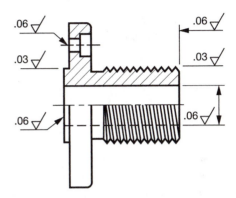

FIGURE 8–5 ■ *The value of the machining allowance as shown on the drawing.*

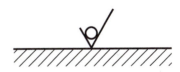

FIGURE 8–6 ■ *Symbol for removal of material not permitted.*

defined and discussed in Unit 12, is used to signify the desired surface-finish quality.

Indicating Machining Allowance

When the value of the machining allowance must be specified, it is indicated to the left of the symbol, Figures 8–4 and 8–5. This value is expressed in inches or millimeters depending on which units of measurement are used on the drawing.

Removal of Material Prohibited

A surface from which the removal of material is prohibited is indicated by the symbol shown in Figure 8–6. This symbol indicates that a surface must be left the

way it is affected by a preceding manufacturing process, regardless of the removal of material or other changes. This machining symbol is part of the surface texture symbol described in Unit 11.

NOT-TO-SCALE DIMENSIONS

When a dimension on a drawing is altered, making it not to scale, a straight freehand line is drawn below the dimension to indicate that the dimension is not drawn to scale, Figure 8–7(A).

This is a change from earlier methods of indicating not-to-scale dimensions. Formerly, a wavy line below the dimension or the letters NTS beside the dimension were used to indicate not-to-scale dimensions, Figure 8–7(B).

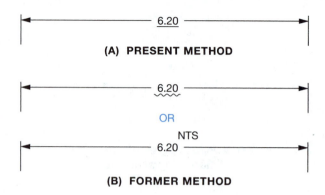

FIGURE 8–7 ■ *Indicating dimensions that are not to scale.*

DRAWING REVISIONS

Drawing revisions are made to accommodate improved manufacturing methods, reduce costs, correct errors, and improve design. A clear record of these revisions must be registered on the drawing.

All drawings should carry a change or revision table located at the bottom or down the top right-hand side of the drawing. The revision number, enclosed in a circle or triangle, should be located near the revised dimension for easy identification, Figure 8–8(A). The revision block should include a revision number or symbol, the date, the drafter's name or initials, and approval of the change.

The recommended method for showing dates on engineering drawings consists of three two-digit values separated by a slash. The first two digits represents the shortest time (day), followed by the next shortest time (month), then the year. Thus, January 12, 2004 would be shown as 12/01/04.

Should the drawing revision cause a dimension or dimensions to be different from the scale indicated, the dimensions that are not to scale should be indicated. Typical revision tables are shown in Figure 8–8(B) and (C).

When many revisions are needed, a new drawing is often made. The words REDRAWN AND REVISED should appear in the revision column of the new drawing when this is done.

After a revision to a drawing is made, a new set of prints of that drawing is distributed to the appropriate departments and the prints of the original drawing are destroyed.

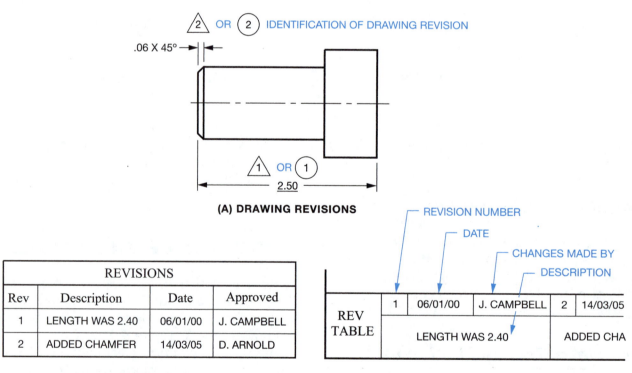

(A) DRAWING REVISIONS

REVISIONS

Rev	Description	Date	Approved
1	LENGTH WAS 2.40	06/01/00	J. CAMPBELL
2	ADDED CHAMFER	14/03/05	D. ARNOLD

**(B) TYPICAL VERTICAL REVISION BLOCK
(SIZE MAY VARY)**

REVISION NUMBER
DATE
CHANGES MADE BY
DESCRIPTION

REV TABLE	1	06/01/00	J. CAMPBELL	2	14/03/05
	LENGTH WAS 2.40				ADDED CHA

**(C) TYPICAL HORIZONTAL REVISION BLOCK
(SIZE MAY VARY)**

FIGURE 8–8 ■ *Drawing revisions.*

BREAK LINES

Break lines, as shown in Figure 8–9, are used to shorten the view of long uniform sections. They are also used when only a partial view is required. Such lines are used on both detail and assembly drawings. The short break lines, shown in Figure 8–9(A)

were introduced in Unit 2. The thin line with freehand zigzags is recommended for long breaks, and the jagged line for wood parts. The special breaks shown for cylindrical and tubular parts are useful when an end view is not shown; otherwise, the thick break line is adequate.

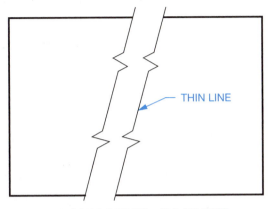

THIN LINE

(A) LONG BREAK - ALL SHAPES

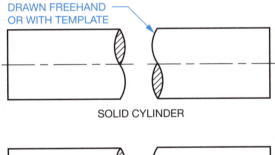

DRAWN FREEHAND OR WITH TEMPLATE

SOLID CYLINDER

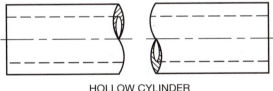

HOLLOW CYLINDER
USEFUL WHEN END VIEW IS NOT SHOWN

(B) CYLINDERS

FIGURE 8–9 ■ *Conventional break lines.*

REFERENCES

ASME Y14.36M-1996 (R2002) Surface Texture Symbols

ASME Y14.2M (R2003) Line Conventions and Lettering

ASME Y14.35M-1997 (R2003) Revision of Engineering Drawings

ASME Y14.5M-1994 (R1999) Dimensioning and Tolerancing

INTERNET RESOURCES

American Society of Mechanical Engineers. For information on surface texture, machining symbols, and conventional breaks, refer to ASME Y14.36M-1996 *(Surface Texture Symbols)* at: http://www.asme.org

American Society of Mechanical Engineers. For information on drawing revisions, refer to ASME Y14.35M-1997 *(Revision of Engineering Drawings and Associated Documents)* at: http://www.asme.org

Drafting Zone. For information on conventional breaks and drawing revisions, see: http://www.draftingzone.com

Integrated Publishing. For information on break lines, see: www.tpub.com/engbas/3-21htm

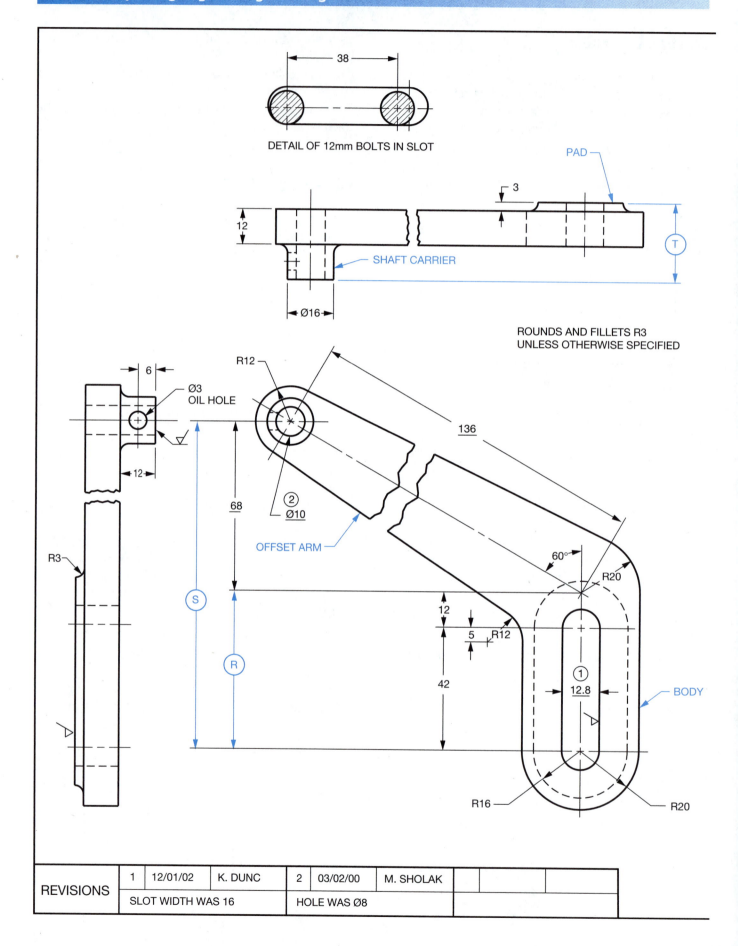

DETAIL OF 12mm BOLTS IN SLOT

38

PAD

3

12

T

Ø16

SHAFT CARRIER

ROUNDS AND FILLETS R3
UNLESS OTHERWISE SPECIFIED

6

Ø3
OIL HOLE

12

R12

136

R3

Ø10
②

OFFSET ARM

60°

R20

S

R

12

5   R12

42

①
12.8

BODY

R16

R20

68

REVISIONS	1	12/01/02	K. DUNC	2	03/02/00	M. SHOLAK			
	SLOT WIDTH WAS 16			HOLE WAS Ø8					

QUESTIONS:

1. At what angle is the offset arm to the body of the piece?

2. What is the center-to-center measurement of the length of the offset arm?

3. Which radius forms the upper end of the offset arm?

4. Which radii form the lower end of the offset arm where it joins the body?

5. What is the width of the bolt slot in the body of the bracket?

6. What is the center-to-center length of this slot?

7. What was the slot width before revision?

8. Which radii forms the ends of the pad?

9. What is the overall length of this pad?

10. What is the overall width of this pad?

11. What is the radii of the fillet between the pad and the body?

12. What is the diameter of the shaft carrier body?

13. What is the diameter of the shaft carrier hole?

14. What is the distance from the face of the shaft carrier to the face of the pad?

15. What is the radii of the inside fillet between the arm and the body of the piece?

16. If M12 bolts are used in holding the bracket to the machine base, what is the clearance on each side of the slot?

17. If the center-to-center distance of the two M12 bolts which fit the slot is 38mm, how much play is there lengthwise in the slot?

18. What size oil hole is in the shaft carrier?

19. How far is the center of the oil hole from the face of the shaft carrier?

20. How thick is the combined body and pad?

21. Calculate distance Ⓡ, Ⓢ and Ⓣ.

22. The hole in the shaft carrier was revised. What is the difference in size between the new and old hole?

23. How many dimensions indicate that they are not drawn to scale?

24. If 2mm is allowed for each surface to be machined, what would be the overall thickness of the original casting?

25. How many conventional breaks are shown?

26. When was the last drawing revision made?

METRIC
DIMENSIONS IN MILLIMETERS

MATERIAL	MI	
SCALE	1:1	
DRAWN	T. Logan	DATE 15/12/99

OFFSET BRACKET   **A-27M**

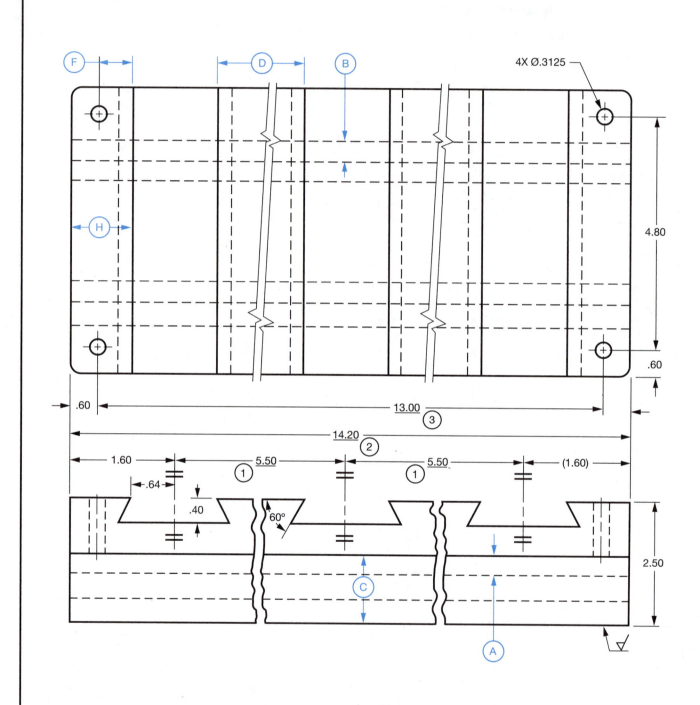

4X Ø.3125

4.80

.60

.60

13.00
3

14.20
2

1.60

5.50
1

5.50
1

(1.60)

.64

.40

60°

2.50

C

A

NOTE:

ROUNDS AND FILLETS R.10

√ SHOWN TO BE .10 √

REV TABLE	1	18/07/05	R. HINES	2	18/07/05	R. HINES	3	18/07/05	R. HINES
	5.50 WAS 6.25			14.20 WAS 15.70			13.00 WAS 14.50		

QUESTIONS:

1. What is the drawing scale?
2. How many machine slots are shown?
3. How many reference dimensions are shown?
4. How many conventional breaks are shown?
5. How many not-to-scale dimensions are shown?
6. What was the total number of dimensions altered due to the drawing revisions?
7. How many symmetrical shapes are indicated by means of the symmetry symbol?
8. What machine allowance is called for on the machined surfaces?
9. How many fillets are shown?
10. How many rounds are shown?
11. What size bolts would be used in the holes?
12. What was the original length of the casting?
13. What was the overall height of the casting before the top and bottom surfaces were machined?
14. What is the overall height of the T-slot?
15. What is the change in distance between the centers of dovetail slots from the present and the original drawing?
16. Determine distances Ⓐ to Ⓗ.

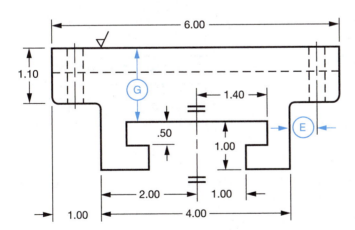

MATERIAL	GRAY IRON	
SCALE	1 : 2	
DRAWN	P. JENSEN	DATE 30/11/04

GUIDE BAR   **A-28**

9 UNIT

SECTIONAL VIEWS

Sectional views, commonly called *sections,* are used to show interior detail too complicated to be shown clearly and dimensioned by outside views and hidden lines. A sectional view is obtained by supposing the nearest part of the object has been cut or broken away on an imaginary cutting plane. The exposed or cut surfaces are identified by section lining or crosshatching. Hidden lines and details behind the cutting-plane line are usually omitted unless they are required for clarity. It should be understood that only in the sectional view is any part of the object shown as having been removed.

A sectional view frequently replaces one of the regular views. For example, a regular front view is replaced by a front view in section, as shown in Figure 9–1.

The Cutting-Plane Line

A *cutting-plane line* indicates where the imaginary cutting takes place. The position of the cutting plane is indicated, when necessary, on a view of the object or assembly by a cutting-plane line, as shown in Figure 9–2. The ends of the cutting-plane line are bent at 90 degrees and terminated by arrowheads to indicate the direction of sight for viewing the section. Cutting planes are not shown on sectional views. The cutting-plane line may be omitted when it corresponds to the center line of the part or when only one sectional view appears on a drawing.

If two or more sections appear on the same drawing, the cutting-plane lines are identified by two identical large, single-stroke, Gothic letters. One letter is placed at each end of the line near the arrowhead. Sectional view subtitles are given when identification letters are used and appear directly below the view, incorporating the letters at each end of the cutting-plane line, thus: SECTION A-A or, abbreviated, SECT A-A. See Assignment A-31M.

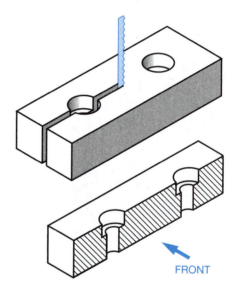

FRONT

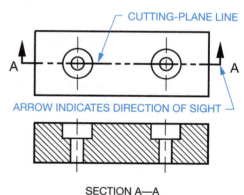

CUTTING-PLANE LINE

ARROW INDICATES DIRECTION OF SIGHT

SECTION A—A

FIGURE 9–1 ■ *A section drawing.*

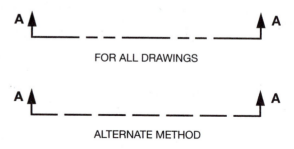

FOR ALL DRAWINGS

ALTERNATE METHOD

NOTE: LETTERS PLACED BESIDE ARROWS

FIGURE 9–2 ■ *Cutting-plane lines.*

Section Lining

Section lining identifies the surface that has been cut and makes it stand out clearly. Section lines usually consist of thin parallel lines, Figure 9–3, drawn at an angle of approximately 45 degrees to the principal edges or axis of the part.

Because the exact material specifications for a part are usually given elsewhere, the general-use section lining, Figure 9–4, is recommended for general use.

When it is desirable to indicate differences in materials, other symbolic section lines are used, such as those shown in Figure 9–4. If the part shape would cause section lines to be parallel or nearly parallel to one of the sides or features of the part, an angle other than 45 degrees is chosen.

The spacing of the hatching lines is uniform to give a good appearance to the drawing. The pitch, or distance, between lines varies from .06 to .18 inch, depending on the size of the area to be sectioned. Section lining is uniform in direction and spacing in all sections of a single component.

Wood and concrete are the only two materials usually shown symbolically. When wood symbols are used, the direction of the grain is shown.

TYPES OF SECTIONS

Full Sections

When the cutting plane extends entirely through the object in a straight line and the front half of the object is theoretically removed, a *full section* is obtained, Figure 9–5(B). This type of section is used for both detail and assembly drawings. When the cutting plane divides the object into two identical parts,

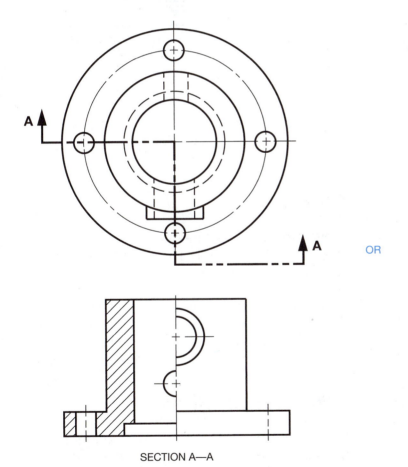

OR

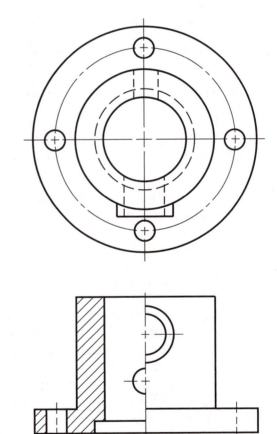

SECTION A—A

LETTERS, SUBTITLE AND CUTTING-PLANE LINE USED WHEN MORE THAN ONE SECTION APPEARS ON A DRAWING OR WHEN THEY MAKE THE DRAWING CLEARER.

LETTERS, SUBTITLE AND CUTTING-PLANE LINE MAY BE OMITTED WHEN THEY CORRESPOND WITH THE CENTER LINE OF THE PART AND WHEN THERE IS ONLY ONE SECTION VIEW ON THE DRAWING.

FIGURE 9–3 ■ *Identification of cutting plane and sectional view.*

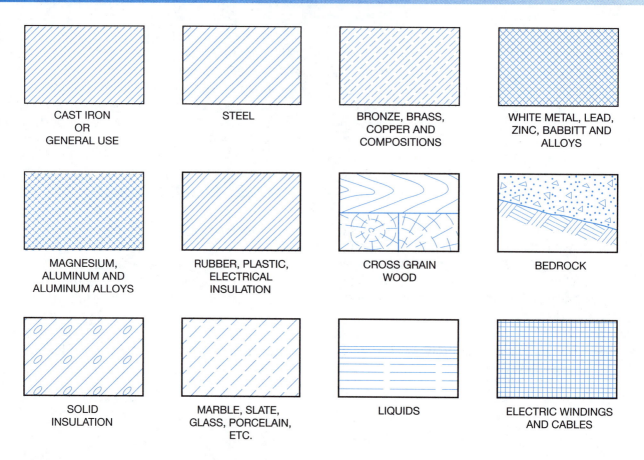

FIGURE 9–4 ■ *Symbolic section lining.*

it is not necessary to indicate its location. However, the cutting plane may be drawn and labeled in the usual manner to increase clarity.

Half Sections

A symmetrical object or assembly may be drawn as a *half section,* Figure 9–5(C), showing one half in section and the other half in full view. A normal center line is used on the section view.

The half section drawing is not normally used where the dimensioning of internal diameters is required. This is because many hidden lines would have to be added to the portion showing the external features. This type of section is used mostly for assembly drawings where internal and external features are clearly shown and only overall and center-to-center dimensions are required.

Offset Sections

In order to include features that are not in a straight line, the cutting-plane line may be offset or bent, so as to include several planes or curved surfaces, Figures 9–6 and 9–7.

An offset section is similar to a full section in that the cutting plane extends through the object from one side to the other. The change in direction of the cutting-plane line is not shown on the sectional view.

Other Types of Sections

Other types of sections, such as **revolved, removed, partial** or **broken-out,** and **sectioned assembly** drawings, are explained in detail in other Units within this text. In addition, sectioning techniques for showing ribs, spokes, shafts, rivets, and keys are also explained.

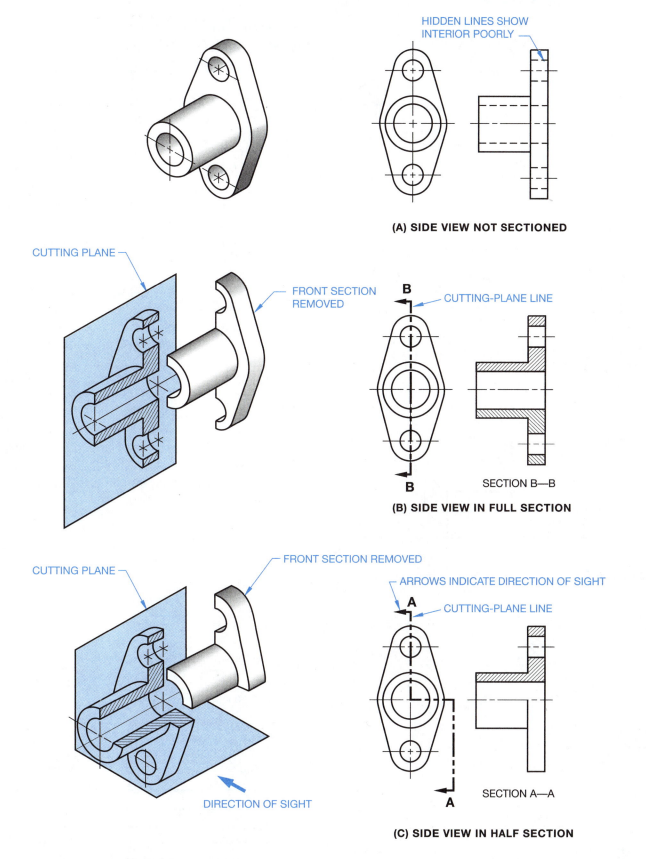

HIDDEN LINES SHOW INTERIOR POORLY

(A) SIDE VIEW NOT SECTIONED

CUTTING PLANE

FRONT SECTION REMOVED

B

CUTTING-PLANE LINE

B

SECTION B—B

(B) SIDE VIEW IN FULL SECTION

CUTTING PLANE

FRONT SECTION REMOVED

ARROWS INDICATE DIRECTION OF SIGHT

A

CUTTING-PLANE LINE

A

SECTION A—A

DIRECTION OF SIGHT

(C) SIDE VIEW IN HALF SECTION

FIGURE 9–5 ■ *Full and half sections.*

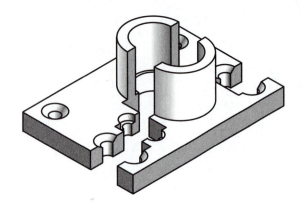

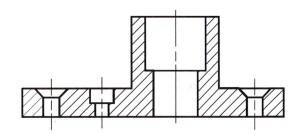

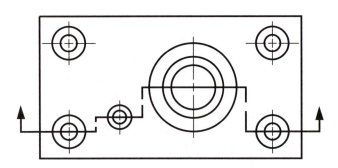

FIGURE 9–6 ■ *An offset section.*

COUNTERSINKS, COUNTERBORES, AND SPOTFACES

A *countersunk* hole is a conical depression cut in a piece to receive a countersunk type of flathead screw or rivet, as illustrated in Figure 9–8. The size is usually shown by a note listing the diameter of the hole first, followed by the diameter of the counter-sink, the abbreviation CSK, and the angle, A *counterbored hole* is one which has been machined larger to a given depth to receive a fillister, hex-head, or similar type of bolt head. Counterbores are specified by a note giving the diameter of the hole first, followed by the counterbore diameter, the abbreviation CBORE, and depth of the counterbore. The counterbore and depth may also be indicated by direct dimensioning. A *spotface* is an area where the surface is machined just enough to provide a level seating surface for a bolt head, nut, or washer. A spotface is specified by a note listing the diameter of the hole first, followed by the spotface diameter, and the abbreviation SFACE. The depth of the spotface is not usually given.

The symbolic means of specifying a counterbore or spotface, a countersink, and the depth of a feature are shown in Figure 9–9. In each case, the symbol precedes the dimension.

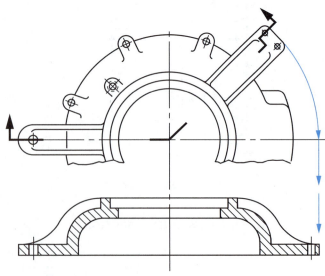

FIGURE 9–7 ■ *Bent cutting-plane line for offset sections.*

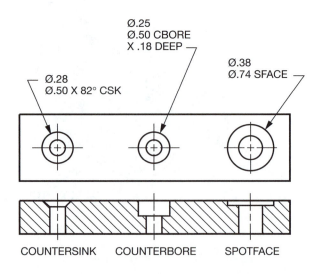

FIGURE 9–8 ■ *Using words to dimension countersinks, counterbores, and spotfaces.*

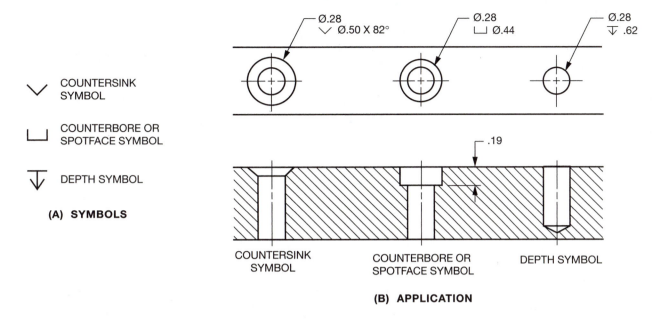

COUNTERSINK
SYMBOL

COUNTERBORE OR
SPOTFACE SYMBOL

DEPTH SYMBOL

(A) SYMBOLS

COUNTERSINK
SYMBOL

COUNTERBORE OR
SPOTFACE SYMBOL

DEPTH SYMBOL

(B) APPLICATION

FIGURE 9–9 ■ *Using symbols to indicate shape and depth of holes.*

INTERSECTION OF UNFINISHED SURFACES

The intersection of unfinished surfaces that are rounded or filleted at the point of theoretical intersection is indicated by a line coinciding with the theoretical point of intersection. The need for this convention is shown by the examples in Figure 9–10. For a large radius, Figure 9–10(C), no line is drawn.

Members such as ribs and arms that blend into other features end in curves called *runouts*.

REFERENCES

ASME Y14.3M-1994 (R1999) Multi- and Sectional-View Drawings

ASME Y14.2M-1992 (R2003) Line Conventions and Lettering

INTERNET RESOURCES

American Society of Mechanical Engineers. For information on sectional views and related topics, refer to ASME Y14.3M-1994 (R1999) (*Multi- and Sectional-View Drawings*) at: http://www.asme.org

Drafting Zone. For information on sectioning, see: http://www.draftingzone.com

IDS Development-Nebraska Education. For information on the various line types used on engineering drawings, see: http://idsdev.mccneb.edu/djackson/lineintro.htm

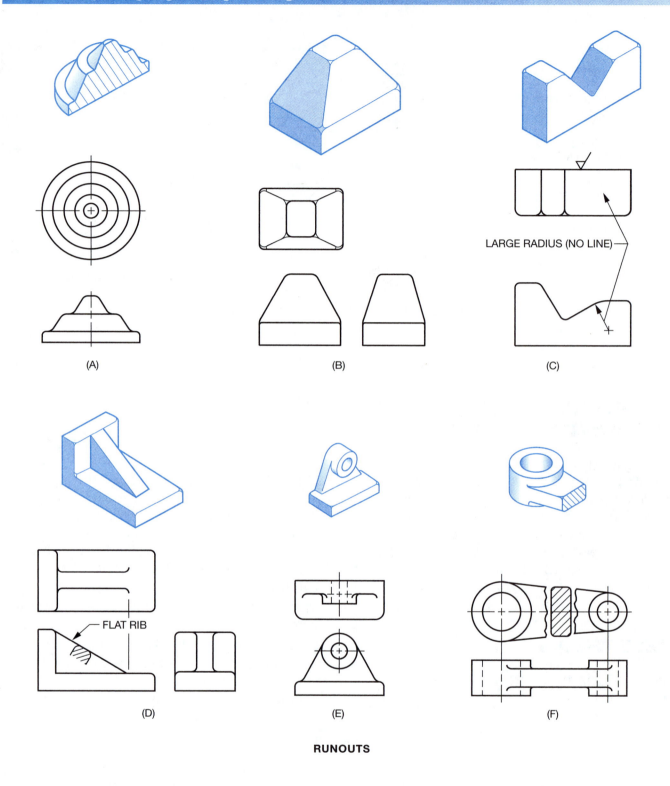

(A)

(B)

LARGE RADIUS (NO LINE)

(C)

FLAT RIB

(D)

(E)

(F)

RUNOUTS

FIGURE 9–10 ■ *Rounded and filleted intersections.*

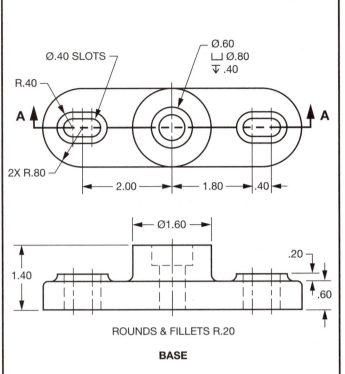

Ø.40 SLOTS

R.40

A

2X R.80

Ø.60
⊔ Ø.80
▽ .40

A

2.00   1.80   .40

Ø1.60

1.40   .20   .60

ROUNDS & FILLETS R.20

BASE

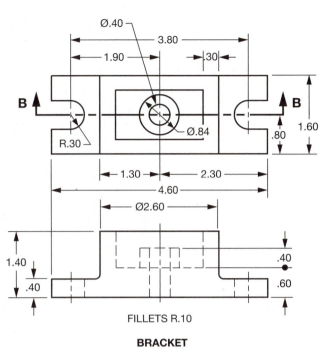

Ø.40   3.80

1.90   30

B   **B**

R.30   Ø.84   1.60   .80

1.30   2.30

4.60

Ø2.60

1.40   .40   .40   .60

FILLETS R.10

BRACKET

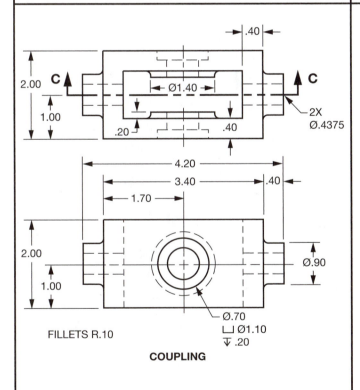

.40

C   **C**

2.00   Ø1.40

1.00   2X Ø.4375

.20   .40

4.20

3.40   .40

1.70

2.00   Ø.90

1.00

Ø.70
⊔ Ø1.10
▽ .20

FILLETS R.10

COUPLING

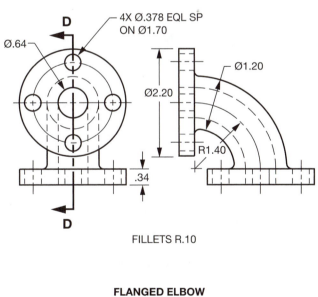

4X Ø.378 EQL SP
ON Ø1.70

D

Ø.64

Ø1.20

Ø2.20

R1.40

.34

D

FILLETS R.10

FLANGED ELBOW

ASSIGNMENT:
USE ONE-INCH GRID SHEETS (.10 IN. SQUARES).
SKETCH THE FOUR FULL-SECTION VIEWS. SCALE 1 : 1.

**SKETCHING
FULL SECTIONS**

A-29

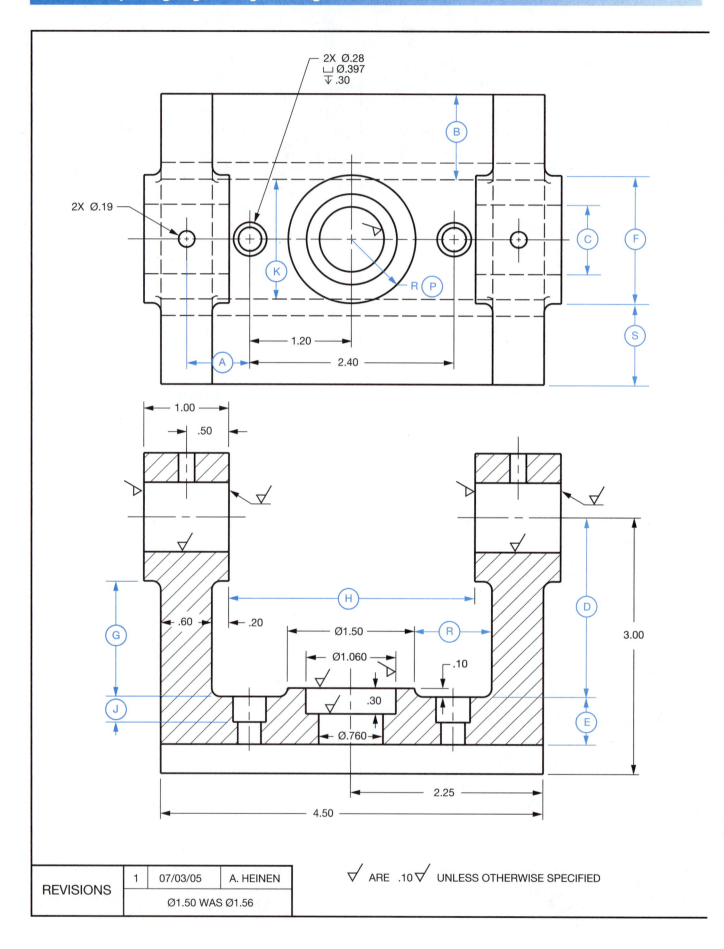

2X Ø.28
⌴ Ø.397
▼ .30

B

2X Ø.19

C   F

K

R P

S

1.20

A

2.40

1.00

.50

D   3.00

H

.60   .20   Ø1.50   R

G

Ø1.060   .10

J   .30

E

Ø.760

2.25

4.50

| REVISIONS | 1 | 07/03/05 | A. HEINEN |
| | Ø1.50 WAS Ø1.56 | | |

∇ ARE .10 ∇ UNLESS OTHERWISE SPECIFIED

QUESTIONS:

1. What is the overall width?

2. What is the overall height?

3. What is the center-to-center distance of the Ø.19 holes?

4. Calculate the width of the casting before machining.

5. At what angle to the vertical are the sides of the dovetail slot?

6. How many different surfaces require finishing?

7. Which type of lines in the top view represent the dovetail?

8. What was the original size of the Ø1.50?

9. How wide is the opening in the dovetail?

10. How high is the dovetail?

11. When was the Ø1.50 dimension altered?

12. How many reference dimensions are shown?

13. Calculate the distances (A) to (S).

14. What is the (A) size and (B) type of cap screw required to fasten the slide bracket to its mating part?

15. How many not-to-scale dimensions are shown?

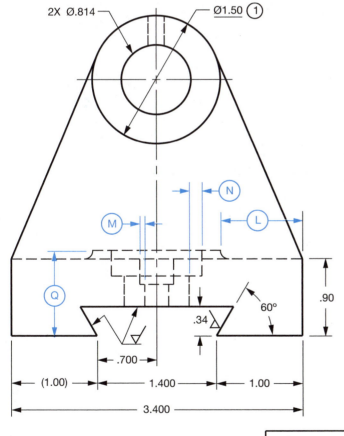

ROUNDS AND FILLETS R.10

MATERIAL	GRAY IRON	
SCALE	1:1	
DRAWN	D. SMITH	DATE 16/10/04

SLIDE BRACKET

A-30

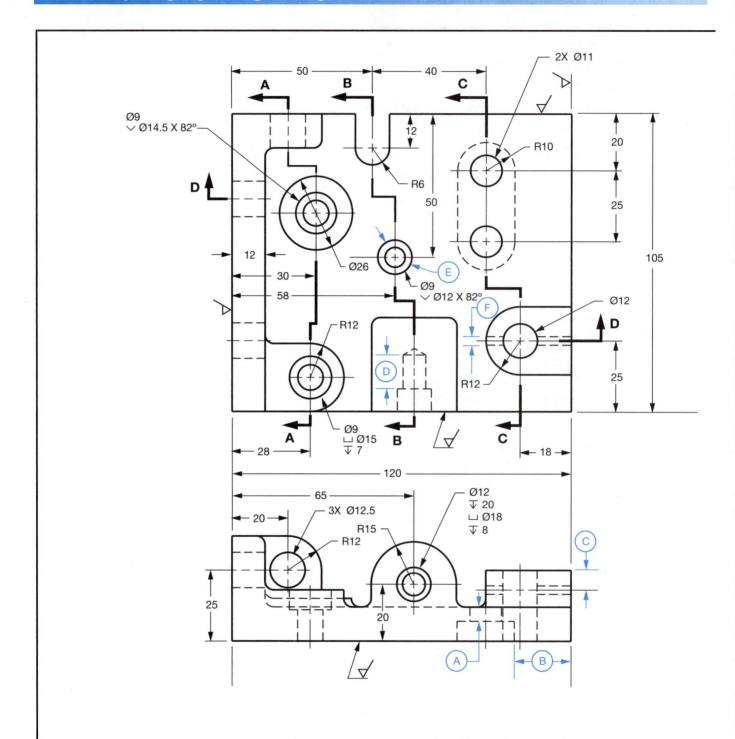

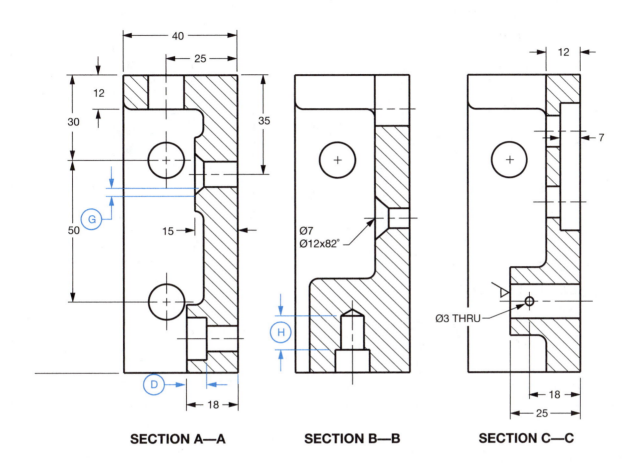

SECTION A—A   **SECTION B—B**   **SECTION C—C**

ROUNDS AND FILLETS R3

✓ SHOWN TO BE 2 ✓

QUESTIONS:

1. What type of sectional view is used?

2. How many surfaces require finishing?

3. What is the total number of holes in the part?
   Note: A THRU hole is considered as one hole.

4. How many countersunk holes are there?

5. What was the (A) width, (B) height and (C) depth of the casting before finishing?

6. With reference to the counterbored hole, what is the (A) size and (B) type of cap screw required if the head of the cap screw must not protrude above the surface of the part? Refer to Table 7 of the Appendix.

7. What is the total number of offsets used on the cutting-plane lines?

8. Calculate distance A through H.

METRIC
NOTE: DIMENSIONS IN MILLIMETERS

ASSIGNMENT:
ON A CENTIMETER GRID SHEET (1mm SQUARES), SKETCH OFFSET SECTION D-D HAVING THE CUTTING-PLANE PASS THROUGH THE CENTERS OF THE COUNTERSUNK HOLES.

MATERIAL	MALLEABLE IRON	
SCALE		
DRAWN	J. MEANS	DATE 06/04/04

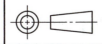 **BASE PLATE**   **A-31M**

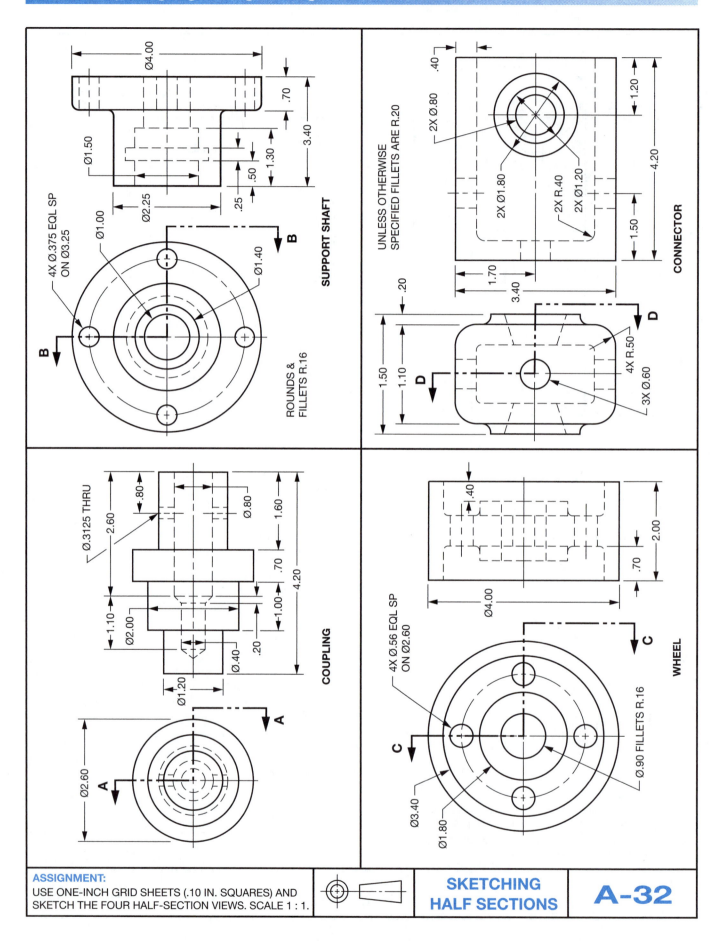

SUPPORT SHAFT

Ø4.00
.70
3.40
1.30
.50
.25
Ø1.50
Ø2.25
4X Ø.375 EQL SP ON Ø3.25
Ø1.00
Ø1.40
B
B
ROUNDS & FILLETS R.16

CONNECTOR

UNLESS OTHERWISE SPECIFIED FILLETS ARE R.20
.40
1.20
4.20
1.50
2X Ø.80
2X Ø1.80
2X R.40
2X Ø1.20
1.70
3.40
.20
1.50
1.10
4X R.50
3X Ø.60
D
D

COUPLING

Ø.3125 THRU
.80
2.60
Ø.80
1.60
.70
4.20
1.00
.20
1.10
Ø2.00
Ø.40
Ø1.20
Ø2.60
A
A

WHEEL

.40
2.00
.70
Ø4.00
4X Ø.56 EQL SP ON Ø2.60
Ø.90 FILLETS R.16
Ø3.40
Ø1.80
C
C

ASSIGNMENT:
USE ONE-INCH GRID SHEETS (.10 IN. SQUARES) AND
SKETCH THE FOUR HALF-SECTION VIEWS. SCALE 1 : 1.

**SKETCHING
HALF SECTIONS**

A-32

CHAMFERS

The process of *chamfering*, that is, cutting away the inside or outside corner of an object, is done to facilitate assembly, Figure 10–1. The recommended method of dimensioning a chamfer is to give an an-gle and the linear length, or an angle and a diameter. For angles of 45 degrees only, a note form may be used. This method is permissible only with 45-degree angles because the size may apply to either the longitudinal or radial dimension. Chamfers are never measured along the angular surface.

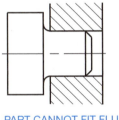
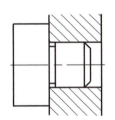
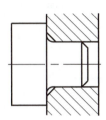

PART CANNOT FIT FLUSH IN HOLE BECAUSE OF SHOULDER

SAME PART WITH UNDERCUT ADDED PERMITS PART TO FIT FLUSH

CHAMFER ADDER TO HOLE TO ACCEPT SHOULDER OF PART

(A) UNDERCUT AND CHAMFER APPLICATION

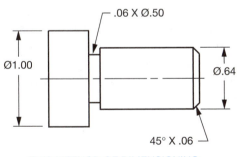

.06 X Ø.50

Ø1.00

Ø.64

45° X .06

THIS METHOD OF DIMENSIONING FOR 45° CHAMFERS ONLY

DIMENSIONING FOR CHAMFERS OTHER THAN 45°

45°

Ø.86

30°

.10

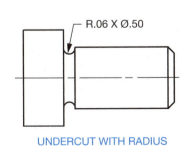

R.06 X Ø.50

UNDERCUT WITH RADIUS

(B) DIMENSIONING CHAMFERS AND UNDERCUTS

FIGURE 10–1 ■ *Chamfers and undercuts.*

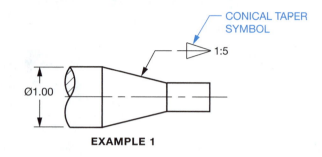

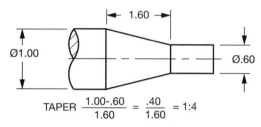

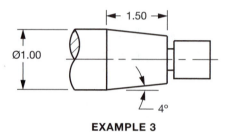

FIGURE 10–2 ∎ *Dimensioning conical tapers.*

UNDERCUTS

The operation of *undercutting*, also referred to as *necking*, is the cutting of a recess in a cylinder. Undercuts permit two parts to join, Figure 10–2. Undercutting is indicated on a drawing by a note listing the width first and then the diameter. If the radius is shown at the bottom of the undercut, it is assumed that the radius will be equal to half the width unless specified differently and the diameter will apply to the center of the undercut. Where the size of the neck is unimportant, the dimension may be omitted from the drawing.

TAPERS

A taper is the ratio of the difference in diameters of two sections along a conical-shape part (perpendicular to the axis).

Conical Tapers

Tapered shanks are used on many small tools such as drills, reamers, counterbores, and spotfaces to hold them accurately in the machine spindle. Conical taper means the difference in diameter or width in a given length. There are many standard tapers; the Morse taper and the Brown and Sharpe taper are the most common.

The following dimensions may be used in suitable combinations, to define the size and form of tapered features:

- The diameter (or width) at one end of the tapered feature

- The length of the tapered feature

- The rate of taper

- The included angle

- The taper ratio

In dimensioning a taper by means of taper ratio, the conical taper symbol should precede the ratio figures, and the vertical leg of the symbol is always shown to the left, Figure 10–2.

Flat Tapers

Flat tapers (slopes) are used as locking devices such as taper keys and adjusting shims. The methods recommended for dimensioning flat tapers are shown in Figure 10–3. The flat taper symbol should precede the ratio figures and the vertical leg of the symbol is always shown to the left.

KNURLS

Knurling is the machining of a surface to create uniform depressions. Knurling permits a better grip. Knurling is shown on drawings as either a straight or diamond pattern. The pitch of the knurl may be specified. It is unnecessary to hatch the whole area to be knurled if enough is shown to clearly indicate the pattern. Knurls are specified on the drawing by a note calling for the type and pitch. The length and diameter of the knurl are shown as dimensions, Figure 10–4.

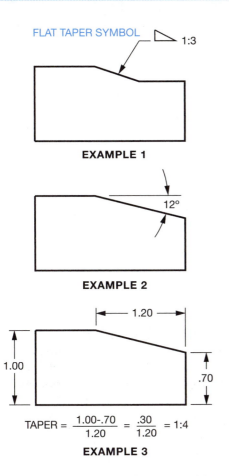

FIGURE 10–3 ■ *Dimensioning flat tapers (slopes).*

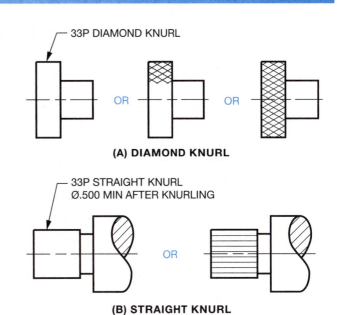

FIGURE 10–4 ■ *Knurling.*

REFERENCE

ASME Y14.5M-1994 (R1999) Dimensioning and Tolerancing

INTERNET RESOURCES

PMPA Designer's Guide. For information on knurls and serrations see: www.pmpa.org/agent/desguide/design/knurls.htm

TechStudent.Com. For information on knurling, see http://www.technologystudent.com (equipment and accessories)

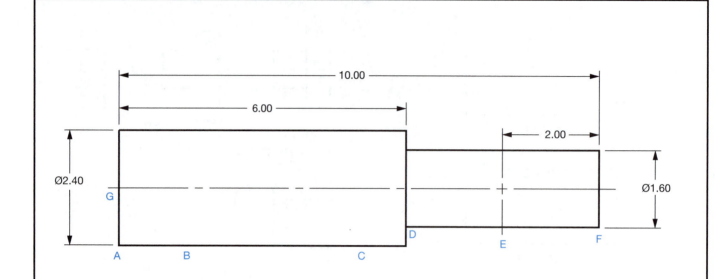

ASSIGNMENT: ON A ONE-INCH GRID SHEET (.10 IN. SQUARES) SKETCH THE HANDLE SHOWN ABOVE AND ADD THE FOLLOWING FEATURES. ADD DIMENSIONS USING SYMBOLS WHEREVER POSSIBLE. USE A CONVENTIONAL BREAK TO SHORTEN THE LENGTH. SCALE 1:1.

A. 45° X .20 CHAMFER
B. 33P DIAMOND KNURL FOR 1.20 IN. STARTING .80 IN. FROM LEFT END
C. 0.1:1 CIRCULAR TAPER FOR 1.20 IN. LENGTH ON RIGHT END OF Ø2.40
D. .20 X Ø1.40 UNDERCUT ON Ø1.60
E. Ø.20 X .50 DEEP, 4 HOLES EQUALLY SPACED
F. 30° X .30 CHAMFER. THE .30 DIMENSION TAKEN HORIZONTALLY ALONG THE SHAFT
G. Ø.60 HOLE, 1.50 DEEP

HANDLE   **A-33**

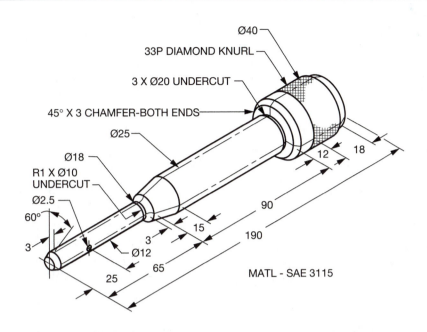

Ø40
33P DIAMOND KNURL
3 X Ø20 UNDERCUT
45° X 3 CHAMFER-BOTH ENDS
Ø25
Ø18
R1 X Ø10 UNDERCUT
Ø2.5
60°
3
18
12
90
190
15
3
Ø12
65
25
Ø18

MATL - SAE 3115

ASSIGNMENT: ON A CENTIMETER GRID SHEET (1mm SQUARES) SKETCH A ONE-VIEW DRAWING, COMPLETE WITH DIMENSIONS, OF THE INDICATOR ROD. USE A CONVENTIONAL BREAK TO SHORTEN THE LENGTH OF THE ROD. SCALE 1:1.

METRIC
DIMENSIONS IN MILLIMETERS

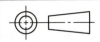

INDICATOR ROD   **A-34M**

ONE- AND TWO-VIEW DRAWINGS

Except for complex objects of irregular shapes, it is seldom necessary to draw more than three views, and for simple parts one- or two-view drawings will often suffice.

In one-view drawings the third dimension is expressed by a note or by symbols or abbreviations, such as ∅, □, HEX ACR FLT, R, Figure 11–1 and Table 2 of the Appendix. The symmetry symbol shown at the ends of the center line indicates that the part is symmetrical. Frequently, the drafter will decide that only two views are necessary to explain the shape of an object fully, Figure 11–2. One or two views usually show the shape of cylindrical objects adequately.

MULTIPLE-DETAIL DRAWINGS

Details of parts may be shown on separate sheets, or they may be grouped together on one or more large sheets. Often the details of parts are grouped according to the department in which they are made. Metal parts to be fabricated in the machine shop may appear on one detail sheet, whereas parts to be made in the wood shop may be grouped on another. Figure 11–3 shows several details used in the assembly of a drafting compass.

FUNCTIONAL DRAFTING

Since the basic function of the drafting department is to produce sufficient information to produce or assemble parts, drafting must embrace every possible means to communicate this information in the least expensive manner.

There are many ways to reduce drafting time in preparing a drawing.

1. Avoid unnecessary views.
2. Use simplified drawing practices.
3. Use explanatory notes to compliment the drawing, thereby eliminating views that are time consuming to draw.
4. Eliminate unnecessary lines.

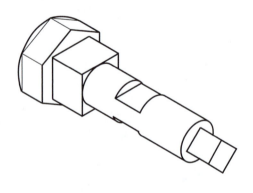

(A) THE PART

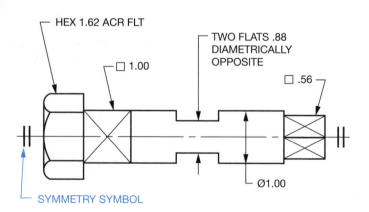

HEX 1.62 ACR FLT

□ 1.00

TWO FLATS .88 DIAMETRICALLY OPPOSITE

□ .56

Ø1.00

SYMMETRY SYMBOL

(B) ONE-VIEW DRAWING OF THE PART

FIGURE 11–1 ■ *Words and symbols used to identify shapes and sizes.*

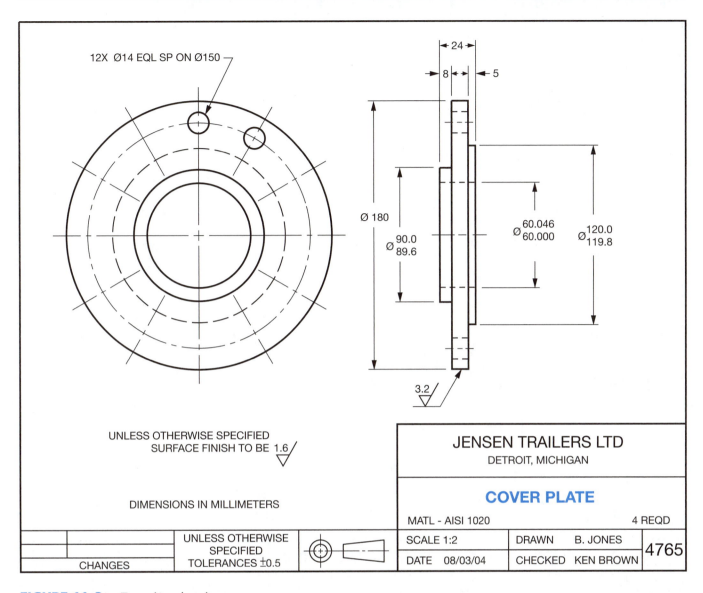

12X Ø14 EQL SP ON Ø150

UNLESS OTHERWISE SPECIFIED
SURFACE FINISH TO BE 1.6

DIMENSIONS IN MILLIMETERS

		UNLESS OTHERWISE SPECIFIED TOLERANCES ±0.5		JENSEN TRAILERS LTD DETROIT, MICHIGAN			
				COVER PLATE			
				MATL - AISI 1020		4 REQD	
				SCALE 1:2	DRAWN	B. JONES	4765
CHANGES				DATE 08/03/04	CHECKED	KEN BROWN	

FIGURE 11–2 ■ *Two-view drawing.*

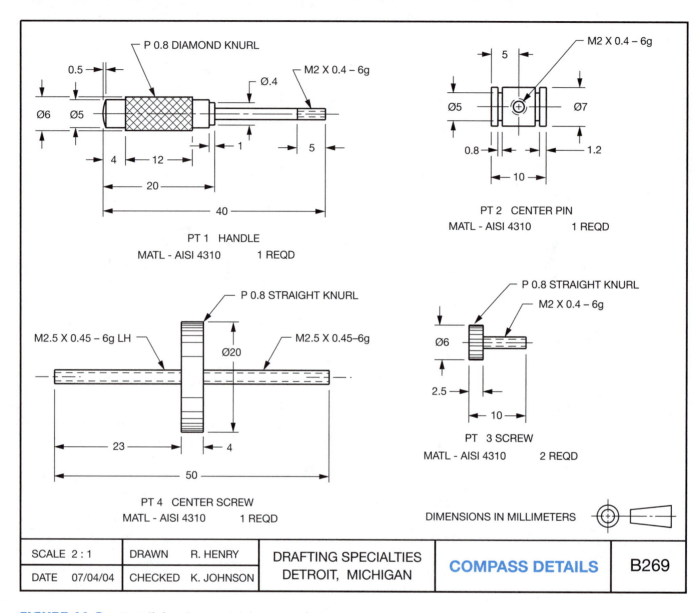

PT 1   HANDLE
MATL - AISI 4310   1 REQD

PT 2   CENTER PIN
MATL - AISI 4310   1 REQD

PT 4   CENTER SCREW
MATL - AISI 4310   1 REQD

PT 3 SCREW
MATL - AISI 4310   2 REQD

DIMENSIONS IN MILLIMETERS

| SCALE 2 : 1 | DRAWN   R. HENRY | DRAFTING SPECIALTIES DETROIT, MICHIGAN | COMPASS DETAILS | B269 |
| DATE   07/04/04 | CHECKED   K. JOHNSON | | | |

FIGURE 11–3 ■ *Detail drawing containing several parts.*

Figure 11–4 shows a simple part described in three different ways, the first example using conventional drawing practices, the second and third examples using simplified drawing techniques.

REFERENCE

ASME Y14.3M-1994 (R1999) Multi- and Sectional-View Drawings

INTERNET RESOURCES

American Society of Mechanical Engineers. For information on multiview drawings, refer to ASME Y14.3M-1994 (R1999) (*Multi- and Sectional-View Drawings*) at: http://www.asme.org

Drafting Zone. For information on functional drafting, see: http://www.draftingzone.com

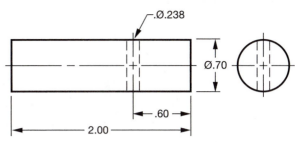

EXAMPLE 1: CONVENTIONAL DRAWING

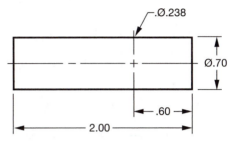

EXAMPLE 2: SIMPLIFIED DRAWING

PT 2   Ø.70 X 2.00 LG
Ø.238 HOLE – .60 FROM END

EXAMPLE 3: PART DESCRIBED BY A NOTE

FIGURE 11–4 ■ *Simplified representation for a simple part.*

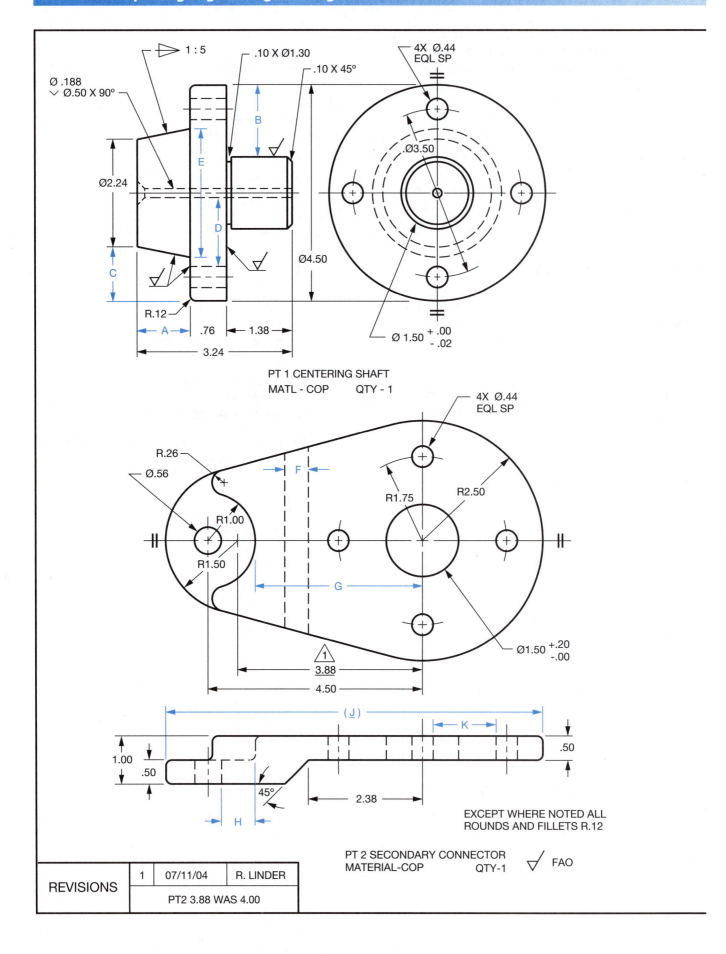

1 : 5

Ø .188
⌵ Ø.50 X 90°

.10 X Ø1.30

.10 X 45°

4X Ø.44
EQL SP

Ø2.24

B

E

D

C

R.12

A   .76   1.38

3.24

Ø4.50

.Ø3.50

Ø 1.50 +.00 / -.02

PT 1 CENTERING SHAFT
MATL - COP   QTY - 1

4X Ø.44
EQL SP

R.26

Ø.56

R1.00

R1.50

F

R1.75   R2.50

G

Ø1.50 +.20 / -.00

⚠1
3.88

4.50

(J)

K

1.00

.50

.50

45°

H

2.38

EXCEPT WHERE NOTED ALL
ROUNDS AND FILLETS R.12

PT 2 SECONDARY CONNECTOR
MATERIAL-COP   QTY-1   ▽ FAO

REVISIONS	1	07/11/04	R. LINDER
	PT2 3.88 WAS 4.00		

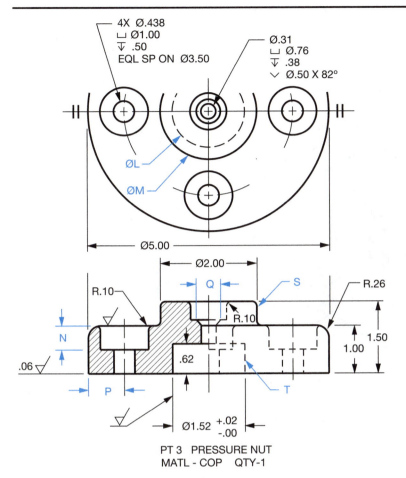

4X Ø.438
⌴ Ø1.00
▽ .50
EQL SP ON Ø3.50

Ø.31
⌴ Ø.76
▽ .38
∨ Ø.50 X 82°

ØL

ØM

Ø5.00

Ø2.00

R.10

Q

S

R.26

R.10

N

.06 ▽

P

.62

1.50

1.00

T

Ø1.52 +.02 / -.00

PT 3   PRESSURE NUT
MATL - COP   QTY-1

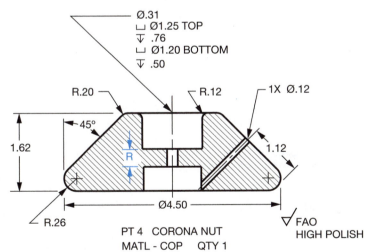

Ø.31
⌴ Ø1.25 TOP
▽ .76
⌴ Ø1.20 BOTTOM
▽ .50

R.20

45°

R.12

1X Ø.12

1.62

R

1.12

Ø4.50

R.26

▽ FAO
HIGH POLISH

PT 4   CORONA NUT
MATL - COP   QTY 1

QUESTIONS:

1. Calculate dimensions A to R. Use nominal sizes. There is no I or O.

Refer to Part 1

2. What is the length of the Ø1.50 shaft? Do not include the undercut or chamfer.

3. What is the length of the Ø.188 hole excluding the countersink?

4. Give two reasons why only part of the end view is drawn.

5. What is the diameter of the undercut?

Refer to Part 2

6. What does FAO mean?

7. What does the line under the 3.88 dimension indicate?

8. What does the abbreviation COP mean?

9. What do the parentheses around dimension J indicate?

10. What is the size of the fillets?

Refer to Part 3

11. How much machining allowance is provided for the bottom of the part?

12. What line in the front view represents line L in the top view?

13. What line in the top view represents line S in the front view?

14. What are the limits of the diameter of the counterbore in the bottom of the part?

15. What type of section view is shown?

Refer to Part 4

16. What is the diameter of the counterbore where the Ø.12 hole terminates?

17. What is the distance between the center points of the .26 radii?

18. What type of section view is shown?

ASSIGNMENT: ON A ONE INCH GRID SHEET (.10 IN. SQUARES) SKETCH THE TOP VIEW OF PART NO. 4.

SCALE	NTS		
DRAWN	J. LONEY	DATE	28/10/04

CENTERING CONNECTOR DETAILS

A-35

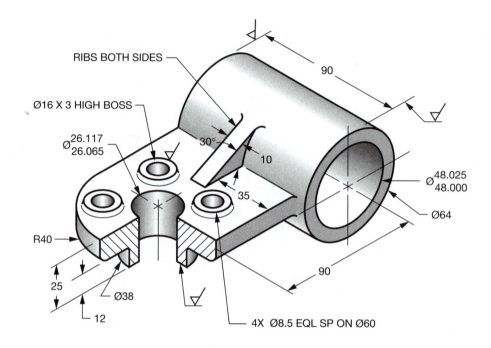

RIBS BOTH SIDES

Ø16 X 3 HIGH BOSS

Ø$^{26.117}_{26.065}$

30°

10

90

35

Ø$^{48.025}_{48.000}$

Ø64

R40

25

Ø38

12

4X Ø8.5 EQL SP ON Ø60

90

MATL - CAST STEEL   ROUNDS AND FILLETS R3

ASSIGNMENT: ON A CENTIMETER GRID SHEET (1mm SQUARES),
SKETCH THE TOP AND FRONT VIEWS OF THE LINK.
SCALE 1:2.

METRIC
DIMENSIONS IN MILLIMETERS

LINK   A-36M

SURFACE TEXTURE

The development of modern, high-speed machines has resulted in higher loadings and faster moving parts. To withstand these more severe operating conditions with minimum friction and wear, a particular surface texture is often essential. This requires the designer to accurately describe the needed texture (sometimes called *finish*) to the persons who are actually making the parts.

Rarely are entire machines designed and manufactured in one plant. They are usually designed in one location, manufactured in another, and perhaps assembled in a third.

All surface finish control begins in the drafting room. The designer is responsible for specifying the correct surface finish for maximum performance and service life at the lowest cost. In selecting the required surface finish for any particular part, the choice is based on the designer's experience, field service data, and engineering tests. Many factors influence the designer's choice. These factors include the function of the parts, the type of loading, the speed and direction of movement, and the operating conditions. Also considered are such factors as the physical characteristics of both materials on contact, whether the part is subjected to stress reversals, the type and amount of lubricant, contaminants, and temperature.

The two principal reasons for surface finish control are friction reduction and the control of wear.

Whenever a lubricating film must be maintained between two moving parts, the surface irregularities must be small enough to prevent penetrating the oil film under even the most severe operating conditions. Such parts as bearings, journals, cylinder bores, piston pins, bushings, pad bearings, helical and worm gears, seal surfaces, and machine ways are objects where this condition must be fulfilled.

Surface finish is also important to the wear service of certain pieces subject to dry friction, such as machine tool bits, threading dies, stamping dies, rolls, clutch plates, and brake drums.

Smooth finishes are essential on certain high-precision pieces. In mechanisms such as injectors and high-pressure cylinders, smoothness and lack of waviness are essential to accuracy and pressure-retaining ability. Smooth finishes are also used on micrometer anvils, gages, gage blocks, and other items.

Smoothness is often important for the visual appeal of the finished product. For this reason, surface finish is controlled on such articles as rolls, extrusions dies, and precision casting dies.

For gears and other parts, surface finish control may be necessary to insure quiet operation.

In cases where boundary lubrication exists or where surfaces are not compatible (for example, two hard surfaces running together), a certain amount of roughness or character of surface will assist in lubrication.

To meet the requirements for effective control of surface quality under diversified conditions, there is a system for accurately describing the surface.

Surfaces are usually very complex in character. Only the height, width, and direction of surface irregularities are covered in this section, because these are of practical importance in specific applications.

Surface Texture Definitions

The following terms relating to surface texture are illustrated in Figure 12–1.

MICROINCH (μin)

A microinch is one millionth of an inch (.000001 inch). For written specifications or reference to surface roughness requirements, microinches may be abbreviated as μin.

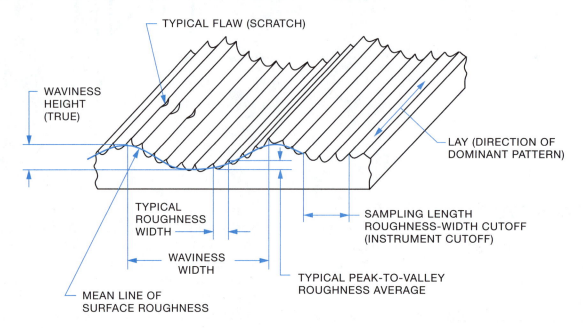

FIGURE 12–1 ■ *Surface texture characteristics.*

MICROMETER (μm)

A micrometer is one millionth of a meter (0.000001 meter.) For written specifications or reference to surface roughness requirements, micrometers may be abbreviated as μm.

ROUGHNESS

Roughness consists of the finer irregularities in the surface texture usually including those irregularities, which result from the inherent action of the production process. These include traverse feed marks and other irregularities within the limits of the roughness-width cutoff.

ROUGHNESS AVERAGE (R_a)

Roughness average is expressed in microinches, micrometers, or roughness grade numbers N1 to N12. The "N" series of roughness grade numbers is often used in lieu of the roughness average values to avoid misinterpretation when drawings are exchanged internationally.

ROUGHNESS WIDTH

Roughness width is the distance parallel to the nominal surface between successive peaks or ridges that constitute the predominant pattern of the roughness. Roughness width is rated in inches or millimeters.

ROUGHNESS-WIDTH CUTOFF

The greatest spacing of repetitive surface irregularities to be included in the measurement of average roughness height is the roughness-width cutoff. Roughness-width cutoff is rated in inches or millimeters and must always be greater than the roughness width in order to obtain the total roughness height rating.

WAVINESS

Waviness is the usually widely spaced component of surface texture and is generally spaced farther apart than the roughness-width cutoff. Waviness may result from machine or work deflections, vibration, chatter, heat treatment or warping strains. Roughness may be considered superimposed on a wavy surface. Although waviness is not currently in ISO standards, it is included as part of the surface texture symbol to follow present industrial practices in the United States.

LAY

The direction of the predominant surface pattern, which is ordinarily determined by the production method used, is the lay. Symbols for the lay are shown in Figure 12–2.

SYMBOL	DESCRIPTION	EXAMPLE
=	LAY PARALLEL TO THE LINE REPRESENTING THE SURFACE TO WHICH THE SYMBOL IS APPLIED	 *DIRECTION OF TOOL MARKS*
⊥	LAY PERPENDICULAR TO THE LINE REPRESENTING THE SURFACE TO WHICH THE SYMBOL IS APPLIED	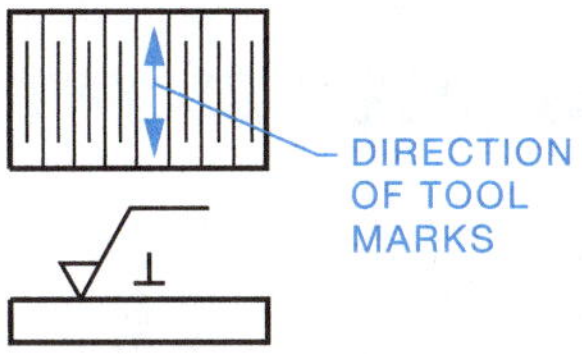 *DIRECTION OF TOOL MARKS*
X	LAY ANGULAR IN BOTH DIRECTIONS TO THE LINE REPRESENTING THE SURFACE TO WHICH THE SYMBOL IS APPLIED	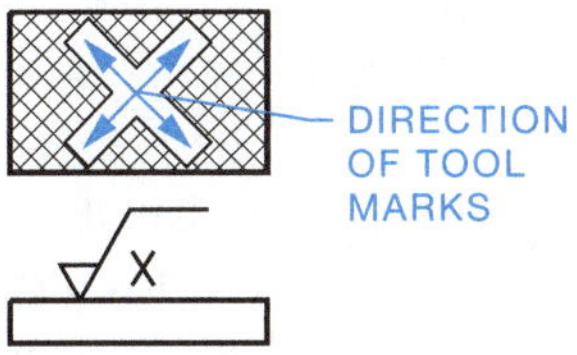 *DIRECTION OF TOOL MARKS*
M	LAY MULTIDIRECTIONAL	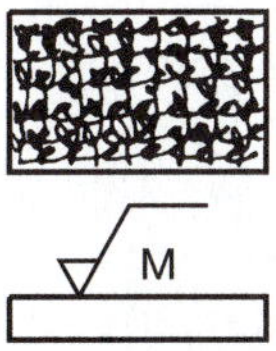
C	LAY APPROXIMATELY CIRCULAR RELATIVE TO THE CENTER OF THE SURFACE TO WHICH THE SYMBOL IS APPLIED	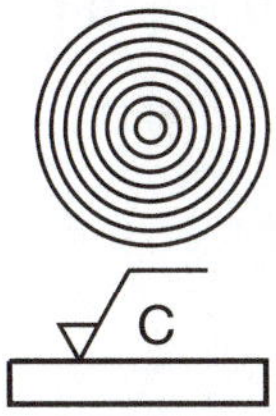
R	LAY APPROXIMATELY RADIAL RELATIVE TO THE CENTER OF THE SURFACE TO WHICH THE SYMBOL IS APPLIED	
P	LAY NONDIRECTIONAL, PITTED, OR PROTUBERANT	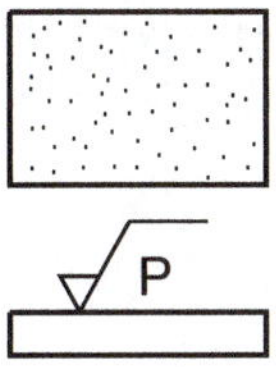

FIGURE 12–2 ■ *Lay symbols.*

FLAWS

Flaws are surface irregularities occurring at one place or at relatively infrequent or widely varying intervals. Flaws include cracks, blow holes, checks, ridges, scratches, and so forth. Unless otherwise specified, the effect of flaws is not included in the roughness height measurements.

SURFACE TEXTURE SYMBOL

The surface texture symbol, Figure 12–3, denotes surface characteristics on the drawing roughness, waviness, and lay controlled by waviness ratings applying the desired values to the surface texture symbol, Figure 12–4, or in a general note. The two methods may be used together. The point of the symbol should be on the line indicating the surface, on an extension line from the surface, or on a leader pointing either to the surface or extension line, Figure 12–5. To be readable from the bottom, the symbol is placed in an upright position when notes or numbers are used. This means the long leg and extension line will be on the right. The symbol applies to the entire surface, unless otherwise specified.

This symbol is the same symbol (machining symbol) that was described in Unit 8. As well as identifying which surfaces require machining, other surface characteristics are defined by this symbol.

Like dimensions, the symbol for the same surface should not be duplicated on other views. They should be placed on the view with the dimensions showing size or location of the surfaces. Surface texture symbols designate surface texture characteristics, which include machining of surfaces. The method of indicating machine finishes on surfaces is covered in Unit 8.

Where all the surfaces are to be machined, a general note such as FAO (finish all over) or ✔ ALL OVER may be used and the symbols on the part may be omitted.

SURFACE TEXTURE RATINGS

Roughness average, which is measured in microinches, micrometers, or roughness grade numbers, is shown to the left of the long leg of the symbol, Figure 12–4. The specification of only one rating defines the maximum value; any lesser value is acceptable. Specifying two ratings defines the minimum and maximum values. Anything within that range is acceptable. The maximum value is placed over the minimum.

Waviness height ratings are indicated in inches or millimeters and positioned as shown in Figure 12–4. Any lesser value is acceptable.

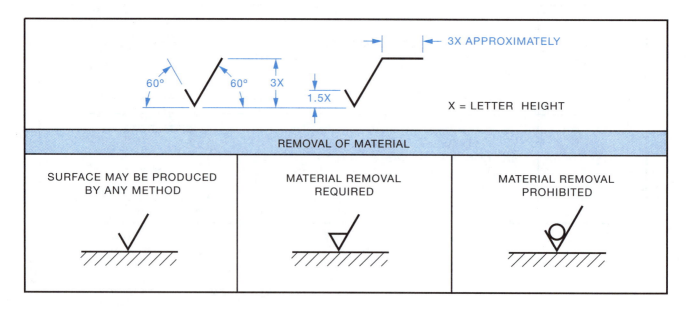

FIGURE 12–3 ■ *Basic surface texture symbol.*

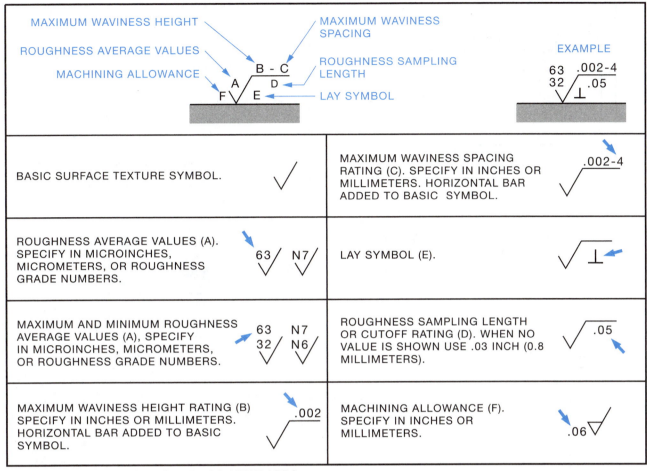

NOTE: WAVINESS IS NOT USED IN ISO STANDARDS.

FIGURE 12–4 ■ *Location of ratings and symbols on surface texture symbol.*

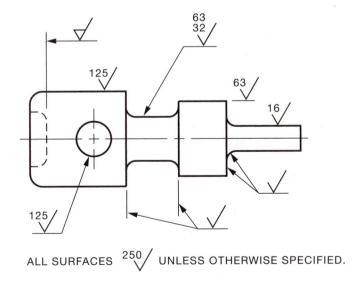

ALL SURFACES ²⁵⁰√ UNLESS OTHERWISE SPECIFIED.

NOTE: VALUES SHOWN ARE IN MICROINCHES.

FIGURE 12–5 ■ *Application of roughness values and waviness ratings.*

Waviness spacing ratings are indicated in inches or millimeters positioned as shown in Figure 12–4. Any lesser value is acceptable.

Lay symbols, indicating the directional pattern of the surface texture, are shown in Figure 12–2. The symbol is on the right of the long leg of the symbol.

Roughness sampling length ratings are given in inches or millimeters and are located below the horizontal extension. Unless otherwise specified, roughness-width cutoff is .03 in. (0.8 mm).

See Figure 12–5 for an application of roughness values and waviness ratings.

NOTES

Usually, a note is used where a given roughness requirement applies to either the whole part or the major portion, or before or after plating. Examples are shown in Figure 12–6.

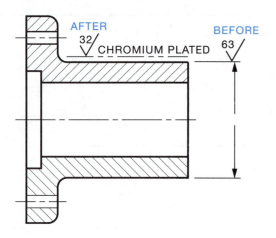

(A) INDICATING SURFACE TEXTURE BEFORE AND AFTER PLATING

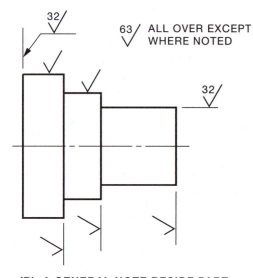

(B) A GENERAL NOTE BESIDE PART

FIGURE 12–6 ■ *The use of notes with surface texture symbol.*

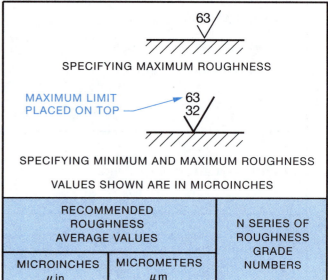

SPECIFYING MAXIMUM ROUGHNESS

MAXIMUM LIMIT PLACED ON TOP

SPECIFYING MINIMUM AND MAXIMUM ROUGHNESS

VALUES SHOWN ARE IN MICROINCHES

RECOMMENDED ROUGHNESS AVERAGE VALUES		N SERIES OF ROUGHNESS GRADE NUMBERS
MICROINCHES μ in.	MICROMETERS μ m	
2000	50	N12
1000	25	N11
500	12.5	N10
250	6.3	N9
125	3.2	N8
63	1.6	N7
32	0.8	N6
16	0.4	N5
8	0.2	N4
4	0.1	N3
2	0.05	N2
1	0.025	N1

FIGURE 12–7 ■ *Recommended roughness average ratings.*

CONTROL REQUIREMENTS

Surface texture control should be specified for surfaces where texture is a functional requirement. For example, most surfaces that have contact with a mating part have a certain texture requirement, especially for roughness. The drawing should reflect the texture necessary for optimum part function without depending on the variables of machining practices.

Many surfaces do not need a specification of surface texture because the function is unaffected by the surface quality. Such surfaces should not receive surface quality designations because they could unnecessarily increase the product cost.

Figures 12–7, 12–8, and 12–9 show recommended roughness average ratings, the machining methods used to produce them, and the application of the ratings to the surface texture symbol.

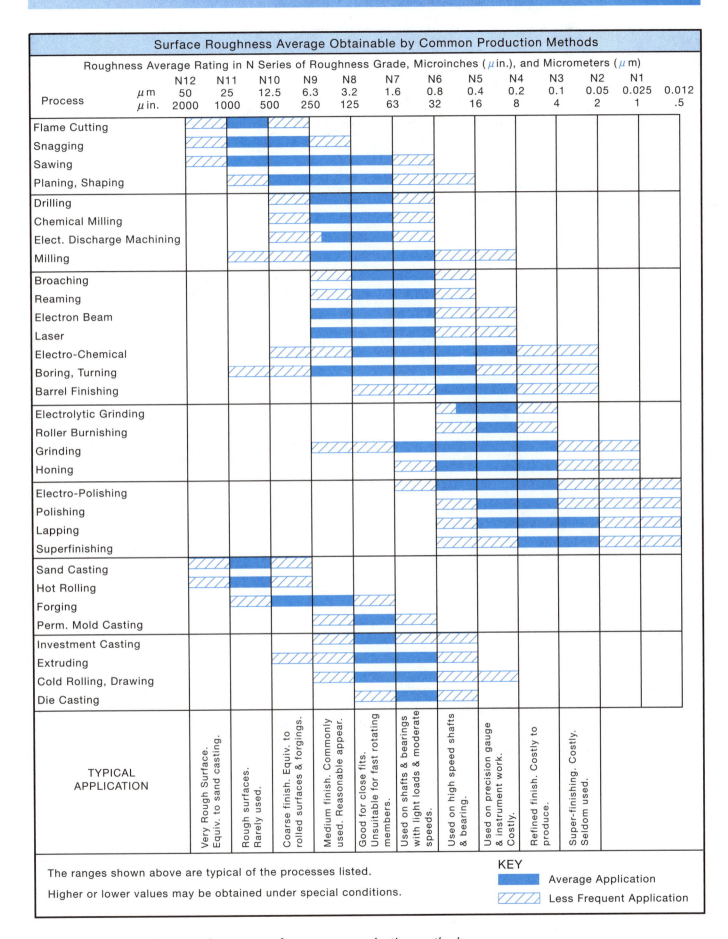

FIGURE 12–8 ■ *Surface roughness range for common production methods.*

MICROMETERS RATING	MICROINCHES RATING	APPLICATION
25	1000	Rough, low grade surface resulting from sand casting, torch or saw cutting, chipping, or rough forging. Machine operations are not required because appearance is not objectionable. This surface, rarely specified, is suitable for unmachined clearance areas on rough construction items.
12.5	500	Rough, low grade surface resulting from heavy cuts and coarse feeds in milling, turning, shaping, boring, rough filing, disc grinding, and snagging. It is suitable for clearance areas on machinery, jigs, and fixtures. Sand casting or rough forging produces this surface.
6.3	250	Coarse production surface, for unimportant clearance and cleanup operation, resulting from coarse surface grind, rough file, disc grind, rapid feeds in turning, milling, shaping, drilling, boring, grinding, etc., where tool marks are not objectionable. The natural surfaces of forgings, permanent mold castings, extrusions, and rolled surfaces also produce this roughness. It can be produced economically and is used on parts where stress requirements, appearance, and conditions of operations and design permit.
3.2	125	The roughest surface recommended for parts subject to loads, vibration, and high stress. It is also permitted for bearing surfaces when motion is slow and loads light or infrequent. It is a medium commercial machine finish produced by relatively high speeds and fine feeds taking light cuts with sharp tools. It may be economically produced on lathes, milling machines, shapers, grinders, etc., or on permanent mold castings, die castings, extrusion, and rolled surfaces.
1.6	63	A good machine finish produced under controlled conditions using relatively high speeds and fine feeds to take light cuts with sharp cuttings. It may be specified for close fits and used for all stressed parts, except fast rotating shafts, axles, and parts subject to severe vibration or extreme tension. It is satisfactory for bearing surfaces when motion is slow and loads light or infrequent. It may also be obtained on extrusions, rolled surfaces, die castings, and permanent mold casting when rigidly controlled.
0.8	32	A high-grade machine finish requiring close control when produced by lathes, shapers, milling machines, etc., but relatively easy to produce by centerless, cylindrical, or surface grinders. Also, extruding, rolling or die casting may produce a comparable surface when rigidly controlled. This surface may be specified in parts where stress concentration is present. It is used for bearings when motion is not continuous and loads are light. When finer finishes are specified, production costs rise rapidly; therefore, such finishes must be analyzed carefully.
0.4	16	A high quality surface produced by fine cylindrical grinding, emery buffing, coarse honing, or lapping, it is specified where smoothness is of primary importance, such as rapidly rotating shaft bearings, heavily loaded bearing and extreme tension members.
0.2	8	A fine surface produced by honing, lapping, or buffing. It is specified where packings and rings must slide across the direction of the surface grain, maintaining or withstanding pressures, or for interior honed surface of hydraulic cylinders. It may also be required in precision gauges and instrument work, or sensitive value surfaces, or on rapidly rotating shafts and on bearings where lubrication is not dependable.
0.1	4	A costly refined surface produced by honing, lapping, and buffing. It is specified only when the design requirements make it mandatory. It is required in instrument work, gauge work, and where packing and rings must slide across the direction of surface grain such as on chrome plated piston rods, etc. where lubrication is not dependable.
0.05 / 0.025	2 / 1	Costly refined surfaces produced by only the finest of modern honing, lapping, buffing, and superfinishing equipment. These surfaces may have a satin or highly polished appearance depending on the finishing operation and material. These surfaces are specified only when design requirements make i tmandatory. They are specified on fine or sensitive instrument parts or other laboratory items, and certain gauge surfaces, such as precision gauge blocks.

FIGURE 12–9 ∎ *Surface roughness description and application.*

REFERENCE

ASME Y14.36M-1996 (R2002) Surface Texture
Symbols

INTERNET RESOURCES

Drafting Zone. For information on surface texture
and surface texture symbols, see: http://www.
draftingzone.com

Precision Devices, Inc. For information on surface
texture and surface texture symbols, see:
http://www.predev.com/smg/intro.htm and
smg/specification.htm

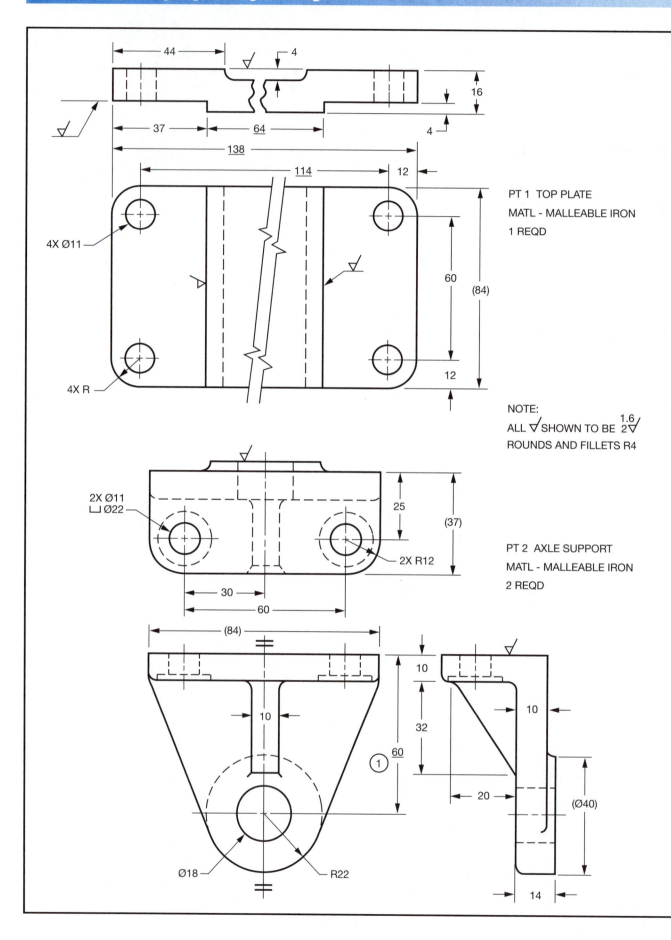

44

4

37

64

16

4

138

114

12

PT 1 TOP PLATE

MATL - MALLEABLE IRON

1 REQD

4X Ø11

60

(84)

4X R

12

NOTE:

ALL ⩗ SHOWN TO BE 2⩗ 1.6

ROUNDS AND FILLETS R4

2X Ø11
⊔ Ø22

25

(37)

2X R12

PT 2 AXLE SUPPORT

MATL - MALLEABLE IRON

2 REQD

30

60

(84)

10

10

10

32

① 60

20

Ø18

R22

(Ø40)

14

QUESTIONS:

1. How many reference dimensions are shown?
2. How many not-to-scale dimensions are shown?
3. What type of section view is shown on part 3?
4. How many machined surfaces are shown? Two surfaces can be on one plane.
5. How many fillets are shown on the drawing?
6. What machining allowance is called for on the machined surfaces?
7. What was the original height of part 2 before machining?
8. What is the height of the caster assembly?
9. What were the overall dimensions of the cast parts before machining?

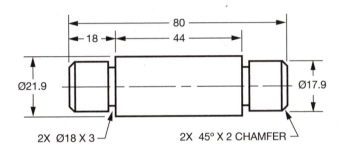

PT 4 AXLE

MATL - CARBON STEEL

1 REQD

Ø21.9 Ø17.9

80 | 18 | 44

2X Ø18 X 3

2X 45° X 2 CHAMFER

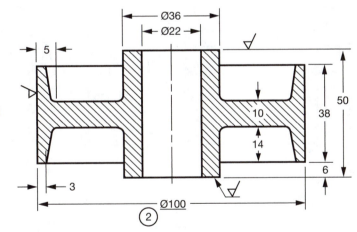

PT 3 WHEEL

MATL - MALLEABLE IRON

1 REQD

Ø36 | Ø22 | 5 | 50 | 38 | 10 | 14 | 6 | 3 | Ø100

②

METRIC
DIMENSIONS IN MILLIMETERS

REV TABLE	1	16/08/04	J. HELSEL	2	16/08/04	J. HELSEL
	60 DIM WAS 65			100 DIM WAS 110	

SCALE 1 : 2

DRAWN J. HELSEL DATE 21/02/04

CASTER DETAILS **A-37M**

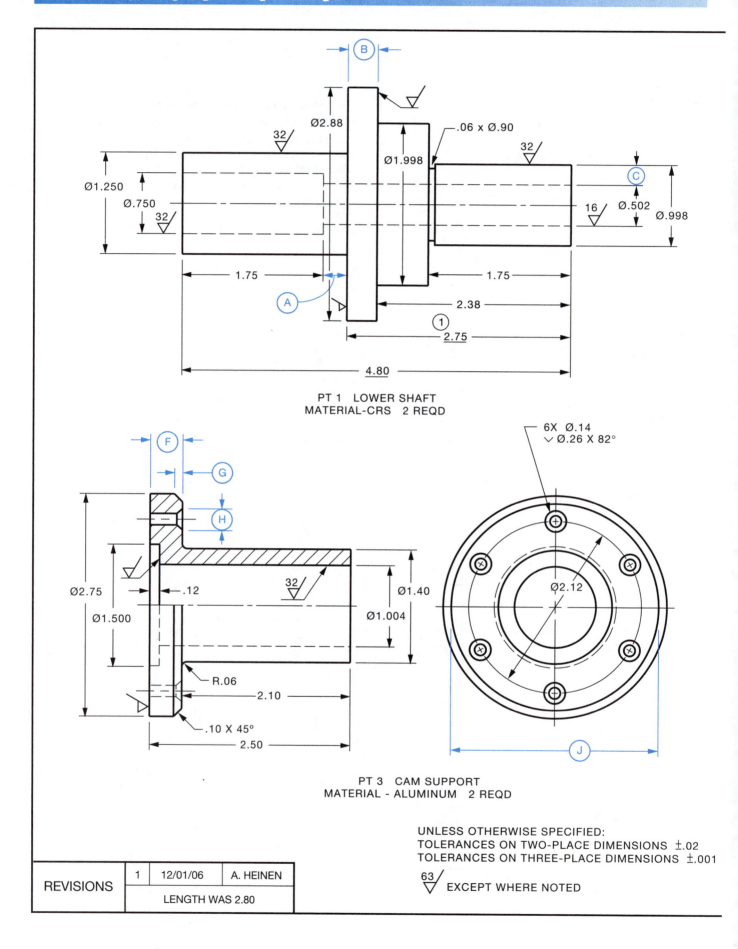

PT 1   LOWER SHAFT
MATERIAL-CRS   2 REQD

PT 3   CAM SUPPORT
MATERIAL - ALUMINUM   2 REQD

UNLESS OTHERWISE SPECIFIED:
TOLERANCES ON TWO-PLACE DIMENSIONS ±.02
TOLERANCES ON THREE-PLACE DIMENSIONS ±.001

63/▽ EXCEPT WHERE NOTED

REVISIONS	1	12/01/06	A. HEINEN
	LENGTH WAS 2.80		

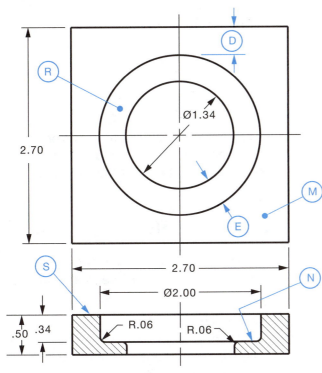

PT 2   WASHER
MATERIAL-MS   4 REQD FAO

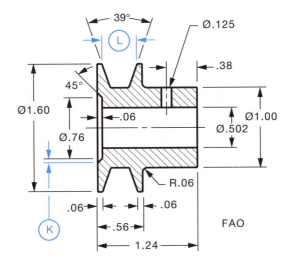

PT 4   V-BELT PULLEY
MATERIAL-CRS   4 REQD

QUESTIONS:

1. Calculate dimensions (A) to (L).

Referring to figure 11-8, what production methods would be suitable to produce the surface texture of

2. Ø.502 hole in Part 1?

3. Ø1.250 on Part 1?

4. Ø1.004 hole in Part 3?

5. How many dimensions indicate that the dimension is not drawn to scale?

Refer to Part 1

6. What is the length of the Ø.998 portion?

7. What was the original length of the 2.75 dimension?

8. How many hidden circles would be seen if the right end view were drawn?

Refer to Part 2

9. Which surface does (R) represent in the front view?

10. Which surface does (S) represent in the top view?

11. If the part that passes through the washer is Ø1.30 what is the clearance per side between the two parts?

12. How many fillets are required?

Refer to Part 3

13. How many degrees apart on the Ø2.12 are the Ø.14 holes?

14. Is the center line of the countersunk holes in the center of the flange?

15. What operation is performed to allow the head of the mounting screws to rest flush with the flange?

16. What type of section view is shown?

17. What is the amount and degree of chamfer?

18. What is the (A) size, (B) type of cap screw required to mount the cam support to its mating parts?

Refer to Part 4

19. How deep is the Ø.125 hole?

20. How deep is the belt groove?

21. What does FAO mean?

22. What type of section view is shown?

SCALE	NOT TO SCALE	
DRAWN	A. HEINEN	DATE 15/10/05

HANGER DETAILS   A-38

TOLERANCES AND ALLOWANCES

The history of engineering drawing as a means for the communication of engineering information spans a period of 6,000 years. It seems inconceivable that such an elementary practice as the tolerancing of dimensions, which is taken for granted today, was introduced for the first time only about 80 years ago.

Apparently, engineers and workers realized only gradually that exact dimensions and shapes could not be attained in the shaping of physical objects. The skilled handicrafters of the past took pride in the ability to work to exact dimensions. This meant that objects were dimensioned more accurately than they could be measured. The use of modern measuring instruments would have shown the deviations from the sizes, which were called exact.

It was soon realized that variations in the sizes of parts had always been present, that such variations could be restricted but not avoided, and that slight variation in the size which a part was originally intended to have could be tolerated without impairment of its correct functioning. It became evident that interchangeable parts need not be identical parts, but rather it would be sufficient if the significant sizes that controlled their fits lay between definite limits. Therefore, the problem of interchangeable manufacture developed from the making of parts to a supposedly exact size, to the holding of parts between two limiting sizes, lying so closely together that any intermediate size would be acceptable.

The concept of limits means essentially that a precisely defined basic condition (expressed by one numerical value or specification) is replaced by two limiting conditions. Any result lying on or between these two limits is acceptable. A workable scheme of interchangeable manufacture that is indispensable to mass production methods had been established.

DEFINITIONS

In order to calculate limits and tolerances, the following definitions must be clearly understood.

Basic Size

The basic size of a dimension is the theoretical size from which the limits for that dimension are derived, by the application of the allowance and tolerance.

Tolerances

The tolerance of a dimension is the total permissible variation in its size. The tolerance is the difference between the limits of size.

Limits of Size

The limits of size are the maximum and minimum sizes permitted for a specific dimension.

Allowance

An allowance is the intentional difference in size of mating parts. It is the minimum clearance (positive allowance) or maximum interference (negative allowance) between such parts. Fits between parts is covered in Units 14 and 15.

All dimensions on a drawing have tolerances. Some dimensions must be more exact than other dimensions and consequently have smaller tolerances.

When dimensions require greater accuracy than the general note provides, individual tolerances or limits must be shown for that dimension.

Where limit dimensions are used and where either the maximum or minimum dimension has digits to the right of the decimal point, the other value should have the zeros added so that both the limits of size are expressed to the same number of decimal places.

When limit dimensions are used for diameter or radial features and the dimensions are placed one above the other, only one diameter or radius symbol is used and located at midheight.

Where one limit alone is important and where any variations away from that limit in the other direction may be permitted, the MAX (maximum) or MIN (minimum) can be specified. Examples are depth of holes, corner radii, and chamfers. Figures 13–1 and 13–2 show applications of limit dimensioning.

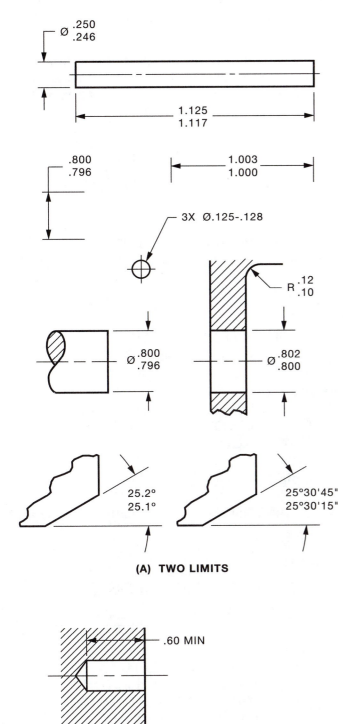

(A) TWO LIMITS

(B) SINGLE LIMITS

FIGURE 13–1 ■ *Limit dimensioning.*

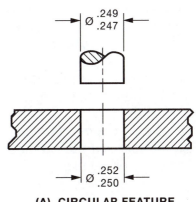

(A) CIRCULAR FEATURE

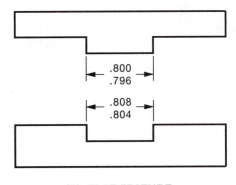

(B) FLAT FEATURE

FIGURE 13–2 ■ *Limit dimensioning application.*

TOLERANCING METHODS

Dimensional tolerances are expressed in either of two ways: limit dimensioning or plus and minus tolerancing.

Limit Dimensioning

In the limit dimensioning method, only the maximum and minimum dimensions are specified, Figure 13–1. When placed above each other, the larger dimension is placed on top. When shown with a leader and placed in one line, the smaller size is shown first. A small dash separates the two dimensions. When limit dimensions are used for diameter or radial features, the ∅ or R symbol is centered midway between the two limits, Figure 13–1(A). These rules apply to both inch and metric drawings.

Plus and Minus Tolerancing

In this method the dimension of the specified size is given first and it is followed by a plus and minus tolerance expression. The tolerance can be bilateral or unilateral, Figure 13–3.

A *bilateral tolerance* is a tolerance that is expressed as plus and minus values. These values need not be the same size.

A *unilateral tolerance* is one that applies in one direction from the specified size, so the permissible variation in the other direction is zero.

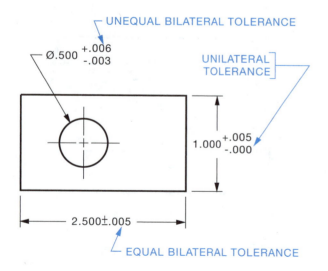

FIGURE 13–3 ■ *Plus and minus tolerancing.*

Inch Tolerances

Where inch dimensions are used on the drawing, both limit dimensions or the plus and minus tolerance and its dimensions are expressed with the same number of decimal places.

Examples

$.500 \pm .005$ not $.50 \pm .005$

$.500 {+.005 \atop -.000}$ not $.500 {+.005 \atop -0}$

$25.0 \pm .2$ not $25 \pm .2$

General tolerance notes greatly simplify the drawing. The following examples illustrate the variety of application in this system. The values given in the following examples are typical only:

Example 1

EXCEPT WHERE STATED OTHERWISE, TOLERANCE ON DIMENSIONS ±.005

Example 2

EXCEPT WHERE STATED OTHERWISE, TOLERANCES ON FINISHED DIMENSIONS TO BE AS FOLLOWS:

Example 3

DIMENSION	TOLERANCE
UP TO 3.00	.01
OVER 3.00 TO 12.00	.02
OVER 12.00 TO 24.00	.04
OVER 24.00	.06

Example 4

UNLESS OTHERWISE SPECIFIED
±.005 TOLERANCE ON MACHINED DIMENSIONS
±.04 TOLERANCE ON CAST DIMENSIONS
ANGULAR TOLERANCE ±30′

Millimeter Tolerances

Where millimeter dimensions are used on the drawings, the following applies:

A. The dimension and its tolerance need not be expressed to the same number of decimal places.

Example

15 ± 0.5 not 15.0 ± 0.5

B. Where unilateral tolerancing is used and either the plus or minus value is nil, a single zero is shown without a plus or minus sign.

Example

$$32 \, {}^{0}_{-0.2} \text{ or } 32 \, {}^{+0.02}_{0}$$

C. Where bilateral tolerancing is used, both the plus and minus values have the same number of decimal places, using zeros where necessary.

Example

$$32 \, {}^{+0.25}_{-0.10} \text{ or } 32 \, {}^{+0.25}_{-0.1}$$

REFERENCE

ASME Y14.5M-1994 (R1999) Dimensioning and Tolerancing

INTERNET RESOURCES

Drafting Zone. For information on tolerances and allowances, see: http://www.draftingzone.com

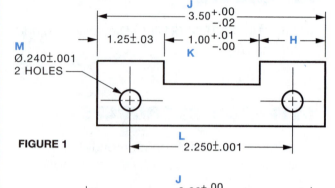

FIGURE 1

M
Ø.240±.001
2 HOLES

J
3.50 +.00 −.02

1.25±.03

1.00 +.01 −.00
K

H

L
2.250±.001

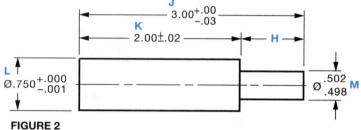

FIGURE 2

J
3.00 +.00 −.03

K
2.00±.02

H

L
Ø.750 +.000 −.001

.502
Ø .498 M

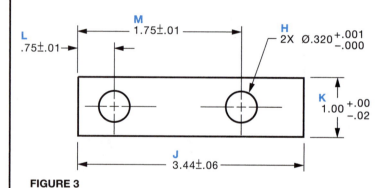

FIGURE 3

L
.75±.01

M
1.75±.01

H
2X Ø.320 +.001 −.000

K
1.00 +.00 −.02

J
3.44±.06

QUESTIONS:

1. With reference to figures 1, 2, and 3, calculate (A) basic size, (B) tolerance, (C) max. limit, and (D) min. limit for dimensions J, K, L, and M.
2. With reference to figures 1, 2, and 3, calculate (A) max size, (B) min. size for dimension H.

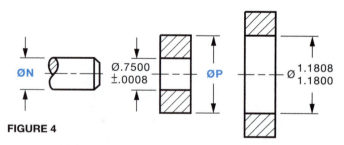

FIGURE 4

ØN

Ø.7500 ±.0008

ØP

Ø 1.1808 1.1800

3. With reference to figure 4, what are the limit dimensions for shaft N if it has a tolerance of .0014 and a min. clearance of .0006?
4. With reference to figure 4, what are the limit dimensions for bushing P if it has a tolerance of .0006 and a min. interference of zero?

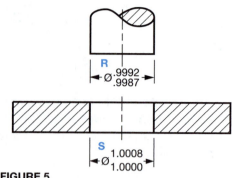

FIGURE 5

R
Ø .9992 .9987

S
Ø 1.0008 1.0000

With reference to figure 5:

5. What is the tolerance on shaft R?
6. What is the tolerance on hole S?
7. What is the min. clearance between the parts?
8. What is the max. clearance between the parts?

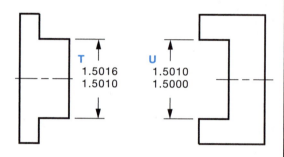

FIGURE 6

T
1.5016
1.5010

U
1.5010
1.5000

With reference to figure 6:

9. What is the tolerance on part T?
10. What is the tolerance on slot U?
11. What is the min. interference between the parts?
12. What is the max. interference between the parts?

INCH TOLERANCES AND ALLOWANCES | **A-39**

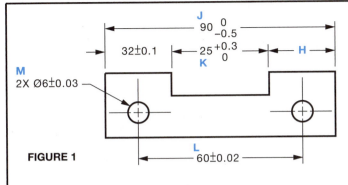

FIGURE 1

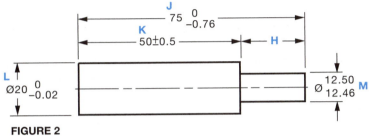

FIGURE 2

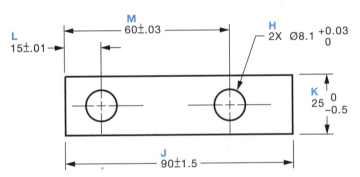

FIGURE 3

QUESTIONS:

1. With reference to figures 1, 2, and 3, calculate (A) basic size, (B) tolerance, (C) max. limit, and (D) min. limit for dimensions J, K, L, and M.
2. With reference to figures 1, 2, and 3, calculate (A) max. size, (B) min. size for dimension H.

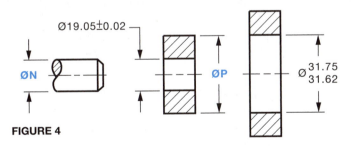

FIGURE 4

3. With reference to figure 4, what are the limit dimensions for shaft N if it has a tolerance of 0.036 and a min. clearance of 0.015?
4. With reference to figure 4, what are the limit dimensions for bushing P if it has a tolerance of 0.016 and a min. interference of zero?

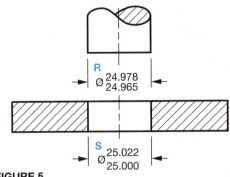

FIGURE 5

With reference to figure 5:

5. What is the tolerance on shaft R?
6. What is the tolerance on hole S?
7. What is the min. clearance between the parts?
8. What is the max. clearance between the parts?

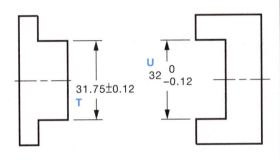

FIGURE 6

With reference to figure 6:

9. What is the tolerance on part T?
10. What is the tolerance on slot U?
11. What is the min. clearance between the parts?
12. What is the max. clearance between the parts?

METRIC
DIMENSIONS IN MILLIMETERS

MILLIMETER TOLERANCES AND ALLOWANCES　**A-40M**

INCH FITS

Fit is the general term used to describe the range of tightness or looseness resulting from the application of a specific combination of allowances and tolerances in the design of mating parts. Fits are of three general types: clearance, interference, and transition. Figures 14–1 and 14–2 illustrate the three types of fits.

Clearance Fits

Refer to Figure 14–2(A). Clearance fits have limits of size prescribed so a clearance always results when mating parts are assembled. Clearance fits are intended for accurate assembly of parts and bearings. The parts can be assembled by hand because the hole is always larger than the shaft.

Interference Fits

Refer to Figure 14–2(C). Interference fits have limits of size so prescribed that an interference always re-

sults when mating parts are assembled. The hole is always smaller than the shaft. Interference fits are for permanent assemblies of parts that require rigidity and alignment, such as dowel pins and bearings in castings. Parts are usually pressed together using an arbor press.

Transition Fits

Refer to Figure 14–2(B). Transition fits have limits of size so prescribed that either a clearance or an interference may result when mating parts are assembled. Transition fits are a compromise between clearance and interference fits.

They are used for applications where accurate location is important, but either a small amount of clearance or interference is permissible.

DESCRIPTION OF FITS

Running and Sliding Fits

These fits, for which tolerances and clearances are given in the Appendix, represent a special type of clearance fit. These are intended to provide a similar running performance, with suitable lubrication allowance, throughout the range of sizes.

Locational Fits

Locational fits are intended to determine only the location of the mating parts; they may provide rigid or accurate location, as with interference fits, or some freedom of location, as with clearance fits. Accordingly, they are divided into three groups: clearance fits, transition fits, and interference fits.

Locational clearance fits are intended for parts that are normally stationary but which can be freely assembled or disassembled.

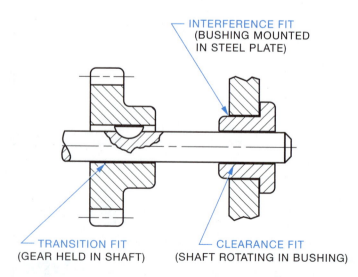

INTERFERENCE FIT
(BUSHING MOUNTED IN STEEL PLATE)

TRANSITION FIT
(GEAR HELD IN SHAFT)

CLEARANCE FIT
(SHAFT ROTATING IN BUSHING)

FIGURE 14–1 ■ *Application of types of fits.*

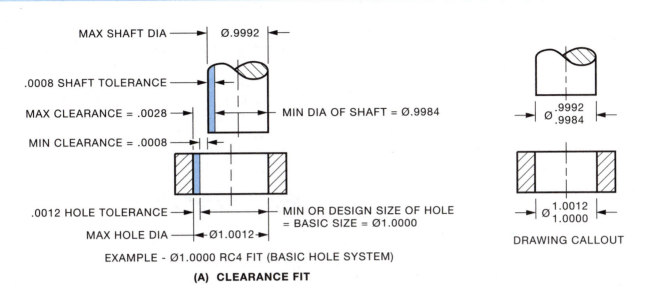

MAX SHAFT DIA — Ø.9992

.0008 SHAFT TOLERANCE

MAX CLEARANCE = .0028 — MIN DIA OF SHAFT = Ø.9984

MIN CLEARANCE = .0008

.0012 HOLE TOLERANCE — MIN OR DESIGN SIZE OF HOLE = BASIC SIZE = Ø1.0000

MAX HOLE DIA — Ø1.0012

Ø .9992 / .9984

Ø 1.0012 / 1.0000

DRAWING CALLOUT

EXAMPLE - Ø1.0000 RC4 FIT (BASIC HOLE SYSTEM)

(A) CLEARANCE FIT

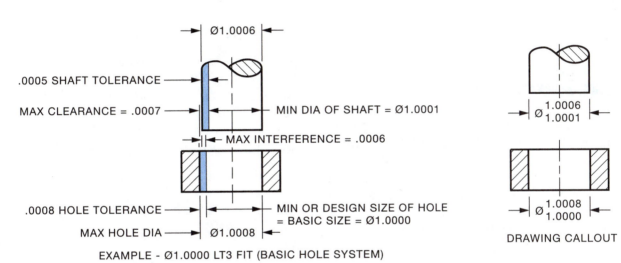

Ø1.0006

.0005 SHAFT TOLERANCE

MAX CLEARANCE = .0007 — MIN DIA OF SHAFT = Ø1.0001

MAX INTERFERENCE = .0006

.0008 HOLE TOLERANCE — MIN OR DESIGN SIZE OF HOLE = BASIC SIZE = Ø1.0000

MAX HOLE DIA — Ø1.0008

Ø 1.0006 / 1.0001

Ø 1.0008 / 1.0000

DRAWING CALLOUT

EXAMPLE - Ø1.0000 LT3 FIT (BASIC HOLE SYSTEM)

(B) TRANSITION FIT

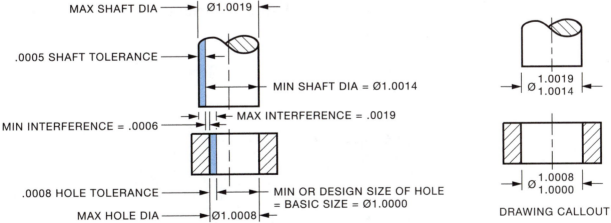

MAX SHAFT DIA — Ø1.0019

.0005 SHAFT TOLERANCE

MIN SHAFT DIA = Ø1.0014

MAX INTERFERENCE = .0019

MIN INTERFERENCE = .0006

.0008 HOLE TOLERANCE — MIN OR DESIGN SIZE OF HOLE = BASIC SIZE = Ø1.0000

MAX HOLE DIA — Ø1.0008

Ø 1.0019 / 1.0014

Ø 1.0008 / 1.0000

DRAWING CALLOUT

EXAMPLE - Ø1.0000 FN2 FIT (BASIC HOLE SYSTEM)

(C) INTERFERENCE FIT

FIGURE 14–2 ■ *Types and examples of inch fits.*

Locational transition fits are a compromise between clearance and interference fits, for application where accuracy of location is important but a small amount of either clearance or interference is permissible.

Locational interference fits are used where accuracy of location is of prime importance and for parts requiring rigidity and alignment.

Drive and Force Fits

Drive and force fits constitute a special type of interference fit, normally characterized by maintenance of constant bore pressures throughout the range of sizes. The interference therefore varies almost directly with diameter, and the difference between its minimum and maximum values is small to maintain the resulting pressures within reasonable limits.

STANDARD INCH FITS

Standard fits are designated for design purposes in specifications and on design sketches by means of the symbols shown in Figure 14–3. These symbols, however, are not intended to be shown directly on shop drawings; instead the actual limits of size are determined, and these limits are specified on the drawings. The letter symbols used are as follows:

RC   Running and sliding fit
LC   Locational clearance fit
LT   Locational transition fit
LN   Locational interference fit
FN   Force or shrink fit

These letter symbols are used in conjunction with numbers representing the class of fit; for example, FN4 represents a class 4, force fit.

Each of these symbols (two letters and a number) represents a complete fit, for which the minimum and maximum clearance or interference and the limits of size for the mating parts, are given directly in Appendix Tables 17 through 21.

Running and Sliding Fits

RC1 PRECISION SLIDING FIT

This fit is intended for the accurate location of parts that must assemble without perceptible play for high precision work such as gages.

RC2 SLIDING FIT

This fit is intended for accurate location, but with greater maximum clearance than class RC1. Parts made to this fit move and turn easily but are not intended to run freely.

RC3 PRECISION RUNNING FIT

This fit is about the closest fit that can be expected to run freely, and is intended for precision work for oil-lubricated bearings at slow speeds and light journal pressures.

RC4 CLOSE RUNNING FIT

This fit is intended chiefly as a running fit for grease- or oil-lubricated bearings on accurate machinery with moderate surface speeds and journal

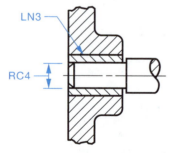
(A) SHAFT IN BUSHED HOLE

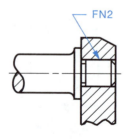
(B) CRANK PIN IN CAST IRON

FIGURE 14–3 ■ *Design sketch showing standard fits.*

pressures, where accurate location and minimum play are desired.

RC5 AND RC6 MEDIUM RUNNING FITS

These fits are intended for higher running speeds and/or where temperature variations are likely to be encountered.

RC7 FREE RUNNING FIT

This fit is intended for use where accuracy is not essential, and/or where large temperature variations are likely to be encountered.

RC8 AND RC9 LOOSE RUNNING FITS

These fits are intended for use where materials made to commercial tolerances are involved such as cold-rolled shafting, tubing, etc.

Locational Clearance Fits

Locational clearance fits are intended for parts that are normally stationary but can be freely assembled or disassembled. These are classified as follows:

LC1 TO LC4

These fits have a minimum zero clearance, but in practice the probability is that the fit will always have a clearance.

LC5 AND LC6

These fits have a small minimum clearance intended for close location fits for non-running parts.

LC7 TO LC11

These fits have progressively larger clearances and tolerances, and are useful for various loose clearances for assembly of bolts and similar parts.

Locational Transition Fits

Locational transition fits are a compromise between clearance and interference fits, for application where accuracy of location is important, but either a small amount of clearance or interference is permissible. These are classified as follows:

LT1 AND LT2

These fits average a slight clearance, giving a light push fit.

LT3 AND LT4

These fits average virtually no clearance, and are for use where some interference can be tolerated. These are sometimes referred to as an easy keying fit, and are used for shaft keys and ball race fits. Assembly is generally by pressure or hammer blows.

LT5 AND LT6

These fits average a slight interference, although appreciable assembly force will be required.

Locational Interference Fits

Locational interference fits are used where accuracy of location is of prime importance, and for parts requiring rigidity and alignment with no special requirements for bore pressure. These are classified as follows: ·

LN1 AND LN2

These are light press fits, with very small minimum interference, suitable for parts such as dowel pins, which are assembled with an arbor press in steel, cast iron, or brass. Parts can normally be dismantled and reassembled.

LN3

This is suitable as a heavy press fit in steel and brass, or a light press fit in more elastic materials and light alloys.

LN4 TO LN6

While LN4 can be used for permanent assembly of steel parts, these fits are primarily intended as press fits for soft materials.

Force or Shrink Fits

Force or shrink fits constitute a special type of interference fit. The interference varies almost directly with diameter, and the difference between its minimum and maximum values is small to maintain the resulting pressures within reasonable limits. These fits are classified as follows:

FN1 LIGHT DRIVE FIT

Requires light assembly pressure and produces more or less permanent assemblies. It is suitable for thin sections or long fits, or in cast-iron external members.

FN2 MEDIUM DRIVE FIT

Suitable for heavier steel parts, or as a shrink fit on light sections.

FN3 HEAVY DRIVE FIT

Suitable for heavier steel parts, or as a shrink fit in medium sections.

FN4 AND FN5 FORCE FITS

Suitable for parts that can be highly stressed.

Basic Hole System

In the basic hole system, which is recommended for general use, the basic size will be the design size for the hole, and the tolerance will be plus. The design size for the shaft will be the basic size minus the minimum clearance, or plus the maximum interference, and the tolerance will be minus, as given in the tables in the Appendix. For example (see Table 17), for a 1-in. RC7 fit, values of +.0020, .0025, and −.0012 are given; hence, tolerances will be:

$$\text{Hole } \varnothing 1.0000 \, {}^{+.0020}_{-.0000}$$

$$\text{Shaft } \varnothing .9975 \, {}^{+.0000}_{-.0012}$$

Basic Shaft System

Fits are sometimes required on a basic shaft system, especially in cases where two or more fits are required on the same shaft. This is designated for design purposes by a letter S following the fit symbol; for example, RC7S.

Tolerances for holes and shafts are identical with those for a basic hole system, but the basic size becomes the design size for the shaft and the design size for the hole is found by adding the minimum clearance or subtracting the maximum interference from the basic size.

For example, for a 1-in. RC7S fit, values of +.0020, .0025, and −.0012 are given; therefore, tolerances will be:

$$\text{Hole } \varnothing 1.0025 \, {}^{+.0020}_{-.0000}$$

$$\text{Shaft } \varnothing 1.000 \, {}^{+.0000}_{-.0012}$$

Additional examples of how to use the inch fits are shown in Tables 17 to 21 in the Appendix.

REFERENCES

ASME Y14.5M-1994 (R1999) Dimensioning and Tolerancing
ASME B4.1-1967 (R1999) Preferred Limits and Fits for Cylindrical Parts

INTERNET RESOURCES

For information and examples of standard decimal inch fits, see: http://www.me.metu.edu.tri/me114/tolerancing.htm

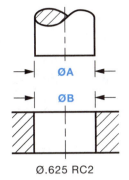

Ø.625 RC2

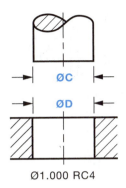

Ø1.000 RC4

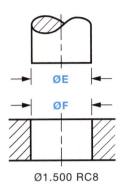

Ø1.500 RC8

RUNNING AND SLIDING FITS

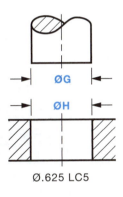

Ø.625 LC5

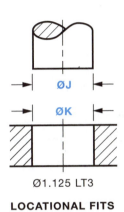

Ø1.125 LT3

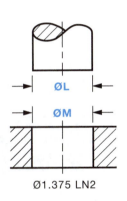

Ø1.375 LN2

LOCATIONAL FITS

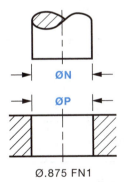

Ø.875 FN1

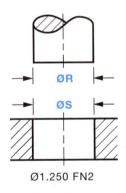

Ø1.250 FN2

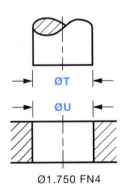

Ø1.750 FN4

FORCE OR SHRINK FITS

ASSIGNMENT: ON A GRID SHEET SKETCH A TABLE SHOWING THE LIMITS OF SIZE FOR EACH OF THE PARTS, AND THE MINIMUM AND MAXIMUM CLEARANCE OR INTERFERENCE FOR EACH OF THE FITS SHOWN.

**INCH FITS–
BASIC HOLE SYSTEM**

A-41

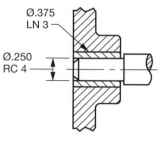

Ø.375
LN 3

Ø.250
RC 4

(A) SHAFT IN BUSHED HOLE

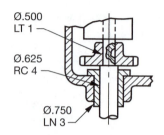

Ø.500
LT 1

Ø.625
RC 4

Ø.750
LN 3

**(B) GEAR AND SHAFT IN
BUSHED BEARING**

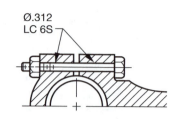

Ø.312
LC 6S

(C) CONNECTING-ROD BOLT

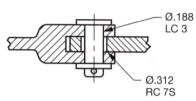

Ø.188
LC 3

Ø.312
RC 7S

(D) LINK PIN

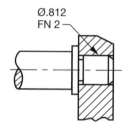

Ø.812
FN 2

(E) CRANK PIN IN CAST IRON

DESIGN SKETCH	BASIC DIAMETER SIZE (IN)	SYMBOL	BASIS	FEATURE	LIMITS OF SIZE		CLEARANCE OR INTERFERENCE	
					MAX	MIN	MAX	MIN
A	.375		HOLE	HOLE				
				SHAFT				
A	.250		HOLE	HOLE				
				SHAFT				
B	.500		HOLE	HOLE				
				SHAFT				
B	.625		HOLE	HOLE				
				SHAFT				
B	.750		HOLE	HOLE				
				SHAFT				
C	.312		SHAFT	HOLE				
				SHAFT				
D	.188		HOLE	HOLE				
				SHAFT				
D	.312		SHAFT	HOLE				
				SHAFT				
E	.812		HOLE	HOLE				
				SHAFT				

ASSIGNMENT: PREPARE A CHART SIMILAR TO THE ONE SHOWN ABOVE AND, USING THE INCH FIT TABLES SHOWN IN THE APPENDIX, ADD THE MISSING INFORMATION.

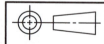

INCH FITS

A-42

METRIC FITS

The ISO (metric) system of limits and fits for mating parts is approved and adopted for general use in the United States. It establishes the designation symbols used to define specific dimensional limits on drawings.

The general terms "hole" and "shaft" can also be taken as referring to the space containing or contained by two parallel faces of any part, such as the width of a slot, or the thickness of a key.

An "International Tolerance Grade" establishes the magnitude of the tolerance zone or the amount of part size variation allowed for internal and external dimensions alike. The smaller the grade number, the smaller the tolerance zone. For general applications of IT grades, see Figure 15–1.

Grades 1 through 4 are very precise grades intended primarily for gage making and similar precision work, although grade 4 can also be used for very precise production work.

Grades 5 through 16 represent a progressive series suitable for cutting operations, such as turning, boring, grinding, milling, and sawing. Grade 5 is the most precise grade, obtainable by fine grinding and lapping, while 16 is the coarsest grade for rough sawing and machining.

Grades 12 through 16 are intended for manufacturing operations such as cold heading, pressing, rolling, and other forming operations.

As a guide to the selection of tolerances, Figure 15–2 has been prepared to show grades that may be expected to be held by various manufacturing processes for work in metals. For work in other materials, such as plastics, it may be necessary to use coarser tolerance grades for the same process.

A fundamental deviation establishes the position of the tolerance zone with respect to the basic size. Fundamental deviations are expressed by "tolerance position letters." Capital letters are used for internal dimensions, and lowercase letters for external dimensions.

Metric Tolerance Symbol

By combining the IT grade number and the tolerance position letter, the tolerance symbol is established, which identifies the actual maximum and minimum limits of the part. The toleranced sizes are thus defined by the basic size of the part followed by the symbol composed of a letter and number, Figure 15–3.

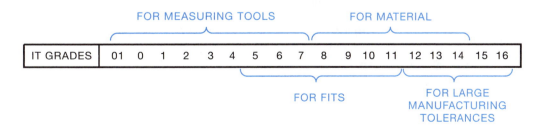

FIGURE 15–1 ■ *Application of international tolerance (IT) grades.*

MACHINING PROCESS	TOLERANCE GRADES									
	4	5	6	7	8	9	10	11	12	13
Lapping & Honing	▓	▓								
Cylindrical Grinding		▓	▓	▓						
Surface Grinding		▓	▓	▓	▓					
Diamond Turning		▓	▓	▓						
Diamond Boring		▓	▓	▓						
Broaching		▓	▓	▓	▓					
Reaming		▓	▓	▓	▓	▓	▓			
Turning				▓	▓	▓	▓	▓	▓	▓
Boring					▓	▓	▓	▓	▓	▓
Milling							▓	▓	▓	▓
Planing & Shaping							▓	▓	▓	▓
Drilling							▓	▓	▓	▓

FIGURE 15–2 ■ *Tolerancing grades for machining processes.*

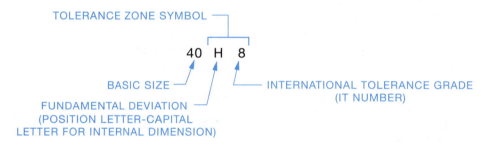

TOLERANCE ZONE SYMBOL

40 H 8

BASIC SIZE

FUNDAMENTAL DEVIATION
(POSITION LETTER-CAPITAL
LETTER FOR INTERNAL DIMENSION)

INTERNATIONAL TOLERANCE GRADE
(IT NUMBER)

(A) INTERNAL DIMENSION (HOLES)

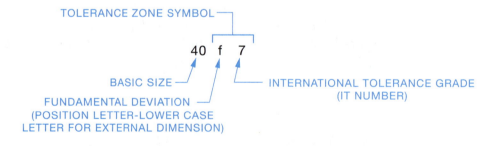

TOLERANCE ZONE SYMBOL

40 f 7

BASIC SIZE

FUNDAMENTAL DEVIATION
(POSITION LETTER-LOWER CASE
LETTER FOR EXTERNAL DIMENSION)

INTERNATIONAL TOLERANCE GRADE
(IT NUMBER)

(B) EXTERNAL DIMENSION (SHAFTS)

FIGURE 15–3 ■ *Tolerance symbol (hole basis fit).*

Hole basis fits have a fundamental deviation of "H" on the hole, and shaft basis fits have a fundamental deviation of "h" on the shaft. Normally, the hole basis system is preferred.

Fit Symbol

A fit is specified by the basic size common to both components, followed by a symbol corresponding to each component, with the internal part symbol preceding the external part symbol, Figure 15–4.

Figure 15–5 shows examples of three common fits.

Hole Basis Fits System

In the hole basis fits system (see Tables 22 and 24 of the Appendix) the basic size will be the minimum size of the hole. For example, for a ⌀25 H8/f7 fit, which is a Preferred Hole Basis Clearance Fit, the limits for the hole and shaft will be as follows:

> Hole limits = ⌀25.000 and ⌀25.033
> Shaft limits = ⌀24.959 and ⌀24.980
> Minimum clearance = 0.020
> Maximum clearance = 0.074

If a ⌀25 H7/s6 Preferred Hole Basis Interference Fit is required, the limits for the hole and shaft will be as follows:

> Hole limits = ⌀25.000 and ⌀25.021
> Shaft limits = ⌀25.035 and ⌀25.048
> Minimum interference = −0.014
> Maximum interference = −0.048

Additional examples of how to use the Hole Basis Fits System are shown in Table 24 of the Appendix.

Shaft Basis Fits System

Where more than two fits are required on the same shaft, the shaft basis fits system is recommended. Tolerances for holes and shafts are identical with those for a basic hole system; however, the basic size becomes the maximum shaft size. For example, for a ⌀16 C11/h11 fit, which is a Preferred Shaft Basis Clearance Fit, the limits for the hole and shaft will be as follows:

> Hole limits = ⌀16.095 and ⌀16.205
> Shaft limits = ⌀15.890 and ⌀16.000
> Minimum clearance = 0.095
> Maximum clearance = 0.315

Additional examples of how to use the Shaft Basis Fits System are shown in Table 25 of the Appendix.

Descriptions of preferred metric fits are described in Figure 15–6.

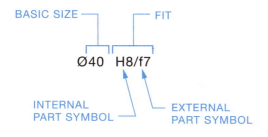

(A) HOLE BASIS

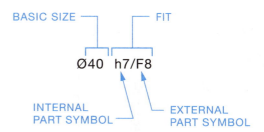

(B) SHAFT BASIS

FIGURE 15–4 ■ *Fit symbol.*

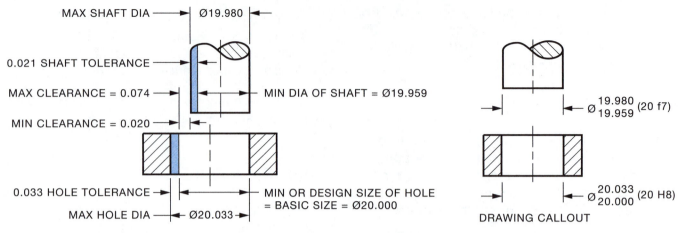

MAX SHAFT DIA ⟶ Ø19.980

0.021 SHAFT TOLERANCE ⟶

MAX CLEARANCE = 0.074 ⟶   MIN DIA OF SHAFT = Ø19.959

MIN CLEARANCE = 0.020 ⟶

0.033 HOLE TOLERANCE ⟶   MIN OR DESIGN SIZE OF HOLE
                                  = BASIC SIZE = Ø20.000

MAX HOLE DIA ⟶ Ø20.033

Ø $^{19.980}_{19.959}$ (20 f7)

Ø $^{20.033}_{20.000}$ (20 H8)

DRAWING CALLOUT

EXAMPLE - H8/f7 PREFERRED HOLE BASIS FIT FOR A Ø20 HOLE (SEE APPENDIX, TABLE 24)

(A) CLEARANCE FIT

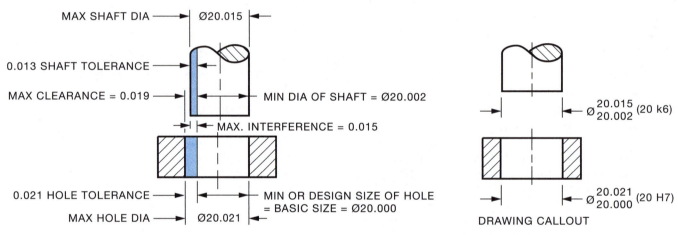

MAX SHAFT DIA ⟶ Ø20.015

0.013 SHAFT TOLERANCE ⟶

MAX CLEARANCE = 0.019 ⟶   MIN DIA OF SHAFT = Ø20.002

MAX. INTERFERENCE = 0.015

0.021 HOLE TOLERANCE ⟶   MIN OR DESIGN SIZE OF HOLE
                                  = BASIC SIZE = Ø20.000

MAX HOLE DIA ⟶ Ø20.021

Ø $^{20.015}_{20.002}$ (20 k6)

Ø $^{20.021}_{20.000}$ (20 H7)

DRAWING CALLOUT

EXAMPLE - H7/k6 PREFERRED HOLE BASIS FIT FOR A Ø20 HOLE (SEE APPENDIX, TABLE 24)

(B) TRANSITION FIT

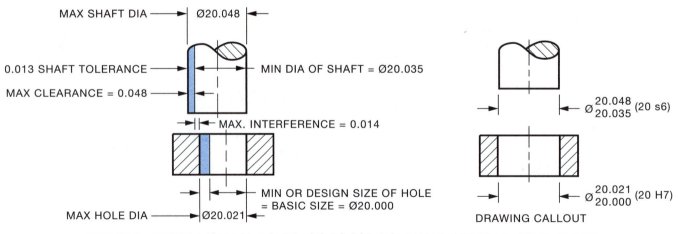

MAX SHAFT DIA ⟶ Ø20.048

0.013 SHAFT TOLERANCE ⟶   MIN DIA OF SHAFT = Ø20.035

MAX CLEARANCE = 0.048 ⟶

MAX. INTERFERENCE = 0.014

MIN OR DESIGN SIZE OF HOLE
= BASIC SIZE = Ø20.000

MAX HOLE DIA ⟶ Ø20.021

Ø $^{20.048}_{20.035}$ (20 s6)

Ø $^{20.021}_{20.000}$ (20 H7)

DRAWING CALLOUT

EXAMPLE - H7/s6 PREFERRED HOLE BASIS FIT FOR A Ø20 HOLE (SEE APPENDIX, TABLE 24)

(C) INTERFERENCE FIT

FIGURE 15–5 ■ *Types and examples of millimeter fits.*

	ISO SYMBOL			
	HOLE BASIS	SHAFT BASIS	DESCRIPTION	
CLEARANCE FITS	H11/c11	C11/h11	LOOSE RUNNING FIT FOR WIDE COMMERCIAL TOLERANCES OR ALLOWANCES ON EXTERNAL MEMBERS.	MORE CLEARANCE
	H9/d9	D9/h9	FREE RUNNING FIT NOT FOR USE WHERE ACCURACY IS ESSENTIAL, BUT GOOD FOR LARGE TEMPERATURE VARIATIONS, HIGH RUNNING SPEEDS, OR HEAVY JOURNAL PRESSURES.	
	H8/f7	F8/h7	CLOSE RUNNING FIT FOR RUNNING ON ACCURATE MACHINES AND FOR ACCURATE LOCATION AT MODERATE SPEEDS AND JOURNAL PRESSURES.	
	H7/g6	G7/h6	SLIDING FIT NOT INTENDED TO RUN FREELY, BUT TO MOVE AND TURN FREELY AND LOCATE ACCURATELY.	
TRANSITION FITS	H7/h6	H7/h6	LOCATIONAL CLEARANCE FIT PROVIDES SNUG FIT FOR LOCATING STATIONARY PARTS, BUT CAN BE FREELY ASSEMBLED AND DISASSEMBLED.	
	H7/k6	K7/h6	LOCATIONAL TRANSITION FIT FOR ACCURATE LOCATION, A COMPROMISE BETWEEN CLEARANCE AND INTERFERENCE.	
	H7/n6	N7/h6	LOCATIONAL TRANSITION FIT FOR MORE ACCURATE LOCATION WHERE GREATER INTERFERENCE IS PERMISSIBLE.	
INTERFERENCE FITS	H7/p6	P7/h6	LOCATIONAL INTERFERENCE FIT FOR PARTS REQUIRING RIGIDITY AND ALIGNMENT WITH PRIME ACCURACY OF LOCATION BUT WITHOUT SPECIAL BORE PRESSURE REQUIREMENTS.	MORE INTERFERENCE
	H7/s6	S7/h6	MEDIUM DRIVE FIT FOR ORDINARY STEEL PARTS OR SHRINK FITS ON LIGHT SECTIONS, THE TIGHTEST FIT USABLE WITH CAST IRON.	
	H7/u6	U7/h6	FORCE FIT SUITABLE FOR PARTS THAT CAN BE HIGHLY STRESSED OR FOR SHRINK FITS WHERE THE HEAVY PRESSING FORCES REQUIRED ARE IMPRACTICAL.	

FIGURE 15–6 ■ *Description of preferred metric fits.*

Drawing Callout

The method shown in Figure 15–7(A) is recommended when the system is first introduced. In this case, limit dimensions are specified and the basic size and tolerance symbol are identified as reference.

As experience is gained, the method shown in Figure 15–7(B) may be used. When the system is established and standard tools, gages, and stock materials are available with size and symbol identification, the method shown in Figure 15–7(C) may be used.

This would result in a clearance between 0.020 and 0.074 mm. A description of the preferred metric fits is shown in Tables 22 and 23 of the Appendix.

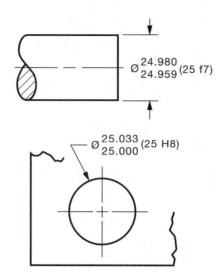

$\varnothing\,^{24.980}_{24.959}$ (25 f7)

$\varnothing\,^{25.033}_{25.000}$ (25 H8)

(A) WHEN SYSTEM IN FIRST INTRODUCED

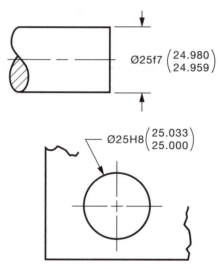

$\varnothing25f7\left(^{24.980}_{24.959}\right)$

$\varnothing25H8\left(^{25.033}_{25.000}\right)$

(B) AS EXPERIENCE IS GAINED

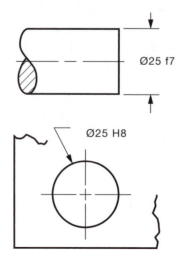

$\varnothing25\ f7$

$\varnothing25\ H8$

(C) WHEN SYSTEM IS ESTABLISHED

FIGURE 15–7 ■ *Metric tolerance symbol shown on drawings.*

REFERENCES

ASME B4.2-1978 (R1999) Preferred Metric Limits and Fits

ASME Y14.5M-1994 (R1999) Dimensioning and Tolerancing

INTERNET RESOURCES

Maryland Metrics. For information and examples of standard millimeter fits, see: http://mdmetric.com/k0k2.htm

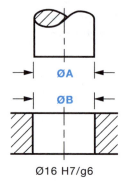

ØA

ØB

Ø16 H7/g6

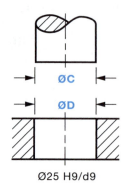

ØC

ØD

Ø25 H9/d9

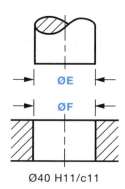

ØE

ØF

Ø40 H11/c11

RUNNING AND SLIDING FITS

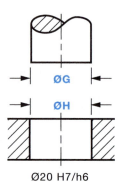

ØG

ØH

Ø20 H7/h6

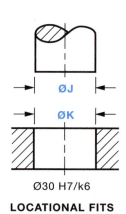

ØJ

ØK

Ø30 H7/k6

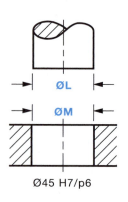

ØL

ØM

Ø45 H7/p6

LOCATIONAL FITS

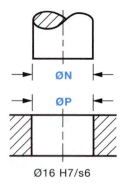

ØN

ØP

Ø16 H7/s6

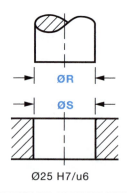

ØR

ØS

Ø25 H7/u6

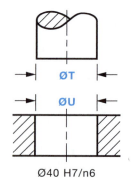

ØT

ØU

Ø40 H7/n6

FORCE OR SHRINK FITS

ASSIGNMENT: ON A GRID SHEET SKETCH A TABLE SHOWING THE
LIMITS OF SIZE FOR EACH OF THE PARTS, AND
THE MINIMUM AND MAXIMUM CLEARANCE OR
INTERFERENCE FOR EACH OF THE FITS SHOWN.

**METRIC FITS–
BASIC HOLE SYSTEM**

A-43M

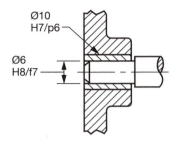

(A) SHAFT IN BUSHED HOLE

Ø10
H7/p6

Ø6
H8/f7

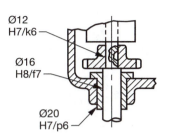

(B) GEAR AND SHAFT IN BUSHED BEARING

Ø12
H7/k6

Ø16
H8/f7

Ø20
H7/p6

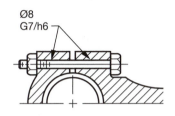

(C) CONNECTING-ROD BOLT

Ø8
G7/h6

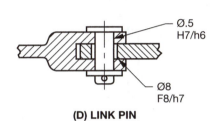

(D) LINK PIN

Ø.5
H7/h6

Ø8
F8/h7

Ø18
H7/u6

(E) CRANK PIN IN CAST IRON

DESIGN SKETCH	BASIC DIAMETER SIZE (mm)	SYMBOL	BASIS	FEATURE	LIMITS OF SIZE		CLEARANCE OR INTERFERENCE	
					MAX	MIN	MAX	MIN
A	10		HOLE	HOLE				
				SHAFT				
A	6		HOLE	HOLE				
				SHAFT				
B	12		HOLE	HOLE				
				SHAFT				
B	16		HOLE	HOLE				
				SHAFT				
B	20		HOLE	HOLE				
				SHAFT				
C	8		SHAFT	HOLE				
				SHAFT				
D	5		HOLE	HOLE				
				SHAFT				
D	8		SHAFT	HOLE				
				SHAFT				
E	18		HOLE	HOLE				
				SHAFT				

ASSIGNMENT: PREPARE A CHART SIMILAR TO THE ONE SHOWN ABOVE AND, USING THE METRIC FIT TABLES SHOWN IN THE APPENDIX, ADD THE MISSING INFORMATION.

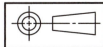

METRIC FITS   **A–44M**

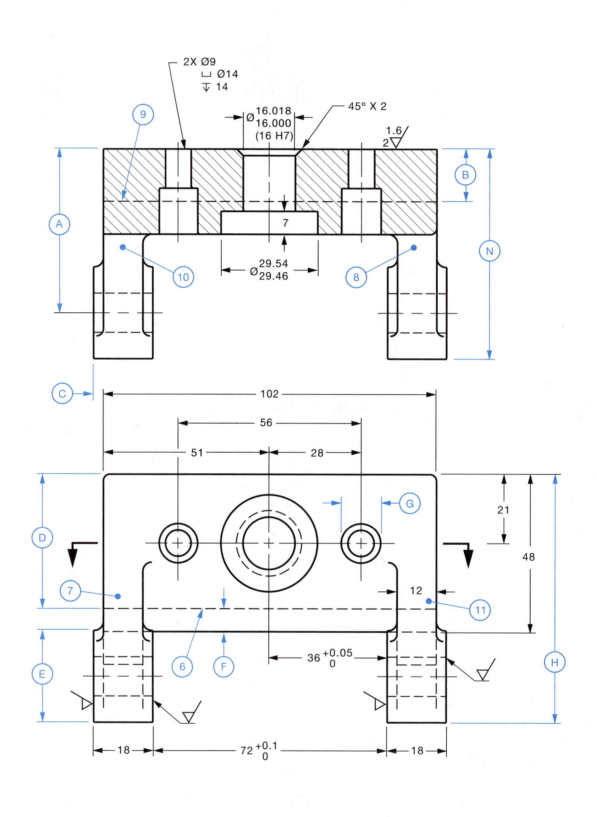

QUESTIONS:

1. How many surfaces are to be finished?

2. Except where noted otherwise, what is the tolerance on all dimensions?

3. What is the tolerance on the Ø12.000-12.018 holes?

4. What are the limit dimensions for the 40.64 dimension shown on the side view?

5. What are the limit dimensions for the 26 dimension shown on the side view?

6. What is the maximum distance permissible between the centers of the Ø9 hole?

7. Express the Ø12.000-12.018 holes as a plus-and-minus tolerance dimension.

8. How many surfaces require a $\frac{6.3}{2\sqrt{}}$ finish?

9. What are the limit dimensions for the 7 dimension shown on the top view?

10. Locate surface ④ on the top view.

11. How many bosses are there?

12. Locate line ③ in the top view.

13. Locate line ⑥ in the side view.

14. Which surface in the front view indicates line ④ in the side view?

15. Calculate distances Ⓐ to Ⓝ.

16. What is the (A) size, and (B) type of cap screw used to fasten the bracket?

17. Can standard lockwashers be used with the cap screws?

18. What type of tolerance is shown on the: (A) 40.64 vertical dimension, and (B) 72 horizontal dimension?

19. What type of sectional view is used?

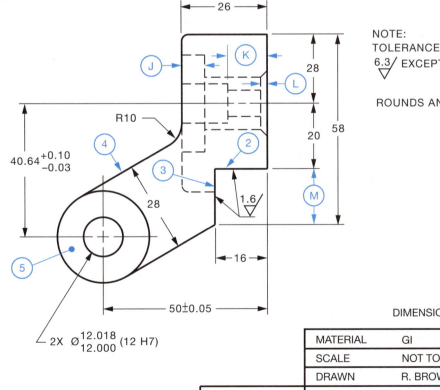

NOTE:
TOLERANCES ON DIMENSIONS ±0.5
$\frac{6.3}{\sqrt{}}$ EXCEPT WHERE NOTED

ROUNDS AND FILLETS R5

METRIC
DIMENSIONS IN MILLIMETERS

MATERIAL	GI	
SCALE	NOT TO SCALE	
DRAWN	R. BROWN	DATE 22/11/04

BRACKET　　**A-45M**

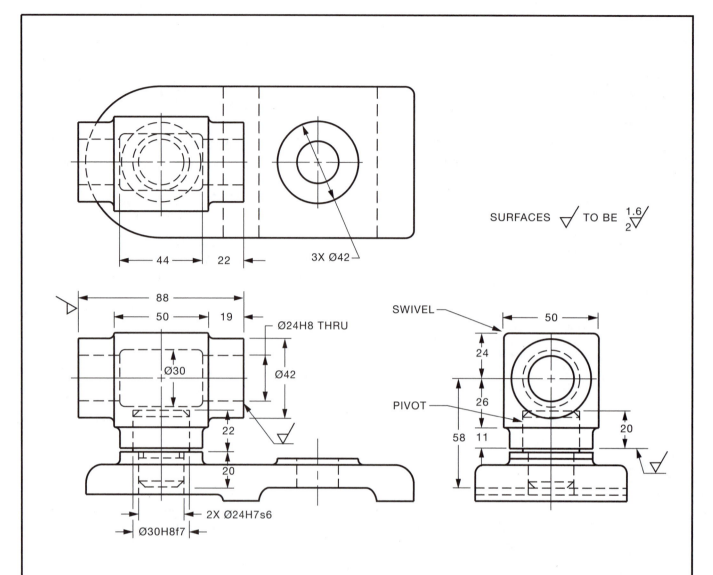

SURFACES ∇ TO BE $\frac{1.6}{2}$∇

44   22   3X Ø42

88
50   19
Ø24H8 THRU
Ø30
Ø42
22
20
2X Ø24H7s6
Ø30H8f7

SWIVEL   50
24
PIVOT   26
58   11   20

ASSIGNMENT:

1. ON A CENTIMETER GRID SHEET (1 mm SQUARES) SKETCH A FULL SECTION VIEW OF THE SWIVEL SHOWN IN THE ABOVE ASSEMBLY. DIMENSION AND SHOW THE HOLE SIZES AS LIMITS. SCALE 1 : 1.

2. ON A CENTIMETER GRID SHEET (1 mm SQUARES) MAKE A ONE-VIEW DRAWING WITH DIMENSIONS OF THE PIVOT. ADD CHAMFERS AT THE END OF THE PIVOT FOR EASE OF ASSEMBLY, AND AN UNDERCUT AT THE SHOULDER. REFER TO UNIT 10 FOR INFORMATION ON CHAMFERS AND UNDERCUTS. SCALE 1 : 1.

METRIC
DIMENSIONS ARE IN MILLIMETERS

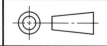

SWIVEL   **A-46M**

THREADED FASTENERS

Fastening devices are vital to most aspects of industry. They are used in assembling manufactured products, the machines and devices used in the manufacturing processes, and in the construction of all types of buildings.

There are two basic types of fasteners: semi-permanent and removable. Rivets are semi-permanent fasteners; bolts, screws, studs, nuts, pins, and keys are removable fasteners. With the progress of industry, fastening devices have become standardized. A thorough knowledge of the design and graphic representation of the common fasteners is essential for interpreting engineering drawings, Figure 16–1.

Thread Representation

True representation of a screw thread is seldom provided on working drawings because of the time involved and the drawing cost. Three types of conventions are generally used for screw thread representation: simplified, detailed, and schematic representation. Simplified representation is used whenever it will clearly indicate the requirements. Schematic and detailed representations require more drafting time, but they are sometimes used to avoid confusion with other parallel lines or to more clearly portray particular aspects of threads. One

method is generally used within any one drawing. When required, however, all three methods may be used. American (ASME) and ISO thread representation vary slightly, Figures 16–2 and 16–3.

When using either simplified or schematic thread representation, on the end views of external threads starting with a chamfer, and the chamfer and minor thread diameter are close to being the same size, for clarity, the minor diameter is not shown.

On the end views showing a countersunk threaded hole, and the countersunk diameter and major thread diameter are close to being the same size, for clarity, the major thread diameter is not shown.

THREADED ASSEMBLIES

Any of the thread representations shown here may be used for assemblies of threaded parts, and two or more methods may be used on the same drawing, Figures 16–2 and 16–3. In sectional views, the externally threaded part is always shown covering the internally threaded part, Figure 16–4.

Thread Standards

With the progress and growth of industry, there is a growing need for uniform, interchangeable, threaded fasteners. Aside from the threaded forms

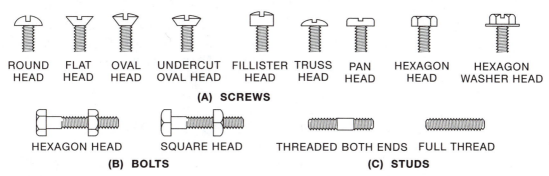

FIGURE 16–1 ■ *Threaded fasteners.*

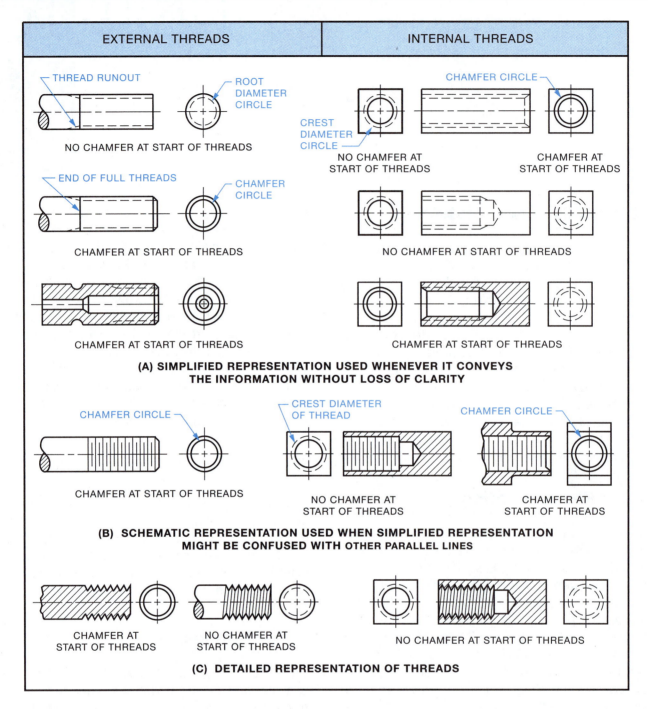

EXTERNAL THREADS	INTERNAL THREADS

(A) SIMPLIFIED REPRESENTATION USED WHENEVER IT CONVEYS THE INFORMATION WITHOUT LOSS OF CLARITY

(B) SCHEMATIC REPRESENTATION USED WHEN SIMPLIFIED REPRESENTATION MIGHT BE CONFUSED WITH OTHER PARALLEL LINES

(C) DETAILED REPRESENTATION OF THREADS

FIGURE 16–2 ■ *Standard thread representations (ASME).*

previously mentioned, the pitch of the thread and the major diameters are factors affecting standards.

THREADED HOLES

When a small threaded hole is required on a part, a common method of producing it is to drill a hole (tap drill size), then add threads to the hole by means of a threading tool called a tap. The tap drill size is not specified on the drawing. For a blind hole, the drilled

hole is made a little deeper than the depth required for the threads. The tap drill sizes for inch and metric threads are shown in Tables 5 and 6 of the Appendix.

INCH THREADS

Until 1976, nearly all threaded assemblies in North America were designed using inch-sized threads. In this system the pitch is equal to the distance be-

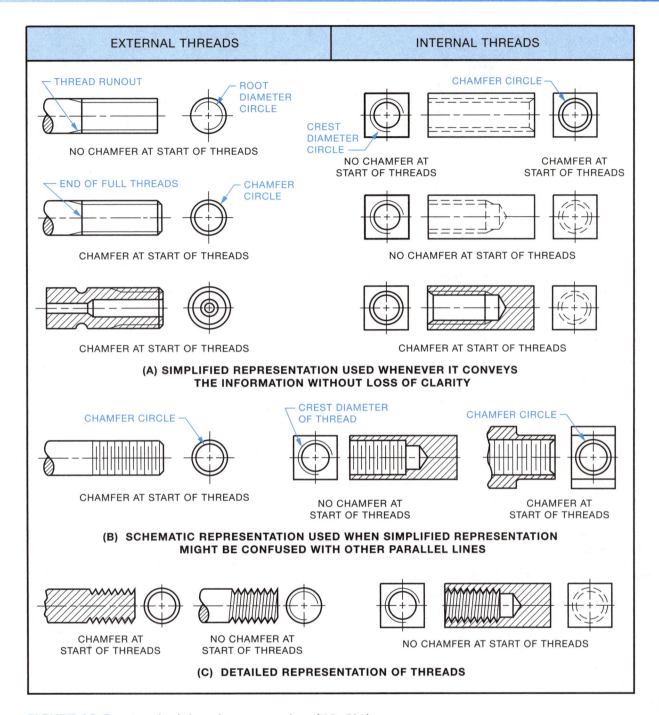

FIGURE 16–3 ∎ *Standard thread representations (ISO, CSA).*

tween corresponding points on adjacent threads and is expressed as:

$$\text{Pitch} = \frac{1}{\text{Number of threads per inch}}$$

The number of threads per inch is set for different diameters in a thread "series." For the Unified National system there are the coarse thread series and the fine thread series.

There is also an extra-fine thread series, UNEF, for use where a small pitch is required, such as on thin-walled tubing. For special work and diameters larger than those specified in the coarse and fine series, the Unified Thread system has three series that allow the same number of threads per inch regardless of the diameter. These are the 8-thread series, the 12-thread series, and the 16-thread series.

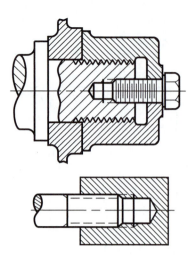

FIGURE 16–4 ■ *Showing threads on assembly drawings.*

Three classes of external thread (Classes 1A, 2A, and 3A) and three classes of internal thread (Classes 1B, 2B, and 3B) are provided.

The general characteristics and uses of the classes are:

CLASSES 1A AND 1B

These classes produce the loosest fit, that is, the most play (free motion) in assembly. They are useful for work where ease of assembly and disassembly is essential, such as for some automotive work and for stove bolts and other rough bolts and nuts.

CLASSES 2A AND 2B

These classes are designed for the ordinary good grade of commercial products, including machine screws and fasteners and most interchangeable parts.

CLASSES 3A AND 3B

These classes are intended for exceptionally high-grade commercial products needing a particularly close or snug fit. Classes 3A and 3B are used only when the high cost of precision tools and machines is warranted.

INCH THREAD DESIGNATION

Thread designation for both external and internal 60° inch threads is expressed in this order: nominal thread diameter in inches, a dash, the number of threads per inch, a space, the letter symbol of the thread series, a dash, the number and letter of the thread class symbol, a space followed by any qualifying information, such as the letters LH for left hand threads, or the gaging system number. To avoid any misunderstanding ASME Y14.6-2001 SCREW THREAD REPRESENTATION Standard recommends the controlling organization

and thread standard be added to the thread designation, or referenced on the drawing, in a general note. See Figure 16–5 and the following examples:

1. Standard Unified External Screw Thread
.250-20 UNC-2A ASME B1.1-2003
2. Standard Unified Internal Screw Thread, Gaging System 21
.500-20 UNF-2B (21) ASME B1.1-2003

For multiple start threads the number of threads per inch is replaced by the following: pitch in inches (P), a dash, lead in inches (L), and the number of starts in inches. Examples:

3. Standard Unified External Multiple Start Screw Thread
.750-.0625P-.1875L(3 STARTS)UNF-2A ASME B1.1-2003
4. Standard Unified Internal Multiple Start Screw Thread, Gaging System 21
.500-.050P-. 100L(2 STARTS)UNF-2B(21) ASME B1.1-2003

Where decimal inch sizes are shown for fractional-inch threads they are shown to 4 decimal places (not showing zero in the fourth place). Examples: a 1/2 " thread would be shown as .500; and a 9/16 " thread as .5625.

Due to established drawing practices, numbered sizes may be shown as the nominal thread diameter. Where a number is used, a three-place decimal inch equivalent, enclosed in parentheses, is placed after the number. Examples:

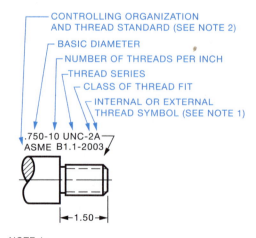

NOTE 1:
LETTER "A" DESIGNATES EXTERNAL THREAD
LETTER "B" DESIGNATES INTERNAL THREAD

NOTE 2:
MAY BE SHOWN AS A GENERAL
NOTE ON THE DRAWING

FIGURE 16–5 ■ *Inch screw thread designation.*

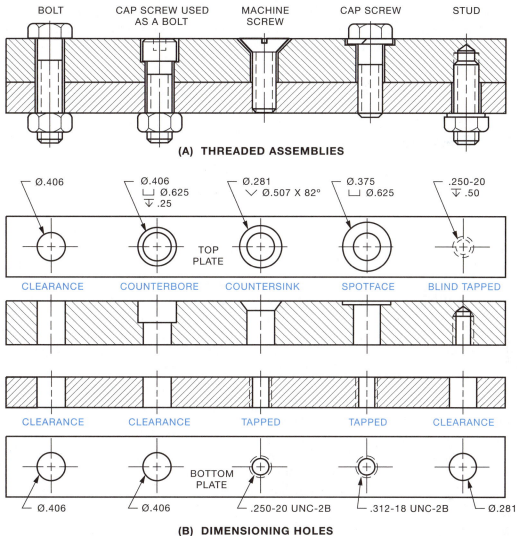

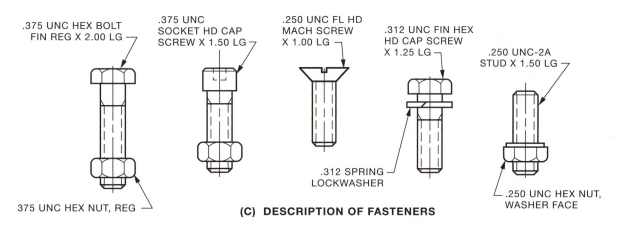

FIGURE 16–6 ■ *Common threaded fasteners.*

5. Standard Unified External Thread
10 (.190)-32 UNF-2A ASME B1.1-2003
6. Standard Unified Internal Thread
5 (.125)-40 UNC-2B ASME B1.1-2003

Typical clearance holes and thread callouts are shown in Figure 16–6.

RIGHT- AND LEFT-HANDED THREADS

Unless designated otherwise, threads are right-handed threads. A bolt being threaded into a tapped hole would be turned in a right-hand (clockwise) direction. For some special uses, such as turnbuckles, left-hand threads are required. When such a thread is necessary, the letters "LH" are added to the thread designation.

Typical hole and thread callouts are shown in Figure 16–6.

METRIC THREADS

Metric threads are grouped into diameter-pitch combinations differentiated by the pitch applied to specified diameters. The pitch for metric threads is the distance between corresponding points on adjacent teeth. In addition to a coarse and fine pitch series, a series of constant pitches is available.

For each of the two main thread elements — pitch diameter and crest diameter — there are numerous tolerance grades. The number of the tolerance grade reflects the tolerance size. For example: Grade 4 tolerances are smaller than Grade 6 tolerances; Grade 8 tolerances are larger than Grade 6 tolerances.

In each case, Grade 6 tolerances should be used for medium quality length of engagement applications. The tolerance grades below Grade 6 are intended for applications involving fine quality and/or short lengths of engagement. Tolerance grades above Grade 6 are intended for coarse quality and/or long lengths of engagement.

In addition to the tolerance grade, positional tolerance is required. The positional tolerance defines the maximum-material limits of the pitch and crest diameters of the external and internal threads and indicates their relationship to the basic profile.

In conformance with current coating (or plating) thickness requirements and the demand for ease of assembly, a series of tolerance positions reflecting the application of varying amounts of allowance has been established.

For External Threads:

Tolerance position "e" (large allowance)
Tolerance position "g" (small allowance)
Tolerance position "h" (no allowance)

For Internal Threads:

Tolerance position "G" (small allowance)
Tolerance position "H" (no allowance)

The two types of metric 60° screw threads in common use are the M and the MJ forms. The MJ form is similar to the standard metric (M) threads, except the sharp V at the root diameter of the external thread has

been replaced with a large radius, which strengthens this stress point. Since the radius increases the minor diameter of the external thread, the minor diameter of the internal thread is enlarged to clear the radius. All other dimensions are the same as the M threads. The MJ thread form is predominately used in applications requiring high fatigue strength, as found in the aerospace and automotive industries.

METRIC THREAD DESIGNATION

Metric 60° screw thread designation is expressed in this order: the metric thread symbol "M", the thread form symbol "J." where applicable, the nominal diameter in millimeters, a lower case "x", the pitch in millimeters, a dash, the pitch diameter tolerance symbol, the crest diameter tolerance symbol (if different from that of the pitch diameter) and a space followed by any qualifying information. To avoid any misunderstanding ASME Y14.6-2001 SCREW THREAD REPRESENTATION standard recommends the controlling organization and thread standard be added to the thread designation or referenced on the drawing. See Figure 16–7 and the following examples:

1. Standard Metric M Screw Thread
   M6×1-4h6h ASME B1.13M-2001
2. Standard Metric MJ Thread, Gaging System 21
   MJ6×1-4H (21) ASME B1.21M-1997

The metric thread size or the pitch should include a zero before the decimal if the value is less than one, and should not show a zero as the last number of the value (e.g. m10×1.5 and MJ 25×0.45).

For multiple start threads the pitch is replaced by the following: L (lead in millimeters), P (pitch in millimeters), and the number of starts in parentheses. Examples:

3. Standard Metric MJ Thread
   MJ20×L7.5P2.5(3 STARTS)-4h6h ASME B1.21M-1997
4. Standard Metric M Multiple Start Thread, Gaging System 21
   M16×L4P2(2 STARTS)-6g (21) ASME B1.13M-2001

An earlier metric thread designation, which was taken from Europian standards is found on many older drawings still in current use. The thread designation was identical for both external and internal threads, and expressed in this order: M denoting the ISO metric thread symbol, the nominal diameter in millimeters, a capital "x" followed by the pitch in

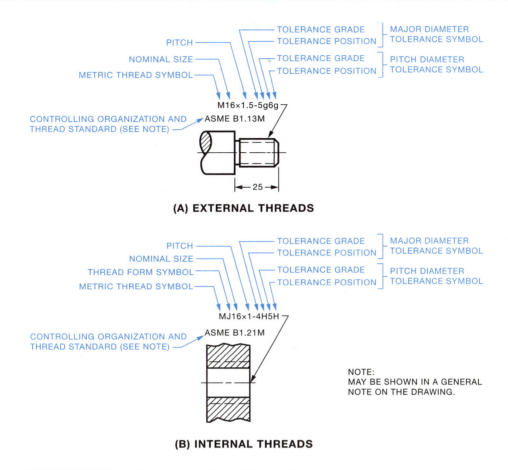

(A) EXTERNAL THREADS

(B) INTERNAL THREADS

NOTE:
MAY BE SHOWN IN A GENERAL
NOTE ON THE DRAWING.

FIGURE 16–7 ■ *Metric screw thread designation.*

millimeters. For the coarse thread series the pitch was not shown. For example, a 10mm diameter, 1.25 pitch, fine thread series, was expressed as M10x1.25. A 10mm diameter, 1.5 pitch, coarse thread series, was expressed as M10. If the length of thread was added to the callout, then the pitch was added to avoid confusion. When specifying the length of the thread in the callout, an "x" separates the length of the thread from the rest of the designation. In the latter example, if a thread length was 25mm, the thread callout would be M10x1.5x25.

In addition to this basic designation, a tolerance class identification was often added. A dash separated the tolerance class identification from the design.

Pipe thread designation is explained in Unit 24.

REFERENCES

ASME Y14.6-2001 SCREW THREAD REPRESENTATION

ASME B1.1-2003 Unified Inch Screw Threads (UN and UNR Thread Form)

ASME B1.13M-2001 Metric Screw Threads – M Profile

ASME B1.21M-1997 Metric Screw Threads – MJ Profile

INTERNET RESOURCES

Drafting Zone. For information on drafting symbols and reference dimensioning, see: http://www.draftingzone.com

Fastener Hut. For information on all types of fasteners, see: http://www.fastenerhut.com

Industrial Fasteners Institute. For both general and specific information on fasteners, see: http://www.ifi-fasteners.org

Machine Design. For information on threaded fasteners and joining processes, see *Machine Design, Fastening/Joining Reference* at: http://www.machinedesign.com

Metrication.com. For information on metric threads, see: http://www.metrication/engineering

Midwest Fasteners, Inc. For information on welded fasteners, see: http://www.studweldmfi.com

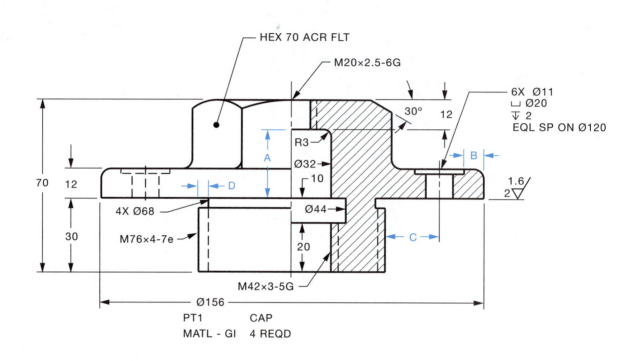

HEX 70 ACR FLT

M20×2.5-6G

6X Ø11
⌴ Ø20
⤓ 2
EQL SP ON Ø120

30°   12

R3

Ø32

10

B

A

70   12

4X Ø68

M76×4-7e

30

Ø44

20

C

1.6
2

M42×3-5G

Ø156

PT1   CAP
MATL - GI   4 REQD

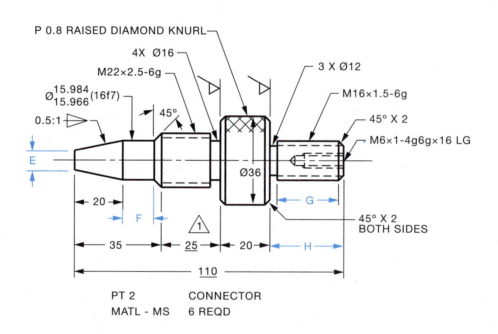

P 0.8 RAISED DIAMOND KNURL

4X Ø16

M22×2.5-6g

3 X Ø12

M16×1.5-6g

Ø 15.984
15.966 (16f7)

45°

45° X 2

M6×1-4g6g×16 LG

0.5:1

E

Ø36

45° X 2
BOTH SIDES

20

F

G

1

35   25   20   H

110

PT 2   CONNECTOR
MATL - MS   6 REQD

ROUNDS AND FILLETS R4
EXCEPT WHERE OTHERWISE SHOWN

REVISIONS	1	27/05/04	R. HINES
	25 IN PT 2 WAS 30		

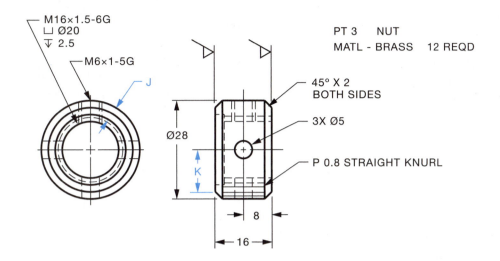

M16×1.5-6G
⊔ Ø20
▽ 2.5

M6×1-5G

J

Ø28

K

8

16

45° X 2
BOTH SIDES

3X Ø5

P 0.8 STRAIGHT KNURL

PT 3   NUT
MATL - BRASS   12 REQD

QUESTIONS:

1. Calculate dimensions A to K.

Refer to Part 1

2. How many holes are in the part?

3. How many external threads are on the part?

4. How many internal threads are in the part?

5. What is the pitch for the M20 thread?

6. How many complete threads are in the M20 hole?

7. How many complete threads are on the M42 section?

8. What is the diameter of the spotface?

9. What is the angle between the Ø11 holes?

10. How much extra metal was provided on the surface that required machining?

Refer to Part 2

11. What is the pitch of the internal thread?

12. What is the pitch on the M22 thread?

13. What provides for better gripping when rotating the part?

14. How many undercuts are shown?

15. How many chamfers are required?

16. If the standard fit of 16f7 was changed to 16g6, what would be the new limits for the size of the shaft?

Refer to Part 3

17. How many holes are in the part?

18. What is the depth of the counterbore?

19. How many full threads has the M16 hole?

20. What surface finish is required for the sides of the knurled portion?

21. What are the tap drill sizes required for the (A) M6 and, (B) M16 threaded holes?

NOTE: UNLESS OTHERWISE SHOWN:
—TOLERANCES ON DIMENSIONS ±0.5
—TOLERANCES ON ANGLES ±30'

SURFACES ✓ TO BE $\frac{3.2}{\nabla}$

THREAD CONTROLLING ORGANIZATION
AND STANDARD – ASME B1.13M-2001

METRIC
DIMENSIONS ARE IN MILLIMETERS

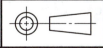

**DRIVE SUPPORT
DETAILS**

A-47M

QUESTIONS:

1. Calculate dimensions Ⓐ to Ⓙ.
2. How many threaded holes or shafts are shown?
3. How many chamfers are shown?
4. How many necks are shown?
5. How many surfaces require finishing?

Refer to Part 1

6. What are the limit sizes for the Ø.64 dimension?
7. What operation provides better gripping when turning the clamping nut?
8. What is the tap drill size?
9. Is the thread right hand or left hand?
10. What is the depth of the threads?

Refer to Part 2

11. What is the length of the thread?
12. Is the thread pitch fine or coarse?
13. What heat treatment does the part undergo?
14. How many threads are there in a one inch length?

NOTE: EXCEPT WHERE NOTED -

TOLERANCE ON TWO-PLACE DIMENSIONS ±.02

TOLERANCE ON THREE-PLACE DIMENSIONS ±.005

TOLERANCE ON ANGLES ±0.5°

32/ EXCEPT WHERE NOTED

THREAD CONTROLLING ORGANIZATION AND STANDARD – ASME B1.1-2003

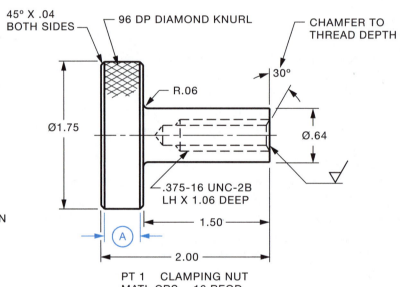

PT 1   CLAMPING NUT
MATL CRS,   16 REQD

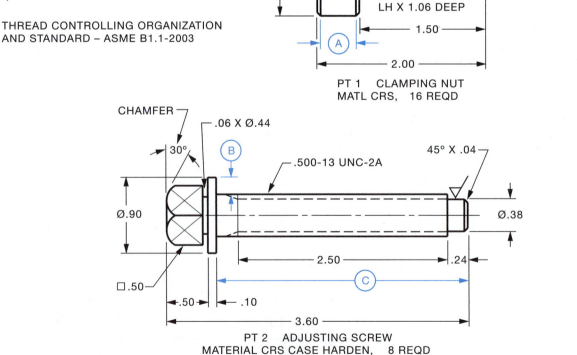

PT 2   ADJUSTING SCREW
MATERIAL CRS CASE HARDEN,   8 REQD

Refer to Part 3

15. What is the maximum permissible width of the part?

16. What size thread is cut on the underside of the piece?

17. What is the distance between the last thread and the flange?

18. What is the tolerance on the center-to-center distance between the tapped holes?

19. What would be the new limit dimensions for the 1.250 diameter hole shown in PT 3 if an RC3 fit is required?

20. What is the smallest diameter to which the hole through the stuffing box can be made?

21. What are the limits of the Ø2.000 dimension?

22. What is the tap drill size for the threaded holes?

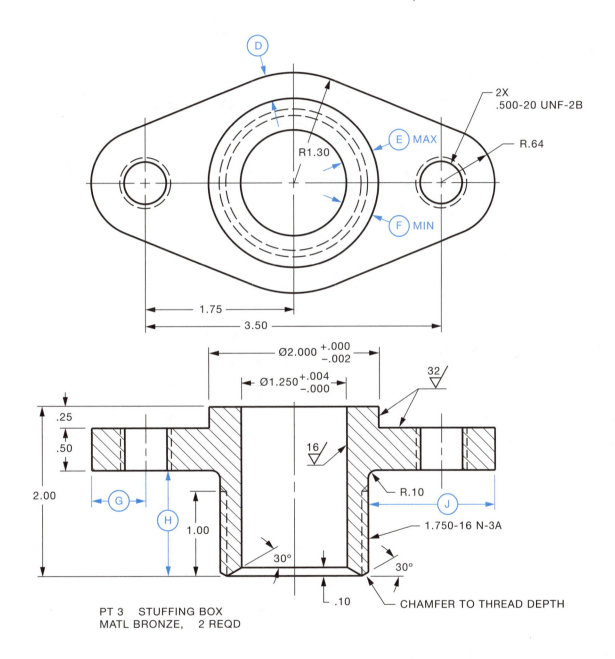

PT 3   STUFFING BOX
MATL BRONZE,   2 REQD

SCALE	NOT TO SCALE		
DRAWN	B. ARMENTI		
		DATE	10/10/04

HOUSING DETAILS   **A-48**

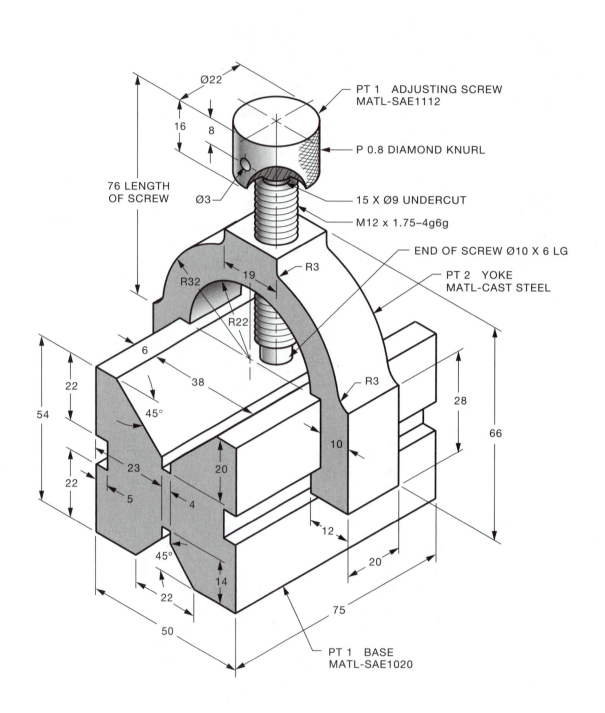

Ø22

16

8

PT 1   ADJUSTING SCREW
MATL-SAE1112

P 0.8 DIAMOND KNURL

76 LENGTH
OF SCREW

Ø3

15 X Ø9 UNDERCUT

M12 x 1.75–4g6g

END OF SCREW Ø10 X 6 LG

R32

19

R3

PT 2   YOKE
MATL-CAST STEEL

R22

6

38

45°

R3

22

54

45°

23

5

20

4

10

28

66

12

20

14

22

50

75

PT 1   BASE
MATL-SAE1020

ASSIGNMENT:
ON A CENTIMETER GRID SHEET (1mm SQUARES),
PREPARE DETAIL DRAWINGS OF THE PARTS
SHOWN IN THE V-BLOCK ASSEMBLY. A CLEARANCE
OF 1mm FOR A SLIDING FIT IS TO BE ADDED TO THE
APPROPRIATE YOKE DIMENSIONS.

METRIC
DIMENSIONS ARE IN MILLIMETERS

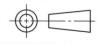

**V-BLOCK
ASSEMBLY**

A-49M

REVOLVED AND REMOVED SECTIONS

In Unit 8 you were introduced to full, half, and off-set sectional views and how they are shown on drawings. For many drawings only a portion of a complete view needs to be shown in section to improve the clarity of the drawing. Two additional types of sectional views are introduced in this unit.

Revolved and removed sections are used to show the cross-sectional shape of ribs, spokes, or arms, when the shape is not obvious in the regular views. End views are often not needed when a revolved section is used.

Revolved Sections

For a *revolved section* a center line is drawn through the shape on the plane to be described, the part is imagined to be rotated 90 degrees, and the view that would be seen when rotated is superimposed on the view. If the revolved section does not interfere with the view on which it is revolved, then the view is not broken unless it would facilitate clearer dimensioning. When the revolved section interferes with or passes through lines on the view on which it is revolved, the view is usually broken. Often the break is used to shorten the length of the object. When superimposed on the view, the outline of the revolved section is a thin continuous line, Figure 17–1.

Removed Sections

The *removed section* differs from the revolved section in that the section is removed to an open area on the drawing instead of being drawn directly on the view. Whenever practical, sectional views should be projected perpendicular to the cutting plane and be placed in the normal position for third-angle projection, Figures 17–2, 17–3, 17–4, and 17–5. Fre-

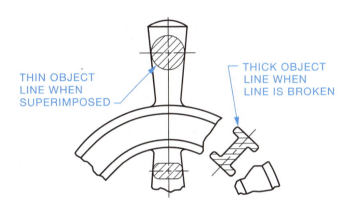

THIN OBJECT LINE WHEN SUPERIMPOSED

THICK OBJECT LINE WHEN LINE IS BROKEN

FIGURE 17–1 ■ *Revolved section.*

quently, the removed section is drawn to an enlarged scale for clarification and easier dimensioning.

Removed sections of symmetrical parts are placed on the extension of the center line where possible.

Broken-out and partial sections are explained and illustrated in Unit 30.

REFERENCE

ASME Y14.3M-1994 (R1999) Multi and Sectional View Drawings

INTERNET RESOURCES

American Society of Mechanical Engineers. For information on revolved and removed sections, refer to ASME Y14.3M-1994 (R1999) (*Multi-and Sectional-View Drawings*) at: http://www.asme.org

Drafting Zone. For information on revolved and removed sections, see: http://www.draftingzone.com

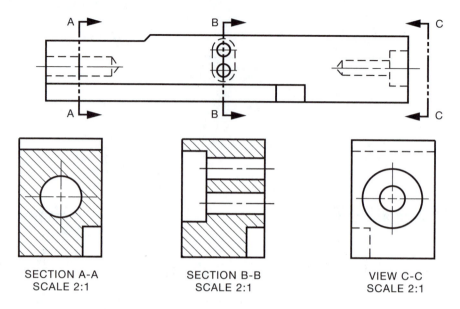

SECTION A-A
SCALE 2:1

SECTION B-B
SCALE 2:1

VIEW C-C
SCALE 2:1

FIGURE 17–2 ■ *Removed sections and removed view.*

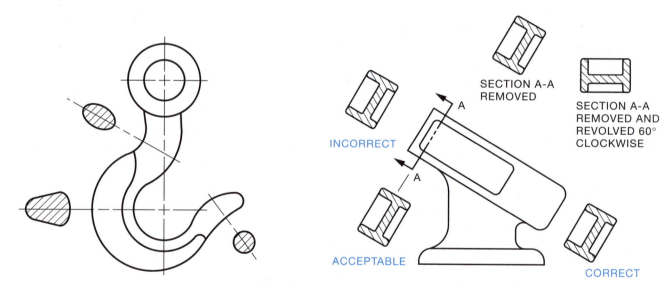

FIGURE 17–3 ■ *Removed sectional view of crane hook.*

SECTION A-A
REMOVED

SECTION A-A
REMOVED AND
REVOLVED 60°
CLOCKWISE

INCORRECT

ACCEPTABLE

CORRECT

FIGURE 17–4 ■ *Placement of removed sectional views.*

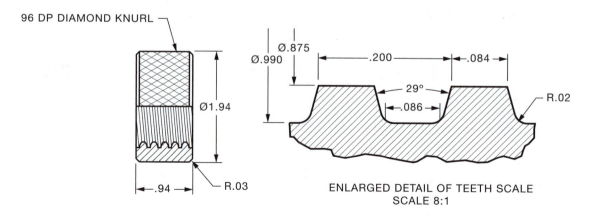

96 DP DIAMOND KNURL

Ø1.94

.94

R.03

Ø.875

Ø.990

.200

.084

29°

.086

R.02

ENLARGED DETAIL OF TEETH SCALE
SCALE 8:1

FIGURE 17–5 ■ *Removed section of thread detail.*

ASSIGNMENT: ON A ONE-INCH GRID SHEET (.10 IN. SQUARES) SKETCH THE BOTTOM VIEW OF THE SHAFT INTERMEDIATE SUPPORT. IF CERTAIN DIMENSIONS CAN BE SHOWN BETTER ON THE BOTTOM VIEW, DUPLICATE THEM. SCALE 1 : 1.

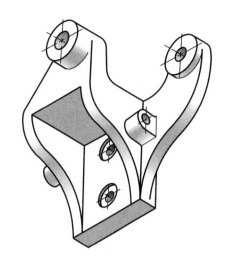

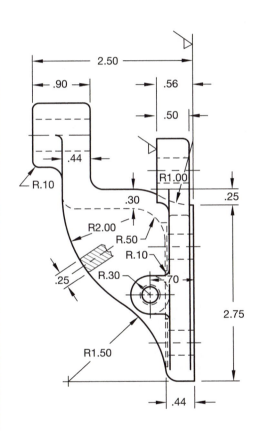

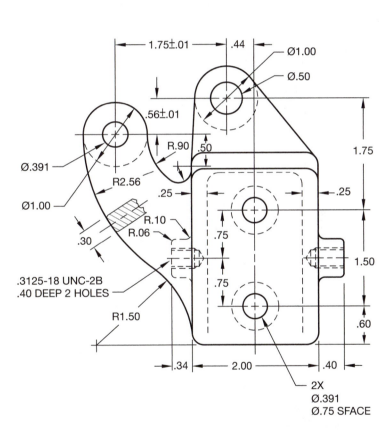

NOTES:
ROUNDS AND FILLETS R.25 EXCEPT WHERE OTHERWISE NOTED.

THREAD CONTROLLING ORGANIZATION AND STANDARD – ASME B1.1-2003

√ TO BE ⁱ²⁵√

MATERIAL	GI		
SCALE			
DRAWN	D. KOLICK	DATE	16/03/04

SHAFT INTERMEDIATE SUPPORT

A-50

QUESTIONS:

1. Calculate distances (A) to (K).

2. How many surfaces require finishing?

3. What tolerance is permitted on dimensions which do not specify a certain tolerance?

4. What type of section view is used on part 1?

5. What type of section view is used on part 2?

6. How many holes are there?

Refer to Part 1

7. What is the maximum center-to-center distance between the holes?

8. Which surface finish is required?

9. Using the smallest permissible hole size as the basic size, replace the limit dimensions for the larger hole with plus-and-minus tolerances.

10. What is the maximum permissible wall thickness at the larger hole?

11. What is the minimum permissible wall thickness at the smaller hole?

12. What is the height of the shaft support casting before machining?

13. The size of the two holes are to be changed to accomodate an LN3 fit with mating bushings. What are the new limit dimensions for these holes?

Refer to Part 2

14. Name two machines that could produce the type of finish required for the slot.

15. What is the nominal size machine screw used in the counterbored holes?

16. What type of cap screw should be used if the heads of these screws must not protrude above the surface of the support?

17. What is the distance between the center of the counterbored holes and the center of the slot?

18. What type of tolerance is used on the (A) Ø1.250 hole and, (B) 3.250 horizontal dimension?

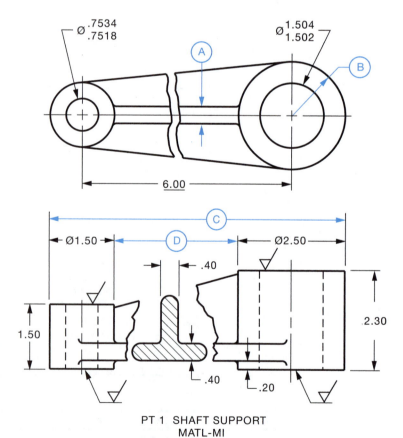

EXCEPT WHERE STATED OTHERWISE:
TOLERANCES ON
    TWO-DECIMAL DIMENSIONS ±.02
    THREE-DECIMAL DIMENSIONS ±.005
ROUNDS AND FILLETS R.10

MACHINE FINISH $\overset{32}{\underset{.05}{\diagdown}}$

PT 1 SHAFT SUPPORT
MATL-MI

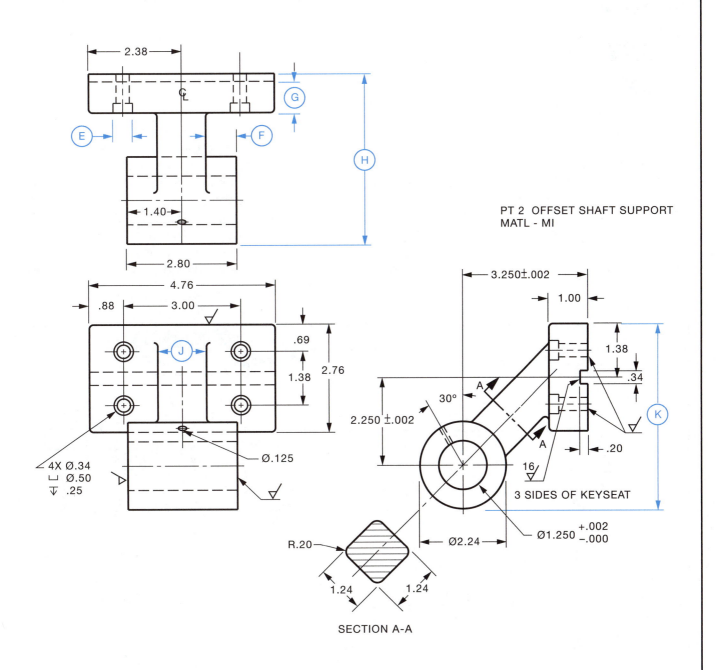

PT 2 OFFSET SHAFT SUPPORT
MATL - MI

2.38

E

F

G

H

1.40

2.80

4.76

.88

3.00

J

.69

1.38

2.76

4X Ø.34
⌴ Ø.50
▽ .25

Ø.125

3.250±.002

1.00

1.38

.34

30°

A

A

2.250 ±.002

16

3 SIDES OF KEYSEAT

.20

K

Ø1.250 +.002 −.000

Ø2.24

R.20

1.24   1.24

SECTION A-A

SCALE	NOT TO SCALE	
DRAWN	S. WOLFE	DATE 07/11/04

SHAFT SUPPORTS

A-51

18 UNIT

KEYS

A *key* is a piece of metal lying partly in a groove in the shaft and extending into another groove in the hub, Figure 18–1. The groove in the shaft is referred to as a keyseat and the groove in the hub or surrounding part is referred to as a keyway. Keys are used to secure gears, pulleys, cranks, handles, and similar machine parts to shafts, so that the motion of the part is transmitted to the shaft or the motion of the shaft to the part, without slippage. The key is also a safety feature. Because of its size, when overloading occurs, the key shears or breaks before the part or shaft breaks.

Common key types are square, flat, and Woodruff. Appendix Tables 13 and 14 give standard square and flat key sizes recommended for various shaft diameters and the necessary dimensions for Woodruff keys.

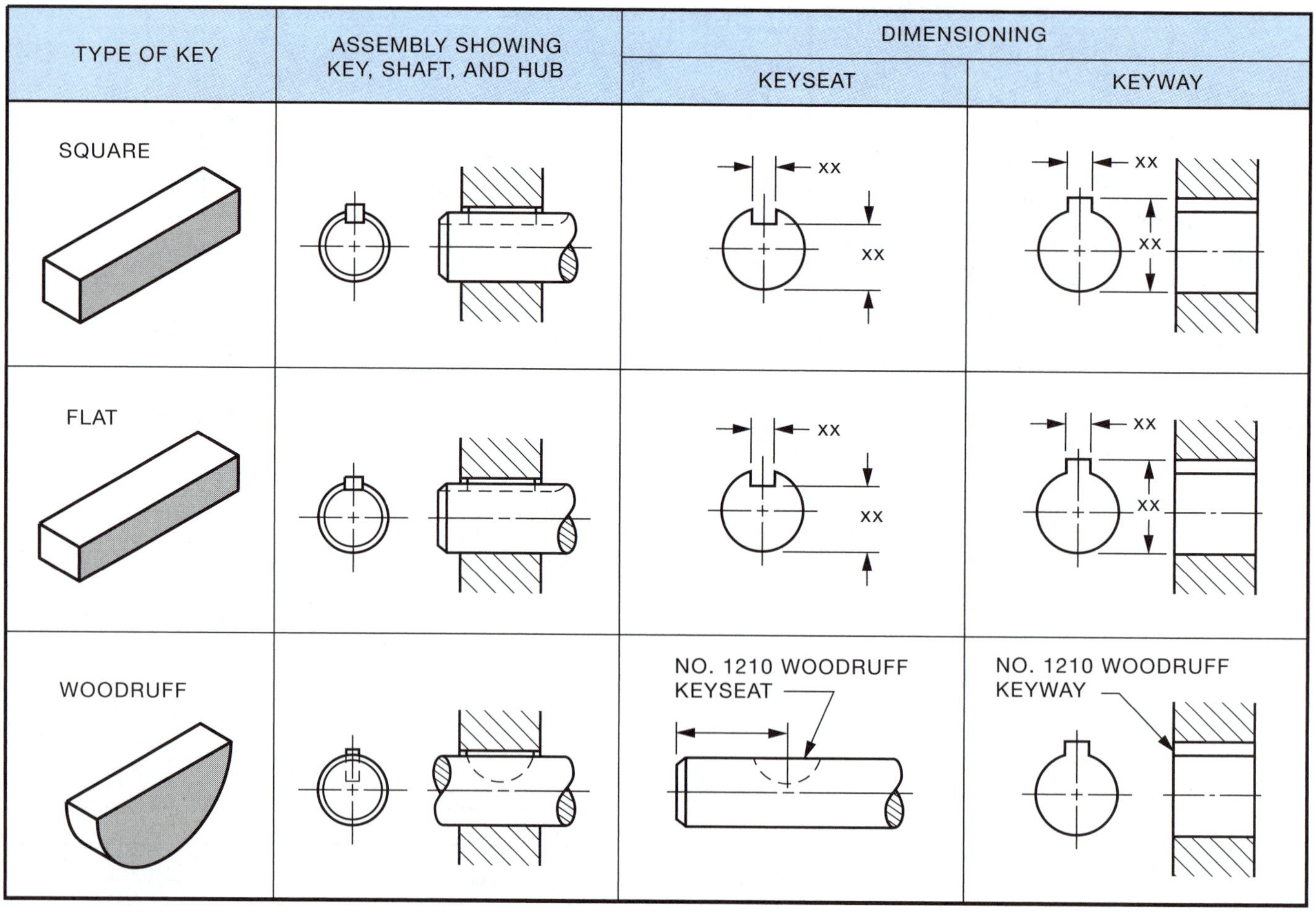

| TYPE OF KEY | ASSEMBLY SHOWING KEY, SHAFT, AND HUB | DIMENSIONING | |
		KEYSEAT	KEYWAY
SQUARE			
FLAT			
WOODRUFF		NO. 1210 WOODRUFF KEYSEAT	NO. 1210 WOODRUFF KEYWAY

FIGURE 18–1 ■ *Common keys.*

The Woodruff key is semicircular and fits into a semicircular keyseat in the shaft and a rectangular keyway in the hub. Woodruff keys are available only in inch sizes, and are identified by a three- or four-digit key number. The last two numbers give the nominal diameter in eighths of an inch. The digit or digits preceding the last two digits gives the key width in thirty-secondths of an inch. For example, a No. 1210 Woodruff key indicates a key 12/32 inch wide by 10/8 (1.25) inches in diameter.

Dimensioning of Keyways and Keyseats

Keyway and keyseat dimensions are usually given in limit dimensions to ensure proper fits and are located from the opposite side of the hole or shaft, Figure 18–2. Alternatively, for unit production where the machinist is expected to fit the key into the keyseat, a leader pointing to the keyseat, specifying first the width and then the depth of the keyseat, may be used. See Part 2, Rack Details, Assignment A-54.

SET SCREWS

Set screws are used as semipermanent fasteners to hold a collar, sheave, or gear on a shaft against rotational or translational forces. In contrast to most fastening devices, the set screw is essentially a compression device. Forces developed by the screw point during tightening produce a strong clamping action that resists relative motion between assembled parts. The basic problem in set screw selection is finding the

best combination of set screw form, size, and point style providing the required holding power.

Set screws are categorized by their forms and the desired point style, Figure 18–3. Each of the standardized set screw forms is available in a variety of point styles. Selection of a specific form or point is influenced by functional as well as other considerations.

The selection of the type of driver and thus, the set screw form, is usually determined by factors other than tightening. Despite higher tightening ability, the protrusion of the square head is a major

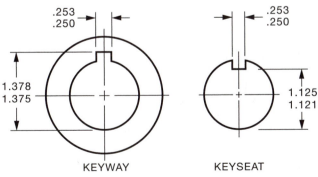

.253
.250

.253
.250

1.378
1.375

1.125
1.121

KEYWAY   KEYSEAT

REFER TO APPENDIX TABLE 14 FOR CALCULATING DIMENSIONS

FIGURE 18–2 ■ *Dimensioning keyways and keyseats.*

STANDARD POINTS	
	CUP Most generally used. Suitable for quick and semipermanent location of parts on soft shafts, where cutting in of edges of cup shape on shaft is not objectionable.
	FLAT Used where frequent resetting is required, on hard steel shafts and where minimum damage to shafts is necessary. Flat is usually ground on shaft for better contact.
	CONE For setting machine parts permanently on shaft, which should be spotted to receive cone point. Also used as a pivot or hanger.
	OVAL Should be used against shafts spotted, splined, or grooved to receive it. Sometimes substituted for cup point.
	HALF DOG For permanent location of machine parts, although cone point is usually preferred for this purpose. Point should fit closely to diameter of drilled hole in shaft. Sometimes used in place of a dowel pin.

STANDARD HEADS			
HEXAGON SOCKET	SLOTTED SOCKET	FLUTED SOCKET	SQUARE HEAD
			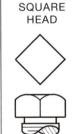

FIGURE 18–3 ■ *Set screws.*

disadvantage. Compactness, weight saving, safety, and appearance may dictate the use of flush-seating socket or slotted headless forms.

The conventional approach to set screw selection is usually based on a rule-of-thumb procedure: the set screw diameter should be roughly equal to one-half the shaft diameter. This rule of thumb often gives satisfactory results, but it has a limited range of usefulness. When a set screw and key are used together, the screw diameter should be equal to the width of the key.

Standard set screws are shown in Table 9 of the Appendix.

FLATS

A *flat* is a slight depression usually cut on a shaft to serve as a surface on which the end of a set screw can rest when holding an object in place, Figure 18–4.

BOSSES AND PADS

A *boss* is a relatively small cylindrical projection above the surface of an object. A *pad* is a slight, non-circular projection above the surface of an object, Figure 18–5. Bosses and pads are generally found

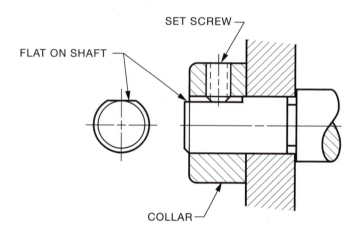

FIGURE 18–4 ■ *Flat and set screw application.*

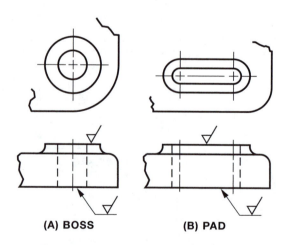

(A) BOSS　　　**(B) PAD**

FIGURE 18–5 ■ *Bosses and pads.*

on castings and provide additional clearance or strength in the area where they are used. They also minimize the amount of machining required.

DIMENSION ORIGIN SYMBOL

This symbol is used to indicate that a toleranced dimension between two features originates from one of these features, Figures 18–6 and 18–7.

RECTANGULAR COORDINATE DIMENSIONING WITHOUT DIMENSION LINES

To avoid having many dimensions extending away from the object, *rectangular coordinate dimensioning* without dimension lines may be used as shown in Figure 18–8. In this system, the "zero" lines are used as reference lines and each of the dimensions

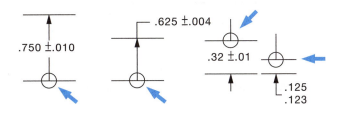

FIGURE 18–6 ■ *Dimension origin symbol.*

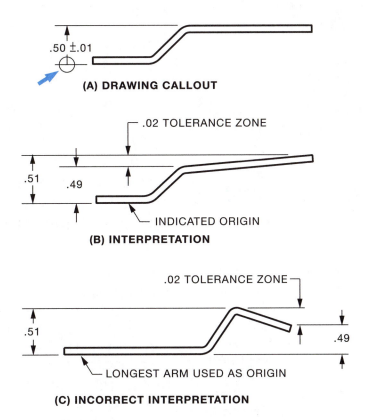

FIGURE 18–7 ■ *Relating dimensional limits to an origin.*

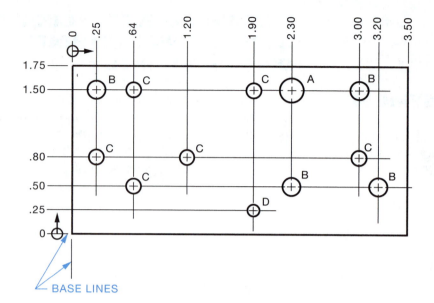

HOLE SYMBOL	HOLE Ø
A	.250
B	.188
C	.156
D	.125

FIGURE 18–8 ■ *Rectangular coordinate dimensioning without dimension lines.*

shown without arrowheads indicates the distance from the zero line. There is never more than one zero line in each direction.

This type of dimensioning is particularly useful when such features are produced on a general-purpose machine, such as a jig borer, a tape-controlled drill, or a turret-type drill press. Drawings for numerical control are covered in Unit 32.

RECTANGULAR COORDINATE DIMENSIONING IN TABULAR FORM

When there are many holes or repetitive features, such as in a chassis or a printed circuit board, and where the multitude of center lines would make a drawing difficult to read, rectangular coordinate dimensioning in tabular form is recommended. In this system each hole or feature is assigned a letter or a letter with a numeral subscript. The feature dimensions and the feature location along the X, Y, and Z axes are given in a table as shown in Figure 18–9.

REFERENCES

ASME Y14.5M-1994 (R1999) Dimensioning and Tolerancing

ASME B17.1-1967 (1998) Keys and Keyseats

ASME B18.25.3M-1998 Square and Rectangular Keys and Keyways

ASME B17.2-1967 (R1998) Woodruff Keys and Keyseats

ASME B18.25.2M-1996 Woodruff Keys and Keyways

ASME B18.6.2-1998 Slotted Headless Set Screws

ASME B18.3.6M-1986 (R2002) Metric Series Socket Set Screws

INTERNET RESOURCES

Drafting Zone. For information on rectangular coordinate dimensioning, see: http://www.draftingzone.com

Machine Design. For information on keys and setscrews, see *Machine Design, Fastening/Joining Reference* at: http://www.machinedesign.com

Maryland Metrics. For information on dimensioning metric and inch keyways and keyseats see: www.mdmetrics.com/tech/cvtchfdm.htm

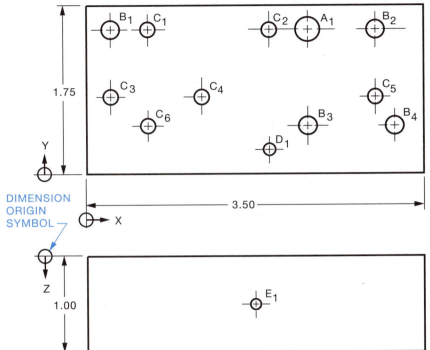

HOLE SYMBOL	Ø HOLE	LOCATION		
		X	Y	Z
A_1	.250	2.30	1.50	.62
B_1	.188	.25	1.50	THRU
B_2	.188	3.00	1.50	THRU
B_3	.188	2.30	.50	THRU
B_4	.188	3.20	.50	THRU
C_1	.156	.64	1.50	THRU
C_2	.156	1.90	1.50	THRU
C_3	.156	.25	.80	THRU
C_4	.156	.120	.80	THRU
C_5	.156	3.00	.80	THRU
C_6	.156	.64	.50	THRU
D_1	.125	1.90	.25	.50
E_1	.109	1.75	1.00	.50

FIGURE 18–9 ■ *Rectangular coordinate dimensioning in tabular form.*

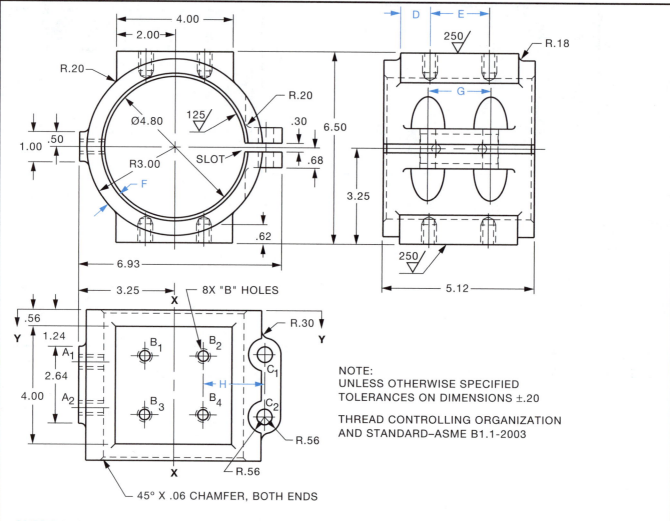

NOTE:
UNLESS OTHERWISE SPECIFIED
TOLERANCES ON DIMENSIONS ±.20

THREAD CONTROLLING ORGANIZATION
AND STANDARD–ASME B1.1-2003

QUESTIONS:

1. What is the center-to-center distance between the following:
   (A) the A_1 and A_2 holes
   (B) the B_1 and B_2 holes
   (C) the C_1 and C_2 holes

2. What finish is required on the 4.00 x 4.00 face?

3. How far apart are the two surfaces of the slot?

4. How many surfaces require finishing?

5. What is the depth of the A holes?

6. What is the length of the threading in the B holes?

7. What size bolts would be used in the C holes if .031 clearance is used?

8. How many chamfers are called for?

9. How many tapped holes are there?

10. Calculate distances D,E,F,G, and H.

11. What is the tap drill size required for the (A) #10 and, (B) .500 threaded holes?

HOLE	HOLE SIZE	LOCATION	
		X-X	Y-Y
A_1	10(.190)-24UNC-2B		1.78
A_2	10(.190)-24UNC-2B		3.32
B_1	.500-13UNC-2B	1.00	1.56
B_2	.500-13UNC-2B	1.00	1.56
B_3	.500-13UNC-2B	1.00	3.56
B_4	.500-13UNC-2B	1.00	3.56
C_1	Ø.531	3.12	1.50
C_2	Ø.531	3.12	3.62

MATERIAL	COPPER	
SCALE	NOT TO SCALE	
DRAWN	J. MILLER	DATE 21/08/04

TERMINAL BLOCK   **A-52**

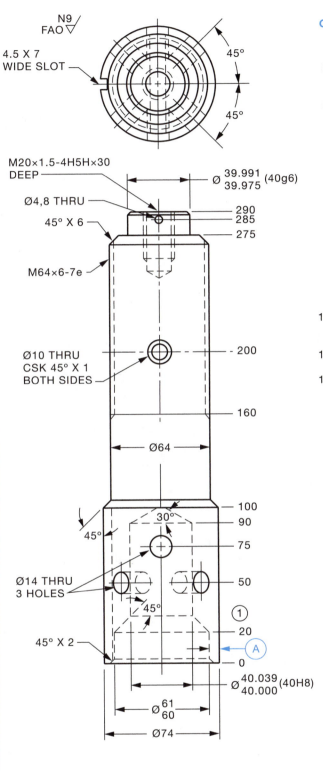

N9
FAO ▽

4.5 X 7
WIDE SLOT

45°

45°

M20×1.5-4H5H×30
DEEP

Ø $\frac{39.991}{39.975}$ (40g6)

Ø4,8 THRU

290
285
275

45° X 6

M64×6-7e

Ø10 THRU
CSK 45° X 1
BOTH SIDES

200

160

Ø64

100
90
30°
75

45°

Ø14 THRU
3 HOLES

50

45°

①

20

45° X 2

A

0

Ø $\frac{40.039}{40.000}$ (40H8)

Ø $\frac{61}{60}$

Ø74

NOTE: HOLES AND THREADS TO BE CLEAN AND BRIGHT

QUESTIONS:

1. How long is the Ø64 threaded section?

2. A. How deep is the M20 thread?
   B. How many full threads are in the tapped hole?

3. How many Ø14 holes are there?

4. What is the maximum thickness of the wall at the Ø14 holes?

5. What series of thread is required for the tapped hole?

6. What is the vertical center distance between the top Ø14 hole and the Ø4.8 hole?

7. What is the length of the Ø64 unthreaded section?

8. What are the minimum and maximum thicknesses at (A)?

9. What is the maximum overall diameter of the finished stud?

10. What is the tolerance on the largest hole at the bottom of the stud?

11. Give the quality of surface texture in micrometers.

12. What is the angle between the slot and the Ø14 hole located 75 from the base of the stud?

NOTE:
UNLESS OTHERWISE SPECIFIED:
—TOLERANCES ON DIMENSIONS ±0.5
—TOLERANCES ON ANGLES ±0.5°

THREAD CONTROLLING ORGANIZATION
AND STANDARD – ASME B1.13M-2001

METRIC
DIMENSIONS ARE IN MILLIMETERS

MATERIAL	COPPER	
SCALE	NOT TO SCALE	
DRAWN	D. THOMPSON	DATE 15/04/04

REVISIONS	1	DN. T. FURMAN	CHK. C. JENSEN
		DIMENSION WAS 24	22/04/04

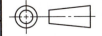

TERMINAL STUD

A-53M

QUESTIONS:

Refer to Part 1

1. What was the original length of the shaft?

2. What symbol is used to indicate that the 1.90 dimension is not to scale?

3. At how many places are threads being cut?

4. Specify for any left-hand threads the thread diameter and the number of threads per inch.

5. What is the pitch for the 1.000 thread?

6. What is the length of that portion of the shaft which includes the (A) .875 thread (include chamfer), (B) 1.250 thread (do not include undercut), and (C) 1.000 thread (include chamfer)?

7. What distance is there between the last thread and the shoulder of the Ø.875 portion of the shaft?

8. How many dimensions have been changed?

9. What type of sectional view is used?

10. What is the largest size to which the Ø1.250 shaft can be turned?

11. What is the minimum permissible size for dimension A?

Refer to Part 2

12. How many holes are there?

13. What are the overall width and depth dimensions of the base?

14. How many surfaces are to be finished?

15. What is the diameter of the bosses on the base?

16. How wide is the pad on the upright column?

17. What is the depth of the keyseat?

18. How far does the horizontal hole overlap the vertical hole? (Use maximum sizes of holes and minimum center-to-center distances.)

19. How much material was added when the change to the bosses was made?

20. Which scale was used on this part?

21. What is the maximum permissible center-to-center distance of the two large holes?

22. If limit dimensions (refer to Tables 13 M and W of the Appendix) were to replace the keyseat dimensions shown, what would they be?

23. What dimension is indicated as not drawn to scale?

24. What was the overall height of the casting before finishing?

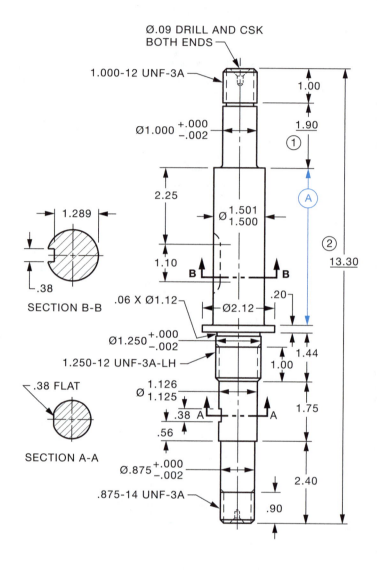

CHAMFER STARTING END OF ALL THREADS 45° TO THREAD DEPTH

NOTE: ALL FILLETS R.10

THREAD CONTROLLING ORGANIZATION AND STANDARD – ASME B1.1-2003

PT 1   SPINDLE SHAFT,   SCALE 1:2,   MATL - CRS
2 REQD

REVISIONS	1	12/01/04	R.H.	2	12/08/04	C.J.	3	15/12/04	G.H.
	1.90 WAS 2.00			13.30 WAS 13.40			.80 WAS .75		

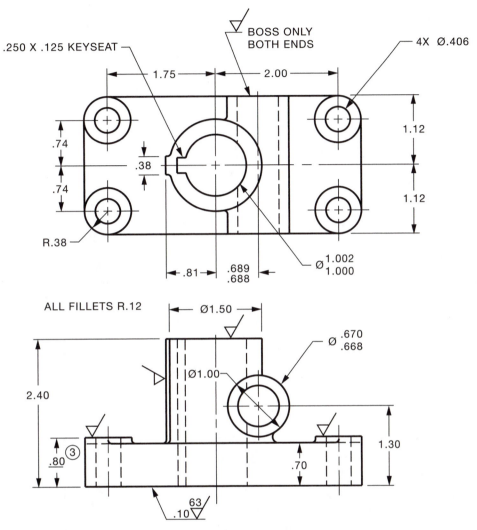

.250 X .125 KEYSEAT

BOSS ONLY
BOTH ENDS

4X Ø.406

1.75

2.00

.74

.38

1.12

.74

R.38

1.12

.81

.689
.688

Ø 1.002
1.000

ALL FILLETS R.12

Ø1.50

Ø .670
.668

2.40

Ø1.00

.80 ③

1.30

.70

.10

63

PT 2   COLUMN BRACKET,   SCALE 1:1
MATL - WROUGHT IRON   4 REQD

UNLESS OTHERWISE SPECIFIED
TOLERANCE ON DIMENSIONS ±.02

∇ TO BE 63/∇

SCALE	AS SHOWN	
DRAWN	C. JENSEN	DATE 15-08-03

RACK DETAILS

A-54

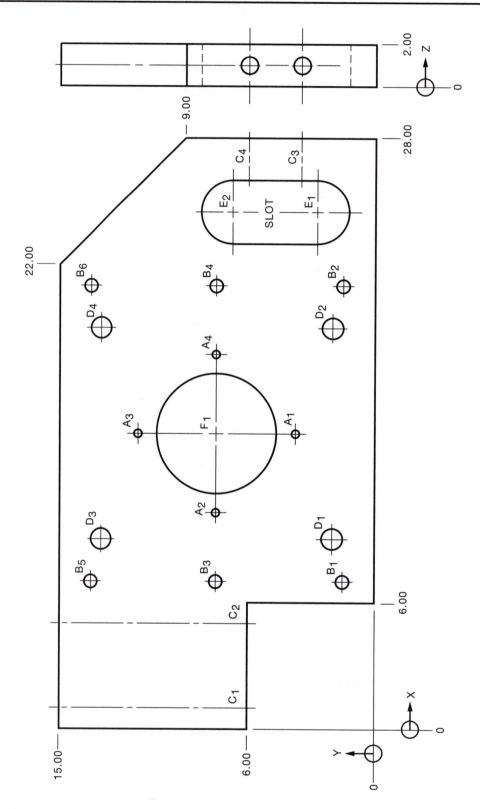

NOTE: TOLERANCE ON LINEAR DIMENSIONS ±.02

QUESTIONS:

1. What is the width of the part?

2. What is the height of the part?

3. What are the width and height of the chamfer on the corner?

4. What is the width of slot E?

5. What is the height of slot E?

6. How much wood is left between slot E and the right-hand edge of the part?

7. What are the center distances between holes?

   (A) B_5 and B_6 (D) D_3 and D_4 (F) C_1 and C_2

   (B) D_2 and D_4 (E) A_1 and A_3 (G) B_3 and A_4

   (C) B_1 and B_5

8. How much wood is left between holes?

   (A) A_1 and E_1 (C) C_1 and C_2

   (B) A_2 and B_3 (D) B_3 and B_4

9. How many places is the origin symbol used?

HOLE SYMBOL	HOLE DIA	LOCATION		
		X	Y	Z
A_1	.375	14.00	3.75	
A_2	.375	10.25	7.50	
A_3	.375	14.00	11.25	
A_4	.375	17.75	7.50	
B_1	.625	7.00	1.50	
B_2	.625	21.00	1.50	
B_3	.625	7.00	7.50	
B_4	.625	21.00	7.50	
B_5	.625	7.00	13.50	
B_6	.625	21.00	13.50	
C_1	.812	1.00		1.00
C_2	.812	5.00		1.00
C_3	.812		3.50	1.00
C_4	.812		6.00	1.00
D_1	1.000	9.00	2.00	
D_2	1.000	19.00	2.00	
D_3	1.000	9.00	13.00	
D_4	1.000	19.00	13.00	
E_1	3.000	24.50	2.75	
E_2	3.000	24.50	6.75	
F_1	5.688	14.00	7.50	

NOTE: ALL HOLES THRU UNLESS
OTHERWISE SPECIFIED

MATERIAL	DRY MAPLE	
SCALE	NOT TO SCALE	
DRAWN	J. HELSEL	DATE 16/10/04

**SUPPORT
BRACKET**

A-55

19 UNIT

OBLIQUE SURFACES

When a surface is sloped so that it is not parallel to any of the six principal orthographic views or planes, it will appear as a surface in all normal views, but never in its true shape. This is referred to as an *oblique* surface, Figure 19–1.

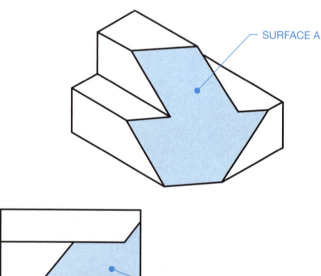

SURFACE A

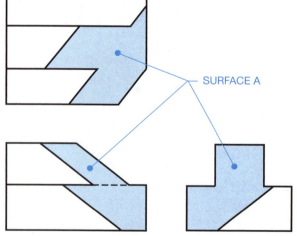

SURFACE A

FIGURE 19–1 ■ *Surface A is not parallel to any of the three normal viewing planes and is therefore considered oblique.*

In many cases, the exact shape of an oblique surface is not necessary. This is especially true if there are no details (holes, slots, etc.) on the oblique surface that require size or location dimensions. However, if a true representation of an oblique surface is required, two successive auxiliary views, primary and secondary, need to be projected and developed.

Auxiliary views are a type of orthographic projection used to develop the true size and shape of inclined and oblique surfaces of objects by projecting them onto imaginary auxiliary planes. These auxiliary planes are not parallel to any of the normal planes, but are parallel to either the primary or secondary auxiliary surface. Auxiliary views are explained in greater detail in Units 20 and 21. Figure 19–2 shows additional examples of parts having oblique surfaces.

REFERENCE

ASME Y14.3M-1994 (R1999] Multi- and Sectional-View Drawings Internet Resources

INTERNET RESOURCE

Michigan Tech. For additional information and practice on oblique surfaces, see: http://www.geneng.mtu.edu/courses/1102/current/g04_teaming_oblique.pdf

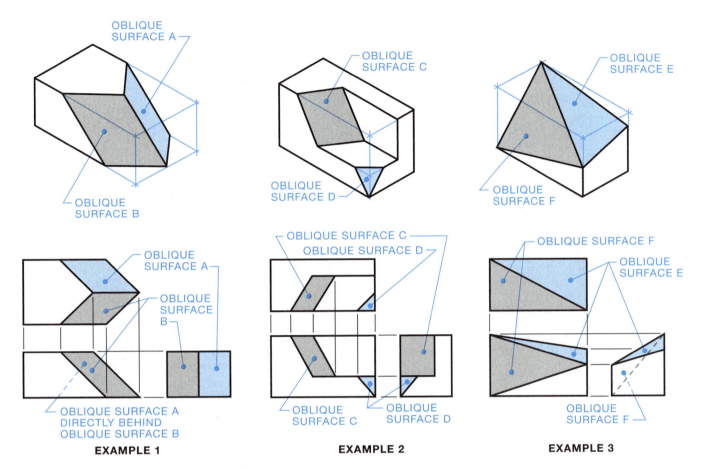

FIGURE 19–2 ■ *Illustration of parts having oblique surfaces.*

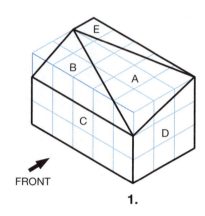

1.

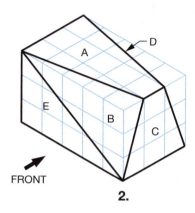

2.

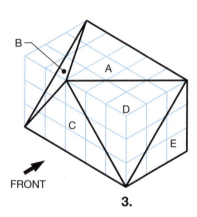

3.

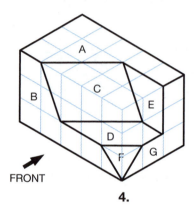

4.

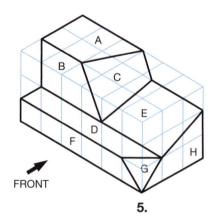

5.

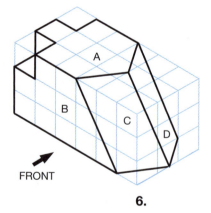

6.

ASSIGNMENT:

ON A ONE-INCH GRID SHEET (.25 IN. SQUARES) AND USING
THE SAME NUMBER OF SQUARES SHOWN ON THE PARTS,
SKETCH THE FRONT, TOP, AND SIDE VIEWS. IDENTIFY THE
OBLIQUE SURFACES ON THE THREE VIEWS WITH THE
APPROPRIATE LETTERS.

**IDENTIFYING
OBLIQUE SURFACES**

A-56

ARROW INDICATES DIRECTION OF FRONT VIEW.

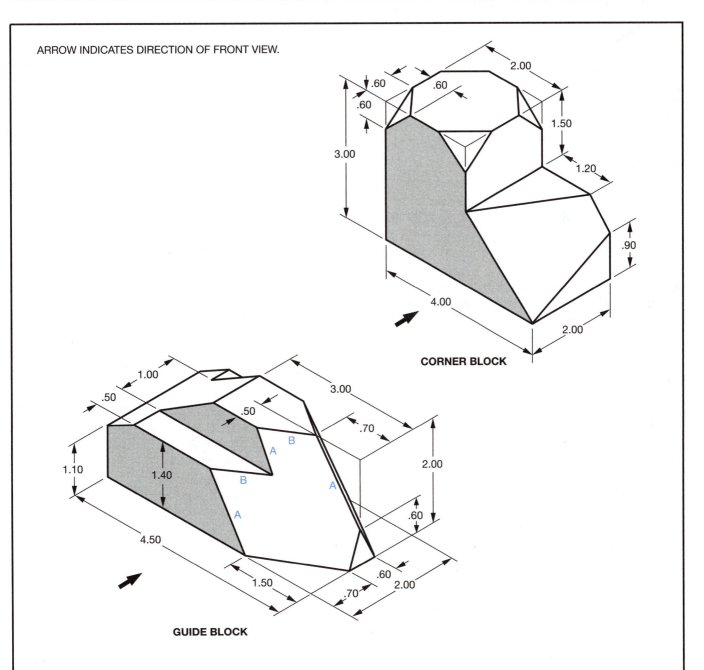

CORNER BLOCK

GUIDE BLOCK

ASSIGNMENT: ON A 1.00 IN. GRID SHEET (.10 IN. SQUARES) MAKE A THREE-VIEW DRAWING OF ONE OF THE PARTS SHOWN ABOVE. ALLOW 1.00 IN. BETWEEN VIEWS. USING LETTERS, IDENTIFY THE OBLIQUE SURFACES ON ALL THREE VIEWS.

NOTE: LINES MARKED "A" ARE PARALLEL
LINES MARKED "B" ARE PARALLEL

COMPLETING OBLIQUE SURFACES

A-57

PRIMARY AUXILIARY VIEWS

Many objects have surfaces that are perpendicular to only one plane of projection. These surfaces are referred to as *inclined sloped surfaces.* In the remaining two orthographic views such surfaces appear to be foreshortened and their true shape is not shown, Figure 20–1. When an inclined surface has important characteristics that should be shown clearly and without distortion, an auxiliary view is used to completely explain the shape of the object.

One of the regular views will have a true-length line representing the edge of the inclined surface. The auxiliary view is projected from this edge line, at right angles, and is drawn parallel to the edge line.

For example, Figure 20–2 clearly shows why an auxiliary view is required. The circular features on the sloped surface on the front view cannot be seen in their true shape on either the top or side view. The auxiliary view is the only view that shows the actual shape of these features, Figure 20–3. Note that only the sloped surface details are shown. Background detail is often omitted on auxiliary views and regular views to simplify the drawing and avoid confusion. A break line is used to signify the break in an incomplete view. The break line is not required if only the exact surface is drawn for either an auxiliary view or a partial regular view. This procedure is recommended for functional and production drafting, when drafting costs are an important consideration. However, complete views of the part are often used on catalog and standard parts drawings.

One of the basic rules for dimensioning is to dimension the feature where it can be seen in its true shape and size. Thus the auxiliary view should only show dimensions pertaining to features for which the auxiliary view was drawn.

REFERENCE

ASME Y14.3M-1994 (R1999) Multi and Sectional View Drawings

INTERNET RESOURCES

For a PowerPoint presentation on primary auxiliary views, see: http://crown.panam.edu/EG/notes/lecture7.ppt

Wikipedia, the Free Encyclopedia. For information on auxiliary views, see: http://en.wikipedia.org/wiki/Orthographic_projection

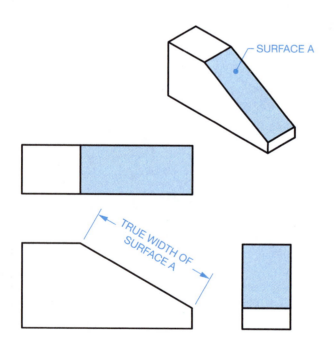

SURFACE A

TRUE WIDTH OF SURFACE A

FIGURE 20–1 ■ *Object having an inclined surface.*

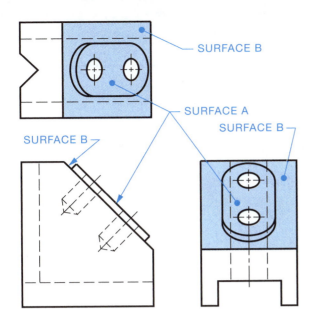

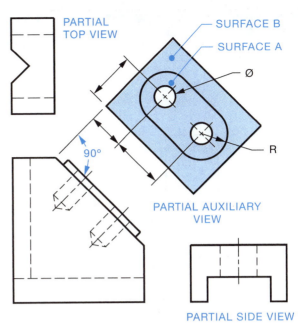

(A) REGULAR VIEWS DO NOT SHOW TRUE FEATURES OF SURFACES A AND B.

(B) AUXILIARY VIEW ADDED TO SHOW TRUE FEATURES OF SURFACE A AND B.

FIGURE 20–2 ■ *The need for auxiliary views.*

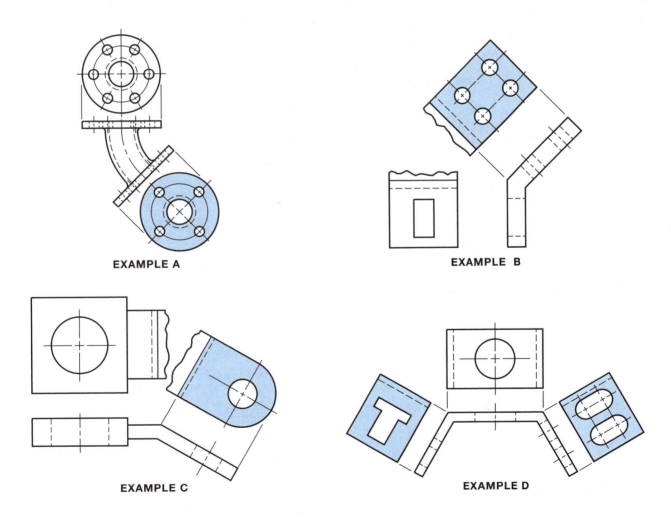

EXAMPLE A

EXAMPLE B

EXAMPLE C

EXAMPLE D

NOTE: CONVENTIONAL BREAK OR PROJECTED SURFACE ONLY NEED BE SHOWN ON PARTIAL VIEWS.

FIGURE 20–3 ■ *Examples of auxiliary-view drawings.*

175

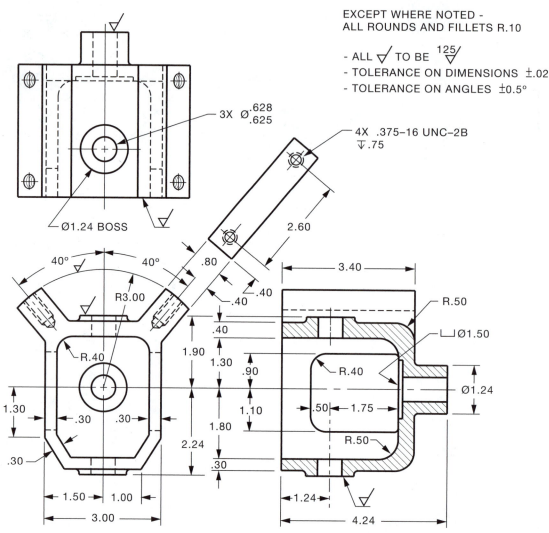

EXCEPT WHERE NOTED -
ALL ROUNDS AND FILLETS R.10

- ALL ▽ TO BE $\overset{125}{\triangledown}$
- TOLERANCE ON DIMENSIONS ±.02
- TOLERANCE ON ANGLES ±0.5°

3X Ø .628 / .625

4X .375–16 UNC–2B ▼.75

Ø1.24 BOSS

QUESTIONS:

1. What is the diameter of the bosses?

2. What is the tolerance on the holes in the bosses?

3. What are the width and height of the cutouts in the sides of the box?

4. What are the width and depth of the legs of the box?

5. How many degrees are there between the legs of the box?

6. What is the maximum surface roughness in microinches permitted on the machined surfaces?

7. How many surfaces are machined?

8. What would be the inside diameter of the mating part that this box fits into?

9. If the mating part is .44 thick and a lockwasher is used, what diameter and longest standard length of socket head cap screws could be used to fasten the parts together? (See Tables 7 and 12 in Appendix.) Lengths available in .25 in. increments.

10. What is the thickness of (A) the side walls of the box, (B) the top of the box, and (C) the bottom of the box? Disregard the bosses.

11. Give overall inside dimensions of the box.

12. What are the overall outside dimensions of the box? (Do not include legs or bosses.)

13. Of what material is the box made?

14. How many screws are required to fasten the box to the mating part?

15. What is the tap drill size required for the four threaded holes?

THREAD CONTROLLING ORGANIZATION
AND STANDARD – ASME B1.13M-2001

MATERIAL	GRAY IRON	
SCALE	NOT TO SCALE	
DRAWN	J. SMITH	DATE 19/10/04

GEAR BOX   **A-58**

ASSIGNMENT:
ON A ONE INCH GRID SHEET (.10 IN. SQUARES) SKETCH A PARTIAL
RIGHT SIDE VIEW AND AN AUXILIARY VIEW OF THE INCLINED STOP.
THE DRAWING BELOW SHOWS THE VIEWING DIRECTION FOR
THESE VIEWS. ADD APPROPRIATE DIMENSIONS. SCALE 1:1.

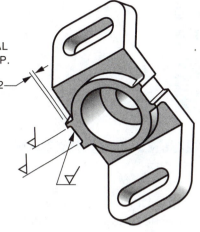

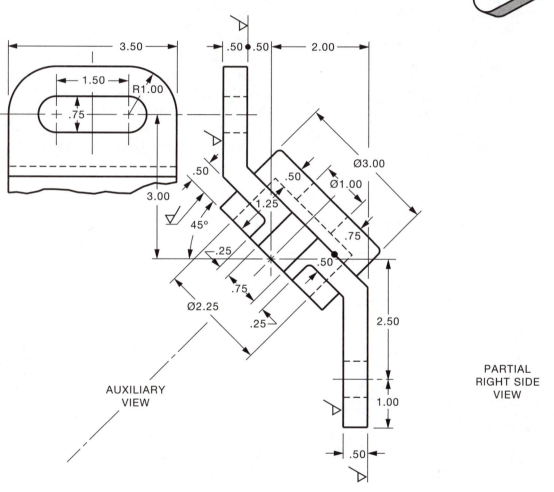

AUXILIARY
VIEW

PARTIAL
RIGHT SIDE
VIEW

MATERIAL	GRAY IRON		
SCALE	1 : 2		
DRAWN	J. HELSEL		
		DATE	12/06/03

INCLINED STOP   **A-59**

21 UNIT

SECONDARY AUXILIARY VIEWS

As mentioned in Unit 20, auxiliary views show the true lengths of lines and the true shapes of surfaces that cannot be described in the ordinary views.

A primary auxiliary view is drawn by projecting lines from a regular view where the inclined surface appears as an edge.

The auxiliary view in Figure 21–1(A), which shows the true projection and true shape of surface X, is called a *primary auxiliary view* because it is projected directly from the regular front view.

Some surfaces are inclined so that they are not perpendicular to any of the three viewing planes. In this case, they appear as a surface in all three views but never in their true shape. These are referred to as *oblique surfaces*. Surface Z shown in Figure 21–1 are oblique surfaces. To show the true shape of surface Z, and the true shape and location of holes M located on surface Z, a second auxiliary view must be shown, as at (C). This auxiliary view is projected from the first or primary auxiliary view, and is known as a *secondary auxiliary view.* The view at (B) is a primary auxiliary view because it is projected from one of the regular views. Notice that the side view is not

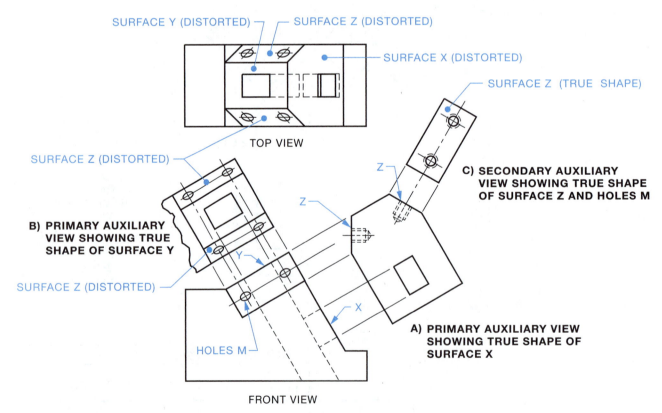

FRONT VIEW

NOTE: MANY HIDDEN LINES ARE OMITTED FOR CLARITY

FIGURE 21–1 ■ *Primary and secondary auxiliary views.*

drawn because the auxiliary views provide the information usually shown on this view.

REFERENCE

ASME Y14.3M-1994 (R1999) Multi and Sectional-View Drawings

INTERNET RESOURCE

For a PowerPoint presentation on secondary auxiliary views, see: http://crown.panam.edu/EG/notes/lecture7.ppt

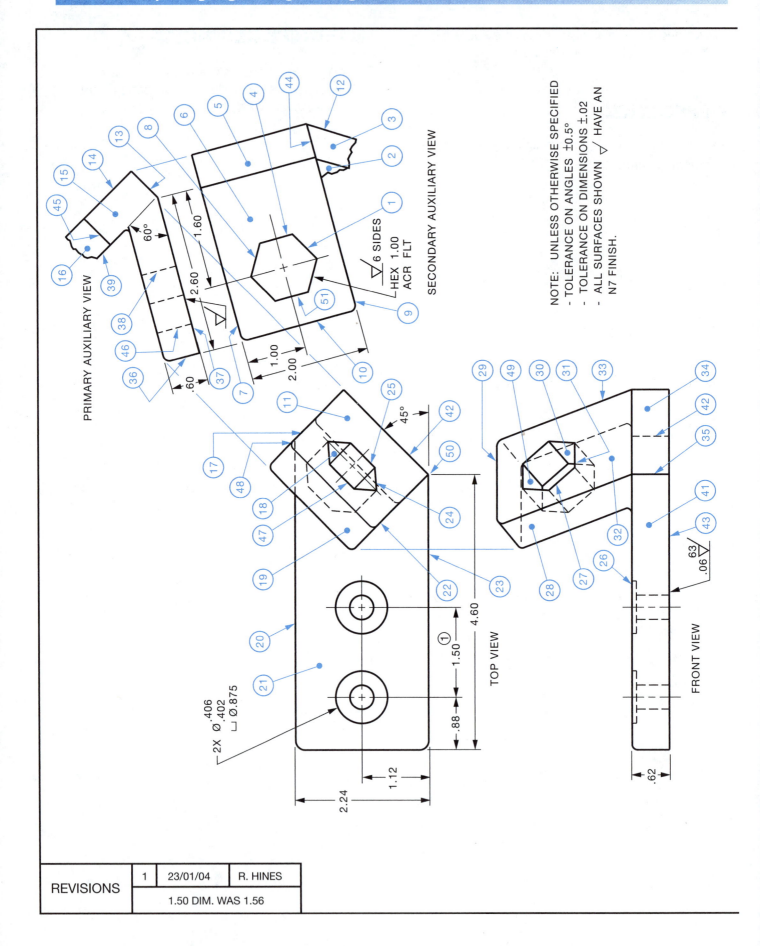

PRIMARY AUXILIARY VIEW

SECONDARY AUXILIARY VIEW

$\sqrt{}$ 6 SIDES
HEX 1.00 ACR FLT

60°

2.60

1.60

1.00

2.00

.60

TOP VIEW

45°

4.60

1.50 ①

.88

1.12

2.24

2X Ø.406
Ø.402
⊔ Ø.875

FRONT VIEW

63 $\triangle$
.06

.62

NOTE: UNLESS OTHERWISE SPECIFIED
- TOLERANCE ON ANGLES ±0.5°
- TOLERANCE ON DIMENSIONS ±.02
- ALL SURFACES SHOWN $\sqrt{}$ HAVE AN N7 FINISH.

REVISIONS	1	23/01/04	R. HINES
		1.50 DIM. WAS 1.56	

ASSIGNMENT: A FEATURE IS IDENTIFIED IN ONE OF THE VIEWS BY A NUMBER. PREPARE A CHART SIMILAR TO THE ONE SHOWN AND IDENTIFY THE FEATURE IN THE OTHER VIEWS BY ADDING THE APPROPRIATE NUMBERS TO THE CHART.

QUESTIONS:

1. Which other view(s) show the true height of the .62 dimension shown in the front view?

2. Which other view(s) show the true width of the .60 dimension shown in the primary auxiliary view?

3. What view(s) show the (A) height, (B) depth, and (C) width of the top portion of the support?

4. How many surfaces require finishing?

5. List the surface(s) or feature(s) which are shown in their true shape or size in the primary auxiliary view but are distorted in all other views.

6. List the surface(s) feature(s) which are shown in their true shape or size in the secondary auxiliary view but are distorted in all other views.

7. What would be the thickness of the base ⑯ before machining?

8. If the hexagon bar support was fastened to another member,
   (Refer to the Appendix when necessary.)
   (A) How many cap screws would be used?
   (B) What would be the cap screw size?
   (C) If the supporting member had tapped holes which were .80 deep and flat washers were used under the cap screw heads, what would be the cap screw length?
   (D) If the cap screws were of the fine-thread series, how would you call out these cap screws?
   (E) What would be the I.D. and O.D. of the flat washers used under these cap screws?

Front View	Top View	Primary Auxiliary View	Secondary Auxiliary View
		—	1
	23	—	
26		38	9
—			—
33	20	—	
	—	14	5
	19	37	
35		—	8
42		—	
	47		—

MATERIAL	GRAY IRON	
SCALE	NOT TO SCALE	
DRAWN	S. KLINGER	DATE 15/10/03

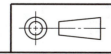

HEXAGON BAR SUPPORT

A-60

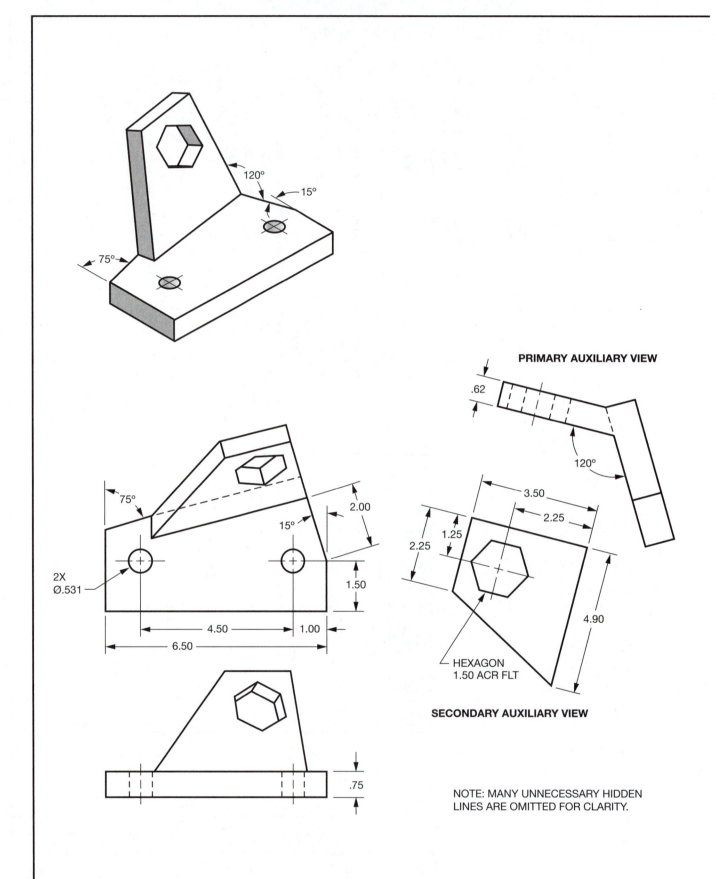

PRIMARY AUXILIARY VIEW

.62

120°

3.50

2.25

1.25

2.25

4.90

HEXAGON
1.50 ACR FLT

SECONDARY AUXILIARY VIEW

75°

120°

15°

2.00

1.50

2X
Ø.531

4.50

1.00

6.50

.75

NOTE: MANY UNNECESSARY HIDDEN
LINES ARE OMITTED FOR CLARITY.

ASSIGNMENT: THE HEXAGONAL HOLE HAS BEEN REPLACED BY THE TRIANGULAR HOLE AS SHOWN IN THE CONTROL BLOCK BELOW. MAKE A PHOTOCOPY OF THE DRAWING AND ACCURATELY SKETCH ON THIS DRAWING THE SIZE AND POSITION OF THE HOLE IN THE OTHER VIEWS. LEAVE THE CONSTRUCTION LINES ON YOUR DRAWING.

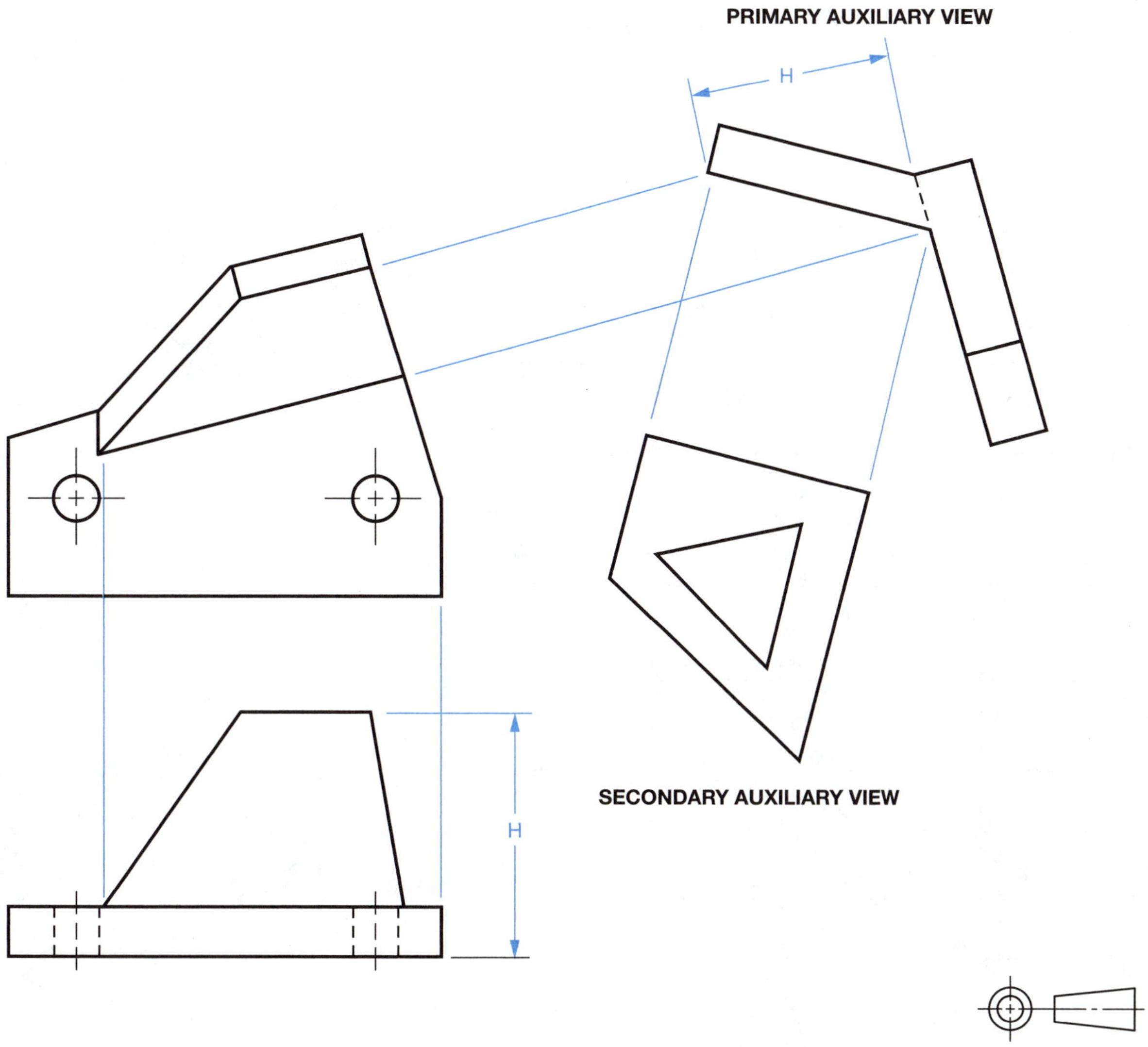

PRIMARY AUXILIARY VIEW

H

SECONDARY AUXILIARY VIEW

H

CONTROL BLOCK   A-61

22 UNIT

DEVELOPMENT DRAWINGS

Many objects, such as metal and cardboard boxes, duct work for heating, funnels, gutters and downspouts, are made from flat sheet material that is cut so that, when folded, formed, or rolled, it will take the shape of the object. Because a definite shape and size is desired, a regular orthographic drawing of the object is first made; then a development drawing is made to show the complete surface or surfaces laid out in a flat plane, Figure 22–1.

A development drawing is sometimes referred to as a pattern drawing, because the layout, when made of heavy cardboard, metal, or wood, is used

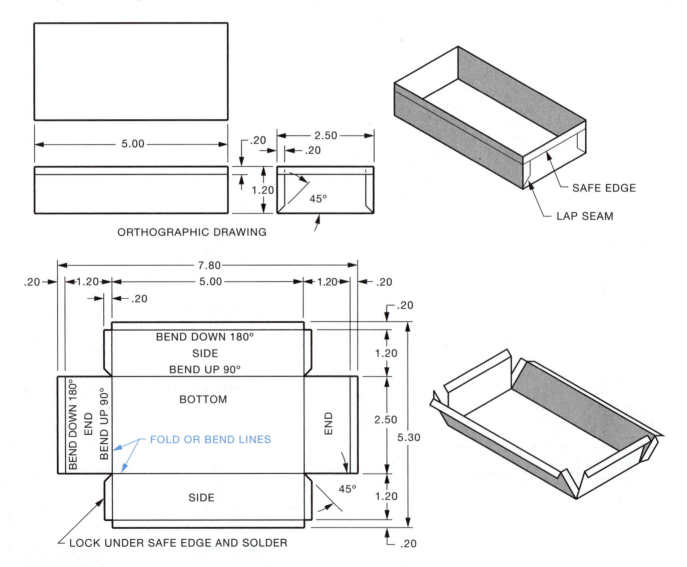

ORTHOGRAPHIC DRAWING

SAFE EDGE

LAP SEAM

BEND DOWN 180°
SIDE
BEND UP 90°

BOTTOM

FOLD OR BEND LINES

BEND DOWN 180°
END
BEND UP 90°

END

SIDE

45°

LOCK UNDER SAFE EDGE AND SOLDER

FIGURE 22–1 ■ *Development drawing with a complete set of folding instructions.*

184

as a pattern for tracing out the developed shape on flat material. Such patterns are used extensively in sheet metal shops.

JOINTS, SEAMS, AND EDGES

Additional material is required for assembly and design purposes. When two or more pieces of material or surfaces are joined, extra material must be provided for the joint or seam. The type of joint or seam for joining metal, Figure 22–2, is dependent on design criteria such as strength, waterproofing, and appearance. Rivets or solder can be added to these joints if required. Exposed edges of metal parts may also be reinforced by the addition of extra material for hemming or for containing a wire. A round metal wastebasket, Figure 22–3, is one example where extra material is provided for the joint, seam, and edge.

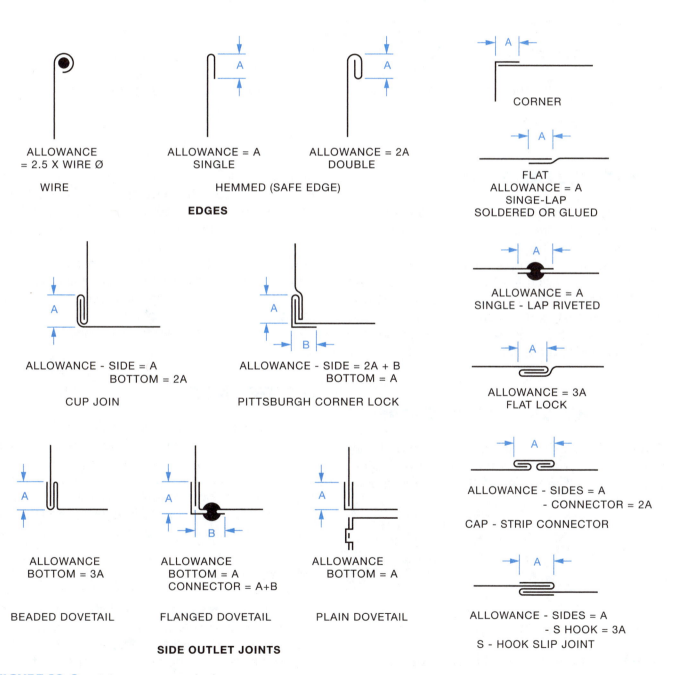

FIGURE 22–2 ■ *Joints, seams, and edges.*

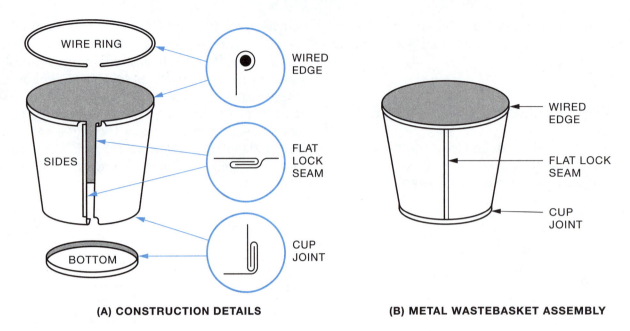

(A) CONSTRUCTION DETAILS   **(B) METAL WASTEBASKET ASSEMBLY**

FIGURE 22–3 ■ *Wastebasket construction.*

SHEET METAL SIZES

Metal thicknesses up to .25 in. (6 mm) are usually designated by a series of gage numbers. The more common gages are shown in Table 16 of the Appendix. Metal .25 in. and over is given in inch or millimeter sizes. In calling for the material size of sheet metal developments, customary practice is to give the gage number, and its inch or millimeter equivalent in brackets followed by the type of gauge, and the developed width and length, Figure 22–4.

STRAIGHT LINE DEVELOPMENT

This is the term given to the development of an object that has surfaces on a flat plane of projection. The true size of each side of the object is known,

FIGURE 22–4 ■ *Callout of sheet metal material.*

and these sides can be laid out in successive order. Figure 22–1 shows the development of a simple rectangular box having a bottom and four sides. Note that in the development of the box, an allowance is made for lap seams at the corners, and for folded edges around the top. All lines for each surface are parallel or perpendicular to the other surfaces. The bottom corners of the lap joints are chamfered to facilitate assembly. The fold lines on the development are shown as thin unbroken lines.

STAMPINGS

Stamping is the art of pressworking sheet metal to change its shape by the use of punches and dies. It may involve punching out a hole or the product itself from a sheet of metal. It may also involve bending or forming, Figures 22–5 and 22–6.

Stamping may be divided into two general classifications: *forming* and *shearing*.

Forming

Forming includes stampings made by forming sheet metal to the shape desired without cutting or shearing the metal. For thicker sheet metal plates, bending allowances must be taken into consideration.

Shearing

Shearing includes stampings made by shearing the sheet metal either to change the outline or to cut holes in the interior of the part. Punching forms a hole or opening in the part.

Height and width dimensions should normally be given to the same side of the metal, either the punch side or the die side. Because larger radii facilitate production, inside radii on stampings should not be less than 1.5 times stock thickness.

The following formula may be used for blank development, Figure 22–7:

$$\text{Total length} = A + B + \text{Bend Allowance}$$

where length for 90° bend =

$$\frac{\pi}{2}(R + .33T \text{ min}) = 1.57 (R + .33T \text{ min})$$

It is not general practice to show the blank development on production drawings.

REFERENCE

ASME Y14.3M-1994 (R1999) Multi- and Sectional-View Drawings

FIGURE 22–5 ■ *Punch press used to punch and form metal.* (Courtesy of Whitney Metal Tool Co.)

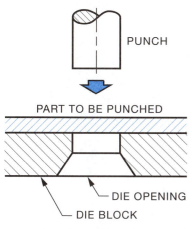 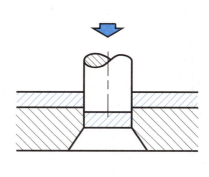 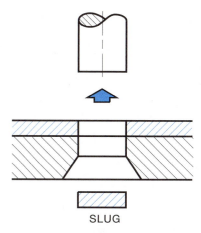

(A) **PUNCH AND DIE COMPONENTS**   (B) **HOLE SHEARED BY LOWERING PUNCH INTO DIE OPENING**   (C) **HOLE COMPLETED IN PART**

FIGURE 22–6 ■ *Punching a hole in sheet metal.*

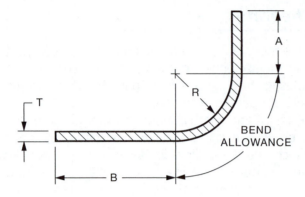

FIGURE 22–7 ■ *Dimensions for calculating blank development.*

INTERNET RESOURCES

Design & Technology Online. For information on packaging, see: http://www.dtonline.org/ (packaging)

Drafting Zone. For information on sheet metal practices, see: http://www.draftingzone.com

eFunda. For information on sheet metal and sheet metal processes, see: http://www.efunda.com/home.cfm

Sheetmetal Shop. For information on sheetmetal layout, see: http://www.thesheetmetalshop.com

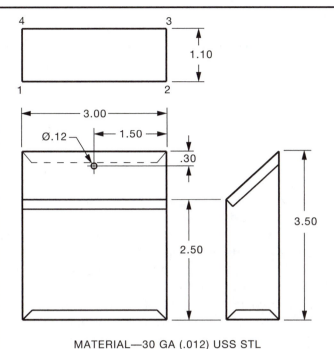

1.10

3.00

Ø.12 — 1.50

.30

2.50

3.50

MATERIAL—30 GA (.012) USS STL

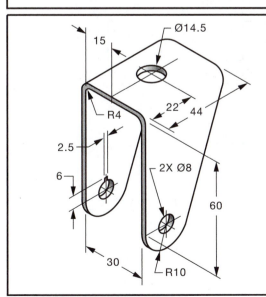

3

4 BACK 2

1

QUESTIONS:

1. What are the overall dimensions of the back?

2. What are the overall dimensions of the side?

3. What are the overall dimensions of the bottom?

4. What are the overall dimensions of the front?

5. How much has the width of the development been increased due to the seam allowance?

6. How much has the height of the development been increased due to the seam allowance?

7. What are the overall sizes of the development?

ASSIGNMENT:
ON A 1.00 IN. GRID SHEET (.10 IN. SQUARES) SKETCH THE DEVELOPMENT DRAWING OF THE LETTER BOX AND SHOW THE OVERALL DIMENSIONS. SCALE 1:2

.20 SINGLE HEMMED EDGES AND 90° LAP SEAMS

SIDE SEAM AT CORNER 1

FOLD LINE BETWEEN BOTTOM AND BACK

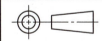

LETTER BOX | **A-62**

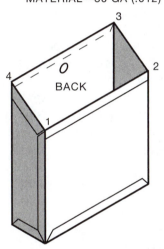

Ø14.5

15

R4

22 44

2.5

6

2X Ø8

60

30

R10

ASSIGNMENT:
ON A CENTIMETER GRID SHEET (1 mm SQUARES) SKETCH THE FRONT AND SIDE VIEWS PLUS THE DEVELOPMENT DRAWING OF THE BRACKET BEFORE THE THREE HOLES ARE PRODUCED. THE BRACKET IS PART OF THE CASTER ASSEMBLY SHOWN IN ASSIGNMENT A-73M. SCALE 1:1. ADD DIMENSIONS TO ALL VIEWS.

NOTE:
MATL - 13 GA (2.38) USS STL

METRIC
DIMENSIONS ARE IN MILLIMETERS

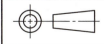

BRACKET | **A-63M**

ARRANGEMENT OF VIEWS

The shape of an object and its complexity influence the possible choices and arrangement of views for that particular object. Because one of the main purposes of making drawings is to furnish the worker with enough information to be able to make the object, only the views that will aid in the interpretation of the drawing should be drawn.

The drafter chooses the view of the object that gives the viewer the clearest idea of the purpose and general contour of the object, and then calls this the front view. This choice of the front view may have no relationship to the actual front of the piece when it is used. The front view does not have to be the actual front of the object.

The selection of views to best describe the mounting plate shown in Assignment A-64 is shown in Figure 23-1. These views could be called front, right side, and bottom views as illustrated in Arrangement A. These views could also be designated top, front, and right side views, as shown in Arrangement B. Note that in Arrangement B the right side view is projected from the top view and not the front view as shown in Arrangement A. The designation of names is not of major importance. What is important is that these views, in the opinion of the drafter or designer, give the necessary information in the most understandable way.

Once the basic views that best describe the mounting plate have been established, auxiliary (helper) views, when required, are added in order to completely describe the part. See Assignment A-64. The No. 1 auxiliary view is added by projecting it from the front view. This is the only one of the five views that shows this surface and the slot-

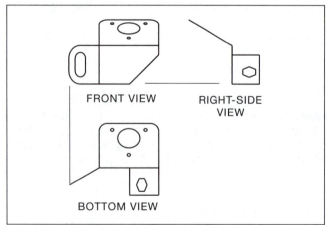

ARRANGEMENT A

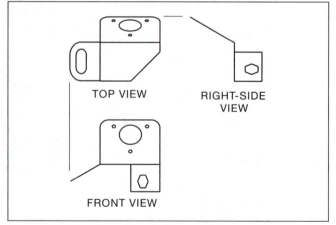

ARRANGEMENT B

FIGURE 23–1 ■ *Naming of views for mounting plate, Assignment A-64.*

ted hole in their true shape. Next the No. 2 auxiliary view is added by projecting it from the top view. This is the only view that shows this surface and the hexagon hole in their true shape. Dimensions are then added to the views or surfaces that are not shown distorted.

A variety of arrangements and naming of views to describe the index pedestal in Assignment 65 is shown in Figure 23–2. Arrangements A and B are identical except for the naming of the views. Although Arrangements C and D are acceptable, they are not as easily read.

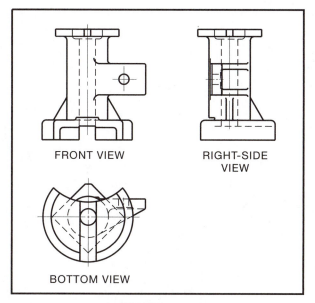

ARRANGEMENT A

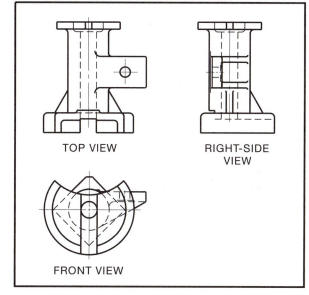

ARRANGEMENT B

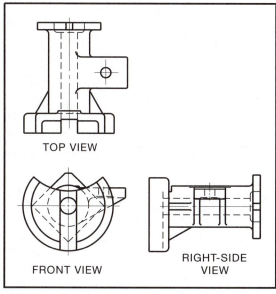

ARRANGEMENT C

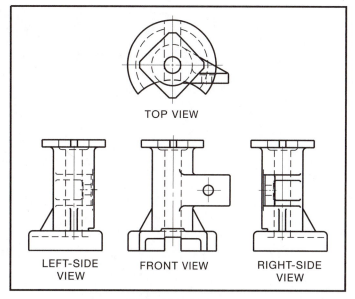

ARRANGEMENT D

FIGURE 23–2 ■ *Arrangements and naming of views for index pedestal, Assignment A-64.*

REFERENCE

ASME Y14.3M-1994 (R1999) Multi- and Sectional-
View Drawings

INTERNET RESOURCE

American Society of Mechanical Engineers. For
information on the arrangement of views, refer
to ASME Y14.3M-1994 (R1999) (*Multi- and
Sectional-View Drawings*) at: http://www.asme.org

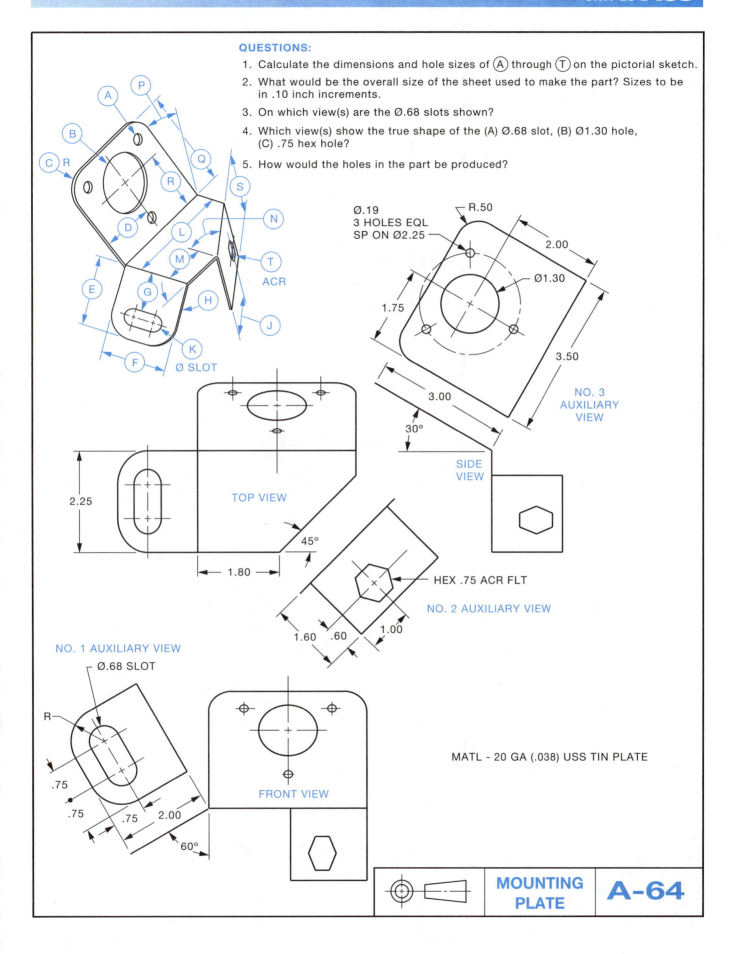

QUESTIONS:

1. Calculate the dimensions and hole sizes of Ⓐ through Ⓣ on the pictorial sketch.
2. What would be the overall size of the sheet used to make the part? Sizes to be in .10 inch increments.
3. On which view(s) are the Ø.68 slots shown?
4. Which view(s) show the true shape of the (A) Ø.68 slot, (B) Ø1.30 hole, (C) .75 hex hole?
5. How would the holes in the part be produced?

MATL - 20 GA (.038) USS TIN PLATE

MOUNTING PLATE

A-64

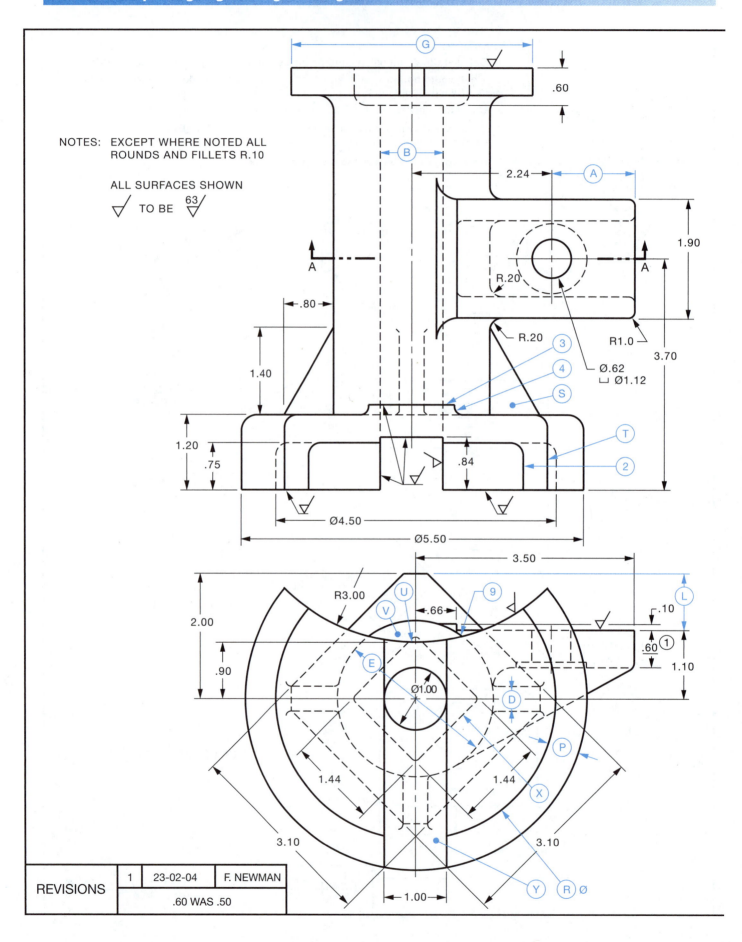

NOTES: EXCEPT WHERE NOTED ALL
ROUNDS AND FILLETS R.10

ALL SURFACES SHOWN
TO BE 63

REVISIONS | 1 | 23-02-04 | F. NEWMAN
.60 WAS .50

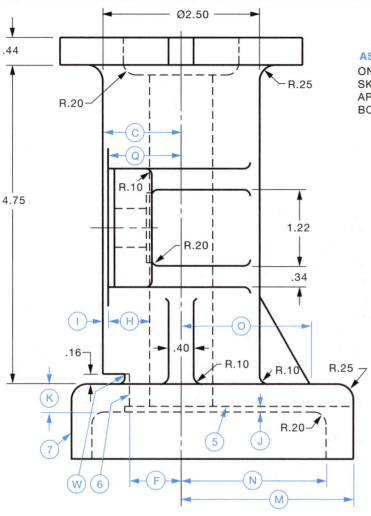

ASSIGNMENT:
ON A 1.00 INCH GRID SHEET (.10 IN. SQUARES)
SKETCH SECTION A-A. SCALE 1 : 1. PLACE THE
APPROPRIATE DIMENSIONS SHOWN ON THE
BOTTOM VIEW OF THIS SECTION.

QUESTIONS:

1. Determine distances Ⓐ through Ⓡ.

2. Which line in the front view does line ⑦ represent?

3. Which line or surface in the front view represents the surface at Ⓥ?

4. Locate line ④ in the right side view.

5. Locate line ④ in the bottom view.

6. How deep is the square Ⓧ hole?

7. How many different finished surfaces are indicated?

8. From which point on the bottom view is line ⑥ projected?

9. Determine overall height of the pedestal.

10. Which line or surface in the right view represents surface Ⓨ?

FRONT VIEW	RIGHT SIDE VIEW
BOTTOM VIEW	

ARRANGEMENT OF VIEWS

MATERIAL	WROUGHT IRON	
SCALE		
DRAWN F. NEWMAN	DATE 27/07/03	

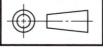

INDEX PEDESTAL

A-65

PIPING

Until one hundred years ago, water was the only important fluid conveyed from place to place through pipe. Today nearly every conceivable fluid is handled in pipe during its production, processing, transportation, or utilization. See Figure 24–1. During the age of atomic energy and rocket power, liquid metals, sodium, and nitrogen have been added to the list of more common fluids such as oil, water, and acids being transported through pipe. Many gases are also being stored and delivered through piping systems.

Pipe is also used for hydraulic and pneumatic mechanisms and used extensively for the controls of machinery and other equipment. Piping is also used as a structural element for columns and handrails.

Pipe is designated in fractional-inch sizes, which signify a nominal diameter only.

The nominal size of pipe and the inside diameter, outside diameter, and wall thickness are given in inches.

Kinds of Pipe

STEEL AND WROUGHT IRON PIPE

This pipe carries water, steam, oil, and gas and is commonly used under high temperatures and pressures. Standard steel or cast iron pipe is specified by the nominal diameter, which is always less than the actual inner diameter (ID) of the pipe. Until recently, this pipe was available in only three weights—standard, extra strong, and double extra strong, Figure 24–2. In order to use common fittings with these different pipe weights, the outside diameter (OD) of each of the different pipes remained the same.

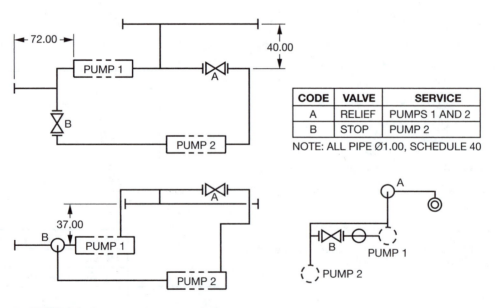

CODE	VALVE	SERVICE
A	RELIEF	PUMPS 1 AND 2
B	STOP	PUMP 2

NOTE: ALL PIPE Ø1.00, SCHEDULE 40

FIGURE 24–1 ■ *Sample piping drawing.*

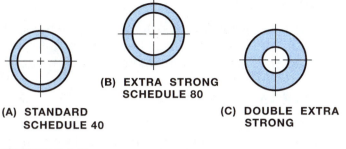

FIGURE 24–2 ■ *Comparison of wall thicknesses.*

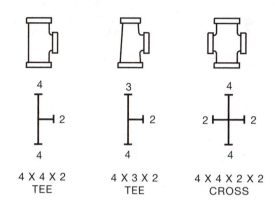

NOTE: NOMINAL PIPE SIZES IN INCHES

FIGURE 24–3 ■ *Order of specifying the openings of reducing fittings.*

The extra metal was added to the ID to increase the wall thickness of the extra strong and double extra strong pipe.

The demand for a greater variety of pipe for use under increased pressure and temperature led to the introduction of ten different pipe weights, each designated by a schedule number. Standard pipe is now called schedule 40 pipe. Extra strong pipe is schedule 80.

CAST IRON PIPE

This is often installed underground to carry water, gas, and sewage.

SEAMLESS BRASS AND COPPER PIPE

These pipes are used extensively in plumbing because of their ability to withstand corrosion.

COPPER TUBING

This is used in plumbing and heating and where vibration and misalignment are factors, such as in automotive, hydraulic, and pneumatic design.

PLASTIC PIPE

This pipe or tubing, because of its resistance to corrosion and chemicals, is often used in the chemical industry. It is easily installed. However, it is not recommended where heat or pressure is a factor.

PIPE JOINTS AND FITTINGS

Parts joined to pipe are called *fittings.* They may be used to change size or direction and to join or provide branch connections. There are three general classes of fittings: screwed, welded, and flanged. Other methods such as soldering, brazing, and gluing are used for cast iron pipe, and copper and plastic tubing.

Pipe fittings are specified by the nominal pipe size, the name of the fitting, and the material. Some fittings, for example, tees, crosses, and elbows, connect different sizes of pipe. These are called *reducing fittings.* Their nominal pipe sizes must be specified. The largest opening of the through run is given first, followed by the opposite end and the outlet. Figure 24–3 illustrates the method of designating sizes of reducing fittings.

SCREWED FITTINGS

Screwed fittings are generally used on small pipe design of 2.50-inch nominal pipe size or less.

There are two types of American Standard Pipe Thread: tapered and straight. The tapered thread is more common. Straight threads are used for special applications, which are listed in the ANSI Handbook.

Tapered threads are designated on drawings as NPT (National Pipe Thread) or whichever standard is used and may be drawn either with or without the taper, Figure 24–4. When drawn in tapered form, the taper is exaggerated. Straight pipe threads are designated on drawings as NPTS and standard thread symbols are used. Pipe threads are assumed to be tapered unless specified otherwise.

Pipe thread designation for drawings is covered in the following sequence: the nominal size in fractional inches, a dash, the number of threads per inch, a space, the thread series symbol, and the thread class if applicable. See Figure 24–4(c).

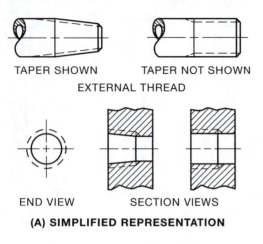

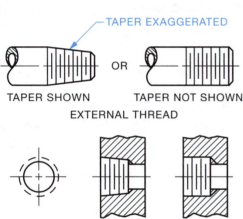

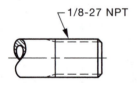

(C) PIPE THREAD DESIGNATION

FIGURE 24–4 ■ *Pipe thread conventions and designations.*

WELDED FITTINGS

Welded fittings are used where connections will be permanent and on high pressure and temperature lines. The ends of the pipe and pipe fittings are usually beveled to accommodate the weld.

FLANGED FITTINGS

Flanged joint fittings provide a quick way to disassemble pipe. Flanges are attached to the pipe ends by welding, screwing, or lapping.

VALVES

Valves are used in piping systems to stop or regulate the flow of fluids and gases. The following information describes a few of the more common types.

GATE VALVES

These are used to control the flow of liquids. The wedge, or gate, lifts to allow full, unobstructed flow and lowers to stop the flow. They are generally used where operation is infrequent and are not intended for throttling or close control.

GLOBE VALVES

These are used to control the flow of liquids or gases. The design of the globe valve produces two changes in the direction of flow, slightly reducing the pressure in the system. The globe valve is recommended for the control of air, steam, gas, and other compressibles where instantaneous on-and-off operation is essential.

CHECK VALVES

Check valves permit flow in one direction, but check all reverse flow. They are operated by pressure and velocity of line flow alone and have no external means of operation.

PIPING DRAWINGS

Piping drawings show the size and location of pipes, fittings, and valves. Because of the detail required to accurately describe these items, there is a set of symbols to represent them on drawings.

There are two types of piping drawings in use, single-line and double-line drawings, Figure 24–5. Double-line drawings take longer to draw and are therefore not recommended for production drawings. They are, however, suitable for catalogs and other applications where the appearance is more important than the extra drafting time.

Single-Line Drawings

Single-line drawings, also known as simplified representation, of pipe lines provide substantial savings without loss of clarity or reduction of comprehensiveness of information. Therefore, the simplified method is used whenever possible.

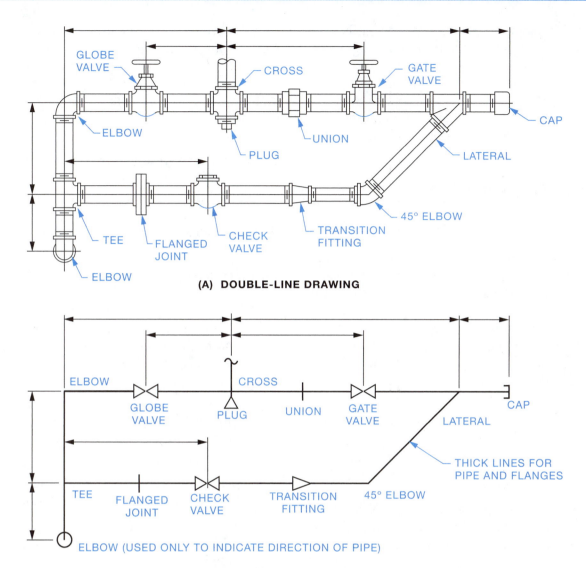

(A) DOUBLE-LINE DRAWING

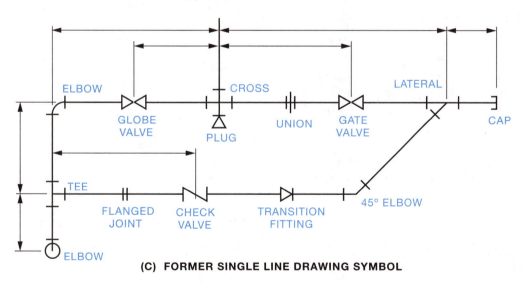

(B) SINGLE-LINE DRAWING

(C) FORMER SINGLE LINE DRAWING SYMBOL

FIGURE 24–5 ■ *Piping drawing symbols.*

Single-line piping drawings use a single line to show the arrangement of the pipe and fittings. The center line of the pipe, regardless of pipe size, is drawn as a thick line to which symbols are added. The size of the symbol is left to the discretion of the drafter. When pipe lines carry different liquids, such as cold or hot water, a coded line symbol is often used.

Drawing Projection

Two methods of projection are used, orthographic and isometric, Figure 24–6. Orthographic projection is recommended for the representation of single pipes that are either straight or bent in one plane only. However, this method is also used for more complicated pipings.

Isometric projection is recommended for all pipes bent in more than one plane and for assembly and layout work because the finished drawing is easier to understand.

CROSSINGS

The crossing of pipes without connections is usually drawn without interrupting the line representing the hidden line, Figure 24–7(A). But when it is desirable to show that one pipe must pass behind the other, the line representing the pipe farthest away from the viewer will be shown with a break or interruption where the other pipe passes in front of it.

CONNECTIONS

Permanent connections or junctions, whether made by welding or other processes, are indicated on the drawing by a heavy dot, Figure 24–7(F). A general note or specification may describe the process used.

Detachable connections or junctions are represented by a single thick line. Specifications, a general note, or the Bill of Material will indicate the type of fitting, for example, flanges, union, or coupling. The specifications will also indicate whether the fittings are flanged, threaded, or welded.

PIPE DRAWING SYMBOLS

If specific symbols are not standardized, fittings such as tees, elbows, crosses, etc., are not specially drawn but are represented, like pipe, by a continuous line. The circular symbol for a tee or elbow may

be used when necessary to indicate whether the piping is viewed from the front or back, as shown in Figure 24–7(H). Elbows on isometric drawings may be shown without the radius. However, if this method is used, the direction change of the piping must be shown clearly.

Adjoining Apparatus

If needed, adjoining apparatus, such as tanks, machinery, etc., not belonging to the piping itself, are shown by outlining them with a thin phantom line.

Dimensioning

- Dimensions of pipe and pipe fittings are always given from center to center of pipe and to the outer face of the pipe end or flange, Figure 24–7(C).

- Individual pipe lengths are usually cut to suit by the pipe fitter. However, the total length of pipe required is usually called for in the Bill of Material.

- Pipe and fitting sizes and general notes are placed on the drawing beside the part concerned, or where space is restricted, with a leader.

- A Bill of Material is usually provided with the drawing.

- Pipes with bends are dimensioned from vertex to vertex.

- Radii and angles of bends are dimensioned as shown in Figure 24–7(C) and (D). Whenever possible, the smaller of the supplementary angles is specified.

- The outer diameter and wall thickness of the pipe are indicated on the line representing the pipe, or in the Bill of Material, general note, or specifications, Figure 24–7(E).

Orthographic Piping Symbols

PIPE SYMBOLS

If flanges are not attached to the ends of the pipe lines when drawn in orthographic projection, pipe line symbols indicating the direction of the pipe are required. If the pipe line direction is toward the

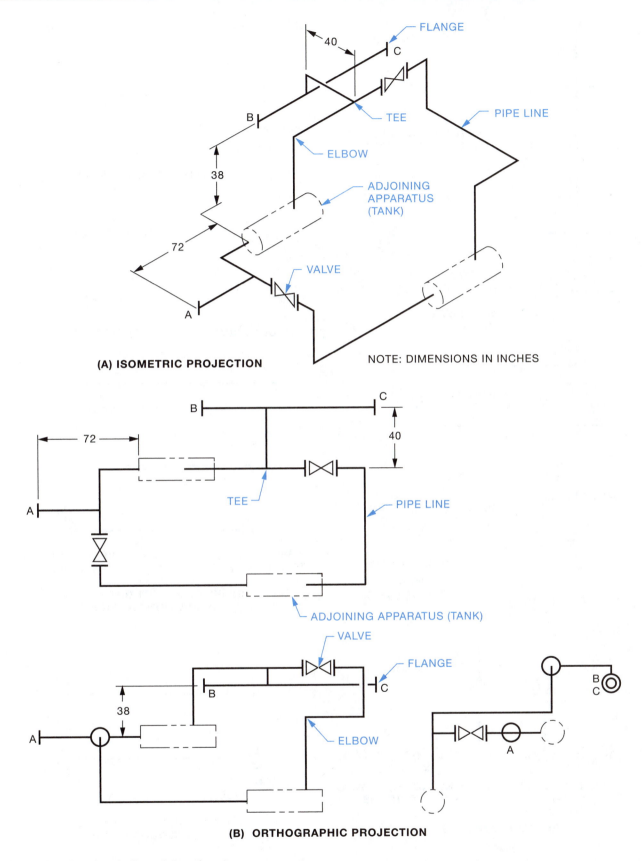

FLANGE

40

C

B

TEE

PIPE LINE

ELBOW

38

ADJOINING
APPARATUS
(TANK)

72

VALVE

A

(A) ISOMETRIC PROJECTION   NOTE: DIMENSIONS IN INCHES

B   C

72

40

A

TEE

PIPE LINE

ADJOINING APPARATUS (TANK)

VALVE

FLANGE

B

38

A

ELBOW

B
C

A

(B) ORTHOGRAPHIC PROJECTION

FIGURE 24–6 ■ *Single-line piping drawing.*

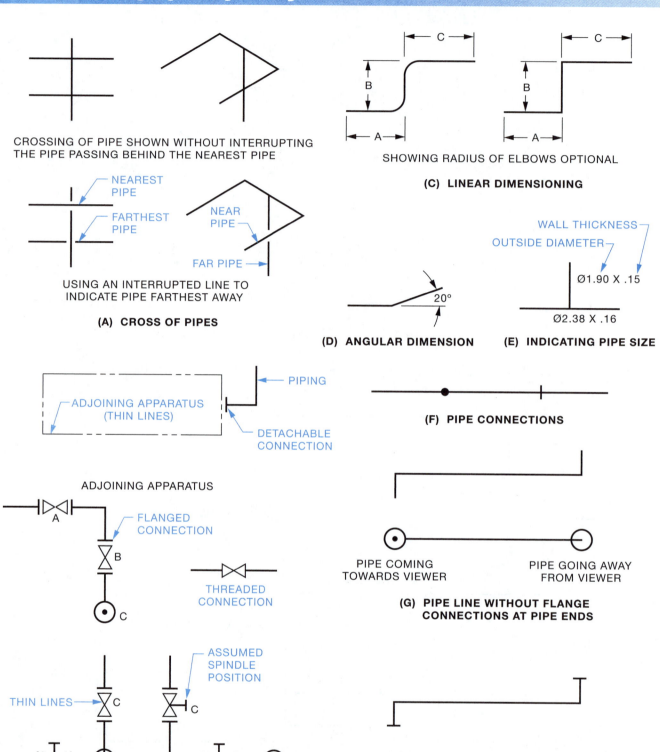

FIGURE 24–7 ■ *Single-line piping drawing symbols.*

front (or viewer), it is shown by two concentric circles, the smaller one of which is a large solid dot, Figure 24–7(G). If the pipe line direction is toward the back (or away from the viewer), it will be shown by one solid circle. No extra lines are required on the other views.

FLANGE SYMBOLS

Irrespective of their type and sizes, flanges are to be represented by:

- two concentric circles for the front view,
- one circle for the rear view,
- a short stroke for the side view,

while using lines of equal thickness as chosen for the representation of pipes, Figure 24–7(H).

VALVE SYMBOLS

Symbols representing valves are drawn with continuous thin lines (not thick lines as for piping and flanges). The valve spindles should only be shown if it is necessary to define their positions. It will be assumed that unless otherwise indicated, the valve spindle is in the position shown in Figure 24–7(B).

REFERENCES

ASME Y32.2.3-1994 (R1999) Graphic Symbols for Pipe Fittings, Valves, and Piping

ASME Y14.6-2001 Screw Thread Representation

INTERNET RESOURCES

American Design and Drafting Association. For information on piping drafting practices in their *Drafting Reference Guide,* see: http://www.adda.org/

Crane Valve Group. For information on valves and links to related sites, see: http://www.cranevalve.com/

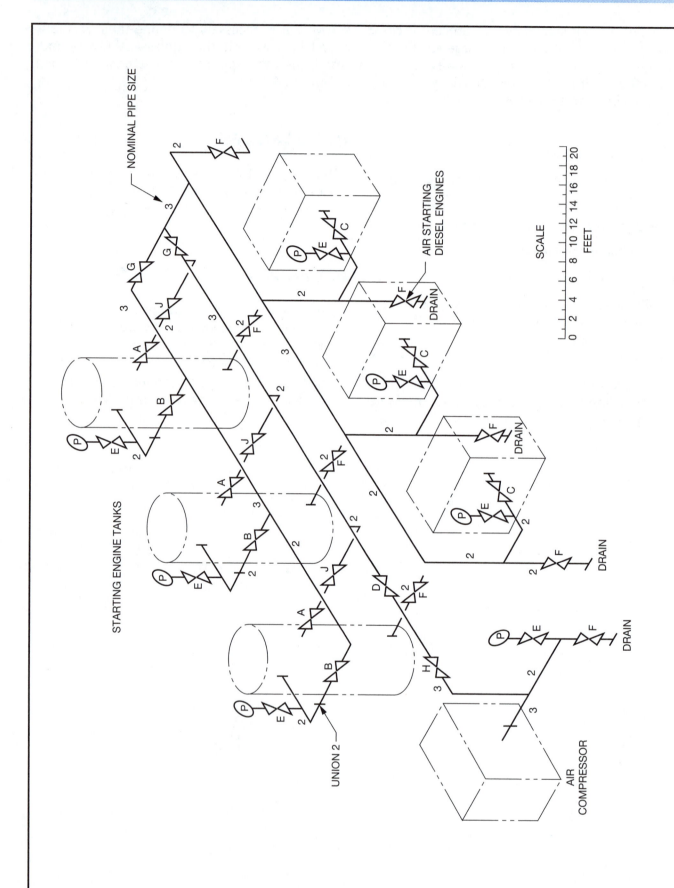

Safety valves are provided for the compressor and the air storage tanks. Check valves are installed on the air storage tank feed lines and the compressor discharge lines to prevent accidental discharge of the tanks.

Piping is arranged so that the compressor will either fill the storage tanks and/or pump directly to the engines. Any of the three storage tanks may be used for starting, and pressure gauges indicate their readiness. The engines are fitted with quick-opening valves to admit air quickly at full pressure and shut it off at the instant rotation is obtained. A bronze globe valve is installed to permit complete shut-down of the engine for repairs, and regulation of the flow of air. Drains are provided at low points to remove condensate from the air storage tanks, lines, and engine feed.

Globe valves are recommended throughout this hookup except on the main shutoff lines where gate valves are used because of infrequent operation.

QUESTIONS (Use scale provided where necessary):

1. What size pipe is used for the main feed line?
2. Disregarding the length of the valves and fittings, what is the total approximate length of
   (A) the 3-inch pipe?   (B) the 2-inch pipe?
3. What is the approximate center line to center line spacing of the diesel engines?
4. Calculate the approximate number of cubic feet of each of the air tanks.
5. Why are diesel engines used instead of public power supply?
6. What is the purpose of the air compressor?
7. Why is a gate valve used instead of a globe valve in the main line shutoff?
8. State the uses of the following parts.
   (A) Valve C   (C) Gauge P   (E) Valve D
   (B) Valve G   (D) Valve F   (F) Valve H

CODE	VALVE	SERVICE
A	BRONZE GLOBE	AIR STORAGE TANK FEED LINES
B	BRONZE GLOBE	AIR STORAGE TANK DISCHARGE LINES
C	BRONZE GLOBE	DIESEL ENGINE SHUTOFF CONTROL
D	BRONZE GLOBE	AIR COMPRESSOR DISCHARGE
E	BRONZE GLOBE	PRESSURE GAUGE SHUTOFF
F	BRONZE GLOBE	DRAIN VALVES
G	SPINDLE GATE	MAIN LINE SHUTOFF
H	BRONZE CHECK	AIR COMPRESSOR CHECK
J	BRONZE CHECK	AIR STORAGE TANK FEED LINES
P	PRESSURE GAUGE	DISCHARGE OR FEED LINES

SCALE	AS SHOWN
DRAWN K. MILLER	DATE 15/06/04

ASSIGNMENT:
PREPARE A BILL OF MATERIAL SHOWING ALL THE VALVES AND FITTINGS.

ISOMETRIC PROJECTION

ENGINE STARTING AIR SYSTEM

A-66

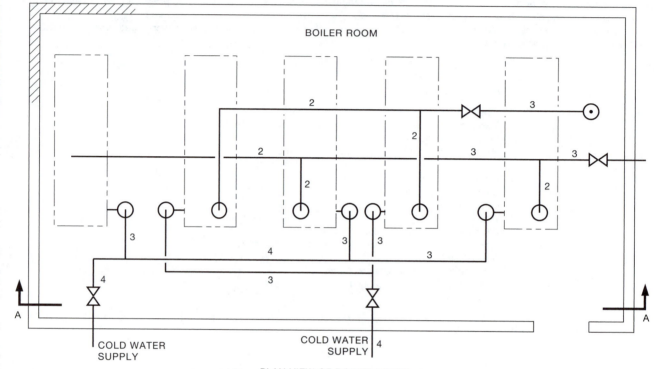

PLAN VIEW OF BOILER ROOM

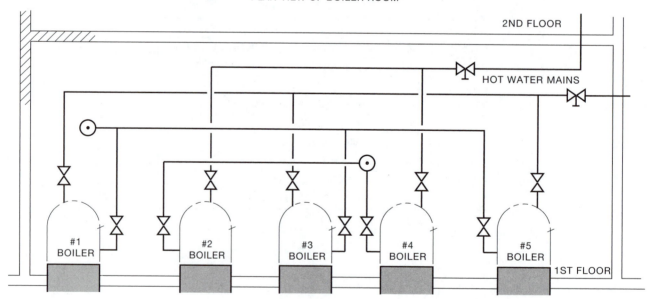

SECTIONAL ELEVATION A-A

ASSIGNMENT:

1. ON A .25 INCH ISOMETRIC GRID SHEET MAKE A SKETCH OF THE BOILER ROOM
   PIPING SHOWN ON THE OPPOSITE PAGE. THE PARTIAL ISOMETRIC VIEW BELOW
   SHOWS THE VIEWING DIRECTION AND THE POSITIONING OF THE BOILERS ON THE
   GRID SHEET. A SCALE IS PROVIDED FOR MEASURING THE DISTANCES ON THE
   DRAWING. THE .25 IN. SQUARES ON THE GRID SHEET REPRESENTS ONE FOOT ON
   THE DRAWING (SCALE 1:48).

2. ON A SEPARATE SHEET, PREPARE A BILL OF MATERIAL CALLING FOR ALL VALVES,
   FITTINGS, AND PIPE. GIVE THE APPROXIMATE TOTAL LENGTH OF EACH SIZE OF PIPE.

3. FROM THE BILL OF MATERIAL, ADD PART NUMBERS TO THE ISOMETRIC SKETCH
   FOR THE VALVES AND FITTINGS.

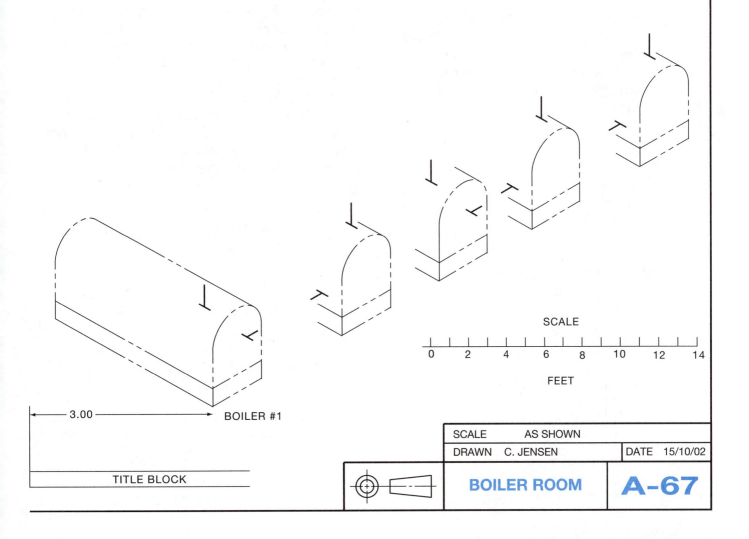

SCALE

0   2   4   6   8   10   12   14

FEET

|← 3.00 →| BOILER #1

SCALE	AS SHOWN	
DRAWN C. JENSEN		DATE 15/10/02
BOILER ROOM		**A-67**

TITLE BLOCK

25 UNIT

BEARINGS

All rotating machinery parts are supported by *bearings.* Each bearing type and style has its particular advantages and disadvantages. Bearings are classified into two groups: plain bearings and anti-friction bearings.

PLAIN BEARINGS

Plain bearings have many uses. They are available in a variety of shapes and sizes, Figure 25–1. Because of their simplicity, plain bearings are versatile. There are several plain bearing categories. The most common are journal (sleeve) bearings and thrust bearings. These are available in a variety of standard sizes and shapes.

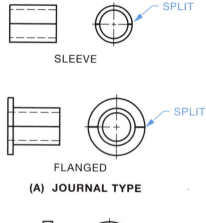

SLEEVE

FLANGED

(A) JOURNAL TYPE

(B) THRUST TYPE

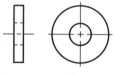

FIGURE 25–1 ■ *Plain bearings.*

Journal or Sleeve Bearings

Journal bearings are the simplest and most economical means of supporting moving parts. *Journal bearings* are usually made of one or two pieces of metal enclosing a shaft. They have no moving parts. The journal is the supporting portion of the shaft.

Speed, mating materials, clearances, temperature, lubrication and type of loading affect the performance of bearings. The maintenance of an oil film between the bearing surfaces is important. The oil film reduces friction, dissipates heat, and retards wear by minimizing metal-to-metal contact, Figure 25–2. Starting and stopping are the most critical periods of operation, because the load may cause the bearing surfaces to touch each other.

The shaft should have a smooth finish and be harder than the bearing material. The bearing will perform best with a hard, smooth shaft. For practical reasons, the length of the bearing should be between one and two times the shaft diameter. The

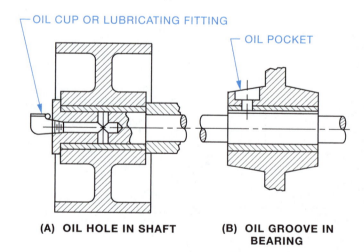

(A) OIL HOLE IN SHAFT

(B) OIL GROOVE IN BEARING

FIGURE 25–2 ■ *Common methods of lubricating journal bearings.*

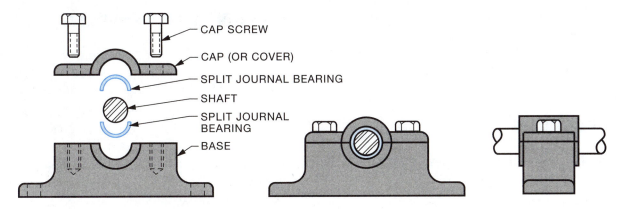

FIGURE 25–3 ■ *Pillow block with split journal bearing.*

outside diameter should be approximately 25 percent larger than the shaft diameter.

Cast bronze and porous bronze are usually used for journal bearings.

Bearings are sometimes split. This design feature facilitates assembly and permits adjustment and replacement of worn parts. Split bearings allow the shaft to be set in one half of the bearing while the other half, or cover, is later secured in position, Figure 25–3.

If the bearings shown in Figure 25–3 are to be made from two parts, they must be fastened together before the hole is bored or reamed. This will facilitate the machining operation and make a perfectly round bearing. For an incorrectly assembled bearing, see Figure 25–4.

One method that gives longer life to the bearing is the insertion of very thin strips of metal between the base and cover halves before boring. These thin strips of varying thickness are called *shims,* Figure 25–5.

When a bearing is shimmed, the same number of pieces of corresponding thickness are used on both sides of the bearing.

As the hole wears, one or more pairs of these shims may be removed for wear compensation.

Thrust Bearings

Plain thrust bearings or thrust washers are available in various materials, including: sintered metal, plastic, woven TFE fabric on steel backing, sintered Teflon-bronze-lead on metal backing, aluminum alloy on steel, aluminum alloy, and carbon-graphite.

Antifriction Bearings

Ball, roller, and needle bearings are classified as antifriction bearings as friction has been reduced to a minimum. These are covered in detail in Unit 42.

PREMOUNTED BEARINGS

Premounted bearing assemblies consist of a bearing element and a housing, usually assembled to permit convenient adaptation to a machine frame. All components are incorporated within a single unit to ensure proper protection, lubrication, and operation of the bearing. Both plain and roller element

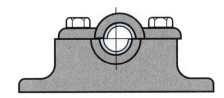

FIGURE 25–4 ■ *Bearing halves incorrectly matched.*

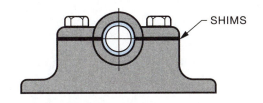

FIGURE 25–5 ■ *"Shimmed" bearing.*

bearing units are available in a wide variety of housing design and shaft sizes. An example of an adjustable premounted bearing is shown in Assignment A-68M.

REFERENCES

A.O. De Hart, "Basic Bearing Types" *Machine Design,* 40, No. 14

W.A. Glaeser, "Plain Bearings" *Machine Design,* 40, No. 14

INTERNET RESOURCES

Howstuffworks. For additional information on the design and use of all types of bearings, see: http://science.howstuffworks.com/bearing.htm

TechSourcer. For information on various types of bearings, see: http://www.techsourcer.com/industrial_supply/bearings.html

Machine Design. For information on the design and applications of various types of bearings, see: http://www.bearings.machinedesign.com/

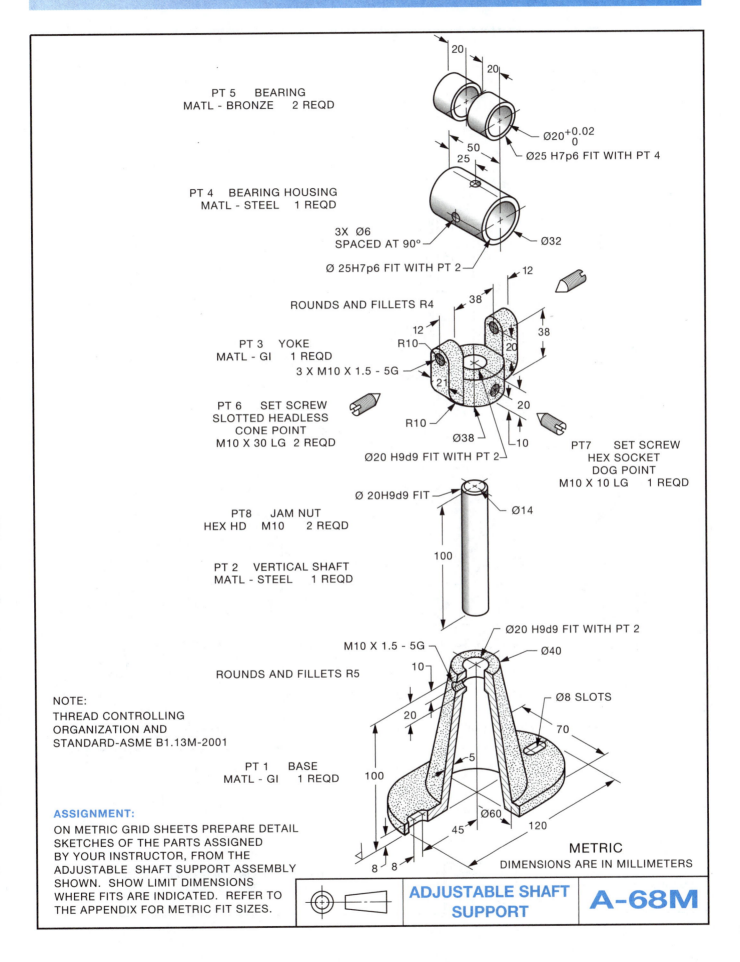

PT 5   BEARING
MATL - BRONZE   2 REQD

20

20

Ø20 +0.02 / 0

Ø25 H7p6 FIT WITH PT 4

50

25

PT 4   BEARING HOUSING
MATL - STEEL   1 REQD

3X Ø6
SPACED AT 90°

Ø32

Ø 25H7p6 FIT WITH PT 2

ROUNDS AND FILLETS R4

12

38

12

R10

PT 3   YOKE
MATL - GI   1 REQD

3 X M10 X 1.5 - 5G

38

20

20

PT 6   SET SCREW
SLOTTED HEADLESS
CONE POINT
M10 X 30 LG 2 REQD

21

R10

20

Ø38

10

Ø20 H9d9 FIT WITH PT 2

PT7   SET SCREW
HEX SOCKET
DOG POINT
M10 X 10 LG   1 REQD

Ø 20H9d9 FIT

Ø14

PT8   JAM NUT
HEX HD   M10   2 REQD

100

PT 2   VERTICAL SHAFT
MATL - STEEL   1 REQD

Ø20 H9d9 FIT WITH PT 2

M10 X 1.5 - 5G

Ø40

10

ROUNDS AND FILLETS R5

20

Ø8 SLOTS

70

NOTE:
THREAD CONTROLLING
ORGANIZATION AND
STANDARD-ASME B1.13M-2001

5

PT 1   BASE
MATL - GI   1 REQD

100

Ø60

45

120

METRIC
DIMENSIONS ARE IN MILLIMETERS

8   8

ASSIGNMENT:

ON METRIC GRID SHEETS PREPARE DETAIL
SKETCHES OF THE PARTS ASSIGNED
BY YOUR INSTRUCTOR, FROM THE
ADJUSTABLE SHAFT SUPPORT ASSEMBLY
SHOWN. SHOW LIMIT DIMENSIONS
WHERE FITS ARE INDICATED. REFER TO
THE APPENDIX FOR METRIC FIT SIZES.

ADJUSTABLE SHAFT
SUPPORT

A-68M

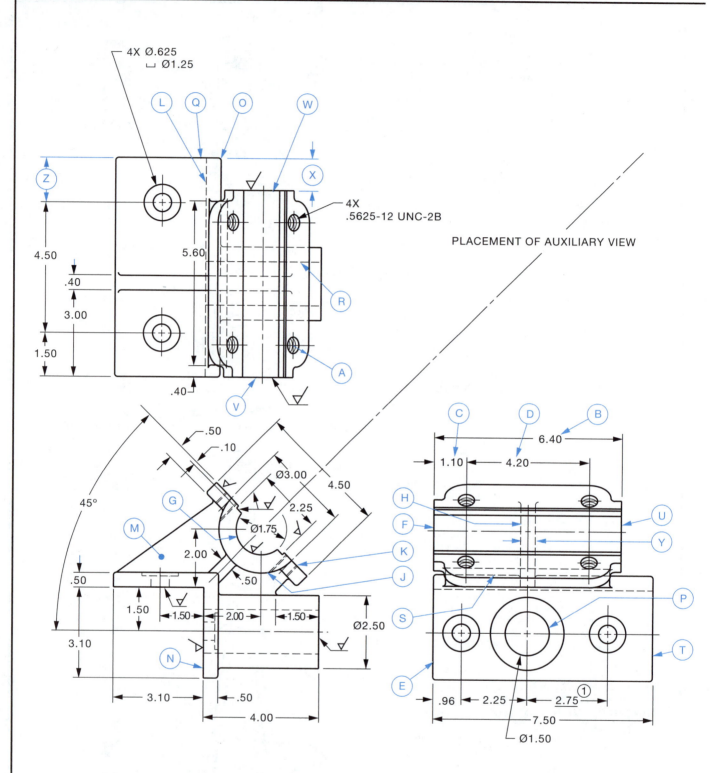

4X Ø.625
⌴ Ø1.25

L Q O W

Z X

4X
.5625-12 UNC-2B

PLACEMENT OF AUXILIARY VIEW

4.50
.40
3.00
1.50
5.60
.40
R
A
V

45°
.50
.10
Ø3.00
4.50
2.25
G
Ø1.75
M
2.00
.50
.50
1.50
1.50
2.00
1.50
Ø2.50
3.10
N
3.10
.50
4.00

C D B
6.40
1.10 4.20
H
F
U
Y
K
J
S
P
E
.96 2.25 2.75 ①
7.50
Ø1.50
T

ROUNDS AND FILLETS R.10

REVISIONS	1	22/01/04	B. JENSEN	THREAD CONTROLLING ORGANIZATION AND STANDARD-ASME B1.1-2003
		2.75 WAS 2.60		

QUESTIONS:

1. How many definite finished surfaces are on the casting? (Note-two or more surfaces could lie on one plane.)
2. What is the size of Ⓟ hole?
3. What size spotface is used on the mounting holes?
4. What are the number and size of the mounting holes?
5. Note that surface Ⓔ is not finished, but surface Ⓕ is to be finished. Allowing .06 in. for finishing, what would be the depth of the rough casting?
6. Would Ⓖ hole be bored before or after bearing cap is assembled?
7. Which line in the side view shows surface Ⓜ?
8. In which view, and by what line, is the surface represented by Ⓝ shown?
9. Which line or surface in the side view shows the projection of point Ⓙ?
10. Which point or surface in the side view does line Ⓡ represent?
11. Locate surface Ⓣ in the top view.
12. Locate surface Ⓤ in the top view.
13. Locate surface Ⓥ in the side view.
14. What are dimensions Ⓧ, Ⓨ, and Ⓩ?
15. What size is the round Ⓞ?
16. Determine dimension Ⓐ and place it correctly on the sketch of the auxiliary view.
17. Place dimensions Ⓑ, Ⓒ, and Ⓓ correctly on the sketch of the auxiliary view.
18. What is the tap drill size required for the four threaded holes?
19. How many dimensions are indicated that they are not drawn to scale?
20. The Ø1.50 hole is to be revised to accomodate a plain bearing with an LN3 fit. What would be the limits of size for the hole?

ASSIGNMENT:

ON A ONE-INCH GRID SHEET (.10 IN. SQUARES), SKETCH THE AUXILIARY VIEW OF THE CORNER BRACKET. ADD THE APPROPRIATE DIMENSIONS TO THE DRAWING. SCALE 1:1.

NOTE: UNLESS OTHERWISE SPECIFIED:

- TOLERANCE ON DIMENSIONS ±.02

- TOLERANCE ON ANGLES ±0.5°

- ▽ TO BE ⁶³▽

MATERIAL	GRAY IRON	
SCALE	NOT TO SCALE	
DRAWN	B. JENSEN	DATE 15/10/03

CORNER BRACKET   **A-69**

26 UNIT

MANUFACTURING MATERIALS

This unit has been expanded to include many of the manufacturing materials that are now readily available. One of the first decisions a designer must make is the choice of material. The choice is influenced by many factors, such as the end use of the product, and the properties of the selected material. Perhaps plastics may be the better choice of material over rubber or metal. Would one choice of metal be better than another? Will the part come in contact with water or chemicals? Is strength a factor? If so, what material will meet the stresses required? What material is in stock or readily obtainable? Is the material the correct choice if a plating or coating is required?

Just as steel composition may vary—tool steel and stainless steel, for example—so do other manufacturing materials. This unit covers some of the commonly used manufacturing materials from which to choose.

CAST IRONS

Iron and the large family of iron alloys called steel are the most frequently specified metals. All commercial forms of iron and steel contain carbon, which is an integral part of the metallurgy of iron and steel.

Types of Cast Iron

GRAY IRON

Gray iron is a supersaturated solution of carbon in an iron matrix. Generally, gray iron serves well in any machinery applications because of its fatigue resistance. Typical applications of gray iron include automotive engine blocks, flywheels, brake disks and drums, machine bases, and gears.

DUCTILE, OR NODULAR IRON

Ductile iron is not as available as gray iron, and is more difficult to control in production. However, ductile iron can be used where higher ductility or strength is required than is available in gray iron. Typical applications of ductile iron include crank shafts, heavy-duty gears, and automotive door hinges.

WHITE IRON

White iron is produced by a process called chilling, which prevents graphic carbon from precipitating out. Because of their extreme hardness, white irons are used primarily for applications requiring wear and abrasion resistance such as mill liners and shot-blasting nozzles. Other uses include brick-making equipment, crushers, and pulverizers. The disadvantage of white iron is that it is very brittle.

MALLEABLE IRON

Malleable iron is a white iron that has been converted to a malleable condition. It is a commercially cast material, which is similar to steel in many respects. It is strong and ductile, has good impact and fatigue properties, and has excellent machining characteristics.

The two basic types of malleable iron are *ferritic* and *pearlite*. Ferritic grades are more machinable and ductile, whereas the pearlite grades are stronger and harder. For design information on the casting process for cast iron see Unit 27.

STEEL

Carbon steels are the workhorses of product design. They account for over 90 percent of total steel production. More carbon steels are used in product manufacturing than all other metals combined. Far more research is going into carbon steel metallurgy and manufacturing technology than into all other steel mill products. Various technical societies and trade associations have issued the specifications covering the composition of metals. They serve as a selection guide and are a way for the buyer to conveniently specify certain known and recognized requirements. The main technical societies and trade associations concerned with metal identification in the United States are the American Iron and Steel Institute (AISI), and the Society of Automotive Engineers (SAE).

SAE and AISI Systems of Steel Identification

The specifications for steel bar are based on a number code listing the composition of each type of steel covered. They include both plain carbon and alloy steels. The code is a 4-number system. The first two figures indicate the alloy series and the last two figures the carbon content in hundredths of a percent, Figure 26–1. Therefore, the numbering code or symbol XX15 indicates 0.15 of 1 percent carbon.

For example, AISI 4830 is a molybdenum-nickel steel containing 0.2–0.3 percent molybdenum, 3.25–3.75 percent nickel, and 0.3 percent carbon. In addition to the 4-number designation, the suffix "H" is used to specify hardenability limits, and the prefix "E" indicates a steel made by the basic electric-furnace method.

Originally, the second figure indicated the percentage of the major alloying element present. This was true of many of the alloy steels. However, this had to be varied in order to classify all the steels that became available.

Alloying materials are added to steel to improve properties such as strength, hardness, machinability, corrosion resistance, electrical conductivity, and ease of forming. Figure 26–2 lists many types of steel, their properties, and their uses.

Effect of Alloys on Steel

CARBON

Increasing the carbon content increases the tensile strength and hardness.

SULPHUR

When sulphur content is over 0.06 percent the metal tends toward *red shortness* (brittleness in steel when it is red hot). Free cutting steel, for threading and screw machine work, is obtained by increasing sulphur content from about 0.075 to 0.1 percent.

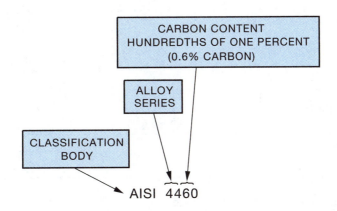

FIGURE 26–1 ■ *Steel designation system.*

TYPE OF STEEL	AISI SYMBOL	PRINCIPAL PROPERTIES	COMMON USES
CARBON STEELS			
- Nonresulfurized (Basic Carbon Steel)	10XX		
- Plain Carbon	10XX		
- Low Carbon Steel (0.06% to 0.2% Carbon)	1006 TO 1020	Toughness and Less Strength	Chains, Rivets, Shafts, Pressed Steel Products
- Medium Carbon Steel (0.2% to 0.5% Carbon)	1020 TO 1050	Toughness and Strength	Gears, Axles, Machine Parts, Forgings, Bolts and Nuts
- High Carbon Steel (Over 0.5% Carbon)	1050 and over	Less Toughness and Greater Hardness	Saws, Drills, Knives, Razors, Finishing Tools, Music Wire
- Resulfurized (Free Cutting)	11XX	Improves Machinability Increases Strength and Hardness but Reduces Ductility	Threads, Splines, Machined Parts
- Phosphorized	12XX		
- Manganese Steels	13XX	Improves Surface Finish	
MOLYBDENUM STEELS			
0.15% - 0.30% Mo	40XX		
0.08% - 0.35% Mo 0.4% - 1.1% Cr	41XX	High Strength	Axles, Forgings, Gears, Cams, Mechanism Parts
1.65% - 2.00% NI 0.2% - 0.3% Mo			
0.4% - 0.9% Cr	43XX		
0.45% - 0.60% Mo	44XX		
0.7% - 2.0% Ni 0.15% - 0.30% Mo	46XX		
0.90% - 1.2% Ni 0.15% - 0.40% Mo			
0.35% - 0.55% Cr	47XX		
3.25% - 3.75% Ni 0.2% - 0.3% Mo	48XX		
CHROMIUM STEELS			
0.3% - 0.5% Cr	50XX	Hardness, Great Strength and Toughness	Gears, Shafts, Bearings Springs, Connecting Rods
0.70% - 1.15%Cr	51XX		
1.00%C 0.90% - 1.15% Cr	E51100		
1.00%C 0.90% - 1.15% Cr	E52100		
CHROMIUM VANADIUM STEELS			
0.5%-1.1% Cr 0.10%-0.15%V	61XX	Hardness and Strength	Punches, Piston Rods, Gears, Axles
NICKEL - CHROMIUM - MOLYBDENUM STEELS			
0.4% - 0.7%Ni 0.4% - 0.6%Cr			
0.15% - 0.25%Mo	86XX	Rust Resistance, Hardness and Strength	Food Containers, Surgical Equipment
0.4% - 0.7%Ni 0.4% - 0.6%Cr			
0.2% - 0.3%Mo	87XX		
0.4% - 0.7%Ni 0.4% - 0.6%Cr			
0.3% - 0.4%Mo	88XX		
SILICON STEELS			
1.8% - 2.2% Si	92XX	Springiness and Elasticity	Springs

FIGURE 26–2 ■ *Designations, uses, and properties of steel.*

PHOSPHORUS

Phosphorus produces brittleness and general cold shortness. It also strengthens low carbon steel, increases resistance to corrosion, and improves machinability.

MANGANESE

Manganese is added during the making of steel to prevent red shortness and increase hardenability.

CHROMIUM

Chromium increases hardenability, corrosion resistance, oxidation, and abrasion.

NICKEL

Nickel strengthens and toughens ferrite and pearlite steels.

SILICON

Silicon is used as a general-purpose deoxidizer. It strengthens low-alloy steels and increases hardenability.

COPPER

Copper is used to increase atmospheric corrosion resistance.

MOLYBDENUM

Molybdenum increases hardenability and coarsening temperature.

BORON

Boron increases hardenability of lower carbon steels and has better machinability than standard alloy steels.

VANADIUM

Vanadium elevates coarsening temperatures, increases hardenability, and is a strong deoxidizer.

STRUCTURAL STEEL

Structural steel shapes and their drawing callouts are covered in Unit 34.

PLASTICS

Plastics may be defined as nonmetallic materials capable of being formed or molded with the aid of heat, pressure, chemical reactions, or a combination of these.

Plastics are strong, tough, durable materials that solve many problems in machine and equipment design. Metals are hard and rigid. This means they can be machined, to a very close tolerance, into cams, bearings, bushings, and gears, which will work smoothly under heavy loads for long periods. Although some come close, no plastic has the hardness or creep resistance of steel, for example. However, metals have many weaknesses that engineering plastics do not. Metals corrode or rust, they must be lubricated, their working surfaces wear readily, they cannot be used as electrical or thermal insulators, they are opaque and noisy, and where they must flex, fatigue rapidly.

Engineering plastics can be run at low speeds and loads and, without lubrication, are amongst the world's slipperiest solids, being compared to ice.

Plastics are a family of materials, not a single material. Each material has its special advantage. Being manufactured, plastic raw materials are capable of being combined to give almost any property desired in an end product.

The advantages of plastics are light weight, range of color, good physical properties, adaptability to mass production methods, and often, lower cost. They are usually classified as either *thermoplastic* or *thermosetting*. See Figure 26–3.

Thermoplastics

Thermoplastics soften or liquefy and flow when heat is applied. Removal of the heat causes these materials to set or solidify. They can be reheated or reformed or reused. This group includes the ABS, acetals, acrylics, the celluloses, fluorocarbons, nylons, polyethylene, polystyrene, the vinyls, and polycarbonates.

Thermosetting Plastics

Thermosetting plastics undergo an irreversible chemical change when heat is applied or when a catalyst

BASE RESIN	OUTSTANDING CHARACTERISTICS	TYPICAL APPLICATIONS
THERMOPLASTICS		
ABS	Colorability, toughness	Instrument panels, telephone housings
Acetal	Impact strength, chemical resistance	Replace zinc or aluminum for castings
Acrylic	Clarity, weather resistance	Windows, lenses, dials, shoe heels, medallions
Cellulosic	Clarity, toughness	Safety glass, film, knobs, bowling balls, electrical parts
Fluorocarbon	Chemical and heat resistance, self-lubrication	Low pressure bearings, chemical linings, gaskets, bushings
Polyamide (Nylon)	Impact strength, cold flow	Gears, cams, bushings, gaskets, cable clamps, housings, electrical insulation
Polycarbonate	Dimensional stability, clarity, good electrical insulation	Switch housings, terminal blocks
Polyethylene	Ease of forming, chemical resistance	Bottles, toys, underground pipe, electrical insulation, ducts
Polypropylene	Ease of forming, chemical resistance	Bottles, toys, replace zinc or aluminum for castings
Polystyrene	Ease of forming, transparent	Toys, display and jewelry cases, foamed insulation, refrigerator parts, containers
Polyvinyl Chloride	Flexibility, toughness, chemical resistance	Toys, upholstery, lawn hose, floor tile, electrical insulation, gaskets, ducts
THERMOSETTING PLASTICS		
Epoxy	Room temperature cure, no pressure	Adhesives, coatings, terminal boards, electrical potting, tooling fixtures
Phenol	Dark colors only, Good electrical insulation	Knobs, terminal boards, adhesives, distributor caps and housings
Polyester	Room temperature cure, on pressure	Car bodies, boat hulls, heater ducts
Silicone	Relative constancy of properties over wide temperature range	Terminal boards, electrical moldings
Urea & Melamine	Indoor use, Pastel colors available	Knobs, jars, buttons, dishes, electrical moldings

FIGURE 26–3 ■ *Common plastics.*

or reactant is added. They become hard, insoluble, and infusible, and they do not soften upon reapplication of heat. Thermosetting plastics include phenolics, amino plastics (melamine and urea), polyesters, epoxies, silicones, alkydes, allylics, and casein.

RUBBER

The use of rubber is advantageous when design considerations involve one or more of the following factors:

- Electrical insulation
- Vibration isolation
- Sealing surfaces
- Chemical resistance
- Flexibility

Elastomers (rubber-like substances) are derived from either natural or synthetic sources. Rubber can be formed into useful rigid or flexible shapes, usually with the aid of heat, pressure, or both. The most outstanding characteristics of rubber are its low modulus of elasticity and its ability to withstand large deformation and to quickly recover its shape when released.

Rubber parts are produced in either mechanical (solid) or cellular form, depending on the desired performance of the part. They are categorized into two kinds of rubber, natural and synthetic. The synthetic rubbers are further classified into several kinds.

Mechanical Rubber

Mechanical rubber is used in either pressure-molded, cast, or extruded form. Typical parts produced by these methods are tires, belts, and bumpers. Mechanical rubber should be preferred to sponge rubber because of its superior physical properties.

Cellular Rubber

Cellular rubber can be produced with "open" or "closed" cells. Open-cell sponge rubber is made by including a gas-forming chemical compound in the mixture before vulcanization. The heat of the vulcanizing process causes a gas to form in the rubber, making a cellular structure. Typical applications are pads and weather stripping. Foam rubber is a specialized type of open cell.

Both open- and closed-cell sponge rubber is available in block or sheet form that can be cut to size and shape.

INTERNET RESOURCES

Animated Worksheets. For information on materials and their properties, see: Animatedworksheets.co.uk (materials)

Bayer MaterialScience AG. For information on industrial plastics, see: http://www.bayermaterialscience.com

eFunda.com. For information on manufacturing materials, see: http://www.efunda.com/home.cfm

Gates Rubber Company. For information on automotive and industrial drive belts, see: http://www.gates.com

Howstuffworks. For information on the drop forging of steel products, see: http://www.science.howstuffworks.com/question376.htm

Machine Design. For information on manufacturing materials and related processes, see: http://www.machinedesign.com

TechStudent.Com. For information on manufacturing materials, see http://www.technologystudent.com (equipment and accessories)

Rubber Cal. For information on various types of rubber and their applications, see: http://www.rubbercal.com

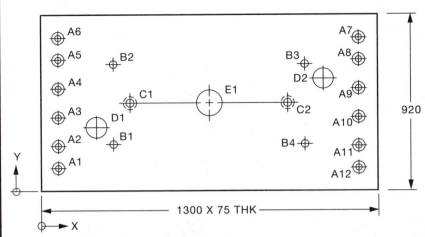

HOLE	LOCATION		HOLE SIZE	HOLE	LOCATION		HOLE SIZE
	X	Y			X	Y	
A1	50	120	12X Ø32 ⊔ Ø64 ⛛ 30 NEAR SIDE ONLY	B1	280	260	4X Ø16
A2	50	235		B2	280	660	
A3	50	385		B3	1020	660	
A4	50	535		B4	1020	260	
A5	50	685		C1	342	460	2X M20 X 2.5-6G
A6	50	800		C2	958	460	
A7	1250	800		D1	210	334	2X Ø 80.30 / 80.00 (80 H7)
A8	1250	685		D2	1090	586	
A9	1250	535		E1	650	460	1X Ø 100.054 / 100.000 (100 H8)
A10	1250	385					
A11	1250	235					
A12	1250	120					

THREAD CONTROLLING ORGANIZATION
AND STANDARD-ASME B1.13M-2001

NOTE: UNLESS OTHERWISE SPECIFIED,
TOLERANCE ON DIMENSIONS ±0.5

QUESTIONS:

1. What is the thickness of the material?

2. In what classification of plastic does the material for the Crossbar belong?

3. What indicates that the drawing is not to scale?

4. How many counterbored holes are there?

5. What is the diameter of the CBORE?

6. How many D holes are there?

7. Which letter specifies the Ø100 hole?

8. How many different sized holes are specified?

9. What is the total number of tapped holes?

10. What is the thread pitch of the C holes?

11. What is the tolerance given for the E hole?

12. What is the high limit of the E hole?

13. What is the maximum size for the D holes?

14. Is the E hole on the center of the part?

15. What is the center distance between the C holes?

16. What type of dimensioning is used to establish the location of the holes?

METRIC
DIMENSIONS ARE IN MILLIMETERS

MATERIAL	ABS PLASTIC	
SCALE	NOT TO SCALE	
DRAWN	B. JENSEN	DATE 16/10/04

CROSSBAR **A-70M**

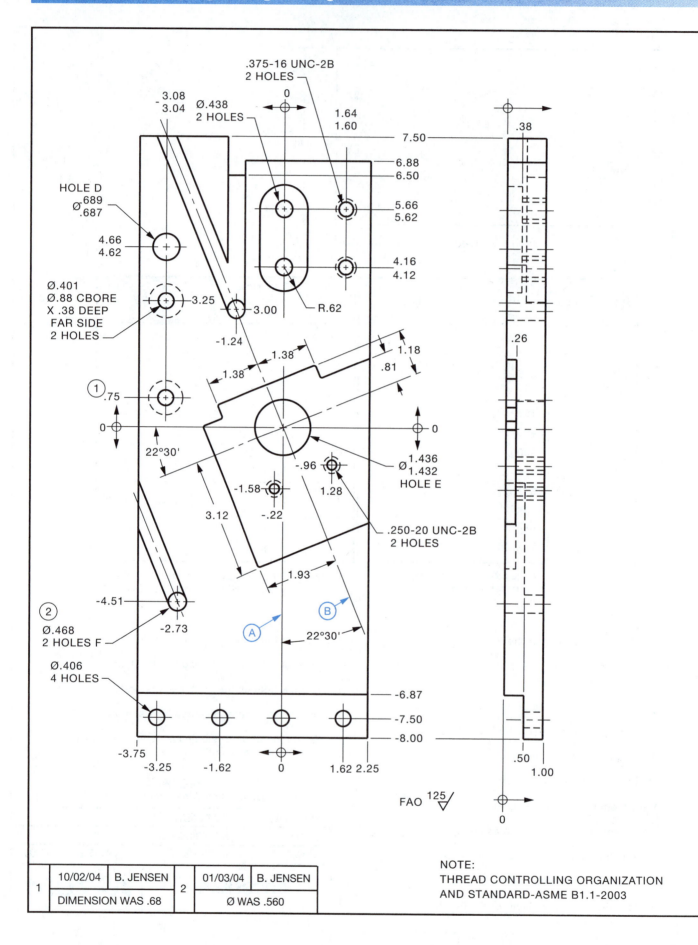

.375-16 UNC-2B
2 HOLES

Ø.438
2 HOLES

3.08
3.04

HOLE D
Ø.689
.687

4.66
4.62

Ø.401
Ø.88 CBORE
X .38 DEEP
FAR SIDE
2 HOLES

3.25

3.00

-1.24

① .75

0

22°30'

R.62

7.50
6.88
6.50
5.66
5.62
4.16
4.12

1.64
1.60

1.38
1.38

1.18
.81

Ø 1.436
1.432
HOLE E

-.96
1.28

-1.58
-.22

3.12

.250-20 UNC-2B
2 HOLES

1.93

② -4.51
-2.73
Ø.468
2 HOLES F

Ⓐ

Ⓑ 22°30'

Ø.406
4 HOLES

-6.87
-7.50
-8.00

-3.75
-3.25

-1.62

0

1.62 2.25

.38

.26

0

0

.50
1.00

FAO 125/▽

	10/02/04	B. JENSEN		01/03/04	B. JENSEN
1	DIMENSION WAS .68		2	Ø WAS .560	

NOTE:
THREAD CONTROLLING ORGANIZATION
AND STANDARD-ASME B1.1-2003

QUESTIONS:

1. What are the overall (A) width, (B) height, and (C) depth of the part?

2. What is the width of the slots that are cut out from the F holes?

3. What is their depth?

4. What class of surface texture is required?

5. What is the depth of the recess adjacent to the (A) E hole, (B) Ø.438 holes?

6. How many degrees are there between line B and the horizontal?

7. What was the size of the F holes before they were changed?

8. What tolerance is required on the E hole?

9. What is the low limit on the E hole?

10. What tolerance is required on the D hole?

11. What is the high limit on the D hole?

12. Determine the maximum vertical center distance from D hole to the four Ø.406 holes located at the bottom of the part.

13. Determine the (A) maximum, (B) minimum, center distance between the two .375 tapped holes.

14. Determine the maximum horizontal center distance between D hole and the .375 tapped holes.

15. What are the width, height, and depth of the cutout at the bottom of the part?

16. How deep are (A) the Ø.438 holes. (B) the Ø.401 holes?

17. How deep are the .250-20 UNC holes?

18. How many full threads do the .250-20 UNC tapped holes have?

19. What are the main alloys in the material?

20. What type of steel is specified for the Oil Chute?

ASSIGNMENT:
ON A ONE-INCH GRID SHEET (.10 IN. SQUARES), SKETCH TWO SECTION VIEWS, ONE ALONG LINE "A", THE OTHER ALONG LINE "B". SCALE 1:2. ADD THE RECTANGULAR COORDINATE DIMENSIONS WITHOUT DIMENSION LINES TO THESE SECTION VIEWS.

NOTE:

- ALL RADII R.06 UNLESS OTHERWISE SPECIFIED.

- DIMENSIONS ARE TAKEN FROM PLANES DESIGNATED 0-0, AND ARE PARALLEL TO THESE PLANES.

- UNLESS OTHERWISE STATED:

±.02 ON DIMENSIONS
±0.5 ° ON ANGLES

MATERIAL	SAE 4020 STEEL	
SCALE	NOT TO SCALE	
DRAWN	B. JENSEN	DATE 15/10/03

OIL CHUTE **A-71**

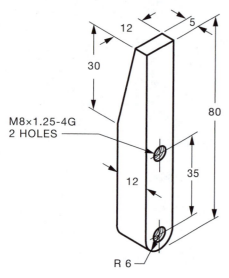

PT 6 CAP SCREW
M3×8 LG
RD HD   1 REQD

M8×1.25-4G
2 HOLES

R 6

PT 1   MOVABLE JAW
1 REQD   MATL - SAE 1020

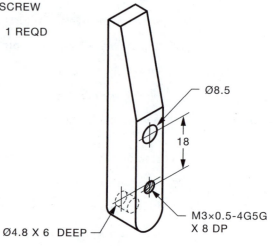

Ø8.5

18

Ø4.8 X 6 DEEP

M3×0.5-4G5G
X 8 DP

AS SHOWN, OTHERWISE SAME
AS PART 1.

PT 2   STATIONARY JAW
1 REQD   MATL - SAE 1020

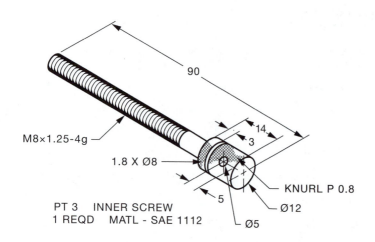

90

M8×1.25-4g

1.8 X Ø8

14

3

KNURL P 0.8

5

Ø12

Ø5

PT 3   INNER SCREW
1 REQD   MATL - SAE 1112

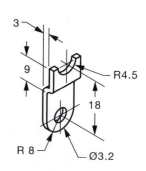

3

9

R4.5

18

R 8

Ø3.2

PT 5   CLIP   1 REQD
MATL - 16 GA (1.6) USS STL

ASSIGNMENT:

ON METRIC GRID PAPER PREPARE WORKING
(DETAIL) DRAWINGS OF THE PARALLEL CLAMP
DETAILS SHOWN. USE SIMPLIFIED THREAD
CONVENTIONS (UNIT 16) AND REFER TO UNIT 11
FOR THE SELECTION OF VIEWS. SCALE 1:1

THREAD CONTROLLING ORGANIZATION
AND STANDARD – ASME B1.13M-2001

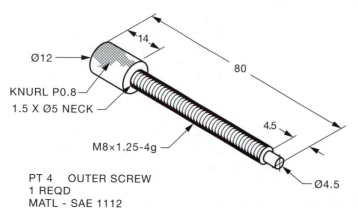

Ø12

14

80

KNURL P0.8

1.5 X Ø5 NECK

M8×1.25-4g

4.5

Ø4.5

PT 4   OUTER SCREW
1 REQD
MATL - SAE 1112

METRIC
DIMENSIONS ARE IN MILLIMETERS

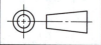

**PARALLEL CLAMP
DETAILS**

A-72M

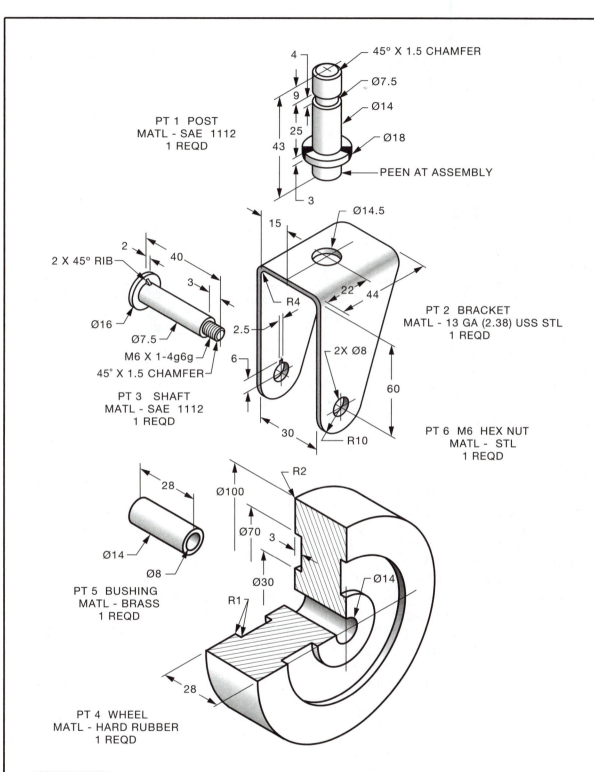

PT 1 POST
MATL - SAE 1112
1 REQD

45° X 1.5 CHAMFER
Ø7.5
Ø14
Ø18
PEEN AT ASSEMBLY

4
9
25
43
3

2 X 45° RIB
2
40
3
Ø16
Ø7.5
M6 X 1-4g6g
45° X 1.5 CHAMFER

PT 3   SHAFT
MATL - SAE 1112
1 REQD

Ø14.5
15
R4
22
44
2.5
6
2X Ø8
60
30
R10

PT 2   BRACKET
MATL - 13 GA (2.38) USS STL
1 REQD

PT 6 M6 HEX NUT
MATL - STL
1 REQD

28
Ø14
Ø8

PT 5 BUSHING
MATL - BRASS
1 REQD

R2
Ø100
Ø70
3
Ø30
Ø14
R1
28

PT 4 WHEEL
MATL - HARD RUBBER
1 REQD

ASSIGNMENT:
ON CENTIMETER GRID SHEETS (1mm SQUARES), PREPARE DETAILED SKETCHES,
COMPLETE WITH DIMENSIONS, OF THE CASTER PARTS SHOWN. SCALE 1:1.

METRIC
DIMENSIONS ARE IN MILLIMETERS

THREAD CONTROLLING ORGANIZATION
AND STANDARD-ASME B1.13M-2001

CASTER ASSEMBLY   A-73M

CASTING PROCESSES

Irregular or odd-shaped parts that are difficult to make from metal plate or bar stock may be cast to the desired shape. Casting processes for metals can be classified by either the type of mold or pattern, or the pressure of force used to fill the mold. Conventional sand, shell, and plaster molds utilize a durable pattern, but the mold is used only once. Permanent molds and die-casting dies are machined in metal or graphite sections and are employed for a large number of castings. Investment casting and the full mold process involve both an expendable mold and pattern.

Sand Mold Casting

The most widely used casting process for metals uses a permanent pattern of metal or wood that shapes the mold cavity when loose molding material is compacted around the pattern. This material consists of a relatively fine sand plus a binder that serves as the adhesive.

Figures 27–1 and 27–2 show a typical sand mold, with the various provisions for pouring the molten metal and compensating for contraction of the solidifying metal, and a sand core for forming a cavity in the casting. Sand molds are prepared in flasks, which consist of two or more sections: bottom (drag), top (cope), and intermediate sections (cheeks) when required.

The cope and drag are equipped with pins and lugs to ensure the alignment of the flask. Molten metal is poured into the sprue, and connecting runners provide flow channels for the metal to enter the mold cavity through gates. Riser cavities are located over the highest section of the casting.

The gating system, besides providing a way for the molten metal to enter the mold, functions as a venting system for the removal of gases from the

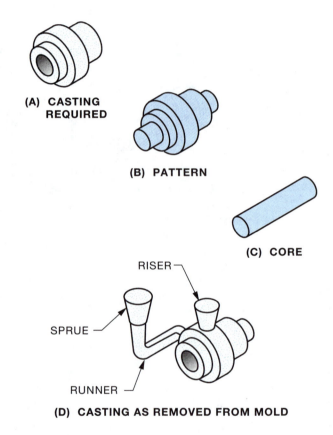

(A) CASTING REQUIRED

(B) PATTERN

(C) CORE

RISER

SPRUE

RUNNER

(D) CASTING AS REMOVED FROM MOLD

FIGURE 27–1 ■ *Sand casting parts.*

mold and acts as a riser to furnish liquid metal to the casting during solidification.

In producing sand molds, a metal or wooden pattern must first be made. The pattern is slightly larger in every dimension than the part to be cast to allow for shrinkage when the casting cools. This is known as *shrinkage allowance*, and the patternmaker allows for it by using a shrink rule for each of the cast metals. Because shrinkage and draft are taken care of by the patternmaker, they are of no concern to the drafter.

Additional metal, known as machining or finish allowance, must be provided on the casting where

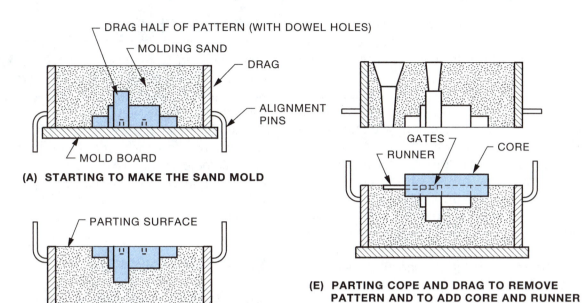

DRAG HALF OF PATTERN (WITH DOWEL HOLES)
MOLDING SAND
DRAG
ALIGNMENT PINS
MOLD BOARD

(A) STARTING TO MAKE THE SAND MOLD

PARTING SURFACE

BOTTOM BOARD

(B) AFTER ROLLING OVER THE DRAG

GATES
RUNNER
CORE

(E) PARTING COPE AND DRAG TO REMOVE PATTERN AND TO ADD CORE AND RUNNER

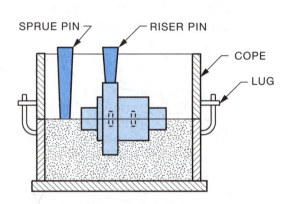

SPRUE PIN
RISER PIN
COPE
LUG

(C) PREPARING TO RAM MOLDING SAND IN COPE

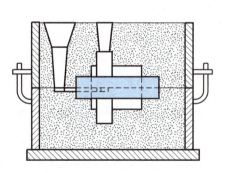

(F) SAND MOLD READY FOR POURING

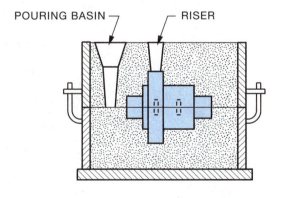

POURING BASIN
RISER

(D) REMOVING RISER AND GATE SPRUE PINS AND ADDING POURING BASIN

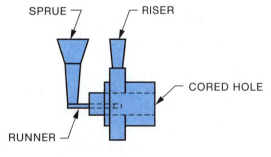

SPRUE
RISER
CORED HOLE
RUNNER

SPRUE, RISER, AND RUNNER TO BE REMOVED FROM CASTING

(G) CASTING AS REMOVED FROM THE MOLD

FIGURE 27–2 ■ *Sequence in preparing a sand casting.*

a surface is to be finished. Depending on the material being cast, from .06 to .12 inch is usually allowed on small castings for each surface requiring finishing.

When casting a hole or recess in a casting, a core is often used. A core is a mixture of sand and a bonding agent that is baked and hardened to the desired shape of the cavity in the casting, plus an allowance to support the core in the sand mold. In addition to the shape of the casting desired, the pattern must be designed to produce areas in the mold cavity to locate and hold the core. The core must be solidly supported in the mold, permitting only that part of the core that corresponds to the shape of the cavity in the casting to project into the mold.

PREPARATION OF SAND MOLDS

The drag portion of the flask is first prepared in an upside-down position with the pins pointing down, Figure 27–2(A). The drag half of the pattern is placed in position on the mold board and a light coating of parting compound is used as a release agent. The molding sand is then rammed or pressed into the drag flask. A bottom board is placed on the drag, the whole unit is rolled over, and the mold board is removed.

The cope half of the pattern is placed over the drag half and the cope portion of the flask is placed in position over the pins, Figure 27–2(C). A light coating of parting compound is sprinkled throughout. Next, the sprue pin and riser pin, tapered for easy removal, are located, and the molding sand is rammed into the cope flask. The sprue pin and riser pin are then removed, Figure 27–2(D). A pouring basin may or may not be formed at the top of the gate sprue.

Now the cope is lifted carefully from the drag and the pattern is exposed. The runner and gate, passageways for the molten metal into the mold cavity, are formed in the drag sand. The pattern is removed and the core is placed in position, Figure 27–2(E). The cope is then put back on the drag, Figure 27–2(F).

The molten metal is poured into the pouring basin and runs down the sprue to a runner and through the gate and into the mold cavity. When the mold cavity is filled, the metal will begin to fill the sprue and the riser. Once the sprue and riser have been filled, the pouring should stop.

When the metal has hardened, the sand is broken and the casting removed, Figure 27–2(G). The excess metal, gates, and risers, are removed and later remelted.

Full Mold Casting

The characteristic feature of the *full mold process* is the use of gasifiable patterns made of foamed plastic. These are not extracted from the mold but are vaporized by the molten metal.

The full mold process is suitable for individual castings, and for small series of up to five castings. The full mold process is very economical, and it reduces the delivery time required for prototypes, articles urgently needed for repair jobs, and individual large machine parts.

CASTING DESIGN

Simplicity of Molding from Flat Back Patterns

Simple shapes such as the one shown in Figure 27–3 are very easy to mold. In this case the flat face of the pattern is at the parting line and lies perfectly

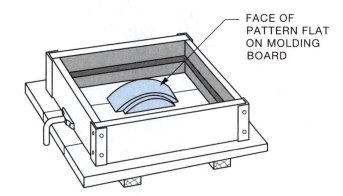

FACE OF PATTERN FLAT ON MOLDING BOARD

(A) PLACING THE PATTERN ON THE MOLDING BOARD

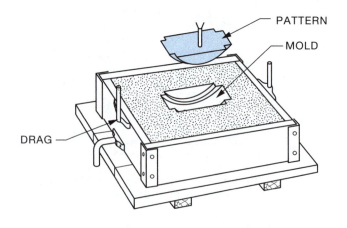

PATTERN

MOLD

DRAG

(B) DRAWING THE PATTERN

FIGURE 27–3 ■ *Making a mold of a flat back pattern.*

flat on the molding board. In this position no molding sand sifts under the flat surface to interfere with the drawing of the pattern. The simplicity with which flat back patterns of this type may be drawn from the mold is illustrated in Figure 27–3.

Irregular or Odd-Shaped Castings

When a casting is to be made for an odd-shaped piece such as the offset bracket, Figure 27–4, it is necessary to make the pattern for the bracket in one or more parts to facilitate the making of the mold. The difficulties with this are mostly due to the removal of the pattern from the sand mold.

The pattern for the bracket is made in two parts as shown in Figure 27–5. The two adjacent flat surfaces of the divided pattern come together at the parting lines.

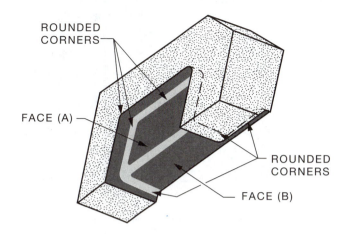

FIGURE 27–6 ■ *Set core for offset bracket.*

Set Cores

In examining the illustration of the offset bracket, it will be noted that there are rounded corners. In order to make it possible to mold these rounded corners, a block must be added to the pattern for ease in its removal from the mold.

This block, which becomes an integral part of the pattern, also acts as a core print for a *set core* (a core that has been baked hard), Figure 27–6. It is made to conform to the shape of the faces of the casting, including the rounded corners.

The face of the core print also forms the parting line for one side of the two-part pattern. When making the mold for the bracket casting, this face, which corresponds to the flat face of a flat back pattern, is laid on the molding board with the drag of the flask in position. The sand is then rammed around the pattern.

When the drag is reversed, the cope, or upper part of the flask, is placed in position. The other half of the pattern is then joined with the first part, and the sand is rammed into the cope to flow around that part of the pattern that projects into it.

After the pattern is removed, the set core, which is formed in a core box and baked hard, is set in the impression in the mold made by the core print of the pattern. When poured into the mold, the molten metal fills the cavity made by the pattern and the faces of the set core as shown in Figure 27–6 at A and B to form the casting.

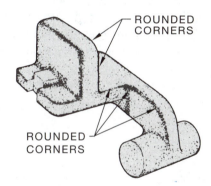

FIGURE 27–4 ■ *Casting of offset bracket shown in Assignment A-74.*

Coping Down

The core print is made as part of the pattern to avoid removing molding sand in the drag, which would correspond to the shape of the core print.

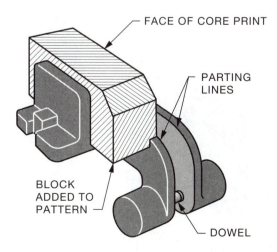

FIGURE 27–5 ■ *Two-piece pattern for offset bracket.*

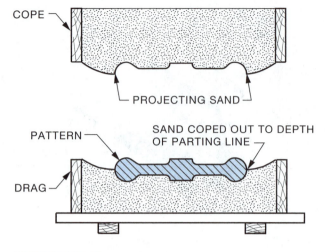

FIGURE 27–7 ■ *Coping down.*

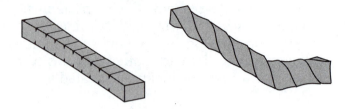

FIGURE 27–8 ■ *Soldiers and gaggers.*

Coping down requires skill and takes time. The set core principle is at times preferred to coping down to avoid delay and ensure a more even parting line on the casting.

Split Patterns

Irregularly shaped patterns that cannot be drawn from the sand are sometimes split so that one half of the pattern may be rammed in the drag as a simple flat back pattern while the cope half, when placed in position on the drag half, forms the mold in the cope. Patterns of this type are called *split patterns* and do not require coping down to the parting line, which would be necessary if the pattern were made solid.

The pattern for the casting shown in Figure 27–9, when made without a print for a core, can be drawn

If this sand were dug out or *coped out,* as shown in Figure 27–7, the remaining cavity would be again filled with molding sand when the cope was rammed. The sand would then hang below the parting line of the cope down into the drag.

When the mold is made by coping down, the hanging portion of the cope is supported by soldiers or gaggers embedded within the sand to hold the projecting part in position for subsequent operations, Figure 27–8.

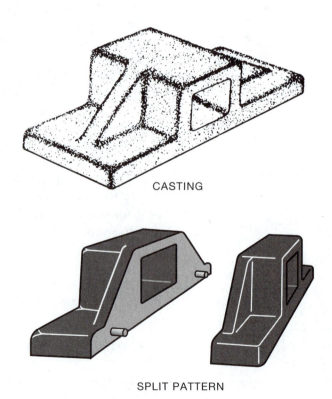

CASTING

SPLIT PATTERN

FIGURE 27–9 ■ *Application of split pattern.*

from the sand only by splitting the pattern on the parting line as illustrated.

The drawing must be examined to determine how the pattern should be constructed. This is important because the parting line must be located in a position that permits the halves of the pattern to be drawn from the sand without interference.

CORED CASTINGS

Cored castings have certain advantages over solid castings. Where practical, castings are designed with cored holes or openings for economy, appearance, and accessibility to interior surfaces.

Cored openings often improve the appearance of a casting. In most instances, cored castings are more economical than solid castings because of the savings in metal. Although cored castings are lighter, they are designed without sacrificing strength. The openings cast in the part eliminate unnecessary machining.

Hand holes may also be formed by coring in order to provide an opening through which the interior of the casting may be reached. These openings also permit machining an otherwise inaccessible surface of the part as shown in Figure 27–10 and Assignment A-76.

MACHINING LUGS

It is difficult to hold and machine certain parts without using lugs. This is because of the design and nature of certain parts. The lugs are an integral part of the casting. They are sometimes removed to avoid interference with the functioning of the part. Lugs are usually represented in phantom outline.

An example of a machining lug is shown in Figure 27–11. This lug is used to provide a flat surface on the end of the casting for centering and also to give a uniform center bearing for other machining operations. Both the function of the lug and its appearance determine whether or not it is removed after machining.

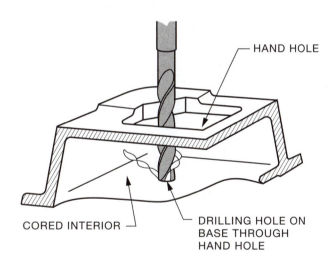

HAND HOLE

CORED INTERIOR

DRILLING HOLE ON BASE THROUGH HAND HOLE

FIGURE 27–10 ■ *Section of cored casting.*

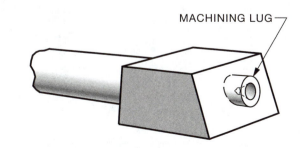

MACHINING LUG

FIGURE 27–11 ■ *Application of machining lug.*

SURFACE COATINGS

Machined parts are frequently finished either to protect the surfaces from oxidation or for appearance. The type of finish depends on the use of the part. The finish commonly applied may be a protective coat of paint, lacquer, or a metallic plating. In some cases only a surface finish such as polishing or buffing may be specified.

The finish may be applied before any machining is done, between the various stages of machining, or after the piece has been completed. The type of finish is usually specified on the drawing in a notation similar to the one indicated on the auxiliary pump base, Assignment A-76, which reads CASTING TO BE PAINTED WITH ALUMINUM BEFORE MACHINING.

The decorative finish on the drive housing should be added before machining so that there will be no accidental deposit of paint on the machined surfaces. This will ensure the desired accuracy when assembled with other parts.

REFERENCE

Machine Design Materials Reference Issue, Mar. 1981.
ASME Y14.8M-1996 (R2002) Castings and Forgings

INTERNET RESOURCES

eFunda. For information on castings, surface coatings, and related topics, see: http://www.efunda.com/home.cfm

Intermet. For links to a wide variety of casting industries and associations, see: http://www.intermet.com

Industrial Coaters List. For a complete list of industrial coating manufacturers with links to specific sites, see: http://www.IndustrialQuickSearch.com

Machine Design. For information on forming processes, see: http://www.machinedesign.com

Meehanite Metal Corp. For information on cast iron and related materials, see: http://www.meehanite.com

TechStudent.Com. For information on cast iron and casting processes, see: http://www.technologystudent.com (equipment and accessories)

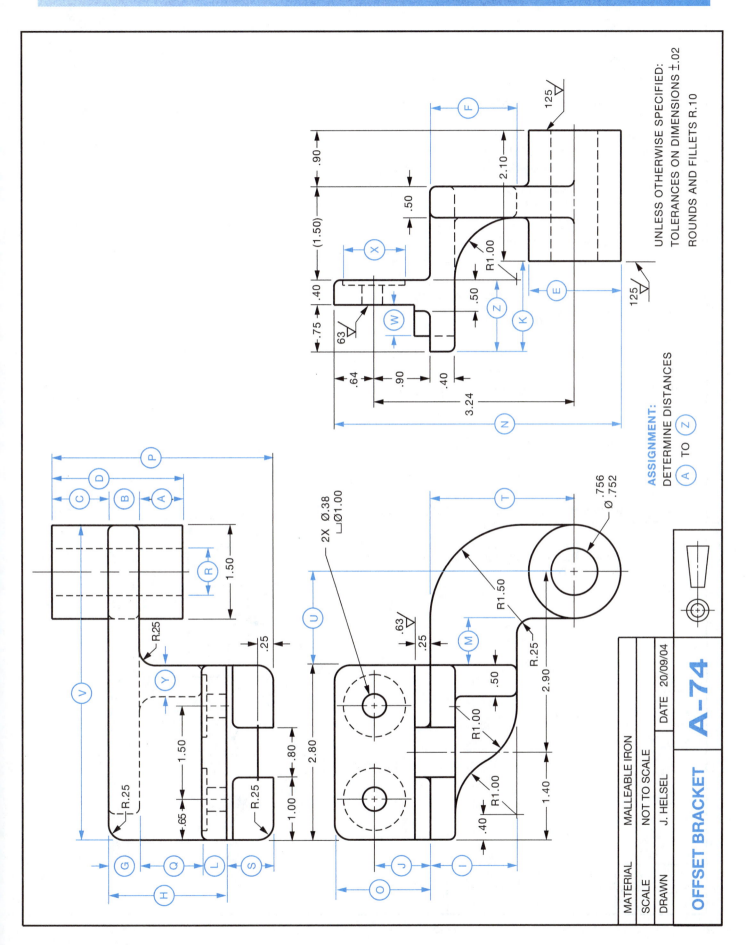

UNLESS OTHERWISE SPECIFIED:
TOLERANCES ON DIMENSIONS ±.02
ROUNDS AND FILLETS R.10

ASSIGNMENT:
DETERMINE DISTANCES
(A) TO (Z)

MATERIAL	MALLEABLE IRON	
SCALE	NOT TO SCALE	
DRAWN	J. HELSEL	DATE 20/09/04
OFFSET BRACKET	**A-74**	

QUESTIONS:

1. Which line in the top view represents surface (1)?

2. Locate surface (A) in the left-side view and the front view.

3. Locate surface (8) in the front view.

4. How many surfaces are to be finished?

5. Which line in the left-side view represents surface (3)?

6. What is the vertical center distance between holes (B) and (O) in the front view?

7. Determine distances at (4) (5) (6) and (11).

8. Locate surface (J) in the top view.

9. Which surface of the left-side view does line (14) represent?

10. What point in the front view is represented by line (15)?

11. What is the thickness of boss (E)?

12. Locate surface (G) in the left-side view.

13. Locate point (K) in the top view.

14. Locate surface (D) in the top view.

15. Determine distances at (M) (N) (S) and (T).

16. What point or line in the top view is represented by point (16)?

17. What is the maximum horizontal center distance between the holes in the front view?

18. What would be the (A) width, (B) height, and (C) depth of the part before machining?

19. Name the machining processes that could provide the surface quality required for this part.

20. What was the vertical distance from the center of the Ø.38 hole and the top of the trip box before the drawing revision was made?

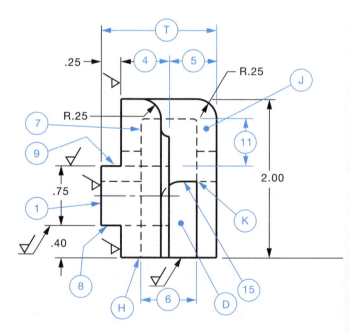

REVISIONS	1	24/10/04	C. JENSEN
		1.15 WAS 1.20	

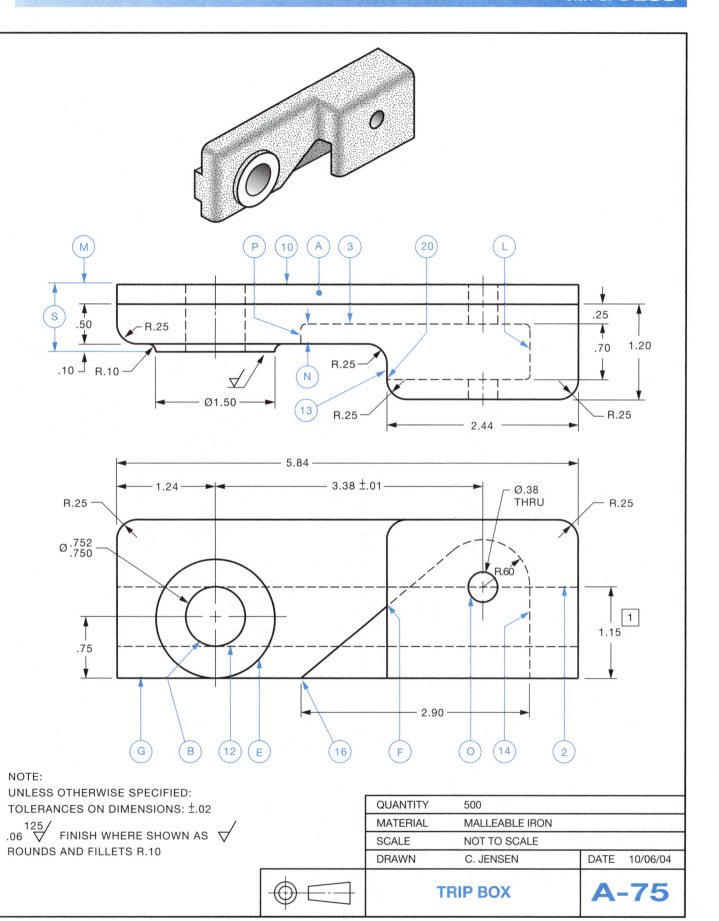

NOTE:
UNLESS OTHERWISE SPECIFIED:
TOLERANCES ON DIMENSIONS: ±.02

.06 $\overset{125}{\diagup}$ ⟍/ FINISH WHERE SHOWN AS ⟍/
ROUNDS AND FILLETS R.10

QUANTITY	500	
MATERIAL	MALLEABLE IRON	
SCALE	NOT TO SCALE	
DRAWN	C. JENSEN	DATE 10/06/04

TRIP BOX   **A-75**

QUESTIONS:

1. Of what material is the base made?
2. How many finished surfaces are indicated?
3. Which circled letters on the drawing indicate the spaces that were cored when the casting was made?
4. Locate the surface in the top view that is represented by line ⑥.
5. What is the height of the cored area Ⓧ?
6. What surface finish is used on the top pad?
7. What is the horizontal length of the pad ⑤?
8. What reason might be given for openings Ⓠ Ⓢ Ⓨ and Ⓩ?
9. Determine distance Ⓦ.
10. What radius fillet would be used at Ⓡ?
11. What is the total allowance added to the height for machining?
12. Determine distances Ⓒ through Ⓝ.
13. What is the size of the tap drill required for the four threaded holes?
14. What size bolts would be used to hold the auxiliary pump base in position?
15. What must be done to the casting before machining?

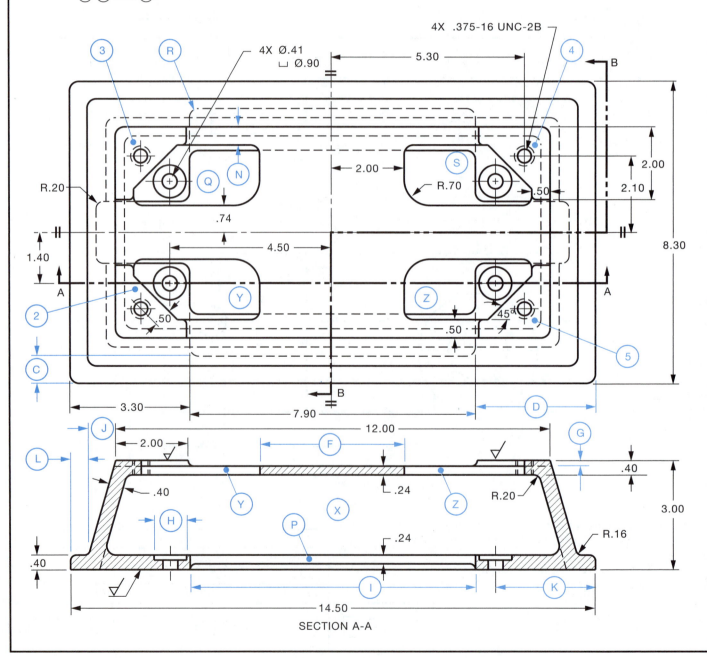

SECTION A-A

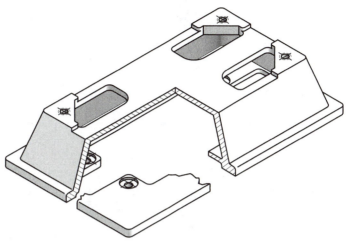

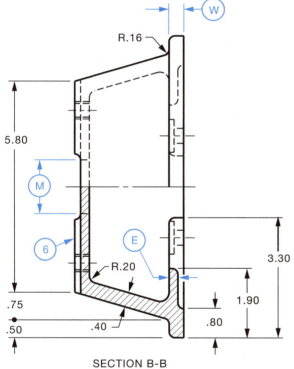

SECTION B-B

R.16

W

5.80

M

6

E

R.20

.75

.50

.40

.80

1.90

3.30

NOTES:

UNLESS OTHERWISE SPECIFIED:

— TOLERANCES ON DIMENSIONS ±.02

— ROUNDS AND FILLETS R.25

— FINISHES $\frac{125}{.06}$∇

— CASTING TO BE PAINTED ALUMINUM
  BEFORE MACHINING

THREAD CONTROLLING ORGANIZATION
AND STANDARD – ASME B1.1-2003

MATERIAL	GRAY IRON		
SCALE	NOT TO SCALE		
DRAWN	J. HELSEL	DATE	22/03/04

AUXILIARY PUMP
BASE

A-76

NOTE:

1. TOLERANCE ON DIMENSIONS ±0.5

2. ▽ TO BE ▽ 1.6

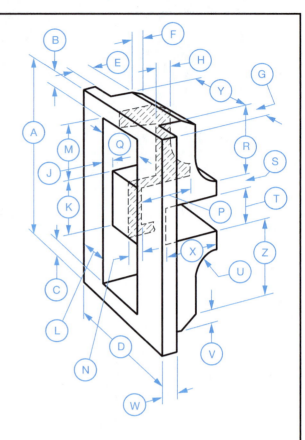

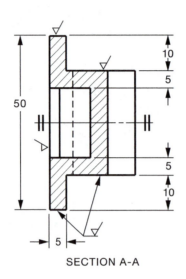

SECTION A-A

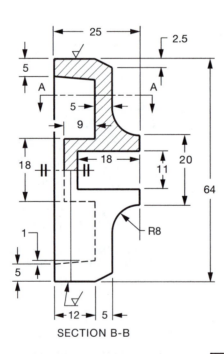

SECTION B-B

METRIC
DIMENSIONS ARE IN MILLIMETERS

ASSIGNMENT:

1. ON A CENTIMETER GRID (1mm SQUARES), SKETCH THE RIGHT-SIDE VIEW OF THE SLIDE VALVE AND SHOW THE POSITION OF THE CUTTING PLANE FOR SECTION B-B.

2. DETERMINE DISTANCES Ⓐ TO Ⓩ.

MATERIAL	GRAY IRON	
SCALE	NOT TO SCALE	
DRAWN	R. DREUCCI	DATE 15/11/04

SLIDE VALVE

A-77M

CHAIN DIMENSIONING

Most linear dimensions are intended to apply on a point-to-point basis. *Chain* dimensioning is applied directly from one feature to another, as shown in Figure 28–1(A). Such dimensions locate surfaces and features directly between the points indicated, or between corresponding points on the indicated surfaces.

For example, a diameter applies to all diameters of a cylindrical surface (not merely to the diameter at the end where the dimension is shown), a thickness applies to all opposing points on the surfaces, and a hole-locating dimension applies from the hole axis perpendicular to the edge of the part on the same center line.

BASE LINE (DATUM) DIMENSIONING

When several dimensions extend from a common data point or points on a line or surface, Figure 28–1(B), this is called *base line* or *datum* dimensioning. This form of dimensioning is preferred for parts to be manufactured by numerical control machines.

A *feature* is a physical portion of a part, that is, a surface, hole, tab, slot, etc.

A *datum* is a theoretically exact point, axis, or plane derived from the true geometric counterpart of a specified datum feature. A datum may be the center of a circle, the axis of a cylinder, or the axis of symmetry.

The location of a series of holes or step features as shown in Figure 28–1(B) is an example of the use of base line dimensioning. Figure 28–2 illustrates more complex uses of base line dimensioning. Without this type of dimensioning the distances between the first and last steps could vary considerably because of the buildup of tolerances permitted between the adjacent holes. Dimensioning from a datum line controls the tolerances between the holes to the basic general tolerance of ±.02 inch.

In this system, the tolerance from the common point to each of the features must be held to half the tolerance acceptable between individual features. For example, in Figure 28–1(C), if a tolerance between two individual holes of ±.02 inch were desired, each of the dimensions shown would have to be held to ±.01 inch.

REFERENCE

ASME Y14.5M-1994 (R1999) Dimensioning and Tolerancing

INTERNET RESOURCE

American Society of Mechanical Engineers. For information on chain and base line dimensioning, refer to ASME Y14,5M-1994 (R1999), see http://www.asme.org

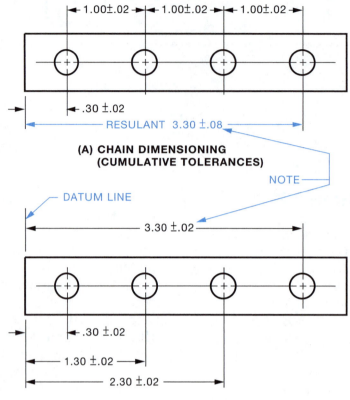

(A) CHAIN DIMENSIONING (CUMULATIVE TOLERANCES)

(B) BASE LINE DIMENSIONING (NON-CUMULATIVE TOLERANCES)

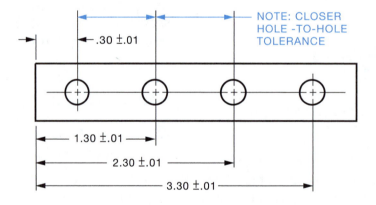

(C) MAINTAINING SAME DIMENSION BETWEEN HOLES AS (A) OR (B) BUT WITH CLOSER TOLERANCE

FIGURE 28–1 ■ *Comparison between chain dimensioning and base line dimensioning.*

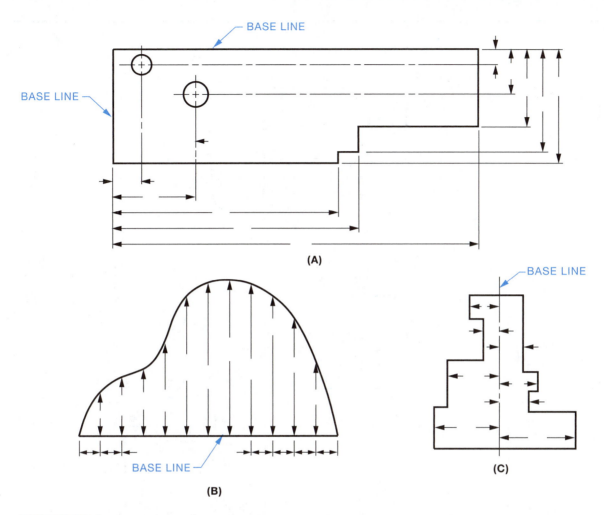

FIGURE 28–2 ■ *Applications of base line dimensioning.*

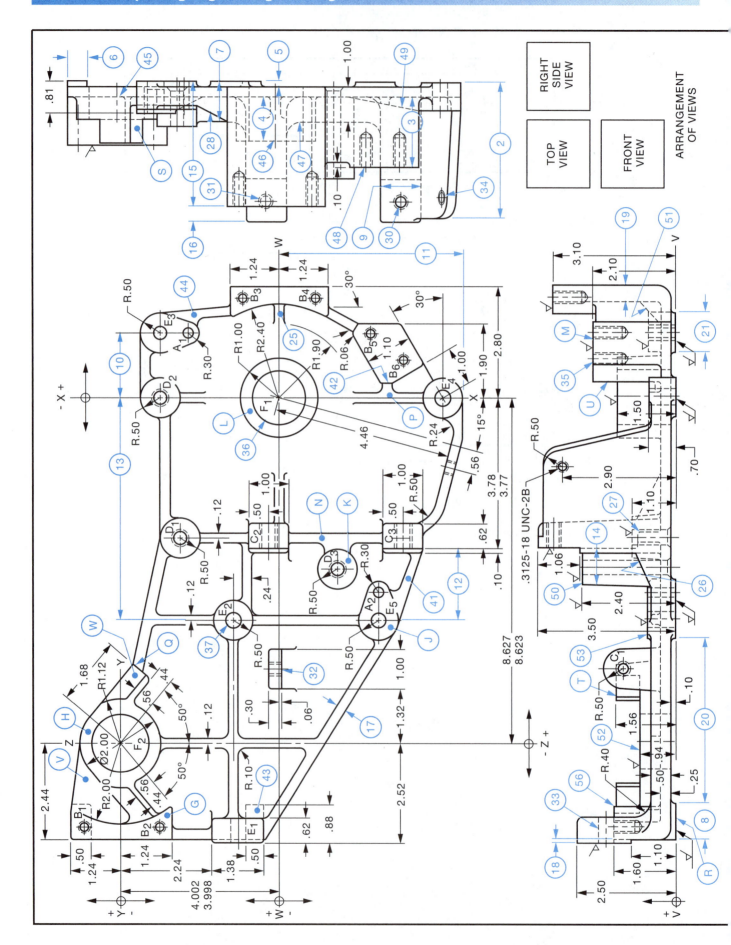

RIGHT SIDE VIEW

TOP VIEW

FRONT VIEW

ARRANGEMENT OF VIEWS

QUESTIONS:

NOTE: Where limit dimensions are given use larger limit.

1. What are the overall dimensions of the casting?

2. Show a roughness symbol meaning the same as the one given on the drawing.

3. Identify surfaces $\textcircled{G}$ to $\textcircled{U}$ on one of the other views.

4. Locate rib $\textcircled{25}$ on the front view.

5. Locate rib $\textcircled{26}$ on the top view.

6. Locate rib $\textcircled{27}$ on the right side view.

7. Locate rib $\textcircled{28}$ on the top view.

8. Give the sizes of the following holes: $\textcircled{30}$, $\textcircled{31}$, $\textcircled{32}$, $\textcircled{33}$, and $\textcircled{34}$.

9. How deep is hole $\textcircled{35}$?

10. What is the tolerance on hole $\textcircled{36}$?

11. How deep is hole $\textcircled{37}$?

12. Determine distances $\textcircled{2}$ to $\textcircled{21}$. (Where limit dimensions are shown, use maximum size.)

13. What is the (A) largest, (B) smallest, permissible hole on the part? Do not include tapped holes.

14. What is the smallest tap drill size used?

HOLE	DISTANCE FROM					SIZE
	V-V	W-W	X-X	Y-Y	Z-Z	
A_1		2.32	1.62			Ø .252
A_2		-2.50	-4.88			Ø .250
B_1				.94	-2.12	
B_2				-.94	-2.12	
B_3		.94	2.50			.312-
B_4		-.94	2.50			18 UNC-2B
B_5		-2.28	1.62			X.75
B_6		-3.08	.96			DEEP
C_1	1.38				-1.82	.375-
C_2	3.00	.28				16 UNC-2B
C_3	3.00	-3.12				
D_1		2.50	-3.50			Ø .3752
D_2		3.00	.00			Ø .3750
D_3		-1.48	-4.28			CSK .06X95°
E_1	1.80			-2.94		
E_2		1.18	-5.56			
E_3		3.00	1.62			Ø.406
E_4		-4.12	.00			
E_5		-2.50	-5.56			
F_1		.00	.00			Ø 1.3765
F_2				.00	.00	Ø 1.3745

HOLE SIZE AND LOCATION

NOTE:
— ALL FILLETS R.06 UNLESS OTHERWISE SHOWN
— ALL RIBS AND WEBS .24 THICK
— TOLERANCE ON DIMENSIONS ±.02
— TOLERANCE ON ANGLES ±0.5°
— SURFACES MARKED $\sqrt{}$ TO BE $\overset{250}{\sqrt{}}$

THREAD CONTROLLING ORGANIZATION
AND STANDARD-ASME B1.1-2003

MATERIAL	GRAY IRON	
SCALE	NOT TO SCALE	
DRAWN	J. HELSEL	

INTERLOCK BASE   A-78

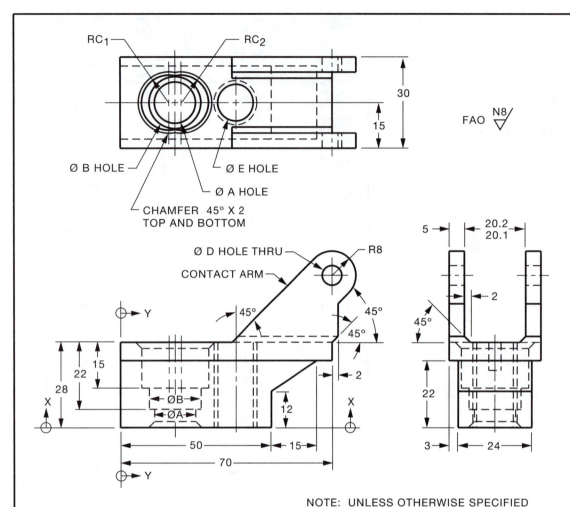

FAO N8

NOTE: UNLESS OTHERWISE SPECIFIED
- TOLERANCE ON DIMENSIONS ±0.5
- TOLERANCE ON ANGLES ±0.5°

QUESTIONS:

1. What is the overall width?
2. What is the overall height?
3. Give the chamfer angle for the C hole.
4. What is the distance from Y-Y to E hole?
5. What is the distance from X-X to D hole?
6. How many complete threads does the tapped hole have?
7. Which thread series is the tapped hole?
8. What is the surface finish in micrometers?
9. How deep is the C counterbore from the top of the surface?
10. How long is the B counterbored hole?
11. Give the distance between the C of C and C radii.
12. What is the nominal thickness of the contact arm at D hole?
13. What is the tolerance on the distance between the contact arms?
14. What is the tolerance on D hole?
15. What is the center distance between A and E holes?
16. What type of dimensioning is shown on the front view?
17. What type of dimensioning is used to locate the holes?

HOLE SYMBOL	HOLE SIZE	LOCATION	
		X-X	Y-Y
A	Ø13.5		18
B	Ø17		18
C$_1$	R9		16
C$_2$	R9		20
D	Ø6.5-6.6	50	70
E	M12x1.25-4G6G		38

METRIC
DIMENSIONS ARE IN MILLIMETERS

MATERIAL	MAGANESE BRONZE	
SCALE	NOT TO SCALE	
DRAWN	C. JENSEN	DATE 16/05/04

THREAD CONTROLLING ORGANIZATION
AND STANDARD-ASME B1.13M-2001

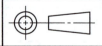

CONTACT ARM   **A-79M**

THREAD CONTROLLING
ORGANIZATION AND
STANDARD–ASME
B1.13M-2001

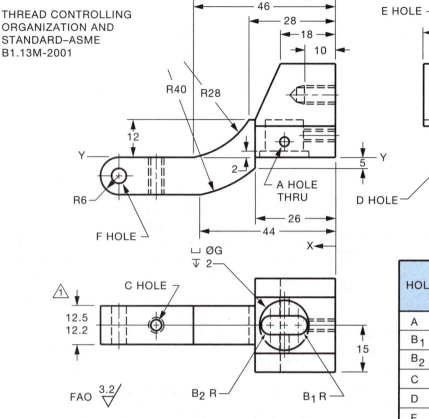

E HOLE

NOTE: UNLESS OTHERWISE SPECIFIED
TOLERANCE ON DIMENSIONS ±0.5

FAO 3.2

HOLE	HOLE SIZE	DISTANCE FROM	
		X-X	Y-Y
A	Ø3	16.5	5
B₁	R3	12	
B₂	R3	21	
C	M5x0.8-5G	58	
D	M4x0.7-5G		7
E	M6x1-5G		20
F	Ø4.78-4.80	70	6
G	Ø16	16.5	

QUESTIONS:

1. What is the overall width?

2. What is the overall height?

3. What is the distance from X-X to the center of hole F?

4. What is the distance from Y-Y to the center of hole E?

5. What is the horizontal distance between the center lines of A and F holes?

6. What is the horizontal distance between the center lines of C hole and B₁ radius?

7. How many full threads does E hole have? (See Appendix.)

8. How deep is the spotface?

9. What is the tolerance on F hole?

10. What type of projection is used on this drawing?

11. How far apart are the D and E holes?

12. What is the tolerance on the thickness of the material at F hole?

13. What is the length of the C hole?

14. What is the distance from line X-X to the termination of both ends of the 28 radius?

15. What does FAO mean?

16. What is the tap drill size required for the (A) C, (B) D, and (C) E threaded holes?

17. What is the length of the slot?

18. What type of dimensioning is shown on the three views?

19. What type of dimensioning is used to locate the holes?

METRIC
DIMENSIONS ARE IN MILLIMETERS

MATERIAL	ALUMINUM	
SCALE	NOT TO SCALE	
DRAWN	F. NEWMAN	DATE 15/02/04

REVISIONS	1	F. NEWMAN 08/06/04	CH A. HEINEN		CONTACTOR	A-80M
		DIMENSION WAS 12.4-12.7				

ALIGNMENT OF PARTS AND HOLES

Two important factors that must be considered when drawing an object are the number of views to be drawn and the time required to draw them. If possible, use time-saving devices, such as templates, for drawing standard features.

To simplify the representation of common features, a number of conventional drawing practices are used, Figure 29–1. Many conventions deviate from true projection for the purpose of clarity; others are used to save drafting time. These conventions must be executed carefully; clarity is even more important than speed.

Foreshortened Projection

When the true projection of ribs or arms results in confusing foreshortening, these parts should be rotated until parallel to the line of the section or projection. See Figure 29–1.

Holes Revolved to Show True Center Distance

Drilled flanges in elevation or section should show the holes at their true distance from center, rather than the true projection. See Figure 29–1.

PARTIAL VIEWS

Partial views, which show only a limited portion of the object with remote details omitted, should be used when necessary, to clarify specific details of the drawing, Figure 29–2. Such views are used to avoid the necessity of drawing many hidden features.

On drawings of objects where two side views can be used to better advantage than one, each need not be complete if together they depict the shape. Show only the hidden lines of features immediately behind the view, Figure 29–2(C).

Another type of partial view is shown in Figure 29–3. However, sufficient information is given in the partial view to complete the description of the object. The partial view is limited by a break line. Partial views are used because:

- They save time in drawing.

- They conserve space that might otherwise be needed for drawing the object.

- They sometimes permit the drawing to be made to scale large enough to bring out all details clearly, whereas if the whole view were drawn, lack of space might make it necessary to draw to a smaller scale, resulting in the loss of detail clarity.

- If the part is symmetrical, a partial view (referred to as a half view) may be drawn on one side of the center line as shown in Figure 29–4. In the case of the coil frame (Assignment A-84) a partial view was used so that the object could be drawn to a larger scale for clarity, thus saving time and space.

NAMING OF VIEWS FOR SPARK ADJUSTER

The drawing of the spark adjuster, Assignment A-81, illustrates several violations of true projection. The names of the views of the space adjuster could

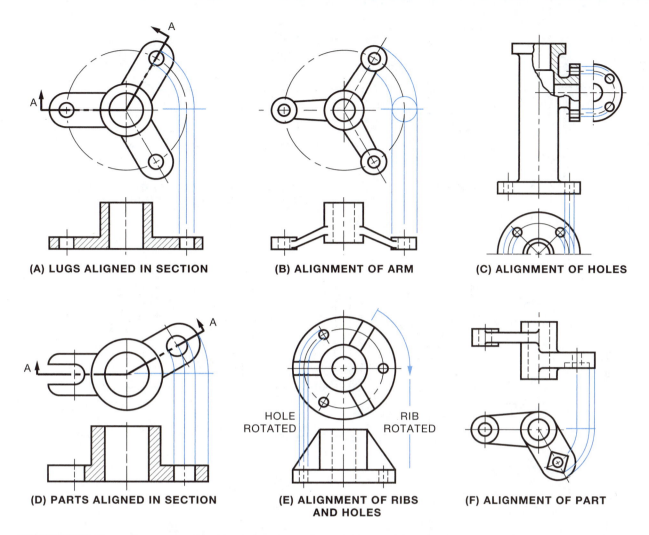

(A) LUGS ALIGNED IN SECTION

(B) ALIGNMENT OF ARM

(C) ALIGNMENT OF HOLES

(D) PARTS ALIGNED IN SECTION

(E) ALIGNMENT OF RIBS AND HOLES

HOLE ROTATED

RIB ROTATED

(F) ALIGNMENT OF PART

FIGURE 29–1 ■ *Alignment of holes and parts to show their true relationship.*

be questionable. The importance lies not in the names, however, but in the relationship of the views to each other. This means that the right view must be on the right side of the front view, the left view must be on the left side of the front view, etc. Any combination in Figure 29–5 may be used for naming the views of the spark adjuster.

DRILL SIZES

Twist drills are the most common tools used in drilling. They are made in many sizes. Inch size twist drills are grouped according to decimal inch sizes; by

number sizes, from 1 to 80, which correspond to the Stubbs steel wire gauge; by letter sizes A to Z; and by fractional sizes from 1/64th up. Twist drill sizes are listed in Table 3 of the Appendix.

Metric twist drill sizes are in millimeters and are classified as *preferred* and *available*. These sizes will eventually replace the fractional-inch, letter, and number size drills that are presently in existence. Metric twist drill sizes are listed in Table 4 of the Appendix.

REFERENCE

ASME Y14.3M-1994 (R1999) Multi and Sectional View Drawings

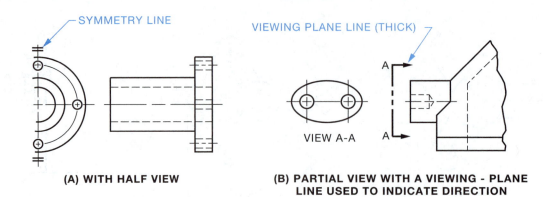

SYMMETRY LINE

VIEWING PLANE LINE (THICK)

VIEW A-A

(A) WITH HALF VIEW

(B) PARTIAL VIEW WITH A VIEWING - PLANE LINE USED TO INDICATE DIRECTION

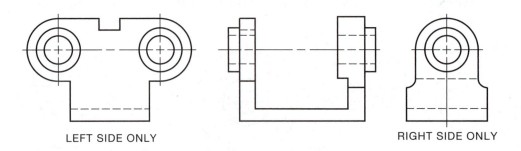

LEFT SIDE ONLY

RIGHT SIDE ONLY

(C) PARTIAL SIDE VIEWS

FIGURE 29–2 ■ *Partial views.*

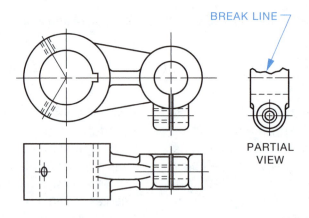

BREAK LINE

PARTIAL VIEW

FIGURE 29–3 ■ *Partial side view.*

INTERNET RESOURCES

Amazon New York, Industrial Press. For information on twist drill sizes, see: http://www.sizes.com/tools/twistdrills/htm

American Society of Mechanical Engineers. For information on multiview drawings, refer to ASME Y14.3M-1994 (R1999) (*Multi- and Sectional-View Drawings*) at: http://www.asme.org

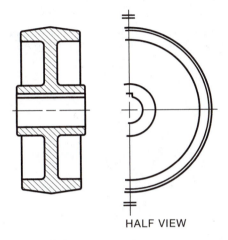

HALF VIEW

FIGURE 29–4 ■ *Half view.*

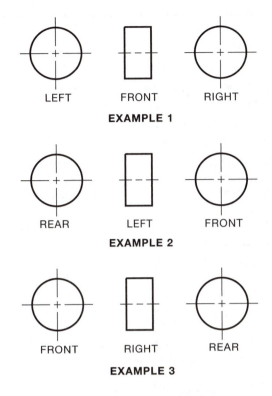

LEFT   FRONT   RIGHT

EXAMPLE 1

REAR   LEFT   FRONT

EXAMPLE 2

FRONT   RIGHT   REAR

EXAMPLE 3

FIGURE 29–5 ■ *Naming of views for spark adjuster, Assignment A-81.*

QUESTIONS:

1. What is radius (A)?
2. What surface is line (2) in the rear view?
3. Locate surface (B) in section A-A.
4. Locate surface (C) in section A-A.
5. Locate line (D) in the rear view.
6. Locate a surface or line in the front view that represents line (F).
7. Which point or line in the front view does line (G) represent?
8. What size cap screw would be used in (N) hole?
9. What is the diameter of (K) hole?
10. Locate the point or line in the rear view from which line (1) is projected.
11. Locate surface (H) in the rear view.
12. Locate (Z) in section A-A.
13. How thick is lug (7)?
14. What are the diameters of holes (L), (M), and (N)?
15. Determine angles (O), and (P).
16. Determine distances (Q) through (U).
17. What standard size drill could be used to produce the Ø.386 hole? Refer to Table 3 in the Appendix.

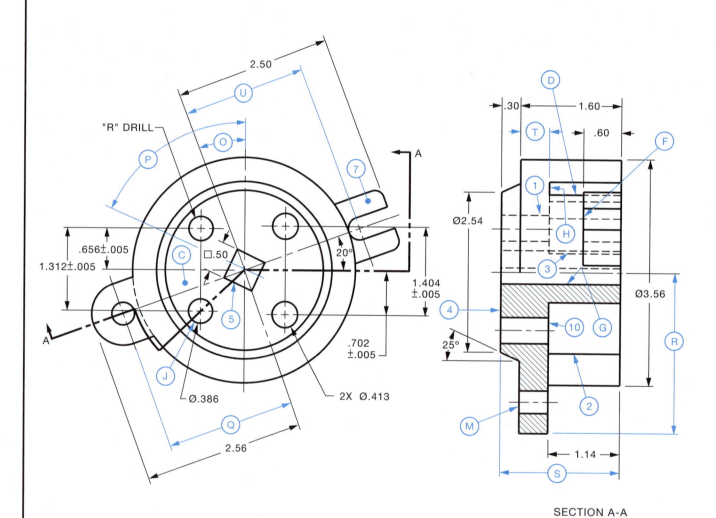

FRONT VIEW

SECTION A-A

RIGHT - SIDE VIEW

NOTE: UNLESS OTHERWISE STATED
TOLERANCE ON TWO-PLACE DIMENSIONS ±.02
TOLERANCE ON THREE-PLACE DIMENSION ±.010
TOLERANCE ON ANGLES ±0.5°

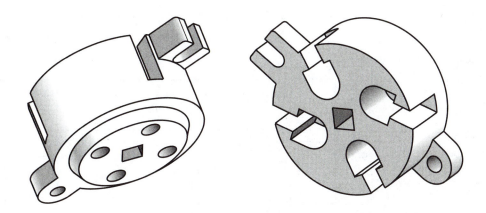

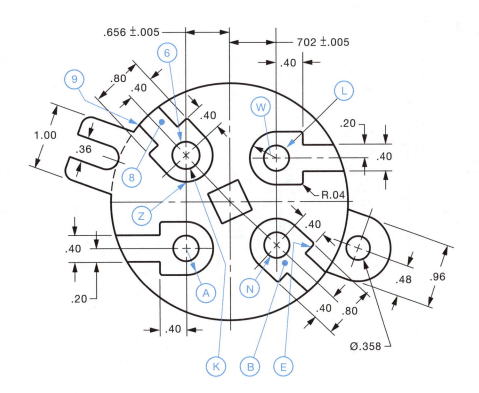

REAR VIEW

MATERIAL	BAKELITE	
SCALE	1:1	
DRAWN	J. HELSEL	DATE 16/10/04

SPARK ADJUSTER   **A-81**

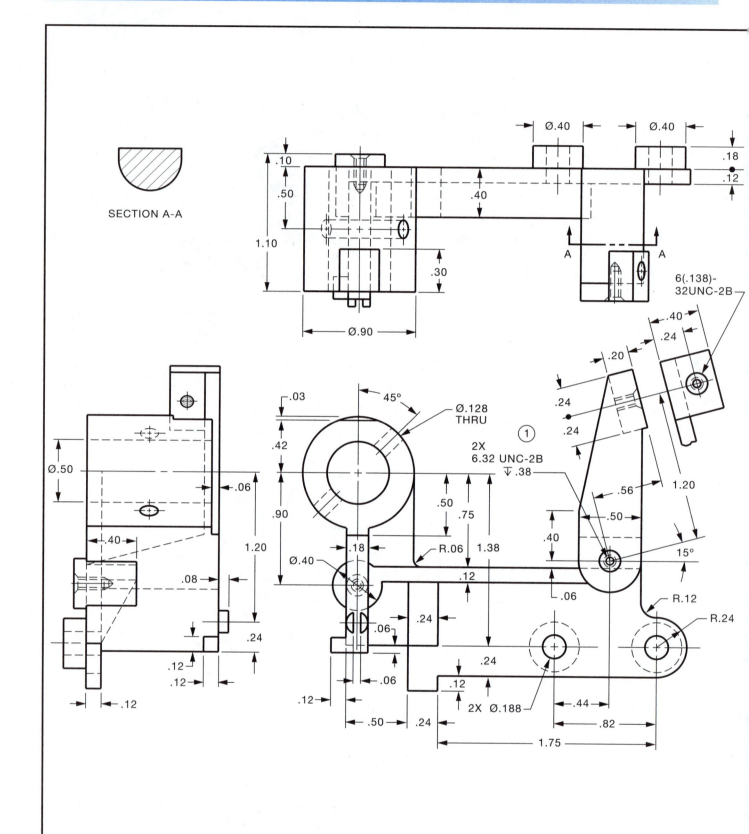

SECTION A-A

6(.138)-
32UNC-2B

①

Ø.128
THRU

2X
6.32 UNC-2B
∇ .38

REVISIONS	1	10/12/04	R.H.
	HOLE WAS .32 DEEP		

ASSIGNMENT: DETERMINE DISTANCES OR DIMENSIONS A TO Z (NO LETTERS I OR O) AND DIMENSIONS 2 TO 47.

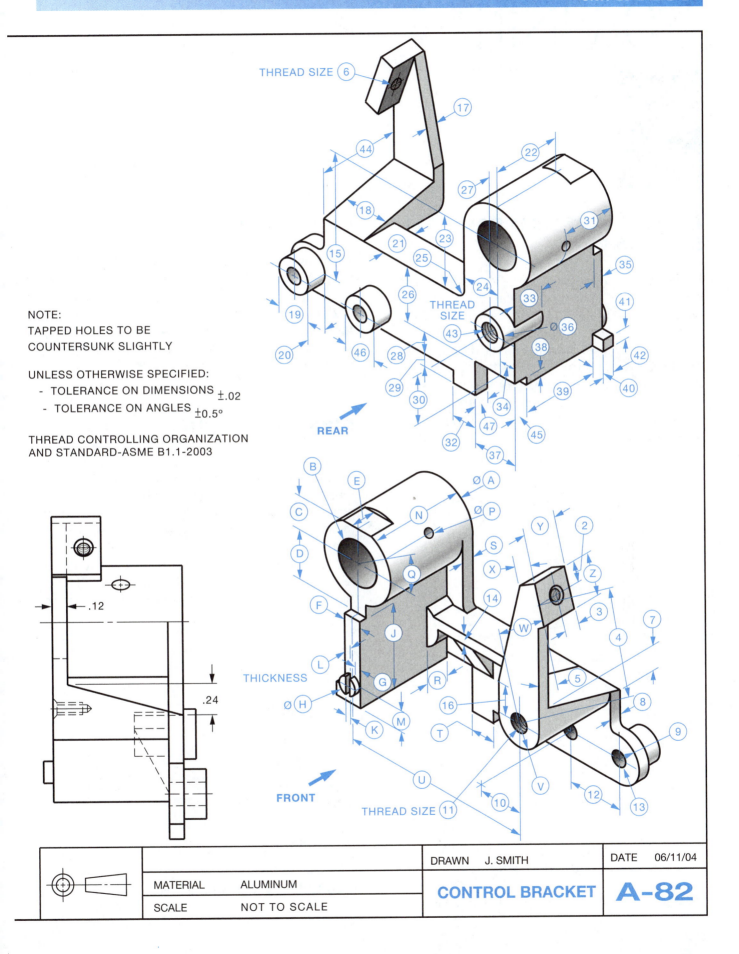

THREAD SIZE ⑥

⑰ ㉒ ㉗ ㉛ ㉟ ㊶ ㊴ ㊵ ㊵

NOTE:
TAPPED HOLES TO BE
COUNTERSUNK SLIGHTLY

UNLESS OTHERWISE SPECIFIED:
- TOLERANCE ON DIMENSIONS ±.02
- TOLERANCE ON ANGLES ±0.5°

THREAD CONTROLLING ORGANIZATION
AND STANDARD-ASME B1.1-2003

THREAD
SIZE

REAR

THICKNESS

.12

.24

FRONT

THREAD SIZE ⑪

		DRAWN	J. SMITH		DATE	06/11/04
MATERIAL	ALUMINUM		**CONTROL BRACKET**		**A-82**	
SCALE	NOT TO SCALE					

BROKEN-OUT AND PARTIAL SECTIONS

Broken-out and partial sections, are used to show certain internal and external features of an object without drawing another view. See Figure 30–1. A break or cutting-plane line is used to indicate where the section is taken. On the raised block, Assignment A-83M, two broken-out sections are shown in the front view, and one broken-out section is shown in the left-side view. Although this method of showing a partial section is not commonly used, it is an accepted practice. See Figure 30–1.

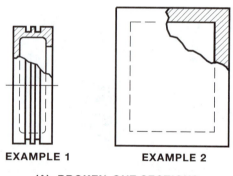

EXAMPLE 1   EXAMPLE 2

(A) BROKEN-OUT SECTIONS

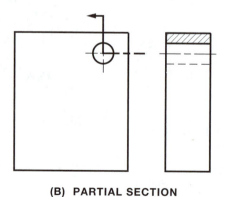

(B) PARTIAL SECTION

FIGURE 30–1 ■ *Broken-out and partial sections.*

WEBS IN SECTION

The conventional (preferred) methods of representing a section of a part having webs or partitions are shown in Figure 30–2. These methods are preferred to drawing the section in true projection. Although the conventional methods are a violation of true projection, they are preferred over true projection for clarity and ease in drawing.

RIBS IN SECTION

A true projection section view, Figure 30–3(A), would be misleading when the cutting plane passes longitudinally through the center of a rib. To avoid this impression of solidity, a preferred section not showing the ribs section-lined or crosshatched is used. When there is an odd number of ribs, Figure 30–3(B), the top rib is aligned with the bottom rib to show its true relationship with the hub and flange. If the rib is not aligned or revolved, it appears distorted on the section view and is misleading.

An alternate method of identifying ribs in a section view is shown in Figure 30–3(C). If rib A of the

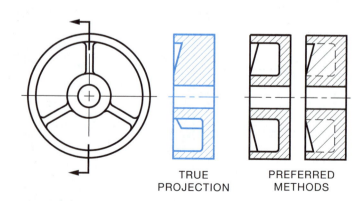

TRUE PROJECTION   PREFERRED METHODS

FIGURE 30–2 ■ *Conventional methods of sectioning webs.*

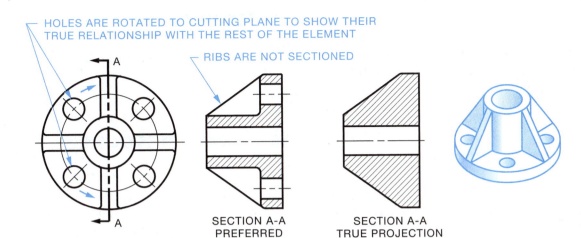

HOLES ARE ROTATED TO CUTTING PLANE TO SHOW THEIR
TRUE RELATIONSHIP WITH THE REST OF THE ELEMENT

RIBS ARE NOT SECTIONED

SECTION A-A
PREFERRED

SECTION A-A
TRUE PROJECTION

(A) CUTTING PLANE PASSING THROUGH TWO RIBS

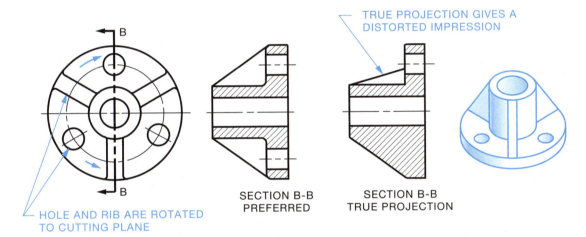

TRUE PROJECTION GIVES A
DISTORTED IMPRESSION

HOLE AND RIB ARE ROTATED
TO CUTTING PLANE

SECTION B-B
PREFERRED

SECTION B-B
TRUE PROJECTION

(B) CUTTING PLANE PASSING THROUGH ONE RIB AND ONE HOLE

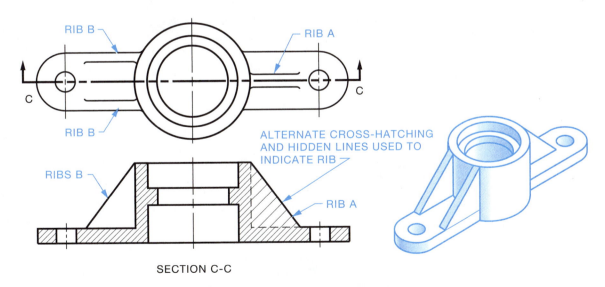

RIB B

RIB A

RIB B

ALTERNATE CROSS-HATCHING
AND HIDDEN LINES USED TO
INDICATE RIB

RIBS B

RIB A

SECTION C-C

(C) ALTERNATE METHOD OF SHOWING RIBS IN SECTION

FIGURE 30–3 ■ *Ribs in section.*

base were not sectioned as previously mentioned, it would appear exactly like B in the section view and would be misleading. To distinguish between the ribs on the base, alternate section lining on the ribs is used. The line between the rib and solid portions is shown as a broken line.

SPOKES IN SECTION

Spokes in section are represented in the same manner as ribs. Figure 30–4 shows the preferred method of representing spokes in section for aligned and unaligned designs. Note that the spokes are not sectioned in either case.

REFERENCE

ASME Y14.3-1994 (R1999) Multi- and Sectional-View Drawings

INTERNET RESOURCE

American Society of Mechanical Engineers. For information on multiview drawings, refer to ASME Y14.3M-1994 (R1999) (*Multi- and Sectional-View Drawings*) at: http://www.asme.org

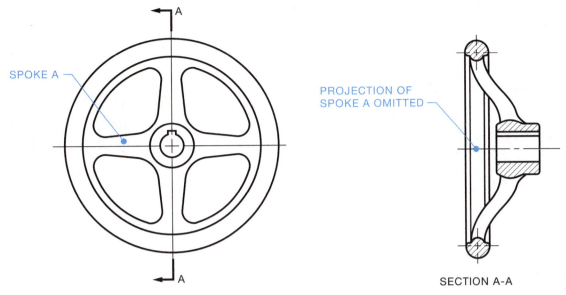

SPOKE A

PROJECTION OF
SPOKE A OMITTED

SECTION A-A

(A) CUTTING PLANE PASSING THROUGH TWO SPOKES

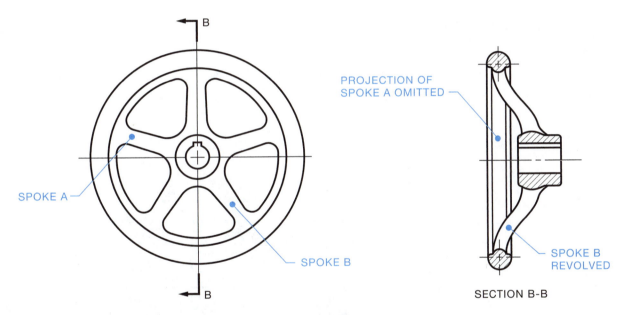

PROJECTION OF
SPOKE A OMITTED

SPOKE A

SPOKE B

SPOKE B
REVOLVED

SECTION B-B

(B) CUTTING PLANE PASSING THROUGH ONE SPOKE

FIGURE 30–4 ■ *Spokes in section.*

QUESTIONS:

1. What is the diameter of the largest unthreaded hole?

2. What size is the smallest threaded hole?

3. What size are the smallest unthreaded holes?

4. Which surface does ⑦ represent in the top view?

5. Which surface does ① represent in the top view?

6. Which line or surface does ⑥ represent in the left view?

7. Which line or surface does Ⓥ represent in the front view?

8. What was the original width of the part?

9. By which line or surface is Ⓗ represented in the left view?

10. Locate in the left view the line or surface that is represented by line Ⓖ.

11. Which line or surface represents Ⓕ in the top view?

12. Which line represents surface Ⓙ in the left view?

13. Determine the overall depth of the raise block.

14. Determine distances Ⓐ through Ⓔ.

15. Determine distances ⑧ through ⑲.

16. What is the tap drill size required for the (A) M10x1.5, (B) M16x2 threaded holes?

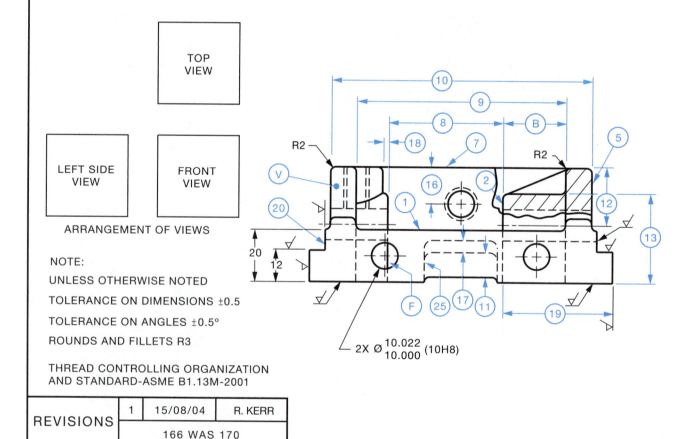

TOP VIEW

LEFT SIDE VIEW

FRONT VIEW

ARRANGEMENT OF VIEWS

NOTE:

UNLESS OTHERWISE NOTED

TOLERANCE ON DIMENSIONS ±0.5

TOLERANCE ON ANGLES ±0.5°

ROUNDS AND FILLETS R3

THREAD CONTROLLING ORGANIZATION
AND STANDARD-ASME B1.13M-2001

2X Ø 10.022 / 10.000 (10H8)

REVISIONS	1	15/08/04	R. KERR
		166 WAS 170	

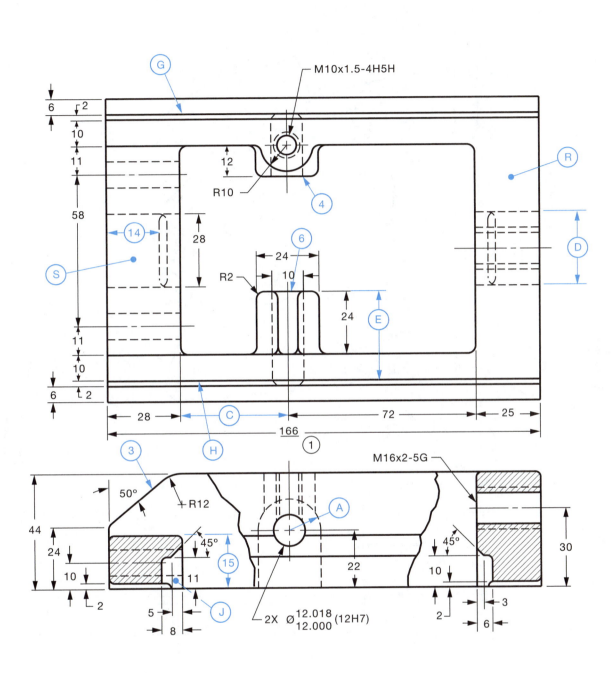

M10x1.5-4H5H

R10

M16x2-5G

2X Ø 12.018 / 12.000 (12H7)

MATERIAL	GRAY IRON		
SCALE	1:1		
DRAWN	R. KERR	DATE	23/10/04

METRIC
DIMENSIONS ARE IN MILLIMETERS

RAISE BLOCK

A-83M

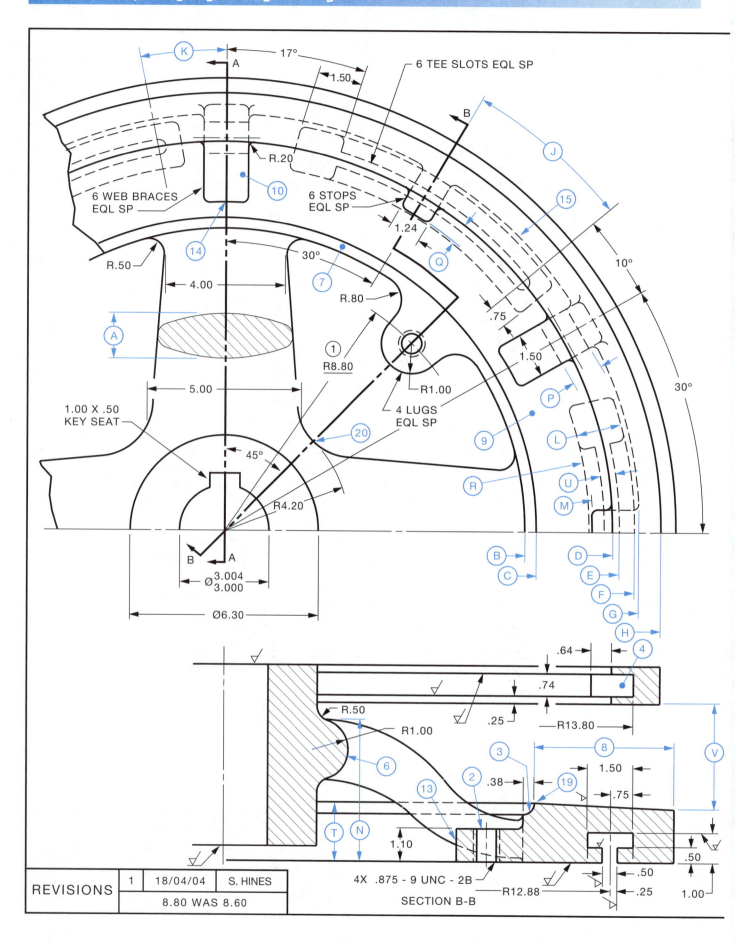

17°

Ⓚ

1.50

6 TEE SLOTS EQL SP

B

Ⓙ

R.20

Ⓙ⑤

10

6 WEB BRACES
EQL SP

6 STOPS
EQL SP

1.24

Ⓠ

⑭

R.50

30°

⑦

10°

4.00

Ⓐ

R.80

.75

Ⓐ

① R8.80

R1.00

1.50

30°

5.00

4 LUGS
EQL SP

Ⓟ

1.00 X .50
KEY SEAT

⑳

Ⓛ

⑨

45°

Ⓡ

Ⓤ

R4.20

Ⓜ

B   A

Ⓑ   Ⓓ

Ø 3.004
3.000

Ⓒ   Ⓔ

Ⓕ

Ø6.30

Ⓖ

Ⓗ

.64   ④

.74

R.50

.25   R13.80

R1.00

③   8

⑥

2   .38   ⑲

13

1.50

.75

1.10

Ⓣ Ⓝ

4X .875 - 9 UNC - 2B

.50

.50

Ⓥ

R12.88

.25

1.00

SECTION B-B

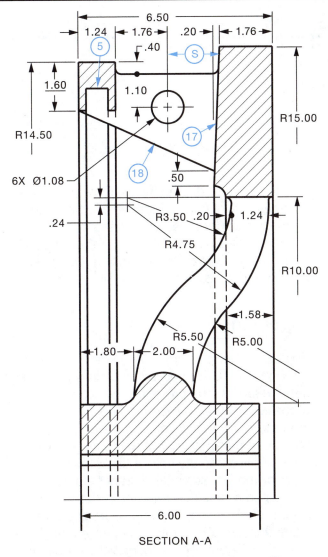

SECTION A-A

EXCEPT WHERE OTHERWISE SPECIFIED:

- FEATURES SYMMETRICAL AROUND
  CENTER POINT
- TOLERANCE ON DIMENSIONS ±.02
- TOLERANCE ON ANGLES ±0.5°

 TO BE ▽125

THREAD CONTROLLING ORGANIZATION
AND STANDARD-ASME B1.1-2003

TOP
VIEW

SECTION A-A

SECTION B-B

ARRANGEMENT OF VIEWS

QUESTIONS:

1. What surface texture is required on the machined surfaces?
2. What is thickness Ⓐ, assuming the revolved section was taken at the middle of the arm?
3. What is the total quantity of feature ④?
4. Locate surface ⑤ in the top view.
5. Locate point ⑥ in the top view.
6. How far is line ⑭ from the center point?
7. How far is surface ⑬ from the center point?
8. Which point on section B-B represents radius Ⓒ?
9. Which line on section B-B represents surface ⑦?
10. What is the radius of surface ③?
11. Determine distance ⑧.
12. Which line on section A-A represents surface ⑩?
13. Which surface in the top view represents line ⑰?
14. What is the angle at Ⓙ?
15. What is the angle at Ⓚ?
16. What is the thickness of the web brace?
17. What would be the length of key used between the shaft and coil frame?
18. What is the thickness of the lugs?
19. What is the distance from the Ø1.00 hole to the center of the coil frame?
20. Determine distances Ⓛ, Ⓝ, Ⓟ, Ⓠ, Ⓢ, Ⓣ, Ⓤ, and Ⓥ.
21. Determine radii Ⓑ, Ⓒ, Ⓓ, Ⓔ, Ⓕ, Ⓖ, Ⓗ, Ⓜ, and Ⓡ.
22. What type of sectional view is used between the shaft and coil frame?
23. What type of section view is section B-B?

MATERIAL	GRAY IRON	
SCALE		
DRAWN	S. HINES	DATE 22/01/04

COIL FRAME **A-84**

31 UNIT

PIN FASTENERS

Pin fasteners offer an inexpensive and effective approach to assembly where loading is primarily in shear. They can be divided into two groups: semipermanent and quick release.

Semipermanent pin fasteners require application of pressure or the aid of tools for installation or removal. Representative types include machine pins (dowel, straight, taper, clevis, and cotter pins) and radial locking pins (grooved surface and spring).

Quick release fasteners are more elaborate self-contained pins that are used for rapid manual assembly or disassembly. They use a form of spring loaded mechanism to provide a locking action in assembly.

Machine Pins

Five types of machine pins are commonly used: ground dowel pins; commercial straight pins; taper pins; clevis pins; and standard cotter pins, Figure 31–1.

Dowel Pins

Dowel pins or small straight pins have many uses. They are used to hold parts in alignment and to guide parts into desired positions. Dowel pins are most commonly used for the alignment of parts that are fastened with screws or bolts and must be accurately assembled.

When two pieces are to be assembled, as in the case of the part in Figure 31–2, one method of alignment is to clamp the two pieces in the desired location, drill and ream the dowel holes, insert the dowel pins, and then drill and tap for the screw holes.

Drill jigs are frequently used when the interchangeability of parts is required or when the nature of the piece does not permit the transfer of the doweled holes from one piece to the other. A drill jig, Figure 31–3, was used in drilling the dowel holes for the spider, Assignment A-85M.

Taper Pins

Holes for taper pins are usually sized by reaming. A through hole is formed by step drills and straight fluted reamers. The present trend is toward the use of helically fluted taper reamers, which provide more accurate sizing and require only a pilot hole the size of the small end of the taper pin. The pin is usually driven into the hole until it is fully seated. The taper of the pin aids hole alignment in assembly.

A tapered hole in a hub and a shaft is shown in Figure 31–4. If the hub and shaft are drilled and reamed separately, a misalignment might occur as shown in Figure 31–5. To prevent misalignment, the hub and shaft should be drilled at the same time as the parts are assembled. Each of the detailed parts should carry a note similar to the following: DRILL AND REAM FOR NO. 1 TAPER PIN AT ASSEMBLY.

Cotter Pins

The cotter pin is a standard machine pin commonly used as a fastener in the assembly of machine parts where great accuracy is not required, Figure 31–6. There is no standard way to represent cotter pins in assembly drawings. The method of representation shown in Figure 31–7 is, however, commonly used to indicate cotter pins.

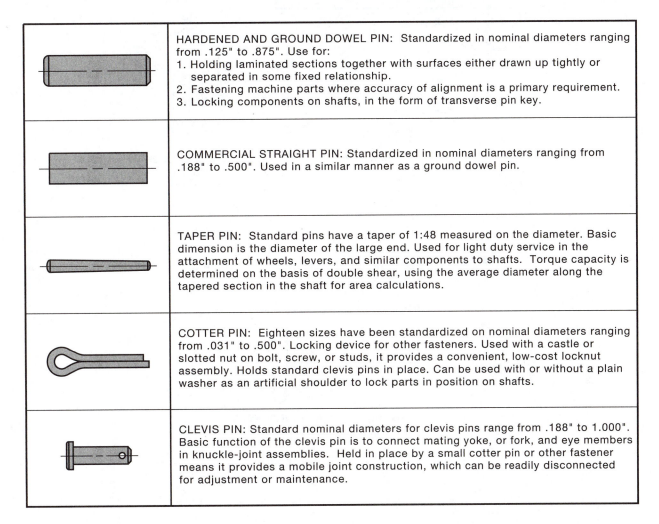

	HARDENED AND GROUND DOWEL PIN: Standardized in nominal diameters ranging from .125" to .875". Use for: 1. Holding laminated sections together with surfaces either drawn up tightly or separated in some fixed relationship. 2. Fastening machine parts where accuracy of alignment is a primary requirement. 3. Locking components on shafts, in the form of transverse pin key.
	COMMERCIAL STRAIGHT PIN: Standardized in nominal diameters ranging from .188" to .500". Used in a similar manner as a ground dowel pin.
	TAPER PIN: Standard pins have a taper of 1:48 measured on the diameter. Basic dimension is the diameter of the large end. Used for light duty service in the attachment of wheels, levers, and similar components to shafts. Torque capacity is determined on the basis of double shear, using the average diameter along the tapered section in the shaft for area calculations.
	COTTER PIN: Eighteen sizes have been standardized on nominal diameters ranging from .031" to .500". Locking device for other fasteners. Used with a castle or slotted nut on bolt, screw, or studs, it provides a convenient, low-cost locknut assembly. Holds standard clevis pins in place. Can be used with or without a plain washer as an artificial shoulder to lock parts in position on shafts.
	CLEVIS PIN: Standard nominal diameters for clevis pins range from .188" to 1.000". Basic function of the clevis pin is to connect mating yoke, or fork, and eye members in knuckle-joint assemblies. Held in place by a small cotter pin or other fastener means it provides a mobile joint construction, which can be readily disconnected for adjustment or maintenance.

FIGURE 31–1 ■ *Machine pins.*

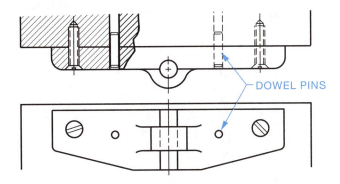

FIGURE 31–2 ■ *Aligning parts with dowel pins.*

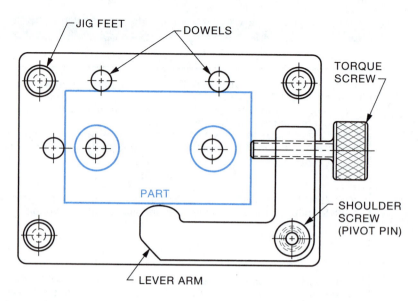

FIGURE 31–3 ■ *Dowel pins used to align part during drilling.*

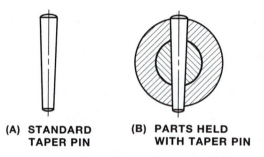

(A) STANDARD TAPER PIN

(B) PARTS HELD WITH TAPER PIN

FIGURE 31–4 ■ *Taper pin application.*

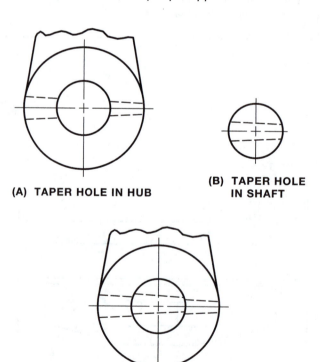

(A) TAPER HOLE IN HUB

(B) TAPER HOLE IN SHAFT

(C) MISALIGNMENT OF TAPER HOLES

FIGURE 31–5 ■ *Possibility of hole misalignment if holes are not drilled at assembly.*

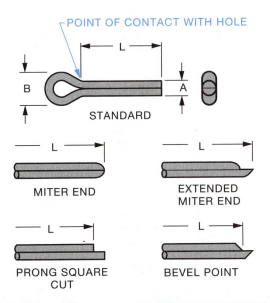

POINT OF CONTACT WITH HOLE

STANDARD

MITER END

EXTENDED MITER END

PRONG SQUARE CUT

BEVEL POINT

NOMINAL THREAD SIZE		NOMINAL COTTER PIN SIZE		COTTER PIN HOLE		END CLEARANCE*	
in.	(mm)	in.	(mm)	in.	(mm)	in.	(mm)
.250	(6)	.062	(1.5)	.078	(1.9)	.12	(3)
.312	(8)	.078	(2)	.094	(2.4)	.12	(3)
.375	(10)	.094	(2.5)	.109	(2.8)	.14	(4)
.500	(12)	.125	(3)	.141	(3.4)	.18	(5)
.625	(14)	.156	(3)	.172	(3.4)	.25	(5)
.750	(20)	.156	(4)	.172	(4.5)	.25	(7)
1.000	(24)	.188	(5)	.203	(5.6)	.31	(8)
1.125	(27)	.188	(5)	.203	(5.6)	.39	(8)
1.250	(30)	.219	(6)	.234	(6.3)	.44	(10)
1.375	(36)	.219	(6)	.234	(6.3)	.44	(11)
1.500	(42)	.250	(6)	.266	(6.3)	.50	(12)
1.750	(48)	.312	(8)	.312	(8.5)	.55	(14)

*DISTANCE FROM EXTREME POINT OF BOLT OR SCREW TO CENTER OF COTTER PIN HOLE.

FIGURE 31–6 ■ *Cotter pin data.*

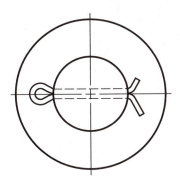

FIGURE 31–7 ■ *Cotter pin in an assembly drawing.*

Radial-Locking Pins

Low cost, ease of assembly, and high resistance to vibration and impact loads are common attributes of this group of commercial pin devices designed primarily for semipermanent fastening service. Two basic pin forms are used: solid with grooved surfaces, and hollow spring pins, which may be either slotted or spiral wrapped, Figure 31–8. In assembly, radial forces produced by elastic action at the pin surface develop a secure, frictional-locking grip against the hole wall. These pins are reusable and can be removed and reassembled many times without appreciable loss of fastening effectiveness. Live spring action at the pin surface also prevents loosening under shock and vibration loads. The need for accurate sizing of holes is reduced because the pins accommodate variations.

Solid Pins with Grooved Surfaces

The locking action of groove pins is provided by parallel, longitudinal grooves uniformly spaced around the pin surface. Rolled or pressed into solid pin stock, the grooves expand the effective diameter of the pin. When the pin is driven into a drilled hole corresponding in size to the nominal pin diameter, elastic deformation of the raised groove edges produces a secure interference fit with the hole wall. Figure 31–8 shows the six standardized constructions of grooved pins.

Hollow Spring Pins

Spiral-wrapped and slotted-tubular pin forms are made to control diameters greater than the holes into which they are pressed. Compressed when driven into the hole, the pins exert spring pressure against the hole wall along their entire engaged length to develop a strong locking action.

SECTION THROUGH SHAFTS, PINS, AND KEYS

Shafts, bolts, nuts, rods, rivets, keys, pins, and similar solid parts, the axes of which lie on the cutting plane, are sectioned only when a broken-out section of the shaft is used to clearly indicate the key, keyway, keyseat, and pin, as shown in Figure 31–9.

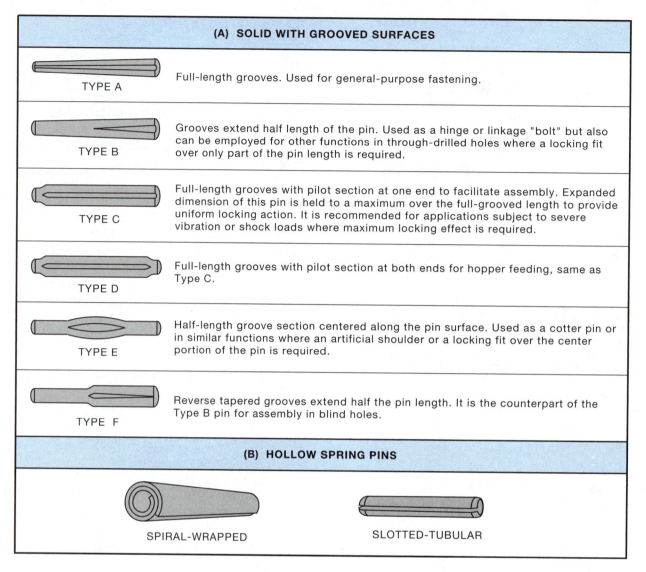

(A) SOLID WITH GROOVED SURFACES

TYPE A — Full-length grooves. Used for general-purpose fastening.

TYPE B — Grooves extend half length of the pin. Used as a hinge or linkage "bolt" but also can be employed for other functions in through-drilled holes where a locking fit over only part of the pin length is required.

TYPE C — Full-length grooves with pilot section at one end to facilitate assembly. Expanded dimension of this pin is held to a maximum over the full-grooved length to provide uniform locking action. It is recommended for applications subject to severe vibration or shock loads where maximum locking effect is required.

TYPE D — Full-length grooves with pilot section at both ends for hopper feeding, same as Type C.

TYPE E — Half-length groove section centered along the pin surface. Used as a cotter pin or in similar functions where an artificial shoulder or a locking fit over the center portion of the pin is required.

TYPE F — Reverse tapered grooves extend half the pin length. It is the counterpart of the Type B pin for assembly in blind holes.

(B) HOLLOW SPRING PINS

SPIRAL-WRAPPED   SLOTTED-TUBULAR

FIGURE 31–8 ■ *Radial locking pins.*

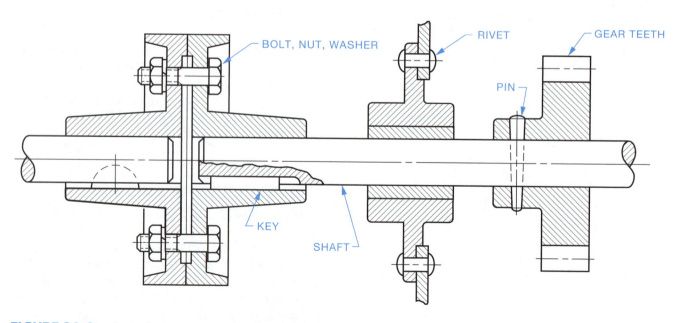

FIGURE 31–9 ■ *Parts that are not section lined in section drawings.*

ARRANGEMENT OF VIEWS OF DRAWING A-85M

Parts that are to be fitted over shafts as a single unit are sometimes made in two or more pieces. This is done for ease in assembly and replacement on the main structure of a machine rather than for ease in manufacture. Drawing A-85M shows two parts that are bolted and doweled together to form one unit.

The arrangement of views of the spider is illustrated by the diagrams in Figure 31–10. By comparing this figure with the drawing of the spider, note that the two halves together represent the top view.

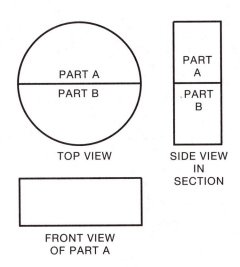

FIGURE 31–10 ■ *Arrangement of views for spider, Assignment A-85M.*

The right view is a full section of each half (**A** and **B**). The front view is a drawing of the front of part **A** only.

Although the front and side views are incomplete, the manner in which they are drawn and the arrangement of the views satisfies the demand for clearness and economy of time and space.

REFERENCES

Machine Design-Fasteners Reference Issue, Nov. 1981.

ASME B18.8.2-2000 Taper Pins, Dowel Pins, Straight Pins, Grooved Pins, and Spring Pins

ASME B18.8.100M-2000 Spring Pins: Coiled Type, Spring Pins: Slotted, Machine Dowel Pins, Grooved Pins (Metric Series)

INTERNET RESOURCES

American Society of Mechanical Engineers. For information on sections through shafts, pins, and keys, refer to ASME Y14.3M-1994 (R1999) (*Multi- and Sectional-View Drawings*) at: http://www.asme.org

Machine Design. For information on pin fasteners, see *Machine Design, Fastening/Joining Reference* at: http://www.machinedesign.com

Vogelsang Corporation. For information on machine pin fasteners, see: http://www.rollpin.com/ ProductInformation/Springpins/Coiled/index.html

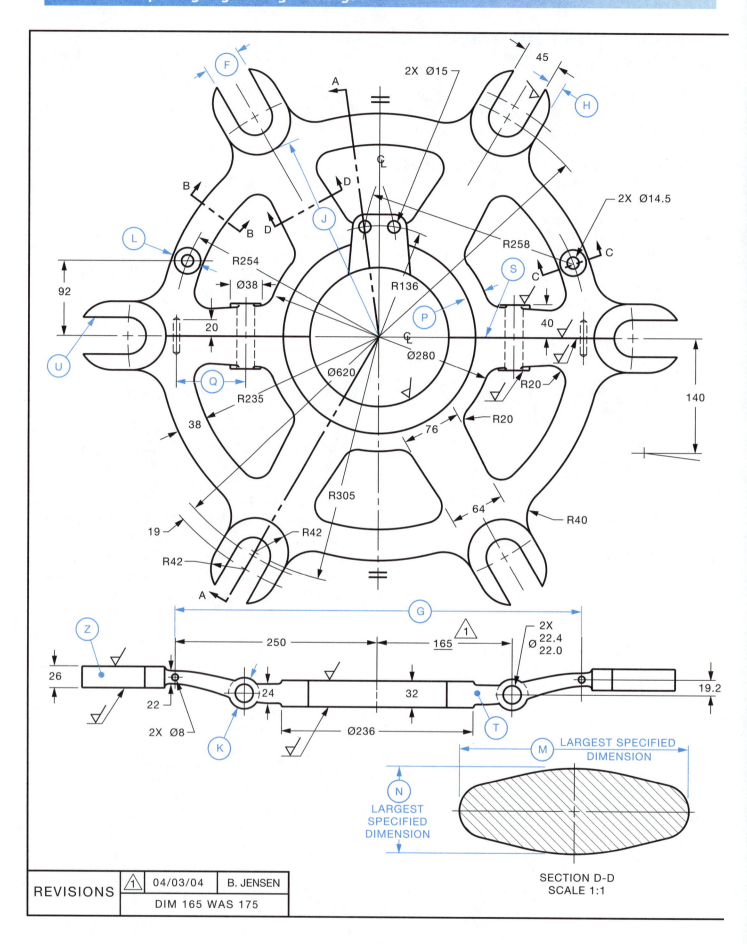

2X Ø15

45

(F)

(H)

2X Ø14.5

A

(D)

(B)

(B) (D)

R258

(J)

C

C

(L)

R254

Ø38

R136

(S)

92

R305

20

(P)

40

Ø280

(U)

(Q)

Ø620

R235

R20

38

76

R20

64

19

R42

R40

R42

A

SECTION D-D
SCALE 1:1

(G)

(Z)

250

165

1

2X
Ø 22.4
22.0

26

24

32

19.2

22

2X Ø8

Ø236

(T)

(K)

(M)

LARGEST SPECIFIED
DIMENSION

(N)

LARGEST
SPECIFIED
DIMENSION

REVISIONS	⚠	04/03/04	B. JENSEN
		DIM 165 WAS 175	

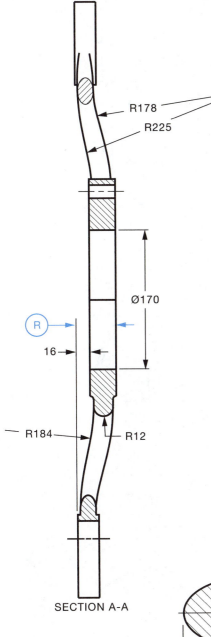

SECTION A-A

R178
R225

Ø170

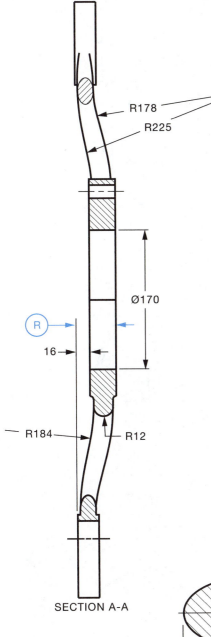

R
16

R184   R12

QUESTIONS:

1. How far were the Ø22 holes moved when the drawing was revised?

2. What type of projection does the ISO projection symbol (E) indicate?

3. How many different size scales were used to make the drawing?

4. Locate surface (T) in the top view.

5. Locate surface (Z) in the top view.

6. What is the approximate outside diameter of the spider?

7. What will be the rough dimension of casting at (F), assuming that 1.5mm have been added for each surface to be finished?

8. What would be used to position both halves of the spider together before bolting? What is their size?

9. What size bolts would be used to fasten the spider together?

10. Calculate distances (G), (H), (J), (K), (L), (M), (N), (P), (Q), and (R).

11. With reference to the scales used on the drawings, what percentage larger are the removed sections over the main drawings?

12. What type of section view is section A-A?

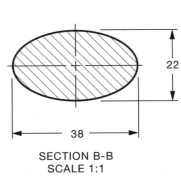

22

38

SECTION B-B
SCALE 1:1

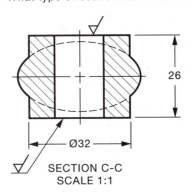

26

Ø32

SECTION C-C
SCALE 1:1

METRIC
DIMENSIONS ARE IN MILLIMETERS

NOTE:
- TOLERANCE ON DIMENSIONS ±0.5
- TOLERANCE ON ANGLES ±0.5°

- ✓ TO BE $^{1.6}$✓

(E)

MATERIAL	GRAY IRON	
SCALE	1:5 EXCEPT WHERE NOTED	
DRAWN	C. JENSEN	DATE 10/06/03

SPIDER   **A-85M**

QUESTIONS:

1. Which views are shown?

2. How many surfaces are to be finished?

3. How many scraped surfaces are indicated?

4. How many holes are to be tapped?

5. What is the purpose of tapped hole (R)?

6. Which surface is (3) in the left-side view?

7. Which surface is (2) in the left-side view?

8. Which surface is (4) in the front view?

9. Which surface is (14) in the left-side view?

10. Which surface in the top view and front view is (9)?

11. Which surface in the top view and front view is (8)?

12. Which surface is (12) in the top view?

13. What is the name of part (V)?

14. What is the purpose of part (V)?

15. What do dotted lines at (W) represent?

16. Which surface is line (6) in the front view?

17. Which top view line indicates point (Z)?

18. What is the depth of the tapped hole at (X)?

19. Which edges or surfaces in the left and front views does line (T) represent?

20. What is the diameter of tap drill (Y)? (See Appendix.)

21. Determine dimensions or operations at (A) to (Q), (20); to (52).

22. What is the largest permissable diameter of the largest hole?

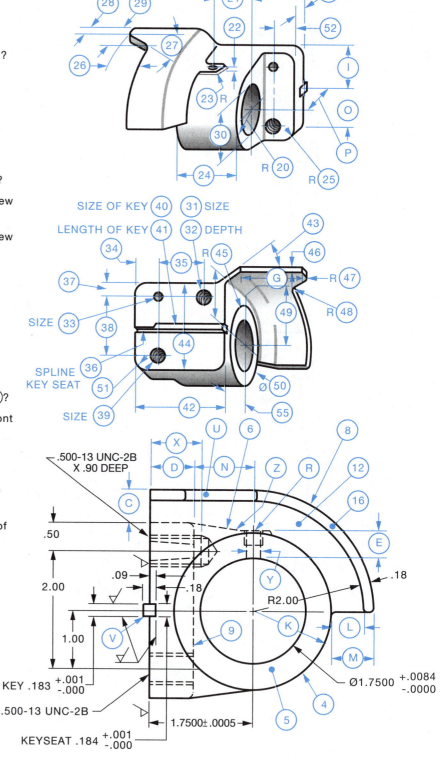

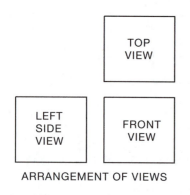

ARRANGEMENT OF VIEWS

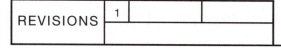

REVISIONS | 1 | | |

NOTES: UNLESS OTHERWISE SPECIFIED:
- TOLERANCE ON DIMENSIONS ±.02
- TOLERANCE ON ANGLES ±0.5°
- FINISH AND SCRAPE SURFACE BEFORE
  CUTTING SPLINE KEY SEAT

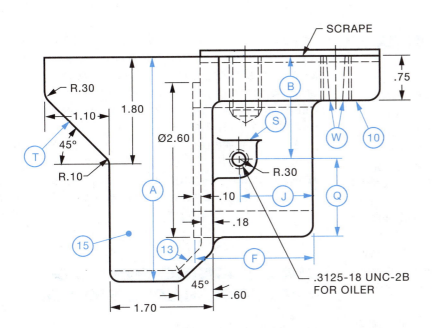

SCRAPE

R.30
1.10
45°
R.10
1.80
Ø2.60
R.30
.10
.18
.75
B
S
W
10
J
Q
T
A
15
13
F
45°
.60
1.70
.3125-18 UNC-2B
FOR OILER

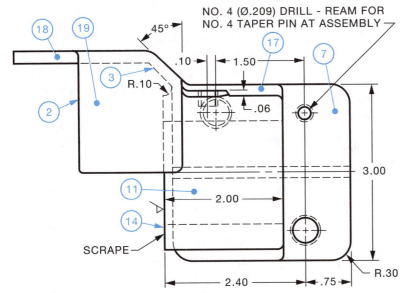

NO. 4 (Ø.209) DRILL - REAM FOR
NO. 4 TAPER PIN AT ASSEMBLY
45°
.10
1.50
.06
R.10
3.00
2.00
2.40
.75
R.30
18
19
3
2
17
7
11
14
SCRAPE

NOTE:
THREAD CONTROLLING ORGANIZATION
AND STANDARD–ASME B1.1-2003

MATERIAL	GRAY IRON	
SCALE	NOT TO SCALE	
DRAWN	D. SMITH	DATE 30/10/03

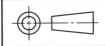

HOOD   **A-86**

32 UNIT

DRAWINGS FOR NUMERICAL CONTROL

Numerical control (NC) is a means of automatically directing the functions of a machine using electronic instructions. Originally, information was fed to NC machines through punched tapes. Improvements in technology have led to the integration of computers with manufacturing machinery, called **computer numerical control (CNC).** The machine interprets digitally coded instructions and directs various operations of the cutting tool.

It has been established that because of the consistent high accuracy of numerically controlled machines, and because human error has been almost entirely eliminated, scrap has been considerably reduced.

Another area where numerically controlled machines are better is in the quality or accuracy of the work. In most cases, a numerically controlled machine can produce parts more accurately at no additional cost, resulting in reduced assembly time and better interchangeability of parts. This latter fact is especially important when spare parts are required.

Computer-aided design and computer-aided manufacturing techniques are now widely used in conjunction with numerical control processes in industry.

DIMENSIONING FOR NUMERICAL CONTROL

Common guidelines have been established for dimensioning for numerical control that enable dimensioning and tolerancing practices to be used effectively for both NC and conventional fabrication. The numerical control concept is based on the system of rectangular or Cartesian coordinates in which any position can be described in terms of distance from an origin point along either two or three mutually perpendicular axes. Each object is pre-

pared using baseline (or coordinate) dimensioning methods as described earlier in Unit 16. First, the selection of an absolute (0, 0, 0) or (0, 0) coordinate origin is made depending on whether the control is three-axis or two-axis. All part dimensions would be referenced from that origin.

After a working drawing is produced, the information is transferred to manufacturing equipment. This allows the NC computer to compile instructions from programs stored within its memory. The result is a detailed program plan for tool-path generation.

DIMENSIONING FOR TWO-AXIS COORDINATE SYSTEM

Two dimensional coordinates (**X, Y**) define points in a plane, Figure 32–1. Examples of parts using rectangular coordinates were shown in Unit 18.

The **X** axis is horizontal and considered the first and basic reference axis. Distances to the right of the origin are considered positive values and those to the left of the origin are negative values.

The **Y** axis is vertical and perpendicular to the **X** axis in the plane of a drawing showing **XY** relationships. Distances above the origin are considered positive **Y** values and below the origin as negative values. The position where the **X** and **Y** axes cross is called the origin, or zero point.

For example, four points lie in a plane, as shown in Figure 32–1. The plane is divided into four quadrants, the origin being in the center. Point **A** lies in quadrant 1 and is located at position (6, 3) with the **X** coordinate first, followed by the **Y** coordinate. Point **B** lies in quadrant 2 and is located at position (−6, 5). Point **C** lies in quadrant 3 and is located at position (−5, −2). Point **D** lies in quadrant 4 and is located at position (2, −3).

Designing for NC would be greatly simplified if all work were done in the first quadrant because all of

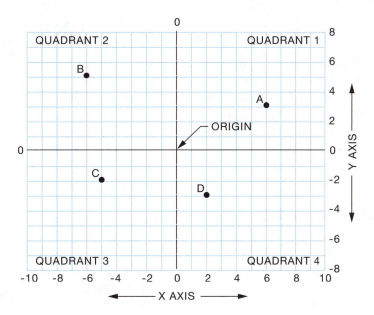

FIGURE 32–1 ■ *Two-dimensional coordinates.*

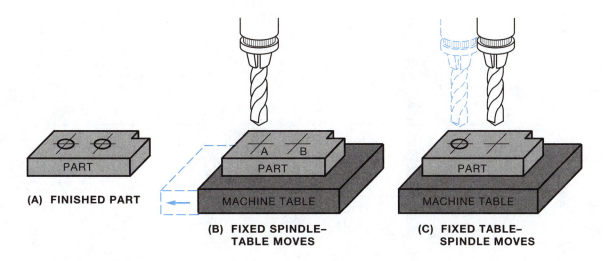

(A) FINISHED PART

(B) FIXED SPINDLE– TABLE MOVES

(C) FIXED TABLE– SPINDLE MOVES

FIGURE 32–2 ■ *Positioning the work.*

the values would be positive, and the plus and minus signs would not be required. For that reason many NC systems place the origin (0, 0) to the lower left of the part. This way only the positive values apply. However, any of the four quadrants may be used.

Some NC machines, called two-axis machines, are designed for locating points in only the X and Y directions. The function of these machines is to move the machine table or tool to a specified position in order to perform work, as shown in Figure 32–2.

With the fixed spindle and movable table, as shown in Figure 32–2(B), hole A is drilled, then the table moves to the left, positioning point B below the drill. This is the most frequently used method. With the fixed table and

movable spindle, as shown in Figure 32–2(C), hole A is drilled, then the spindle moves to the right, positioning the drill above point B. This changes the direction of the motion, but the movement of the cutter as delivered to the work remains the same.

Origin (Zero Point)

This is the position where all coordinate dimensions are measured. A set-up point is located on the part or the fixture holding the part. It may be the intersection of two finished surfaces, the center of a previously machined hole in the part, or a feature of the fixture.

Set-Up Point

The part must be accurately positioned on the fixture before any work is performed. This establishes a set-up point, which is accurately located in relationship to the origin. It may be located on the part, or on the fixture holding the part. It may be the intersection of two finished surfaces, the center of a previously machined hole, or a feature of the fixture.

Relative Coordinate (Point-to-Point) Programming

With point-to-point programming, each new position is given from the last position. To compute the next position wanted, it is necessary to establish the sequence in which the work is to be done.

Absolute Coordinate Programming

Many systems use absolute coordinate programming instead of the point-to-point method of dimensioning. With this type of dimensioning all dimensions are taken from the origin; as such, base line or datum dimensioning is used.

Examples of both of these dimensioning techniques are shown in Figure 32–3. In these examples two of the outer surfaces of the part are positioned on the fixture by means of three locating pins. This establishes the set-up point. It is located 80 mm above and 80 mm to the right of the origin. Regardless of which of the two methods of dimensioning is to be used, the coordinates for the first hole (hole 1) to be machined are the same and are taken from the origin (zero point). The X coordinate is 100 (80 + 20), and the Y coordinate is also 100 (80 + 20).

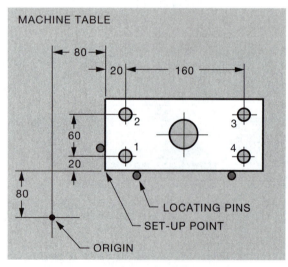

POINT-TO-POINT DIMENSIONING

HOLE	X	Y
1	100	100
2	0	60
3	160	0
4	0	-60

(A) RELATIVE COORDINATE (POINT-TO-POINT) DIMENSIONING FOR 4 HOLES SHOWN ON PART

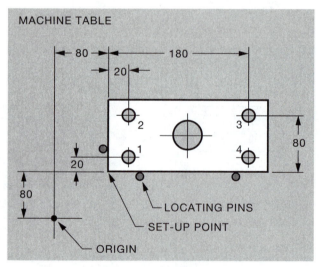

BASE LINE DIMENSIONING

HOLE	X	Y
1	100	100
2	100	160
3	260	160
4	260	100

(B) ABSOLUTE COORDINATE (BASE LINE) DIMENSIONING FOR 4 HOLES SHOWN ON PART

FIGURE 32–3 ■ *Dimensioning for numerical control.*

INTERNET RESOURCES

Machine Design. For information on CAD/CAM, see *Machine Design, CAD/CAM Reference* at: http://www.machinedesign.com

Sandria National Laboratories. For information on NC/CNC, see: http://mfgshop.sandria.gov/ 1400_ext_Num_Control.htm

TechStudent.Com. For information, including illustrations on the basics of NC/CNC, go to: http://www.technologystudent.com/cam/ camex/htm

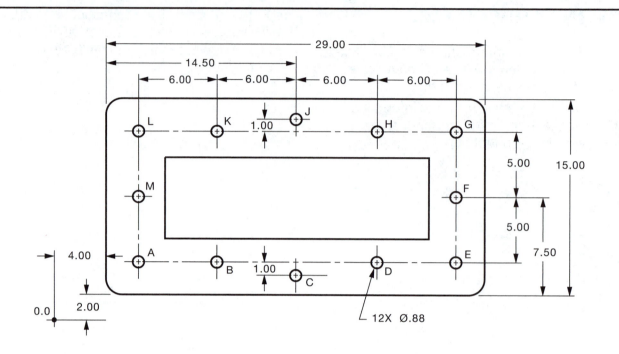

ABSOLUTE COORDINATES		HOLE	RELATIVE COORDINATES	
X	Y		X	Y
		A		
		B		
		C		
		D		
		E		
		F		
		G		
		H		
		J		
		K		
		L		
		M		

ASSIGNMENT:
PREPARE A CHART SIMILAR TO THE ONE SHOWN ABOVE AND PLACE
THE COORDINATES FOR EACH OF THE CIRCULAR HOLES IN THE CHART.
THE LETTERS AT THE HOLES INDICATE THE SEQUENCE IN WHICH THEY
ARE TO BE DRILLED. NOTE THE LOCATION OF THE ORIGIN AND THAT THE
LETTER "I" IS NOT USED TO IDENTIFY A HOLE.

COVER PLATE	A-87

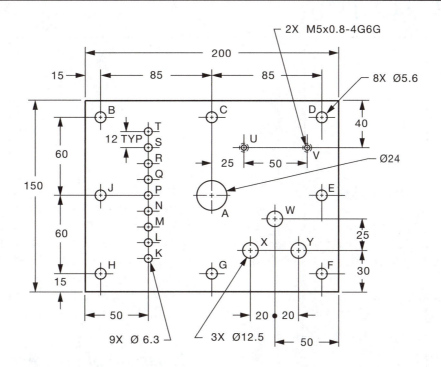

POINT TO POINT PROGRAMMING		
HOLE	X AXIS	Y AXIS
A		
B		
C		
D		
E		
F		
G		
H		
J		
K		
L		
M		

POINT TO POINT PROGRAMMING		
HOLE	X AXIS	Y AXIS
N		
P		
Q		
R		
S		
T		
U		
V		
W		
X		
Y		

ASSIGNMENT:
PREPARE A CHART SIMILAR TO THE ONE SHOWN ABOVE AND PLACE THE X AND Y COORDINATES FOR EACH OF THE HOLES IN THE CHART. POINT-TO-POINT PROGRAMMING IS TO BE USED TO LOCATE EACH HOLE. THE LETTERS AT THE HOLES INDICATE THE SEQUENCE IN WHICH THEY ARE TO BE DRILLED. ORIGIN FOR THE X AND Y COORDINATES IS THE BOTTOM LEFT-HAND CORNER OF THE PART. NOTE THAT THE LETTERS "I"AND "O" ARE NOT USED TO IDENTIFY HOLES.

THREAD CONTROLLING ORGANIZATION AND STANDARD-ASME B1.13M-2001

METRIC
DIMENSIONS ARE IN MILLIMETERS

TERMINAL BOARD   **A-88M**

ASSEMBLY DRAWINGS

The term *assembly drawing* refers to the type of drawing in which the various parts of a machine or structure are drawn in their relative positions in the completed unit.

In addition to showing how the parts fit together, the assembly drawing is used mainly:

- To represent the working relationships of the mating parts of a machine or structure and the function of each.

- To give a general idea of how the finished product should look.

- To aid in securing overall dimensions and center distances in assembly.

- To give the detailer data needed to design the smaller units of a larger assembly.

- To supply illustrations that may be used for catalogs, maintenance manuals, or other illustrative purposes.

In order to show the working relationship of interior parts, the principles of projection may be violated and details omitted for clarity. Assembly drawings should not be overly detailed because precise information describing part shapes is provided on detail drawings.

Detail dimensions that would confuse the assembly drawing should be omitted. Only such dimensions as center distances, overall dimensions, and dimensions showing the relationship of the parts as they apply to the mechanism as a whole should be included. There are times when a simple assembly drawing may be dimensioned so that no other detail drawings are needed. In such a case the assembly drawing becomes a working assembly drawing.

Sectioning is used more extensively on assembly drawings than on detail drawings. The conventional method of section lining is used on assembly drawings to show the relationship of the various parts, Figure 33–1.

Symbolic section lining, as shown in Figure 8–4, may be used for many purposes: (1) to represent the material of the part or parts; (2) to represent conductive or non-conductive materials; and (3) to represent moving and stationary parts.

Subassembly Drawings

Subassembly drawings are often made of smaller mechanical units. When combined in final assembly, they make a single machine. For a lathe, subassembly drawings would be furnished for the headstock, the apron, and other units of the carriage. These units might be machined and assembled in different departments by following the subassembly drawings. The individual units would later be combined in final assembly according to the assembly drawing.

Identifying Parts of an Assembly Drawing

When a machine is designed, an assembly drawing or design layout is first drawn to visualize clearly the method of operation, shape, and clearances of the various parts. From this assembly drawing, the detail drawings are made and each part is given a part number.

To assist in the assembly of the machine, item numbers corresponding with the part numbers of various details are placed on the assembly drawing attached to the corresponding part with a leader. The part number is often enclosed in a small circle, called a balloon, which helps distinguish part numbers from dimensions, Figure 33–1.

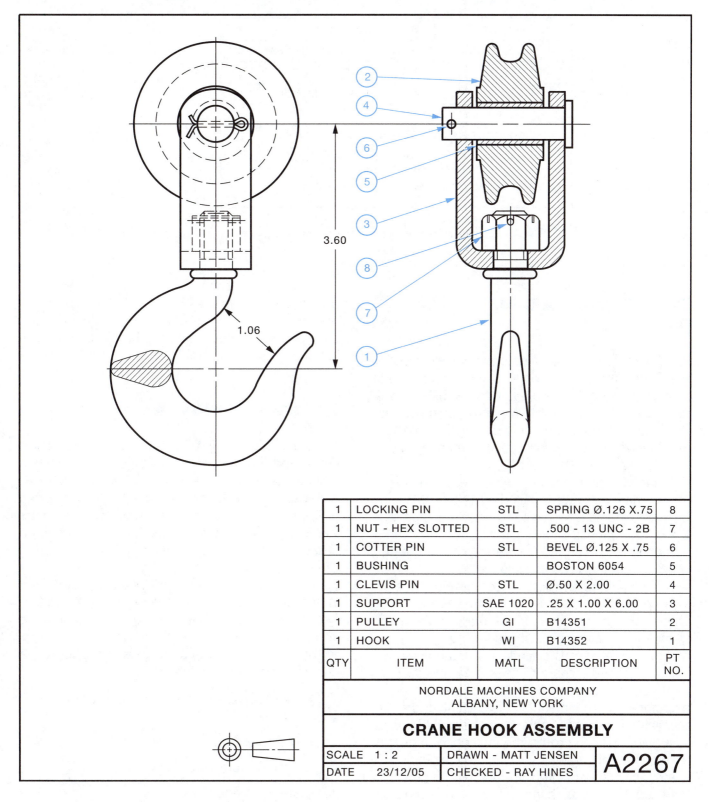

1	LOCKING PIN	STL	SPRING Ø.126 X.75	8
1	NUT - HEX SLOTTED	STL	.500 - 13 UNC - 2B	7
1	COTTER PIN	STL	BEVEL Ø.125 X .75	6
1	BUSHING		BOSTON 6054	5
1	CLEVIS PIN	STL	Ø.50 X 2.00	4
1	SUPPORT	SAE 1020	.25 X 1.00 X 6.00	3
1	PULLEY	GI	B14351	2
1	HOOK	WI	B14352	1
QTY	ITEM	MATL	DESCRIPTION	PT NO.

NORDALE MACHINES COMPANY
ALBANY, NEW YORK

CRANE HOOK ASSEMBLY

SCALE 1 : 2	DRAWN - MATT JENSEN	A2267
DATE 23/12/05	CHECKED - RAY HINES	

FIGURE 33–1 ■ *A typical assembly drawing.*

QTY	ITEM	MATL	DESCRIPTION	PT N0.
4	NUT-HEX REG	STL	.375-16 UNC	7
4	BOLT-HEX REG	STL	.375-16 UNC X 1.50	6
1	KEY	MS	WOODRUFF 608	5
2	BEARINGS	SKF	RADIAL BALL 620	4
1	SHAFT	CRS	Ø1.00 X 6.50 LG	3
1	SUPPORT	MST	.375 X 2.00 X 5.50	2
1	BASE	GI	PATTERN - A3154	1

FIGURE 33–2 ■ *Bill of material for assembly drawing.*

BILL OF MATERIAL (ITEMS LIST)

A *bill of material* or *items list* is an itemized list of all the components shown on an assembly drawing or a detail drawing, Figure 33–2. Often, a bill of material is placed on a separate sheet of paper for handling and duplicating. For castings, a pattern number would appear in the size column instead of the physical size of the part.

Standard components, which are purchased rather than fabricated, including bolts, nuts, and bearings, should have a part number and appear on the bill of material. There should be sufficient information in the descriptive column to enable the purchasing agent to order these parts.

Standard components are incorporated in the design of machine parts for economical production. These parts are specified on the drawing according to the manufacturer's specification. The use of manufacturers' catalogs is essential for determining detailing standards, characteristics of a special part, methods of representation, etc. However, it should be pointed out that manufacturers' catalogs are very unreliable for specifying parts; they should be used as a guide only. To protect the integrity of a design, a purchase part drawing must be made. This overcomes the frequent problem whereby the component supplier makes changes, unknown to the user, which frequently affect the design.

The four-wheel trolley (Assignment A-91) includes many standard parts, grease cups, lockwashers, Hyatt roller bearings, rivets, and nuts, all of which are standard purchased items. These parts are not detailed but are listed in the bill of material. However, the special countersunk head bolts and the taper washers, commonly called Dutchmen, are not standard parts and must therefore be made especially for this particular assembly.

HELICAL SPRINGS

The *coil* or *helical spring* is commonly used in machine design and construction. It may be cylindrical or conical in shape or a combination of the two, Figure 33–3.

Because of the labor and time involved, the true projection of a helical spring is usually not drawn. Instead, a schematic or simplified drawing is preferred because of its simplicity. All the required information can be given on such a drawing.

On assembly drawings, springs are usually shown in section; either crosshatched lines or simplified, symbolic representations recommended, depending on the size of the wire diameter, Figure 33–4.

The following information must be given on a drawing of a spring, Figure 33–5.

- Size, shape, and type of material used in the spring

- Diameter (outside or inside)

- Pitch or number of coils

- Shape of ends

- Length

For example, ONE HELICAL TENSION SPRING 3.00 LG (OR NUMBER OF COILS), .50 ID, PITCH .18, 18 B&S GA. SPRING BRASS WIRE clearly states the required information.

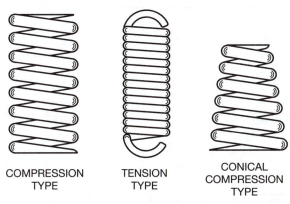

(A) PICTORIAL REPRESENTATION

COMPRESSION TYPE   TENSION TYPE   CONICAL COMPRESSION TYPE

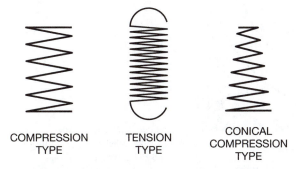

(B) SCHEMATIC OR SIMPLIFIED REPRESENTATION

COMPRESSION TYPE   TENSION TYPE   CONICAL COMPRESSION TYPE

FIGURE 33–3 ■ *Helical springs.*

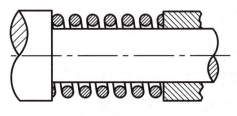

(A) LARGE SPRINGS

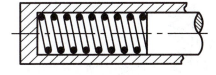

(B) SMALL SPRINGS

FIGURE 33–4 ■ *Showing helical springs on assembly drawings.*

The *pitch* of a coil spring is the distance from the center of one coil to the center of the next. The sizes of spring wires are designated by inch sizes and also in gauge numbers. The tables for these are found in handbooks.

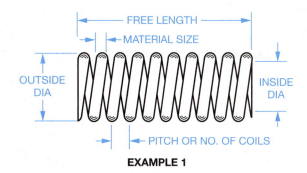

EXAMPLE 1

FREE LENGTH — MATERIAL SIZE — OUTSIDE DIA — INSIDE DIA — PITCH OR NO. OF COILS

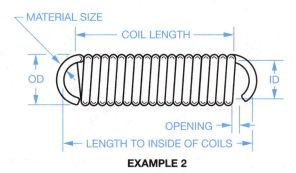

EXAMPLE 2

MATERIAL SIZE — COIL LENGTH — OD — ID — OPENING — LENGTH TO INSIDE OF COILS

FIGURE 33–5 ■ *Information given on spring drawings.*

Springs are made to a dimension of either outside diameter (if the spring works in a hole) or inside diameter (if the spring works on a rod). In some cases the mean diameter is specified for computation purposes.

REFERENCES

ASME Y14.3M-1994 (R1999) Multi- and Sectional-view Drawings

ASME Y14.24-1999 Types and Applications of Engineering Drawings

INTERNET RESOURCES

eFunda. For information on spring applications, see: http://www.efunda.com/home.cfm

Integrated Publishing. For information on Bill of Materials, see: www.tpub.com/engbas/3-16htm

Machine Design. For information on universal joints, see *Machine Design, Mechanical Reference* at: http://www.machinedesign.com

TechStudent.Com. For information on springs and spring applications, see www.technologystudent.com (Mechanisms)

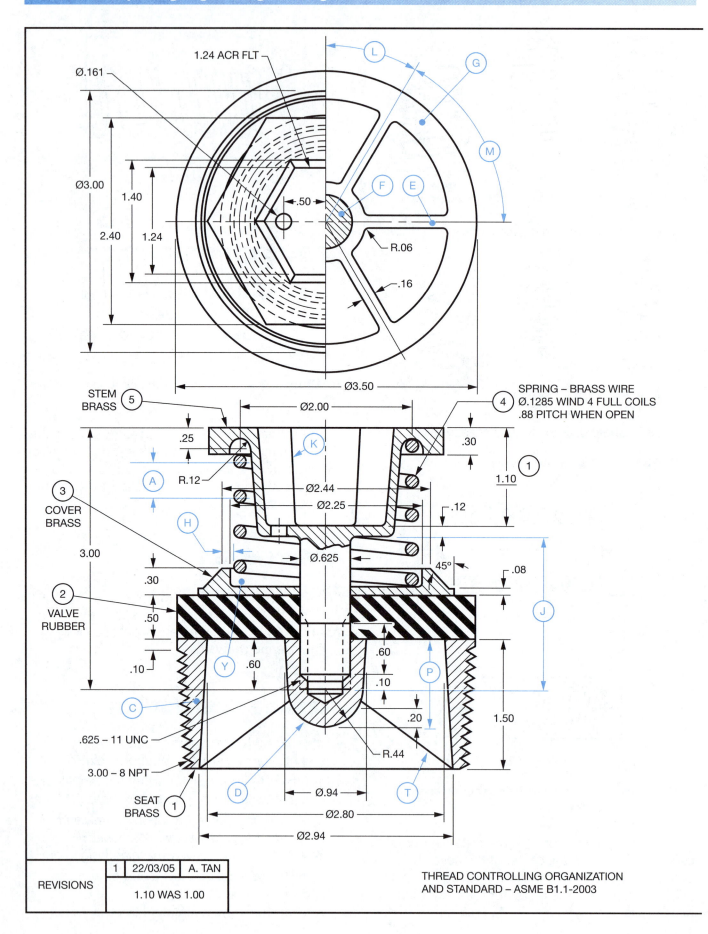

STEM BRASS (5)

COVER BRASS (3)

VALVE RUBBER (2)

SEAT BRASS (1)

Ø.161

Ø3.00

2.40

1.40

1.24

1.24 ACR FLT

.50

R.06

.16

Ø3.50

3.00

.25

R.12

.30

Ø2.00

Ø2.44

Ø2.25

Ø.625

45°

.08

.12

1.10

.30

.50

.10

.60

.60

.10

.20

1.50

.625 – 11 UNC

3.00 – 8 NPT

R.44

Ø.94

Ø2.80

Ø2.94

SPRING – BRASS WIRE
Ø.1285 WIND 4 FULL COILS
.88 PITCH WHEN OPEN

(4)

(1)

REVISIONS	1	22/03/05	A. TAN
		1.10 WAS 1.00	

THREAD CONTROLLING ORGANIZATION
AND STANDARD – ASME B1.1-2003

QUESTIONS

1. How many separate parts are shown on the valve assembly?
2. What is the length of the spring when the valve is closed?
3. Determine distances (A), (J), and (P).
4. What is the overall free length of the spring?
5. Locate (E) in the front view.
6. How many supporting ribs are there connecting (D) to (C)?
7. How thick are these ribs?
8. Identify the part numbers to which features (F) and (G) belong.
9. How many full threads are there on the stem (part 5)?
10. What is the nominal size of the pipe thread?
11. Give the length of the pipe thread.
12. Determine clearance distance (H).
13. Determine angles (L) and (M).

NOTE: UNLESS OTHERWISE SPECIFIED
 – TOLERANCES ON DIMENSIONS ±.02
 – TOLERANCES ON ANGLES ±0.5°

ASSIGNMENT: 1. ON A 1.00 INCH GRID SHEET (.10 IN. SQUARES) SKETCH THE TOP VIEW AND THE FRONT VIEW OF THE STEM IN FULL SECTION, PT 5. USE A CONVENTIONAL BREAK TO SHORTEN THE HEIGHT OF THE FRONT VIEW. ADD DIMENSIONS. SCALE 1 : 1.

2. ON A 1.00 INCH GRID SHEET (.10 IN. SQUARES) SKETCH A PARTIAL TOP VIEW AND THE FRONT VIEW OF THE SEAT IN FULL SECTION, PT 1. ADD DIMENSIONS. SCALE 1 : 1.

SCALE	NOT TO SCALE	
DRAWN	J. MILLER	
		DATE 15/09/04

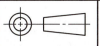

FLUID PRESSURE VALVE

A-89

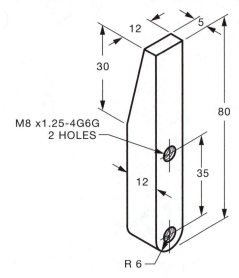

M8 x1.25-4G6G
2 HOLES

PT 1   MOVABLE JAW
1 REQD   MATL - SAE 1020

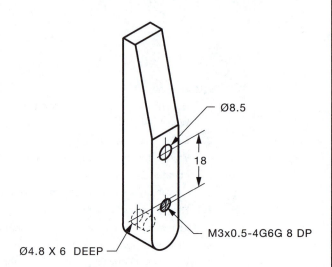

Ø8.5

M3x0.5-4G6G 8 DP

Ø4.8 X 6 DEEP

AS SHOWN, OTHERWISE SAME
AS PART 1.
PT 2   STATIONARY JAW
1 REQD   MATL-SAE 1020

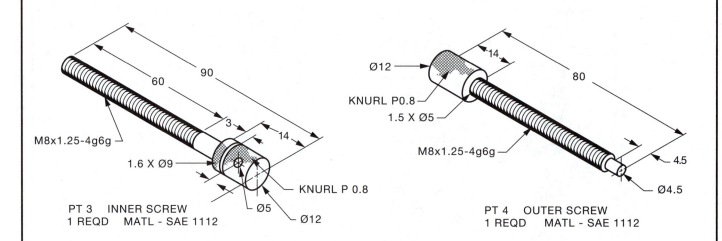

M8x1.25-4g6g

1.6 X Ø9

KNURL P 0.8

Ø5

Ø12

PT 3   INNER SCREW
1 REQD   MATL - SAE 1112

Ø12

KNURL P0.8

1.5 X Ø5

M8x1.25-4g6g

4.5

Ø4.5

PT 4   OUTER SCREW
1 REQD   MATL - SAE 1112

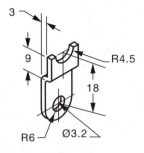

R4.5

R6

Ø3.2

PT 5   CLIP   1 REQD
MATL - 1.60 (16 USS) STL

PT 6   CAP SCREW
1 REQD   M3 x 8 LG RD HD

ASSIGNMENT:
ON A CENTIMETER GRID SHEET (1mm SQUARES) MAKE A ONE-VIEW
ASSEMBLY DRAWING OF THE PARALLEL CLAMP. USE SIMPLIFIED
THREAD CONVENTIONS (UNIT 16). PREPARE A BILL OF MATERIAL
SIMILAR TO THE ONE SHOWN IN FIGURE 33-2 CALLING FOR ALL
THE PARTS. IDENTIFY THE PARTS ON THE ASSEMBLY. THE ONLY
DIMENSION REQUIRED IS THE MAXIMUM OPENING OF THE JAWS.
SCALE 1:1.

METRIC
DIMENSIONS ARE IN MILLIMETERS

NOTE:
THREAD CONTROLLING ORGANIZATION
AND STANDARD-ASME B1.13M-2001

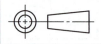

**PARALLEL CLAMP
ASSEMBLY**

A-90M

STRUCTURAL STEEL SHAPES

Structural steel is widely used in the metal trades for the fabrication of machine parts because the many standard shapes lend themselves to many different types of construction. Assignments A-91, A-94 and A-96 show three assemblies where components made from structural steel shapes are used.

Steel produced at the rolling mills and shipped to the fabricating shop comes in a wide variety of shapes and forms (approximately 600). At this stage it is called *plain material*.

The great bulk of this material can be designated as shown in Figure 34–1.

Abbreviations

When structural steel shapes are designated on drawings, a standard method of abbreviating should be followed that will identify the group of shapes without reference to the manufacturer and without the use of inches and pounds per foot. Therefore, it is recommended that structural steel be abbreviated as listed in Figure 34–2.

The abbreviations shown are intended only for use on design drawings. When lists of materials are being prepared for ordering from the mills, the requirements of the respective mills from which the material is to be ordered should be observed.

SYMBOL	WWF	W	M	S	C	MC
SHAPE						
NAME	WELDED WIDE FLANGE SHAPES	WIDE FLANGE SHAPES	MISCELLANEOUS SHAPES	STANDARD BEAMS	STANDARD CHANNELS	MISC. CHANNELS

SYMBOL	WWT	WT OR MT	L
SHAPE			
NAME	STRUCTURAL TEES		EQUAL LEG UNEQUAL LEG ANGLES

(A) INCH DESIGNATION

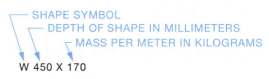

(B) METRIC DESIGNATION

FIGURE 34–1 ■ *Common structural steel shapes and drawing callouts.*

SHAPE	U.S. CUSTOMARY EXAMPLES SEE NOTE 1		METRIC SIZE EXAMPLES SEE NOTE 2
	NEW DESIGNATION	OLD DESIGNATION	
Welded Wide Flange Shapes (WWF Shapes)			
- Beam	WWF48 X 230	48WWF320	WWF1000 X 244
- Columns			WWF350 X 315
Wide Flange Shapes (W Shapes)	W24 X 76	24WF76	W600 X 114
	W14 X 26	14B26	W160 X 18
Miscellaneous Shapes (M Shapes)	M8 X 18.5	8M18.5	M200 X 56
	M10 X 9	10JR9.0	M160 X 30
Standard Beams (S Shapes)	S24 X 100	24I100	S380 X 64
Standard Channels (C Shapes)	C12 X 20.7	12C20.7	C250 X 23
Structural Tees			
- cut from WWF Shapes	WWT24 X 160	ST24WWF160	WWT280 X 210
- cut from W Shapes	WT12 X 38	ST122F38	WT130 X 16
- cut from M Shapes	MT4 X 9.25	ST4M9.25	MT100 X 14
Bearing Piles (HP Shapes)	HP14 X 73	14BP73	HP350 X 109
Angles (L Shapes)	L6 X 6 X .75	L6 X 6 X 3/4	L75 X 75 X 6
(leg dimensions X thickness)	L6 X 4 X .62	L6 X 4 X 5/8	L150 X 100 X 13
Plates (width X thickness)	20 X .50	20 X 1/2	500 X 12
Square Bar (side)	⊘1.00	BAR 1⊘	⊘25
Round Bar (diameter)	Ø1.25	BAR 1-1/4 Ø	Ø30
Flat Bar (width X thickness)	250 X .25	BAR 2-1/2 X 1/4	60 X 6
Round Pipe (type of pipe X OD X wall thickness)	12.75 OD X .375	12-3/4 X 3/8	XS 102 OD X 8
Square and Rectangular Hollow Structural Sections (outside dimensions X wall thickness)	HSS4 X 4 X .375	4X 4RT X 3/8	HSS102 X 102 X 8
	HSS8 X 4 X .375	8 X 4RT X 3/8	
Steel Pipe Piles (OD X wall thickness)			320 OD X 6

Note 1 - Values shown are nominal depth (inches) X weight per foot length (pounds).

Note 2 - Values shown are nominal depth (millimeters) X mass per meter length (kilograms).

Note 3 - Metric size examples shown are not necessarily the equivalents of the inch size examples shown.

FIGURE 34–2 ■ *Abbreviations for shapes, plates, bars, and tubes.*

S-shaped beams and all standard and miscellaneous channel have a slope on the inside flange of 16.67 percent (16.67 percent slope is equivalent to 9° 28' or a bevel of 1:6). All other beams have parallel face flanges.

PHANTOM OUTLINES

At times, a part or mechanism not included in the actual detail or assembly drawing is shown to clarify how the mechanism will connect with or operate from an adjacent part. This part is shown by draw-

ing thin dash lines (one long line and two short dashes) in the operating position. Such a drawing of the extra part is known as a *phantom drawing* or view drawn in *phantom*, Figure 34–3.

On the drawing of the four-wheel trolley (Assignment A-91), the track on which the wheels run on is an S beam. The wheels are set at an angle to the vertical plane in order to ride upon the sloping bottom flange of the S beam. The outline of the beam is shown by dash lines and, while not an integral part of the trolley, the outline or phantom view of the S beam shows clearly how the trolley operates.

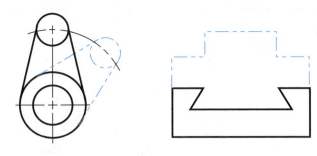

FIGURE 34–3 ■ *Phantom lines.*

SQUARE   OR   ROUND

FIGURE 34–4 ■ *Conical washers.*

CONICAL WASHERS

Conical washers, Figure 34–4, are available in a variety of sizes to accommodate the slopes found on structural steel shapes. A typical application can be found on the four-wheel trolley, parts **D** and **W** (Assignment A-91).

REFERENCES

American Institute of Steel Construction
ASME Y14.2M-1992 (R2003) Line Conventions and Lettering

INTERNET RESOURCES

American Institute of Steel Construction. For information on structural steel design and construction with links to an online library, training CD-ROM, directories, and job postings, see: http://www.aisc.org

American Iron and Steel Institute. For the latest news and information about the use of iron and steel in manufacturing and construction, see: http://www.steel.org

IDS Development-Nebraska Education. For information on the various line types used on engineering drawings, see: http://idsdev.mccneb.edu/djackson/lineintro.htm

QUESTIONS:

1. What does hidden line Ⓔ indicate?

2. Which cutting plane in the primary auxiliary view indicates (A) where the section to the left of line N-N is taken, (B) where the section to the right of line N-N is taken?

3. What is the slope of angle Ⓙ?

4. Locate part Ⓚ in the section view.

5. What is the wheel diameter of the trolley?

6. Locate parts ②, ③, ⑤, Ⓐ, Ⓒ, Ⓥ, and Ⓩ in the primary auxiliary view.

7. What are the names of parts Ⓣ, Ⓤ, Ⓥ, Ⓦ, Ⓧ, and Ⓨ ?

8. What is the diameter of the bearing rollers?

9. Determine distance Ⓛ.

10. How many not-to-scale dimensions are shown?

11. What type of line is used to show the S beam?

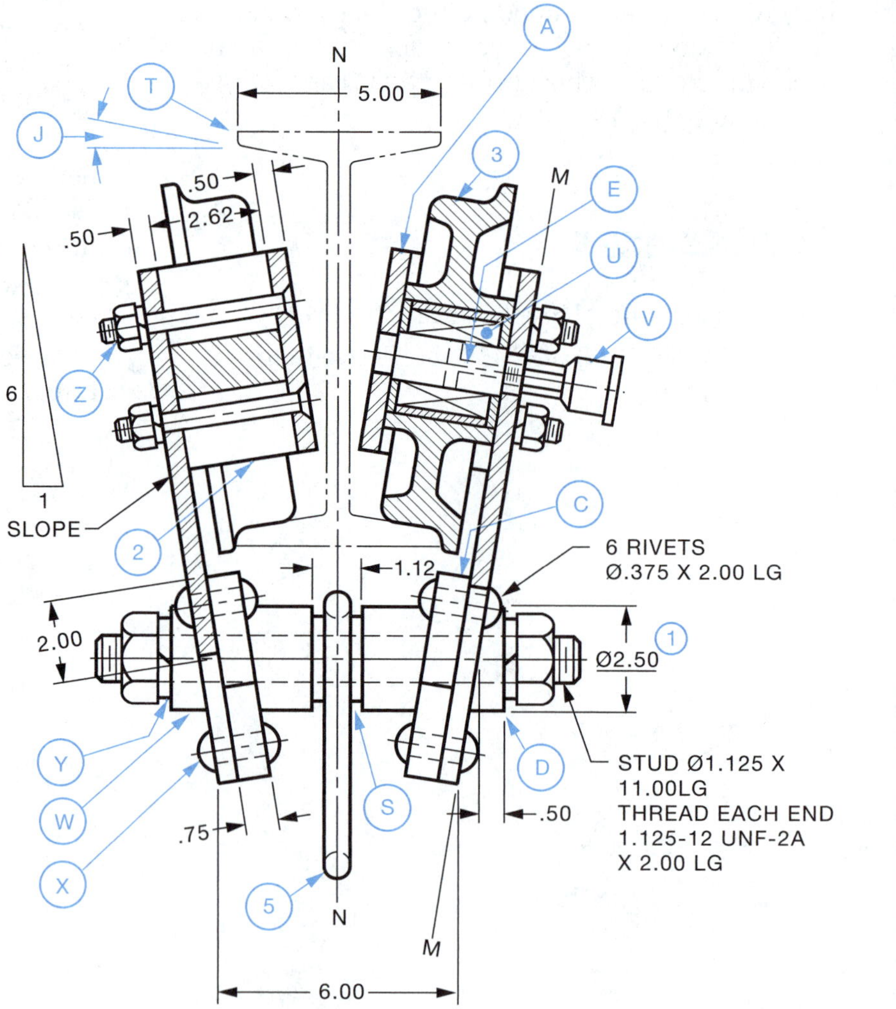

S BEAM - S10 X 35
WHEEL - Ø8,00
SHAFT - Ø1.374
BEARING - 2.835 OD
    -ROLLERS - Ø.562

NOTE:
THREAD CONTROLLING ORGANIZATION
AND STANDARD-ASME B1.13M-2001

REVISIONS	1	10/06/04	F. NEWMAN
		2.50 WAS 2.60	

5.00

.50
2.62
.50

6
1
SLOPE

2.00

.75

6.00

1.12

Ø2.50

6 RIVETS
Ø.375 X 2.00 LG

STUD Ø1.125 X
11.00LG
THREAD EACH END
1.125-12 UNF-2A
X 2.00 LG

.50

ASSIGNMENT:

1. ON A 1.00 INCH GRID SHEET (.10 IN. SQUARES) SKETCH THE SECONDARY AUXILIARY VIEW DRAWING OF PART A. THE SECONDARY AUXILIARY VIEW IS POSITIONED ABOVE AND PROJECTED FROM THE PRIMARY AUXILIARY VIEW. THE WIDTH OF THE PART IS TO BE SHORTENED BY USING CONVENTIONAL BREAKS AND ITS WIDTH TO BE DETERMINED BY THE STUDENT. ADD DIMENSIONS. SCALE 1:2.

2. ON 1.00 INCH GRID SHEETS (.10 IN. SQUARES) SKETCH WORKING DRAWINGS OF PARTS C AND D. SCALE 1:1.

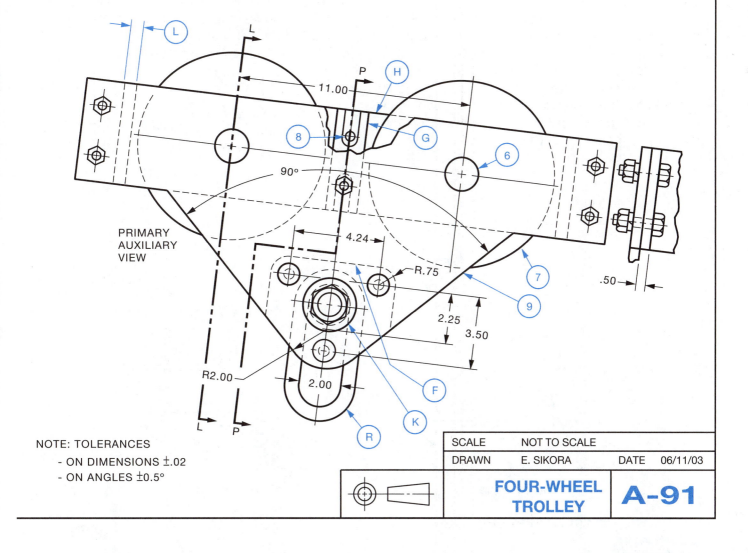

PRIMARY AUXILIARY VIEW

NOTE: TOLERANCES
- ON DIMENSIONS ±.02
- ON ANGLES ±0.5°

SCALE	NOT TO SCALE		
DRAWN	E. SIKORA	DATE	06/11/03

FOUR-WHEEL TROLLEY   **A-91**

35 UNIT

WELDING DRAWINGS

The primary importance of welding is the joining of pieces of metal so they will operate properly as a unit to support the loads to be carried. In order to design and build such a structure, to be economical and efficient, a basic knowledge of welding is essential. Figure 35–1 illustrates many basic welding terms.

The use of welding symbols on a drawing enables the designer to specify clearly the type and size of weld required to meet the design requirements. It is becoming increasingly important for the designer to specify the required type of weld correctly. Basic welding joints are shown in Figure 35–2. Points that must be made clear are the type of weld, the joint penetration, the weld size, and the root opening (if any). These points can be clearly indicated on the drawing by the welding symbol.

WELDING SYMBOLS

Welding symbols are a shorthand language. They save time and money and if used correctly, ensure understanding and accuracy. Welding symbols should be a universal language; for this reason the symbols of the American Welding Society have been adopted.

A distinction between the terms *weld symbol* and *welding symbol* should be understood. The weld symbol indicates the type of weld. The welding symbol is a method of representing the weld on drawings. It includes supplementary information and consists of the following eight elements. Not all elements need be used unless required for clarity.

1. Reference line
2. Arrow
3. Basic weld symbol
4. Dimensions and other data

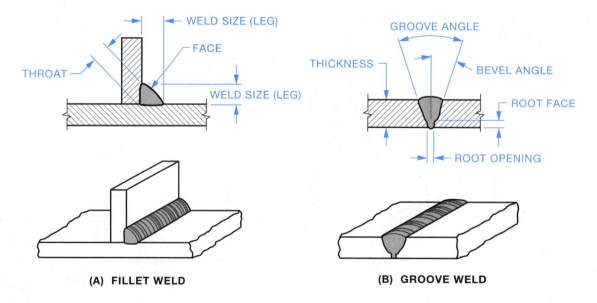

(A) FILLET WELD

(B) GROOVE WELD

FIGURE 35–1 ■ *Basic welding nomenclature.*

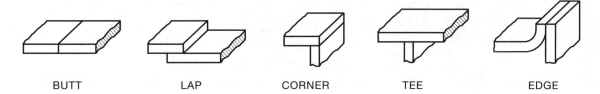

FIGURE 35–2 ■ *Basic welding joints.*

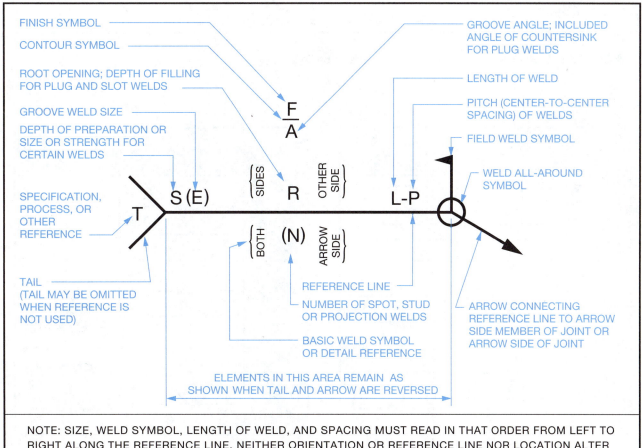

FIGURE 35–3 ■ *Standard location of elements of a welding symbol.*

5. Supplementary symbols
6. Finish symbols
7. Tail
8. Specification, process, or other reference.

The size and spacing of welds shown on the welding symbol are given in inches (U.S. customary) or millimeters (metric).

Figure 35–3 illustrates the position of the weld symbols and other information in relation to the welding symbol. The various weld symbols that may be applied to the basic welding symbol are shown in Figure 35–4. Figure 35–5 shows the actual shape of many of the weld types symbolized in Figure 35–4.

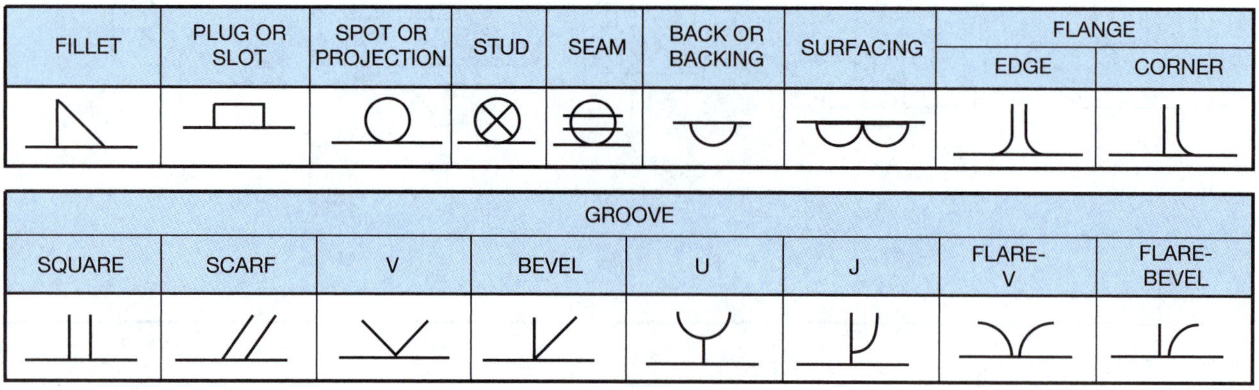

FILLET	PLUG OR SLOT	SPOT OR PROJECTION	STUD	SEAM	BACK OR BACKING	SURFACING	FLANGE	
							EDGE	CORNER

GROOVE								
SQUARE	SCARF	V	BEVEL	U	J	FLARE-V	FLARE-BEVEL

FIGURE 35–4 ■ *Basic weld symbols shown on reference line.*

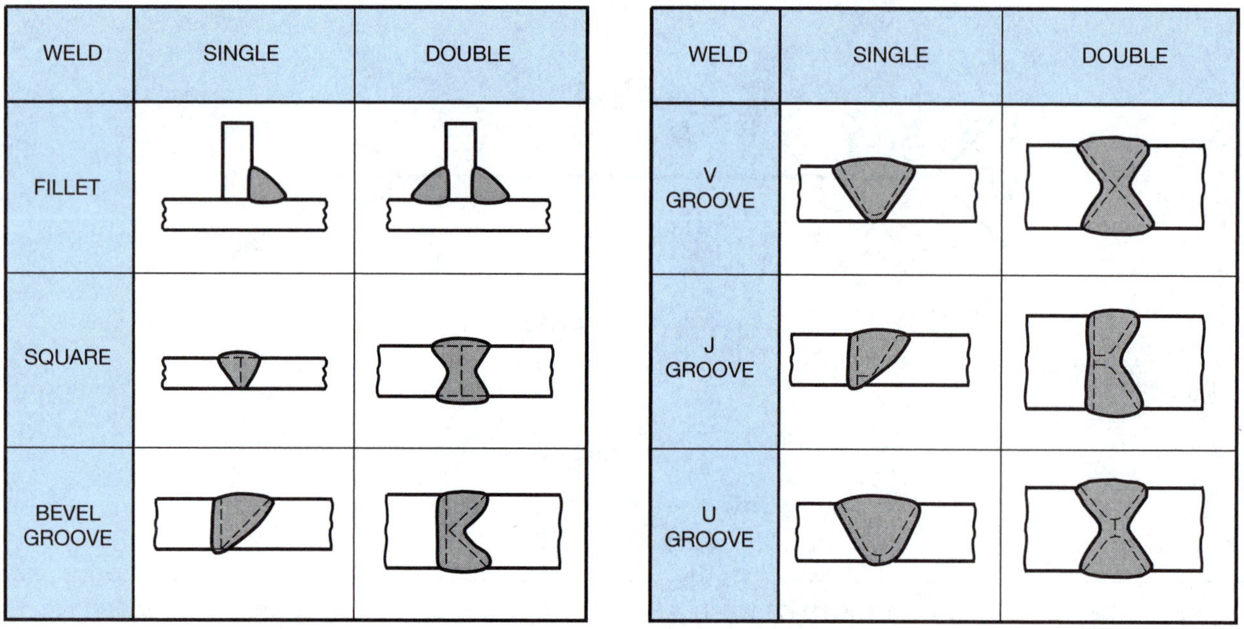

WELD	SINGLE	DOUBLE
FILLET		
SQUARE		
BEVEL GROOVE		

WELD	SINGLE	DOUBLE
V GROOVE		
J GROOVE		
U GROOVE		

FIGURE 35–5 ■ *Types of welds.*

Supplementary symbols may also be added to the welding symbol. The supplementary symbols are illustrated in Figure 35–6.

Any welding joint indicated by a symbol will always have an arrow side and an other side. The words arrow side, other side, and both sides are used accordingly to locate the weld with respect to the joint.

Tail of Welding Symbol

The welding symbol and allied process to be used may be specified by placing the appropriate letter designations from Figure 35–7 in the tail of the welding symbol, Figure 35–8.

WELD ALL AROUND	FIELD WELD	MELT-THRU	BACKING OR SPACER MATERIAL	CONSUMABLE INSERT	CONTOUR		
					FLUSH	CONVEX	CONCAVE
			BACKING / SPACER				

FIGURE 35–6 ■ *Supplementary symbols.*

Welding Process	Welding Process (Specific)	Letter Designation
Brazing (B)	Infrared Brazing	IRB
	Torch Brazing	TB
	Furnace Brazing	FB
	Induction Brazing	IB
	Resistance Brazing	RB
	Dip Brazing	DB
Oxyfuel Gas Welding (OFW)	Oxyacetylene Welding	OAW
	Oxyhydrogen Welding	OHW
	Pressure Gas Welding	PGW
Resistance Welding (RW)	Resistance-Spot Welding	RSW
	Resistance-Seam Welding	RSEW
	Projection Welding	PW
	Flash Welding	FW
	Upset Welding	UW
	Percussion Welding	PEW
Arc Welding (AW)	Stud Arc Welding	SW
	Plasma-Arc Welding	PAW
	Submerged Arc Welding	SAW
	Gas Tungsten-Arc Welding	GTAW
	Gas Metal-Arc Welding	GMAW
	Flux Cored Arc Welding	FCAW
	Shielded Metal-Arc Welding	SMAW
	Carbon-Arc Welding	CAW
Other Processes	Thermit Welding	TW
	Laser Beam Welding	LBW
	Induction Welding	IW
	Electroslag Welding	ESW
	Electron Beam Welding	EBW
Solid State Welding (SSW)	Ultrasonic Welding	USW
	Friction Welding	FRW
	Forge Welding	FOW
	Explosion Welding	EXW
	Diffusion Welding	DFW
	Cold Welding	CW

Cutting Method	Letter Designation
Arc Cutting	AC
Air Carbon-Arc Cutting	AAC
Carbon-Arc Cutting	CAC
Metal-Arc Cutting	MAC
Plasma-Arc Cutting	PAC
Oxygen Cutting	OC
Chemical Flux Cutting	FOC
Metal Powder Cutting	POC
Oxygen-Arc Cutting	AOC

FIGURE 35–7 ■ *Designation of welding process by letters.*

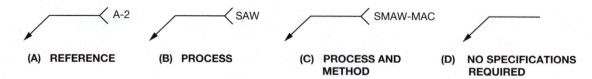

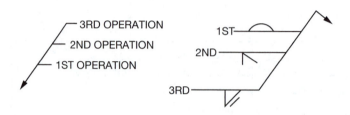

FIGURE 35–8 ▪ *Location of specifications, processes, and other references on welding symbols.*

FIGURE 35–9 ▪ *Multiple reference lines.*

Codes, specifications, or any other applicable documents may be specified by placing the reference in the tail of the welding symbol. Information contained in the referenced document need not be repeated in the welding symbol.

Multiple Reference Lines

Two or more reference lines may be used to describe a sequence of operations. The first operation is specified on the reference line nearest the arrow. Subsequent operations are specified sequentially on other reference lines, Figure 35–9.

Weld Locations on Symbol

Welds on the arrow side of the joint are shown by placing the weld symbol on the bottom side of the reference line. Welds on the other side of the joint are shown by placing the weld symbol on the top side of the reference line. Welds on both sides of the joint are shown by placing the weld symbol on both sides of the reference line. A weld extending completely around a joint is indicated by means of a weld-all-around symbol placed at the intersection of the reference line and the arrow line.

Field welds (welds not made in the shop or at the initial place of construction) are indicated by means of the field weld symbol placed at the intersection of the reference line and the arrow.

All weld dimensions on a drawing may be subject to a general note. Such a note might state: ALL FILLET WELDS .25 UNLESS OTHERWISE NOTED.

Only the basic fillet welds will be discussed in this unit. Figure 35–10 illustrates several typical fillet welding symbols and the resulting welds.

FILLET WELDS

Fillet welds are triangular in shape and are used to join surfaces that are perpendicular to each other. They are used on lapped, tee, and corner joints. No preparation of the surfaces of the metal is required.

1. Fillet weld symbols are drawn with the perpendicular leg always to the left.

2. Dimensions of fillet welds are shown on the same side of the reference line and to the left of the weld symbol.

3. The dimensions of fillet welds on both sides of a joint are shown whether the dimensions are identical or different.

4. The dimension does not need to be shown when a general note is placed on the drawing to specify the dimension of fillet welds.

NOTE: SIZE OF FILLET WELDS .25 UNLESS OTHERWISE SPECIFIED.

5. The *length* of a fillet weld, when indicated on the welding symbol, is shown to the right of the weld symbol.

6. The *pitch* (center-to-center spacing) of an intermittent fillet weld is shown as the distance between centers of increments on one side of the joint. It is shown to the right of the length dimension following a hyphen.

7. Staggered intermittent fillet welds are illustrated by staggering the weld symbols.

8. Fillet welds that are to be welded with approximately flat, convex, or concave faces without postweld finishing are specified by adding the flat, convex, or concave contour symbol to the weld symbol.

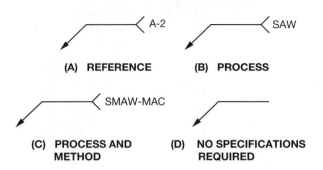

(A) REFERENCE

(B) PROCESS

(C) PROCESS AND METHOD

(D) NO SPECIFICATIONS REQUIRED

9. Fillet welds whose faces are to be finished approximately flat, convex, or concave by postweld finishing are specified by adding both the appropriate contour and finishing symbol to the weld symbol.

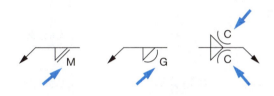

The following finishing symbols may be used to specify the method of finishing, but not the degree of finish:

C —Chipping
G —Grinding
H —Hammering
M—Machining
R —Rolling

10. A weld with a length less than the available joint length whose location is significant is specified on the drawing in a manner similar to that shown in Figure 35–11.

11. Weld-all-around symbol. A continuous weld extending around a series of connected joints may be specified by the addition of the weld-all-around symbol at the junction of the arrow and reference line. The series of joints may involve different directions and may lie on more than one plane, Figure 35–12(A).

Welds extending around a pipe or circular or oval holes do not require the weld-all-around symbol, Figure 35–12(B).

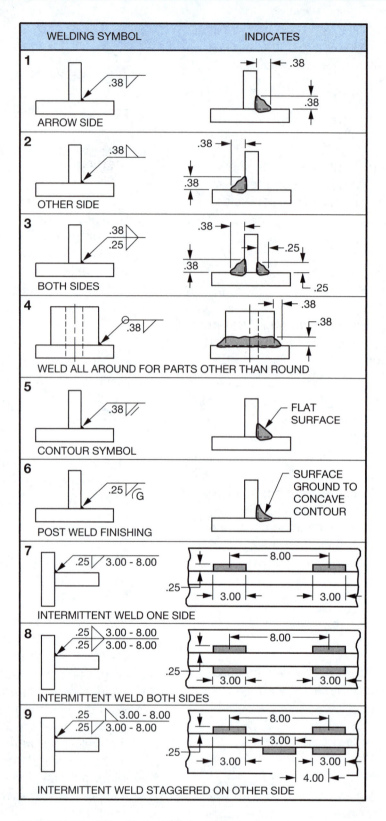

FIGURE 35–10 ■ *Typical fillet welds.*

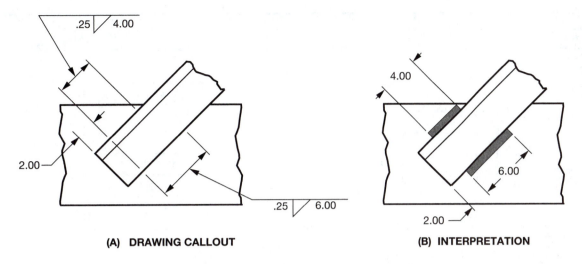

(A) DRAWING CALLOUT   **(B) INTERPRETATION**

FIGURE 35–11 ■ *Weld lengths located by dimensions.*

REFERENCES

ASME/AWS A2.4-86
American Welding Society

INTERNET RESOURCES

American Welding Society. For information on all aspects of welding with links to related organization and materials, see: http://www.aws.org

eFunda. For information on the various types of welding, see http://www.efunda.com/home.cfm

Machine Design. For information on welding and weld joints, see *Machine Design, Fastening/Joining Reference* at: http://www.machinedesign.com

Unified Engineering Inc. For information on fillet welds, see: http://www.unified-eng.com/scitech/weld/weld.html

DRAWING CALLOUT	INTERPRETATION

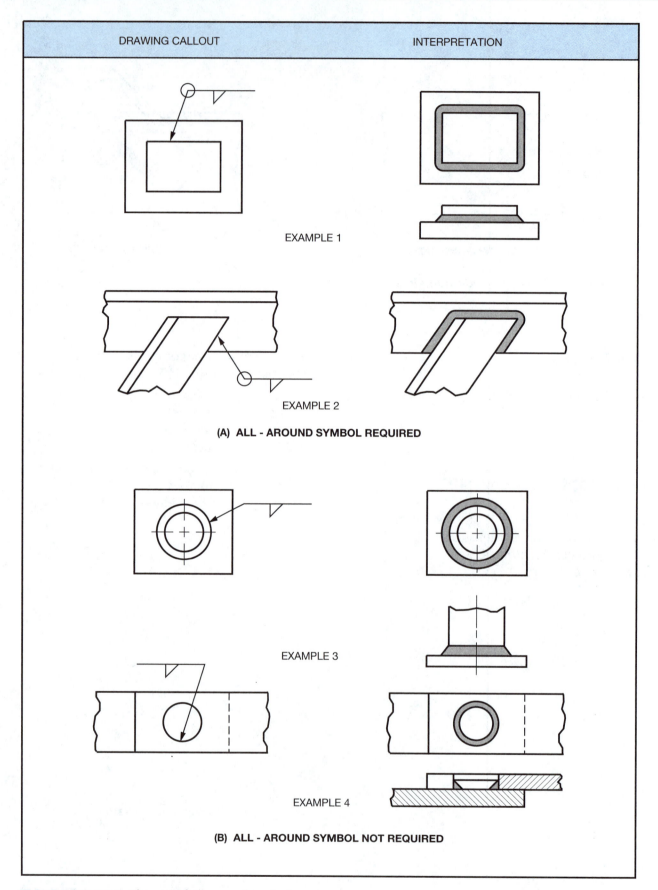

EXAMPLE 1

EXAMPLE 2

(A) ALL - AROUND SYMBOL REQUIRED

EXAMPLE 3

EXAMPLE 4

(B) ALL - AROUND SYMBOL NOT REQUIRED

FIGURE 35–12 ■ *The use of all-around symbols.*

DESIRED WELD	DRAWING CALLOUT

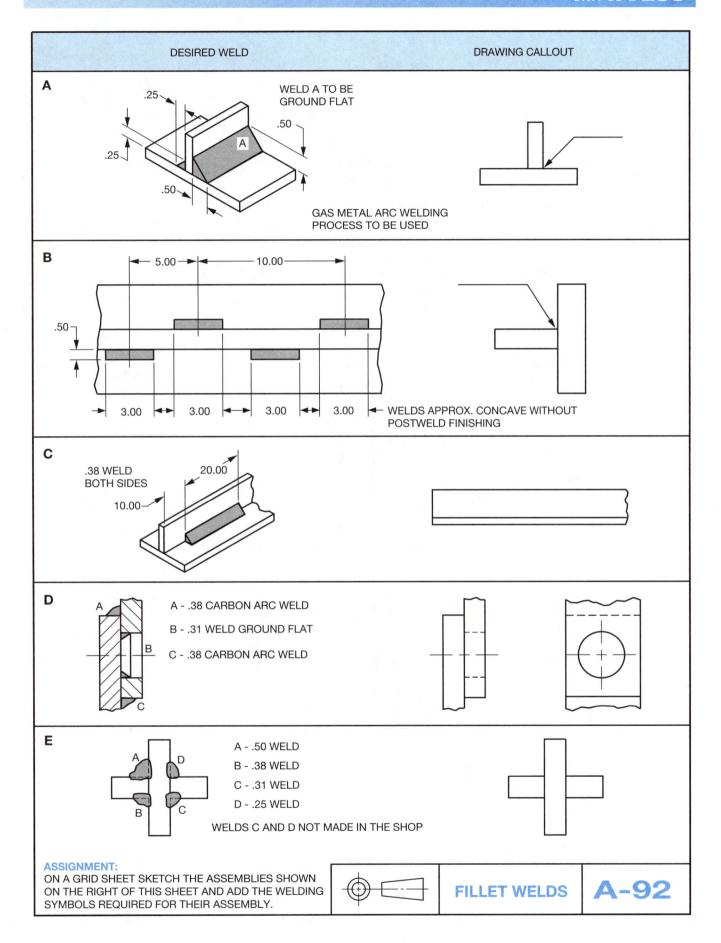

A

.25

.25

.50

.50

WELD A TO BE
GROUND FLAT

A

GAS METAL ARC WELDING
PROCESS TO BE USED

B

5.00   10.00

.50

3.00   3.00   3.00   3.00   WELDS APPROX. CONCAVE WITHOUT
POSTWELD FINISHING

C

.38 WELD
BOTH SIDES

20.00

10.00

D

A

B

C

A - .38 CARBON ARC WELD

B - .31 WELD GROUND FLAT

C - .38 CARBON ARC WELD

E

A

B

C

D

A - .50 WELD

B - .38 WELD

C - .31 WELD

D - .25 WELD

WELDS C AND D NOT MADE IN THE SHOP

ASSIGNMENT:
ON A GRID SHEET SKETCH THE ASSEMBLIES SHOWN
ON THE RIGHT OF THIS SHEET AND ADD THE WELDING
SYMBOLS REQUIRED FOR THEIR ASSEMBLY.

FILLET WELDS

A-92

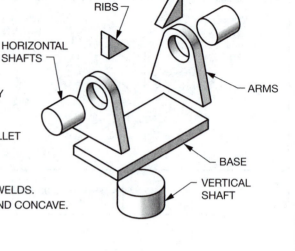

RIBS

HORIZONTAL
SHAFTS

ARMS

BASE

VERTICAL
SHAFT

ASSIGNMENT:

ON A ONE-INCH GRID SHEET (.10 IN. SQUARES) SKETCH THE ASSEMBLY
SHOWN BELOW TO THE SCALE OF 1 : 2. ADD THE FOLLOWING WELD
INFORMATION TO THE DRAWING:

- HORIZONTAL SHAFTS WELDED BOTH SIDES TO ARMS WITH .25 FILLET
  WELDS ON OUTSIDE AND .19 FILLET WELDS INSIDE. WELD TO BE
  FLAT WITHOUT POSTWELD FINISHING.
- ARMS WELDED BOTH SIDES TO BASE WITH .19 FILLET WELDS.
- RIBS WELDED BOTH SIDES TO BASE AND ARMS WITH .12 FILLET WELDS.
- VERTICAL SHAFT WELDED TO BASE WITH .25 FILLET WELD GROUND CONCAVE.

PROCESS - CARBON ARC WELDING.

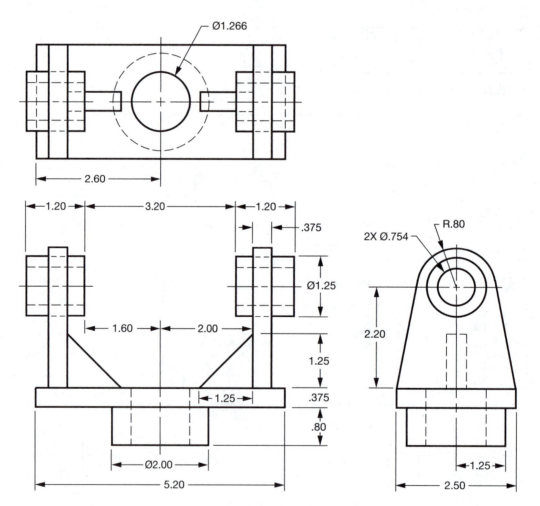

Ø1.266

2.60

1.20   3.20   1.20

.375

Ø1.25

1.60   2.00

1.25

1.25

.375

.80

Ø2.00

5.20

2X Ø.754   R.80

2.20

1.25

2.50

SHAFT SUPPORT   A-93

GROOVE WELDS

Groove welds are used to join two butted parts. The welds are classified as SQUARE, SCARF, BEVEL, V, U, J, FLARE-V, and FLARE-BEVEL. One or more beads (welding passes) may be used to produce the desired weld. For most of these welds the preparation of the metal surfaces before welding is required.

1. Bevel-groove, J-groove, and flare bevel-groove weld symbols are always drawn with the perpendicular leg to the left.

2. Dimensions of single-groove welds are shown on the same side of the reference line as the weld symbol.
3. Each groove of a double-groove joint is dimensioned; however, the root opening need appear only once.
4. For bevel-groove and J-groove welds, a broken arrow is used, when necessary, to identify the member to be prepared, Figure 36–1.
5. The depth of groove preparation "S" and size (E) of a groove weld when specified, is

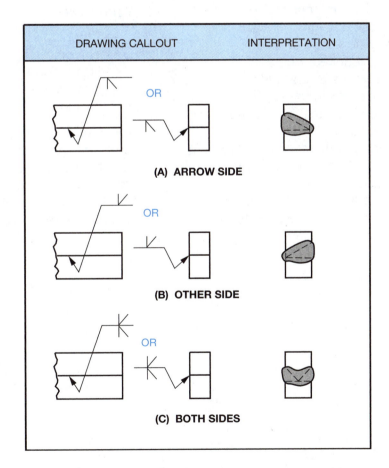

DRAWING CALLOUT	INTERPRETATION

(A) ARROW SIDE

(B) OTHER SIDE

(C) BOTH SIDES

FIGURE 36–1 ■ *Application of break in arrow on welding symbol.*

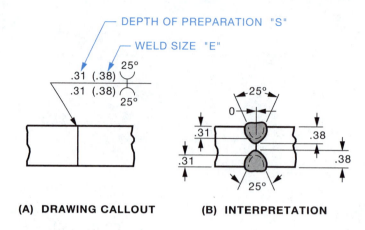

DEPTH OF PREPARATION "S"

WELD SIZE "E"

25°
.31 (.38)
.31 (.38)
25°

25°
0
.31
.38
.31
.38
25°

(A) DRAWING CALLOUT   **(B) INTERPRETATION**

FIGURE 36–2 ■ *Groove weld symbol showing use of combined dimensions.*

placed to the left of the weld symbol. Either or both may be shown. Except for square-groove welds, the groove weld size (E) in relation to the depth of the groove preparation "S" is shown as "S(E)," Figure 36–2.

6. Only the groove weld size is shown for square-groove welds.

7. When no depth of groove preparation and no groove weld size are specified on the welding symbol for single-groove and symmetrical double-groove welds, complete joint preparation is required, Figure 36–3.

8. When the groove welds extend only partly through the member being joined, the size of the weld is shown on the weld symbol, Figures 36–4, 36–5, and 36–6.

9. A dimension not in parentheses placed to the left of bevel, V-, J-, or U-groove weld symbol indicates only the depth of preparation.

10. Groove welds that are to be welded with approximately flush or convex faces without

postweld finishing are specified by adding the flush or convex contour symbol to the weld symbol.

11. Groove welds whose faces are to be finished flush or convex by postweld finishing are specified by adding both the appropriate contour and finishing symbol to the weld symbol. Standard finishing symbols are:

C — Chipping
G — Grinding
H — Hammering
M — Machining
R — Rolling

12. The size of flare-groove welds when no weld size is given is considered as extending only to the tangent points indicated by dimension "S," Figure 36–7. For application of flare-groove welds with partial joint preparation, see Figure 36–7.

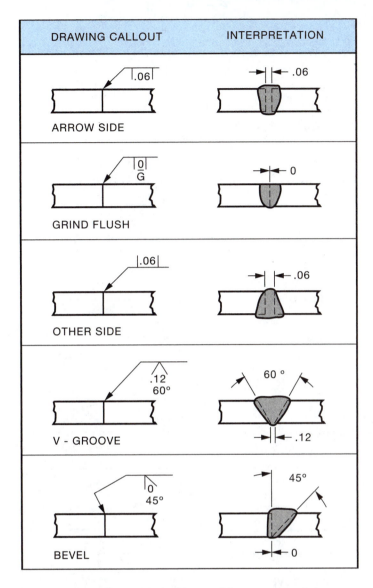

DRAWING CALLOUT	INTERPRETATION
ARROW SIDE	
GRIND FLUSH	
OTHER SIDE	
V - GROOVE	
BEVEL	

FIGURE 36–3 ■ *Single-groove welds—complete joint preparation.*

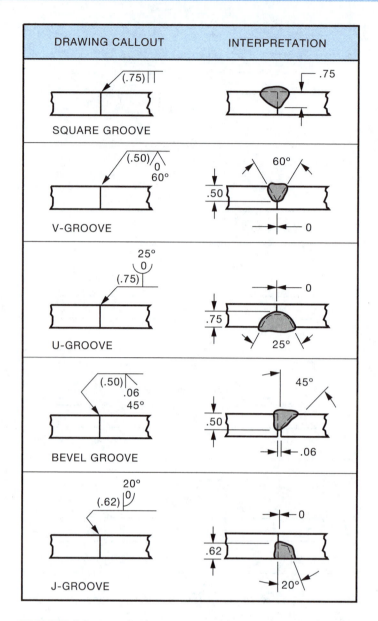

FIGURE 36–4 ■ *Single-groove welds—partial penetration.*

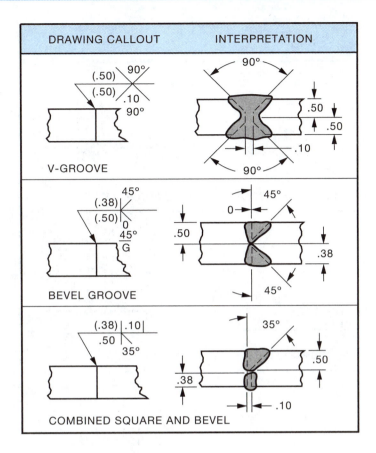

FIGURE 36–5 ■ *Double-groove welds.*

SUPPLEMENTARY SYMBOLS

Back and Backing Welds

The back or backing weld symbol is used to indicate bead-type back or backing welds of single-groove welds.

The back and backing weld symbols are identical. The sequence of welding determines which designation applies. The back weld is made *after* the groove weld and the backing weld is made *before* the groove weld.

1. The back weld symbol is placed on the side of the reference line opposite a groove weld

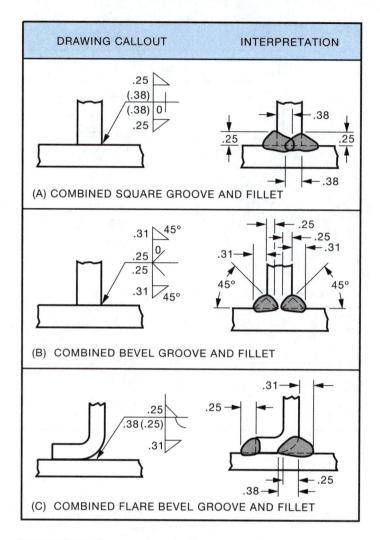

DRAWING CALLOUT	INTERPRETATION

(A) COMBINED SQUARE GROOVE AND FILLET

(B) COMBINED BEVEL GROOVE AND FILLET

(C) COMBINED FLARE BEVEL GROOVE AND FILLET

FIGURE 36–6 ■ *Combined groove and fillet weld.*

symbol. When a single reference line is used, "BACK WELD" is specified in the tail of the symbol. Alternately, if a multiple reference line is used, the back weld symbol is placed on a reference line next to the reference line specifying the groove weld, Figure 36–8(A).

2. The backing weld symbol is placed on the side of the reference line opposite the groove weld symbol. When a single reference line is used, "BACKING WELD" is specified in the tail of the arrow. If a multiple reference line is used, the backing weld sym-

bol is placed on a reference line prior to that specifying the groove weld, Figures 36–8(B) and (C).

Melt-Through Symbol

The melt-through symbol is used only when complete root penetration plus visible root reinforcement is required in welds made from one side.

The melt-through symbol is placed on the side of the reference line opposite the weld symbol, Figure 36–9.

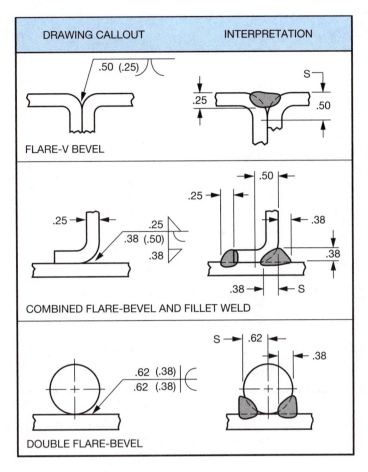

FIGURE 36–7 ■ *Flare-V and flare-bevel groove welds with partial joint preparation.*

The height of root reinforcement may be specified by placing the required dimension to the left of the melt-through symbol. The height of root reinforcement may be unspecified.

REFERENCES

ANSI/AWS A2.4-86
American Welding Society

INTERNET RESOURCES

eFunda. For information on the various types of welding, see: http://www.efunda.com/home.cfm

Machine Design. For information on welding and weld joints, see *Machine Design, Fastening/Joining Reference* at: http://www.machinedesign.com

Unified Engineering Inc. For information on groove welds, see: http://www.unified-eng.com/scitech/weld/weld.html

DRAWING CALLOUT	INTERPRETATION

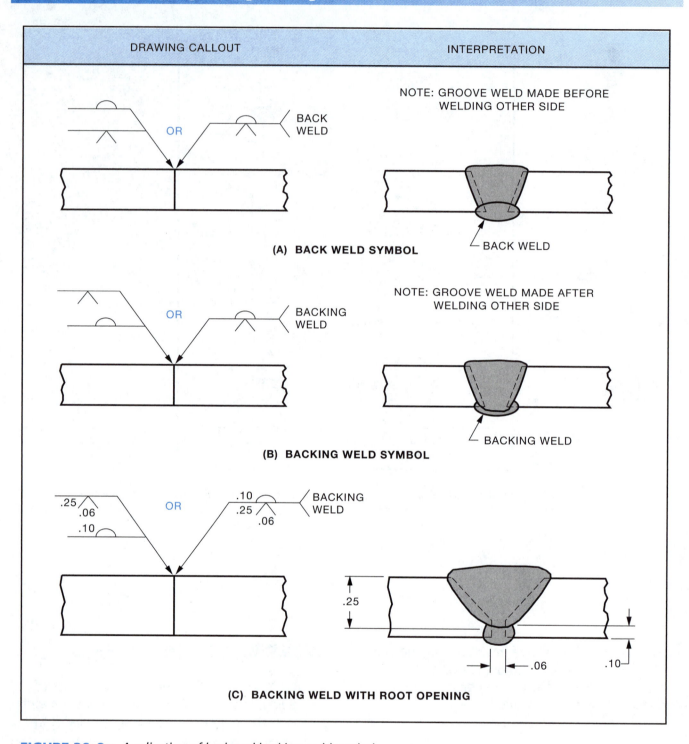

FIGURE 36–8 ■ *Application of back and backing weld symbol.*

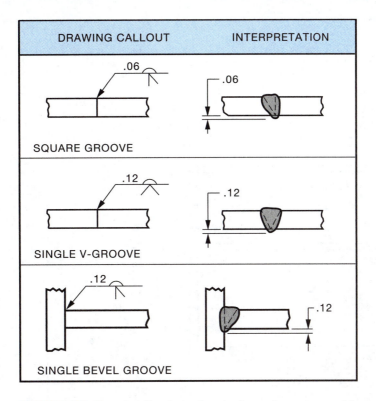

DRAWING CALLOUT	INTERPRETATION
SQUARE GROOVE	
SINGLE V-GROOVE	
SINGLE BEVEL GROOVE	

FIGURE 36–9 ■ *Application of melt-through groove weld symbols.*

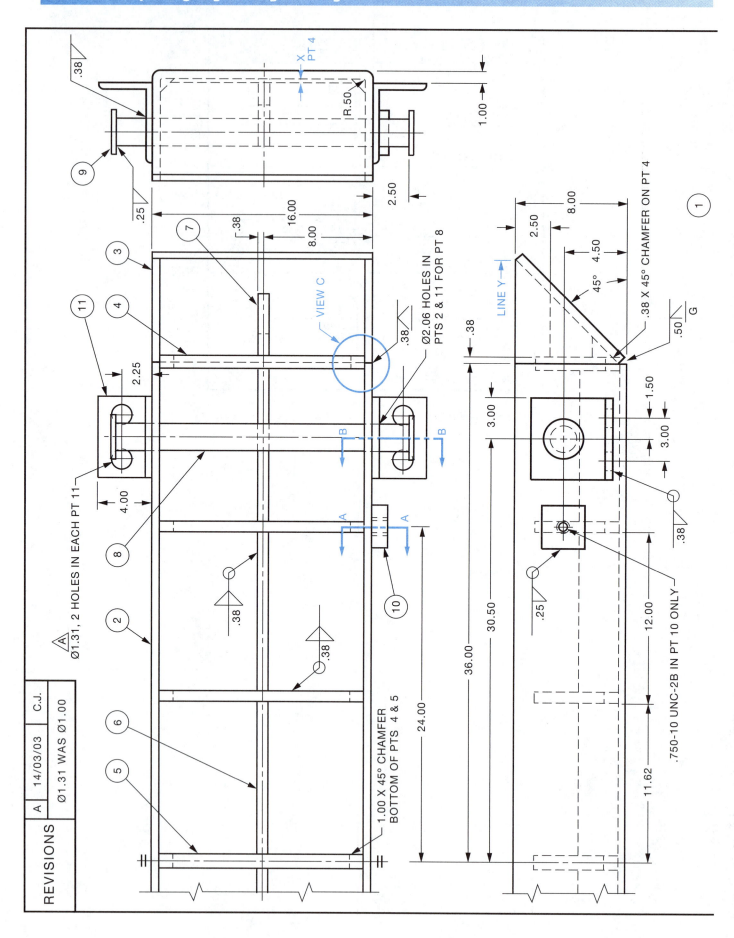

QUESTIONS:

1. How many Ø1.31 holes are there in the complete assembly?

2. How deep is the .750 tapped hole?

3. What was the original size of the Ø1.31 hole?

4. What is the overall height of the assembly?

5. Determine the distance from line Y to the center line of the assembly.

6. Determine distance X.

7. What is the developed width of part 2? Use inside dimensions of channel.

8. What is the clearance for fitting on the length of pt. 5?

9. What is the length of (A) pt. 2, (B) pt. 4, allow .25 for clearance, (C) pt. 7, (D) pt. 8?

10. What is the difference between pt. 4 and pt. 5?

11. How many 1.00 chamfers are needed?

12. What does G mean on .50 bevel weld symbols?

13. What type of weld is used to fasten pt. 11 to pt. 2?

14. What type of weld is used to join pt. 3 to pt. 2 at the sides?

15. How many parts make up the assembly?

16. Complete the missing sizes in the Bill Material. Use inside travel for calculating part 2.

NOTE:

UNLESS OTHERWISE SPECIFIED
- TOLERANCE ON LINEAR DIMENSIONS ±.04
- TOLERANCE ON ANGLES ±0.5°
- TOLERANCE ON HOLES ±.004

ASSIGNMENT:

ON A ONE INCH GRID SHEET (.10 IN. SQUARES) SKETCH SECTION VIEWS AT A-A AND B-B AND AN ENLARGED VIEW AT VIEW C. SHOW ONLY THE PARTS AND THE WELDS. SCALE 1:1.

	TOP VIEW	RIGHT SIDE VIEW
	FRONT VIEW	

ARRANGEMENT OF VIEWS

QTY	ITEM	MATL	DESCRIPTION	PT NO.
4	LOCATING ANGLE	STL	6.00 X 4.00 X .50 X 6.00 LG	11
2	GROUND BAR	STL BAR	1.00 X 3.00 X 3.00	10
4	RETAINER	STL PL	.50 X Ø3.00	9
2	DRAW BAR	STL RD	Ø2.00 X	8
2	GUSSET	STL BAR	.75 X 3.00 X	7
6	GUSSET	STL BAR	.75 X 3.00 X 11.25	6
5	GUSSET	STL BAR	.75 X 6.00 X 14.75	5
2	GUSSET	STL BAR	.75 X 6.00 X	4
2	END PLATE	STL PL	.50 X 10.62 X 25.90	3
1	BASE	STL PL	.50 X X	2
1	SKID ASS'Y			1

BASE SKID

A-94

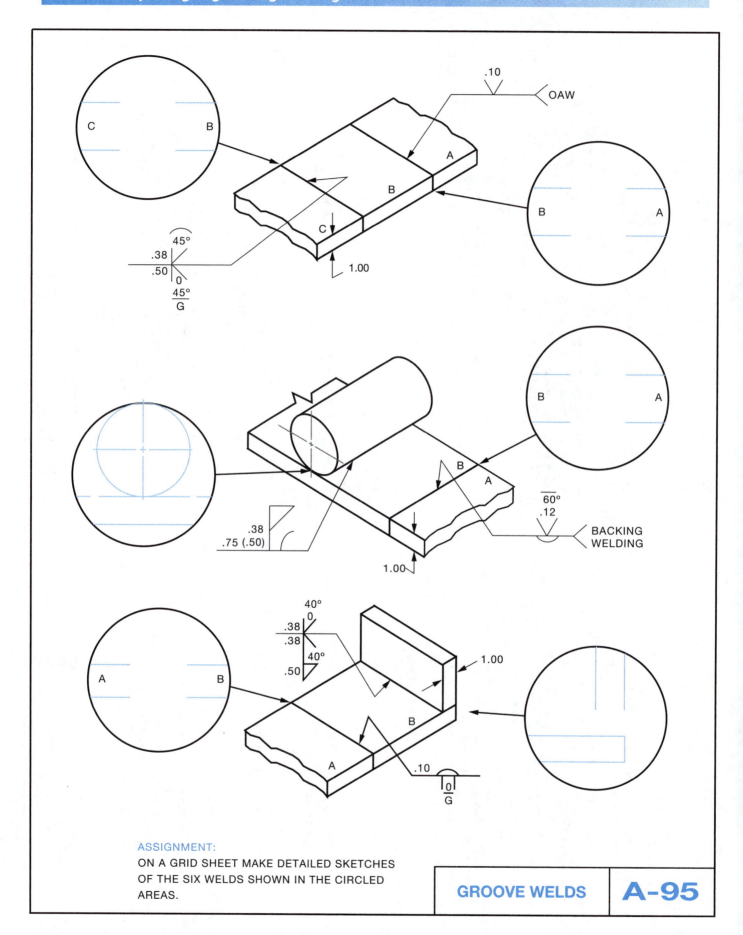

.10

OAW

C   B

A

B

B   A

45°
.38
.50   0
45°
G

C

1.00

.38
.75 (.50)

B
A

B   A

60°
.12

BACKING
WELDING

1.00

1.00

40°
0
.38
.38
40°
.50

1.00

A   B

B

A

.10
0
G

ASSIGNMENT:
ON A GRID SHEET MAKE DETAILED SKETCHES
OF THE SIX WELDS SHOWN IN THE CIRCLED
AREAS.

GROOVE WELDS   **A-95**

OTHER BASIC WELDS

PLUG AND SLOT WELDS

Plug and slot welds are used to join overlapping parts. The top member contains holes (round for plug welds) or slots (elongated for slot welds). Weld metal is deposited in the holes and fuses the two parts together. Theses holes may be completely or partially filled with weld metal.

Plug Welds (Figure 37–1)

1. Holes in the arrow-side member of a joint for plug welding are specified by placing the weld symbol below the reference line.

2. Holes in the other-side member of a joint for plug welding are indicated by placing the weld symbol above the reference line.

3. The size of a plug weld is shown on the same side and to the left of the weld symbol.

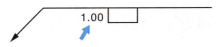

4. The included angle of countersunk or plug welds is the user's standard, unless otherwise specified. Included angle, when not the user's standard, is shown.

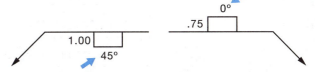

5. The depth of filling of plug welds is complete unless otherwise indicated. When the depth of filling is less than complete, the depth of filling, in inches or millimeters, is shown inside the weld symbol.

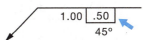

6. Pitch (center-to-center spacing) of plug welds is shown to the right of the weld symbol.

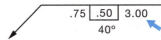

7. Plug welds that are to be welded with approximately flush or convex faces without postweld finishing are specified by adding the flush or convex contour symbol to the weld symbol.

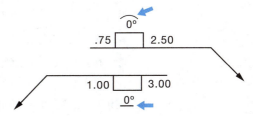

8. Plug welds whose faces are to be finished approximately flush or convex by postweld finishing are specified by adding both the appropriate contour and finishing symbol to the welding symbol. Welds that require a flat but not flush surface require an explanatory note in the tail of the symbol.

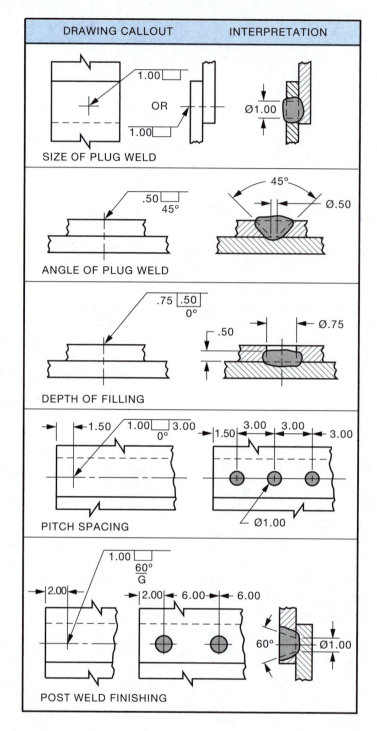

FIGURE 37–1 ■ *Plug welds.*

Slot Welds (Figure 37–2)

1. Slots in the arrow-side member of a joint for slot welding are specified by placing the weld symbol below the reference line. Slot orientation must be shown on the drawing.

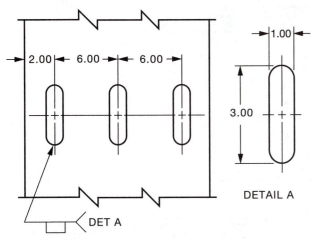

EXAMPLE 1 SLOTS PERPENDICULAR TO LINE OF WELD

DETAIL A

DET A

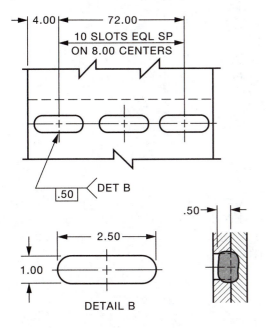

DET B

DETAIL B

EXAMPLE 2 SLOTS PARALLEL TO LINE OF WELD

FIGURE 37–2 ■ *Slot welds.*

4. Length, width, spacing, included angle of countersink, orientation, and location of slot welds cannot be specified on the welding symbol. These data are to be specified on the drawing or by a detail with reference to it on the welding symbol.

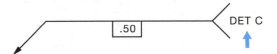

DET C

5. Slot welds that are to be welded with approximately flush or convex faces without postweld finishing are specified by adding the flush or convex contour symbol to the weld symbol.

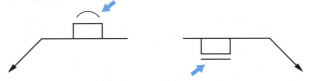

6. Slot welds whose faces are to be finished approximately flush or convex by postweld finishing are specified by adding both the appropriate contour and finishing symbol to the welding symbol. Welds that require a flat but not flush surface require an explanation note in the tail of the symbol.

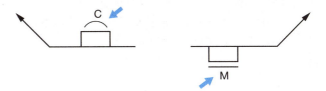

Spot Welds (Figure 37–3)

Spot welding is the most popular type of resistance welding used to join sheet metal parts. The parts to be joined are placed under pressure between two electrodes. An electrical charge is then passed between the two electrodes at controlled interval spacing.

2. Slots in the other-side member of a joint for slot welding are indicated by placing the weld symbol above the reference line.

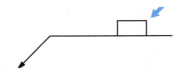

3. Depth of filling of slot welds is complete unless otherwise specified. When the depth of filling is less than complete, the depth of filling, in inches or millimeters, is shown inside the welding symbol.

1. The symbol for all spot or projection welds is a circle, regardless of the welding process used. There is no attempt to provide symbols for different ways of making a spot weld, such as resistance, arc, and electron

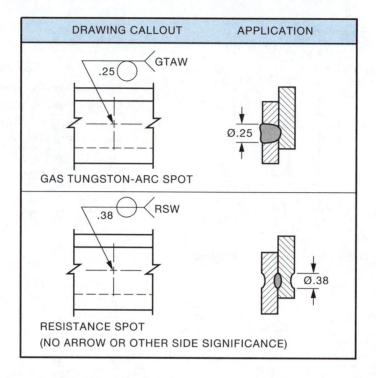

DRAWING CALLOUT	APPLICATION
GTAW .25	Ø.25
GAS TUNGSTON-ARC SPOT	
RSW .38	Ø.38
RESISTANCE SPOT (NO ARROW OR OTHER SIDE SIGNIFICANCE)	

FIGURE 37–3 ■ *Spot welds.*

beam welding. The symbol for a spot weld is a circle placed:

■ Below the reference line, indicating arrow side.

■ Above the reference line, indicating other side.

■ On the reference line, indicating that there is no arrow or other side.

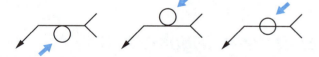

2. Dimensions of spot welds are shown on the same side of the reference line as the weld symbol, or on either side when the symbol is located astride the reference line and has no arrow-side or other-side significance. They are dimensioned by either the size or the strength. The size is designated as the diameter of the weld and is shown to the left of

the weld symbol. The strength of the spot weld is designated in pounds (or newtons) per spot and is shown to the left of the weld symbol.

SPECIFYING DIAMETER OF SPOT

SPECIFYING STRENGTH OF SPOT

3. The process reference is specified in the tail of the welding symbol.

4. When projection welding is used, the spot weld symbol is used and the projection welding process is referenced in the tail of the symbol. The spot weld symbol is located above or below (not on) the reference line to designate on which member the emboss-ment is placed.

5. The pitch (center-to-center spacing) is shown to the right of the weld symbol.

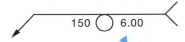

6. When spot welding extends less than the dis-tance between abrupt changes in the direc-tion of the welding or less than the full length of the joint, the extent is dimensioned.

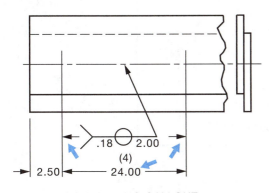

(A) DRAWING CALLOUT

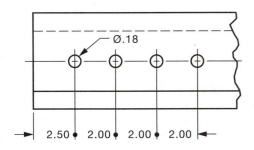

(B) INTERPRETATION

7. Where the exposed surface of either mem-ber of a spot welded joint is to be welded with approximately flush or convex faces without postweld finishing, that surface is specified by adding the flush or convex con-tour symbol to the weld symbol.

8. Spot welds whose faces are to be finished approximately flush, or convex by postweld finishing are specified by adding both the appropriate contour and finishing symbol to the welding symbol. Welds that require a flat but not flush surface require an ex-planatory note in the tail of the symbol.

Seam Welds (Figure 37–4)

Seam welding is similar to spot welding except that the charges between electrodes are more closely spaced which produces a continuous-type weld. Seam welds can be continuous or intermittent.

1. The symbol for all seam welds is a circle tra-versed by two horizontal parallel lines. This symbol is used for all seam welds regardless of the way they are made. The seam weld symbol is placed (1) below the reference line to indicate arrow side, (2) above the refer-ence line to indicate other side, and (3) on the reference line to indicate that there is no arrow or other side significance.

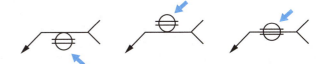

2. Dimensions of seam welds are shown on the same side of the reference line as the weld symbol or all on either side when the sym-bol is centered on the reference line. They are dimensioned by either size or strength. The size of the seam welds is designated as the width of the weld at the faying (fitted) surfaces and is shown to the left of the weld symbol. The strength of seam welds is des-ignated in pounds per linear inch (lb/in.) or newtons per millimeter (N/mm) and is shown to the left of the weld symbols.

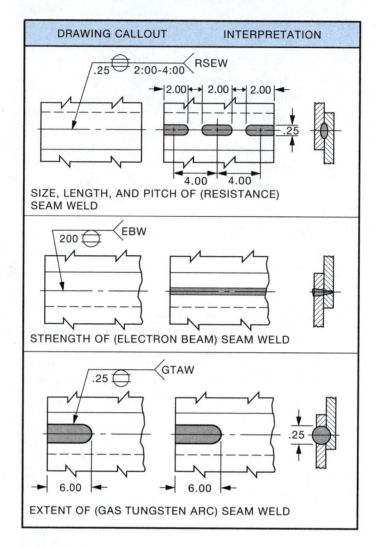

FIGURE 37–4 ■ *Seam welds.*

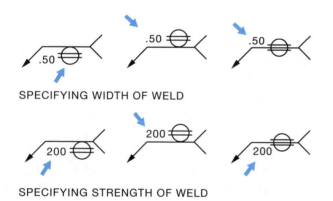

SPECIFYING WIDTH OF WELD

SPECIFYING STRENGTH OF WELD

3. The process reference is specified in the tail of the welding symbol.

4. The length of a seam weld, when indicated on the welding symbol, is shown to the right of the weld symbol. When seam welding extends for the full distance between abrupt changes in the direction of the welding, no length dimension needs to be shown on the welding symbol. When a seam weld extends less than the full length of the joint, the extent of the weld should be shown.

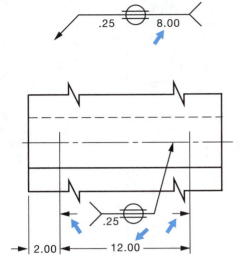

5. The pitch of an intermittent seam weld is shown as the distance between centers of the weld increments. The pitch is shown to the right of the length dimension.

6. When the exposed surface of either member of a seam-welded joint is to be welded with approximately flush or convex faces without postweld finishing, that surface is specified by adding the flush or convex contour symbol to the weld symbol.

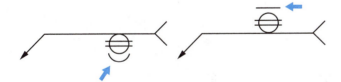

7. Seam welds with faces to be finished approximately flush or convex are specified by adding both the appropriate contour and finish symbol to the welding symbol.

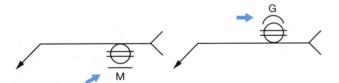

Flange Welds (Figure 37–5)

The following welding symbols are intended to be used for light-gage metal joints involving the flaring or flanging of the edges to be joined.

1. Edge-flange welds are shown by the edge-flange-weld symbol.

2. Corner-flange welds on joints detailed on the drawing are specified by the corner-flange weld symbol. Weld symbols are always drawn with the perpendicular leg to the left.

3. Corner-flange welds on joints not detailed on the drawing are specified by the corner-flange weld symbol. A broken arrow points to the member being flanged.

4. Edge-flange welds requiring complete joint penetration are specified by the edge-flange weld symbol with the melt-through symbol placed on the opposite side of the reference line. The same welding symbol is used for joints either detailed or not detailed on the drawing.

5. Corner-flange welds requiring complete joint penetration are specified by the corner-flange weld symbol with the melt-through symbol placed on the opposite side of the reference line. A broken arrow points to the member to be flanged where the joint is not detailed.

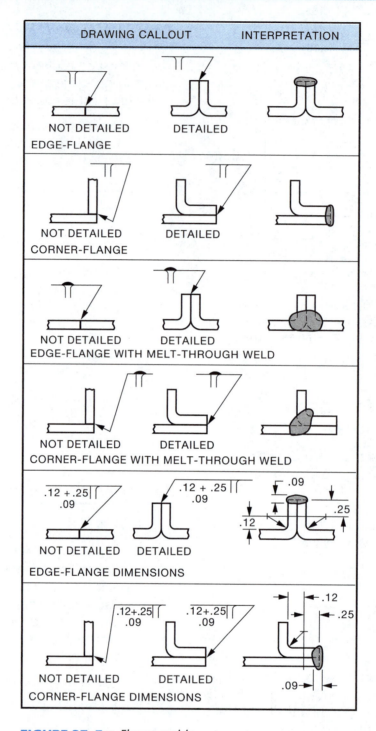

FIGURE 37–5 ■ *Flange welds.*

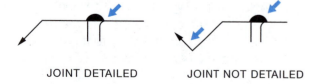

JOINT DETAILED   JOINT NOT DETAILED

6. Dimensions of flange welds are shown on the same side of the reference line as the weld symbol. The radius and the height, separated by a plus (+) are placed to the left of the weld symbol. The radius and the

height read in that order from left to right
along the reference line.

WHERE T = WELD THICKNESS
      H = HEIGHT OF FLANGE
      R = RADIUS OF FLANGE

7. The size (thickness) of flange welds is specified by a dimension placed above or below
the flange dimensions.

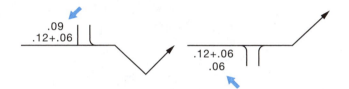

REFERENCES

ASME/AWS A2.4-86
American Welding Society

INTERNET RESOURCES

American Welding Society. For information on all
aspects of welding with links to related
organizations and materials, see:
http://www.aws.org

eFunda. For information on the various types of
welding, see: http://www.efunda.com/
home.cfm

Machine Design. For information on welding and
weld joints, see *Machine Design,
Fastening/Joining Reference* at: http://www.
machinedesign.com

Unified Engineering Inc. For information on plug,
spot, and flange welds, see: http://www.
unified-eng.com/scitech/weld/weld.html

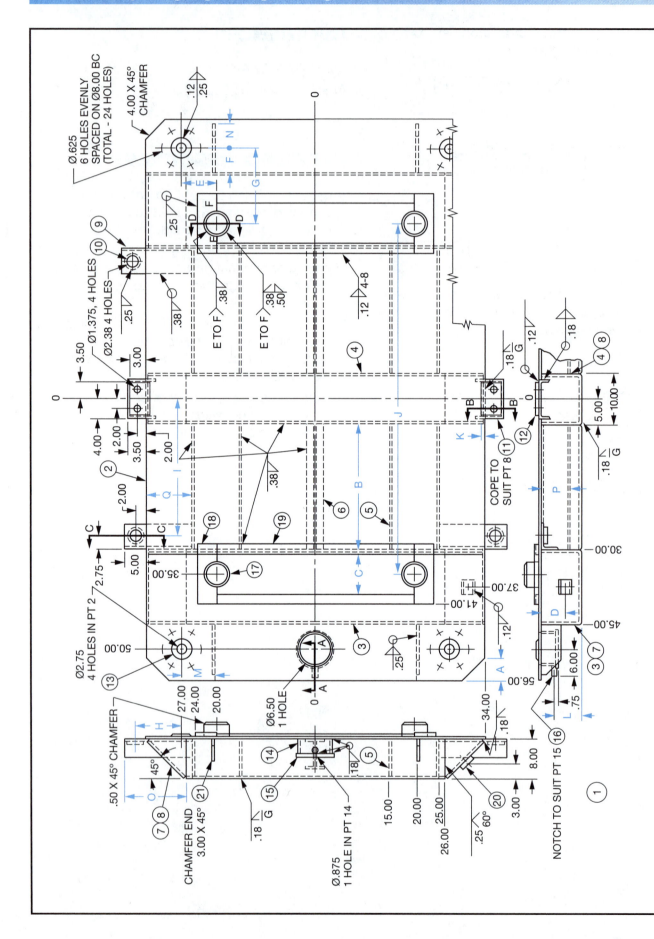

NOTES:
UNLESS OTHERWISE SPECIFIED:
- DIMENSIONS SYMMETRICAL AROUND CENTER LINE
- TOLERANCE ON DIMENSIONS ±.06 EXCEPT HOLES
- TOLERANCE ON HOLES ±.02
- TOLERANCE ON ANGLES ±1.0°

- DIMENSIONS ARE TO CENTER LINES UNLESS OTHERWISE SHOWN.

- REFER TO HANDBOOK FOR STRUCTURAL SIZES AND SHAPES.

QTY	ITEM	MATL	DESCRIPTION	PT NO.
4	RIB	STL	.50 X 4.00 X 10.00	21
1	GROUND PAD		DWG A - 4158	20
4	RETAINER	STL	.50 X 2.00 X LG	19
4	RETAINER	STL	.75 X 4.00 X LG	18
4	BUMPER PIN	STL	Ø5.00 X 3.00	17
1	NIPPLE	STL	.50 IPS X 3.00 LG	16
1	END PLATE	STL	Ø7.20 X .50 THK	15
1	SUMP	STL	6.00 IPS X 3.00 LG	14
4	FLANGE	STL	WELD -IN FLANGE 3.00 IPS	13
2	SUPPORT	STL	.75 X 4.50 X 7.00	12
2	SUPPORT	STL	C8 X 11.5 X LG	11
4	SUPPORT	STL	Ø3.50 X 1.00 THK	10
4	DRAW BAR	STL	L5.00 X 3.50 X .50 X LG	9
2	RIB END	STL	.50 X 10.50 X 20.00 LG	8
4	RIB END	STL	.50 X 10.50 X 25.00	7
2	RIB	STL	S6 X 12.5 X 25.00 LG	6
4	RIB	STL - PL	.50 X X LG	5
1	RIB	STL - PL	.50 X X LG	4
2	RIB	STL - PL	.50 X X LG	3
1	BASE PLATE	STL - PL	.50 X X LG	2
1	BASE ASSEMBLY			1

DRAWN D. SMITH

SCALE   NONE

BASE ASSEMBLY

A-96

LEFT
SIDE
VIEW

TOP
VIEW

FRONT
VIEW

ARRANGEMENT OF VIEWS

QUESTIONS:

1. Determine dimensions A to P.

2. What is the overall (A) width, (B) depth, and (C) height of the complete assembly?

3. Assuming the formed channels to have square inside corners, what is the width of the material used to make pt. 3? (Use inside travel.)

4. What is the (A) width, and (B) depth of the base plate, pt. 2?

5. If steel weighs .281 pound per cu. in., what is the weight of the base plate, pt. 2? (Disregard holes and chamfers.)

6. Determine the distance from the bottom of the S-beam (pt. 6) to the bottom of the base assembly.

7. What is the angle between the Ø.62 holes?

8. If 6-in. pipe has an OD of 6.62 in., what distance does pt. 15 project beyond pt. 14?

9. Determine the distance pt. 16 projects into pt.14.

10. What is the area enclosed by parts 17, 18, and 19?

ASSIGNMENTS:

1. ON ONE-INCH GRID SHEETS (.10 IN. SQUARES), SKETCH SECTION VIEWS AT A-A, B-B, C-C, AND D-D. SHOW ONLY THE PARTS AND THE WELDS. SCALE 1 : 1.

2. ON A ONE-CENTIMETER (1 MM SQUARES) GRID SHEET SKETCH ONE-HALF THE DEVELOPMENT OF PT. 8 SHOWING DIMENSIONS AND INDICATING BEND LINES. USE INSIDE TRAVEL. LET THE ONE-CENTIMETER GRID EQUAL ONE INCH.

DRAWING CALLOUT	INTERPRETATION

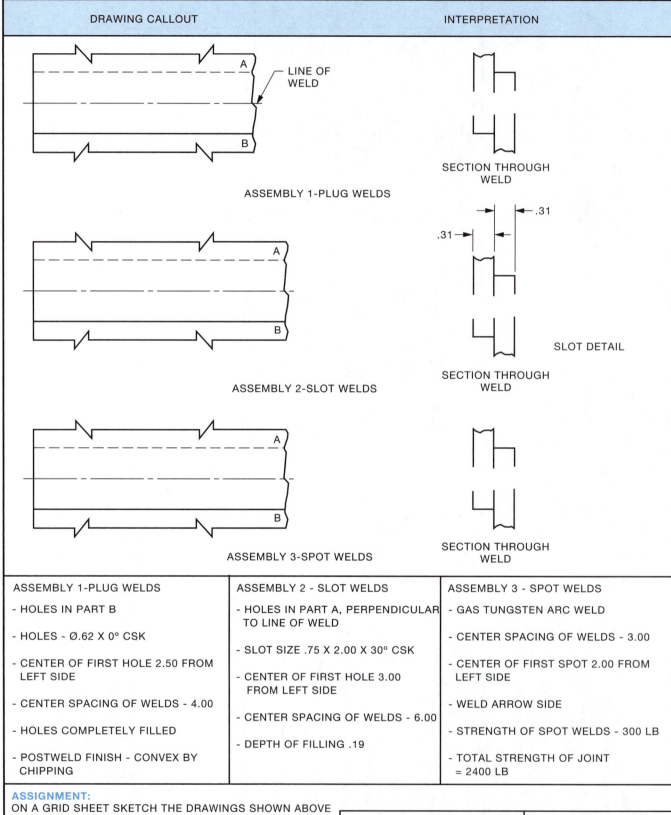

ASSEMBLY 1-PLUG WELDS

SECTION THROUGH WELD

ASSEMBLY 2-SLOT WELDS

.31

.31

SLOT DETAIL

SECTION THROUGH WELD

ASSEMBLY 3-SPOT WELDS

SECTION THROUGH WELD

LINE OF WELD

ASSEMBLY 1-PLUG WELDS	ASSEMBLY 2 - SLOT WELDS	ASSEMBLY 3 - SPOT WELDS
- HOLES IN PART B	- HOLES IN PART A, PERPENDICULAR TO LINE OF WELD	- GAS TUNGSTEN ARC WELD
- HOLES - Ø.62 X 0° CSK	- SLOT SIZE .75 X 2.00 X 30° CSK	- CENTER SPACING OF WELDS - 3.00
- CENTER OF FIRST HOLE 2.50 FROM LEFT SIDE	- CENTER OF FIRST HOLE 3.00 FROM LEFT SIDE	- CENTER OF FIRST SPOT 2.00 FROM LEFT SIDE
- CENTER SPACING OF WELDS - 4.00	- CENTER SPACING OF WELDS - 6.00	- WELD ARROW SIDE
- HOLES COMPLETELY FILLED	- DEPTH OF FILLING .19	- STRENGTH OF SPOT WELDS - 300 LB
- POSTWELD FINISH - CONVEX BY CHIPPING		- TOTAL STRENGTH OF JOINT = 2400 LB

ASSIGNMENT:
ON A GRID SHEET SKETCH THE DRAWINGS SHOWN ABOVE AND ADD THE WELDING SYMBOLS TO THE SKETCHES ON THE LEFT AND SHOW THE DETAIL OF THE WELD ON THE SKETCHES ON THE RIGHT.

PLUG, SLOT, AND SPOT WELDS

A-97

DRAWING CALLOUT	INTERPRETATION

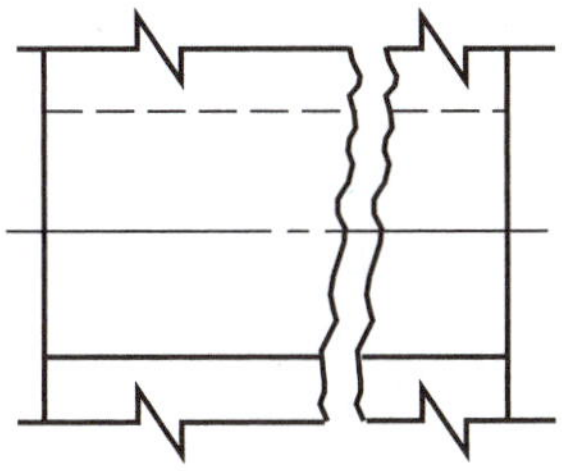

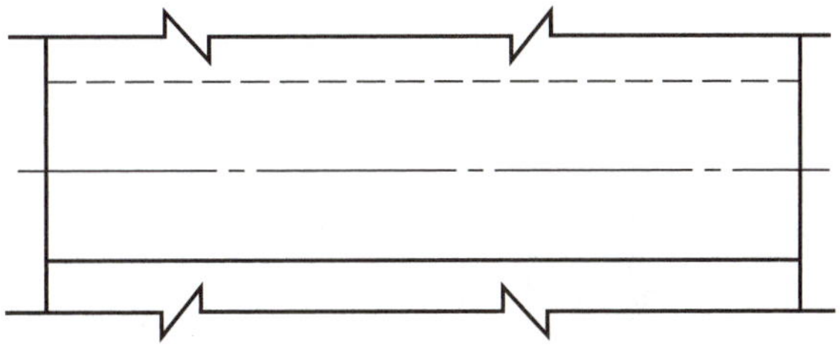

ASSEMBLY 1 - SEAM WELD

SECTION THROUGH WELD

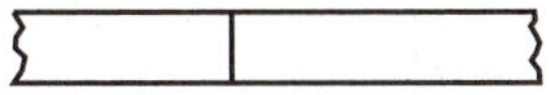

EDGE FLANGE WELD ONLY

COMPLETE JOINT PREPARATION

EDGE FLANGE WELD ONLY

COMPLETE JOINT PREPARATION

ASSEMBLY 2 - EDGE FLANGE WELD

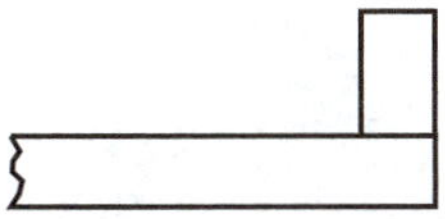

CORNER FLANGE WELD ONLY

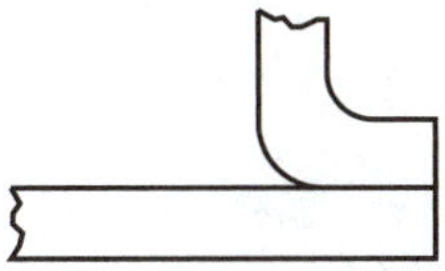

COMPLETE JOINT PREPARATION

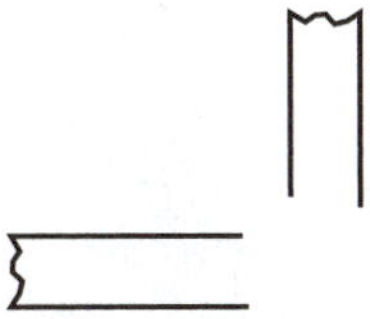

CORNER FLANGE WELD ONLY

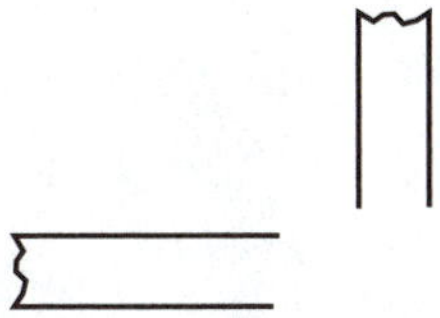

COMPLETE JOINT PREPARATION

ASSEMBLY 3 - CORNER FLANGE WELD

ASSEMBLY 1 RESISTANCE SEAM WELD	ASSEMBLY 2 EDGE FLANGE WELD	ASSEMBLY 3 CORNER FLANGE WELD
- SIZE .25	- WELD THICKNESS .10	- WELD THICKNESS .20
- LENGTH 1.50	- HEIGHT OF FLANGE .20	- HEIGHT OF FLANGE .16
- PITCH 3.00	- RADIUS OF FLANGE .30	- RADIUS OF FLANGE .24

ASSIGNMENT:
ON A GRID SHEET SKETCH THE DRAWINGS SHOWN ABOVE AND ADD THE WELDING SYMBOLS TO THE SKETCHES ON THE LEFT AND SHOW THE DETAIL OF THE WELD ON THE SKETCHES ON THE RIGHT.

SEAM AND FLANGE WELDS

A-98

38 UNIT

GEARS

The function of a gear is to transmit rotary or reciprocating motion from one machine part to another. Gears are often used to reduce or increase the r/min of a shaft. *Gears* are rolling cylinders or cones. They have teeth on their contact surfaces to ensure the transfer of motion, Figure 38–1.

There are many kinds of gears; they may be grouped according to the position of the shafts they connect. *Spur gears* connect parallel shafts; *bevel gears* connect shafts having intersecting axes; and *worm gears* connect shafts having axes that do not intersect. A spur gear with a rack converts rotary motion to reciprocating or linear motion. The smaller of two gears is the *pinion*.

A simple gear drive consists of a toothed driving wheel meshing with a similar driving wheel. Tooth forms are designed to ensure uniform angular rotation of the driven wheel during tooth engagement.

FIGURE 38–1 ■ *Gears.*

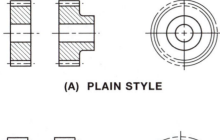

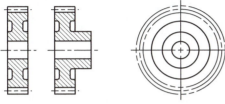

(A) PLAIN STYLE

(B) WEBBED STYLE

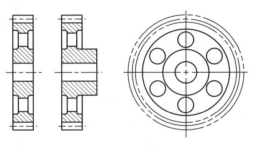

(C) WEBBED WITH CORED HOLES

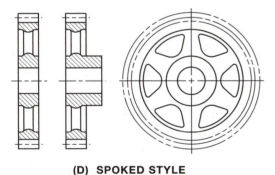

(D) SPOKED STYLE

FIGURE 38–2 ■ *Stock spur gears.*

SPUR GEARS

Spur gears, which are used for drives between parallel shafts, have teeth on the rim of the wheel, Figure 38–2. A pair of spur gears operates as though it consists of two cylindrical surfaces with formed teeth that maintain constant speed ratio between the driving wheel and the driven wheel.

ORIGINAL FRICTIONAL CYLINDRICAL
SURFACE (PITCH CIRCLE)

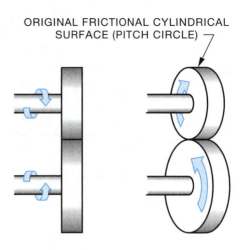

FIGURE 38–3 ■ *The pitch circle of spur gears.*

Gear design is complex, dealing with such problems as strength, wear, noise, and material selection. Usually, a designer selects a gear from a catalog. Most gears are made of cast iron or steel. However, brass, bronze, and plastics are used when factors such as wear or noise must be considered.

Theoretically, the teeth of a spur gear are built around the original frictional cylindrical surface called the *pitch circle,* Figure 38–3.

The angle between the direction of pressure between contacting teeth and a line tangent to the pitch circle is the *pressure angle.*

The 14.5-degree pressure angle has been used for many years and remains useful for duplicate or replacement gearing.

The 20- and 25-degree pressure angles have become the standard for new gearing because of their smoother and quieter operation and greater load-carrying ability.

Gear Terms

The following terms, shown in Figures 38–4 and 38–5, are used in spur gear train calculations.

Pitch Diameter (PD)

The diameter of an imaginary circle on which the gear tooth is designed.

Number of Teeth (N)

The number of teeth on the gear.

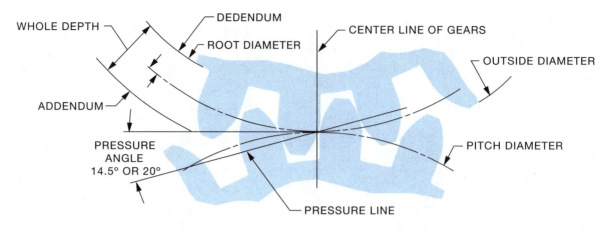

FIGURE 38–4 ■ *Gear teeth terms.*

Diametral Pitch (DP)

The diametral pitch is a ratio of the number of teeth (N) to a unit length of pitch diameter, DP = N/PD.

Outside Diameter (OD)

The overall gear diameter.

Root Diameter (RD)

The diameter at the bottom of the tooth.

Addendum (ADD)

The radial distance from the pitch circle to the top of the tooth.

Dedendum (DED)

The radial distance from the pitch circle to the bottom of the tooth.

Whole Depth (WD)

The overall height of the tooth.

Clearance

The radial distance between the bottom of one tooth and the top of the mating tooth.

Circular Pitch

The distance measured from the point of one tooth to the corresponding point on the adjacent tooth on the circumference of the pitch diameter.

Circular Thickness

The thickness of a tooth or space measured on the circumference of the pitch diameter.

Chordal Thickness

The thickness of a tooth or space measured along a chord on the circumference of the pitch diameter.

Chordal Addendum

Chordal addendum, also known as corrected addendum, is the perpendicular distance from the chord to the outside circumference of the gear.

Chordal Thickness and Corrected Addendum

After the gear teeth have been milled or generated, the width of the tooth space and the thickness of the tooth, measured on the pitch circle, should be equal.

Instead of measuring the curved length of line known as *circular thickness of tooth,* it is more convenient to measure the length of the straight line *(chordal thickness),* which connects the ends of that arc.

TERM AND SYMBOL	FORMULA	
	INCH GEARS	METRIC GEARS
Pitch Diameter - PD	PD = N ÷ DP	PD = MDL X N
Number of Teeth - N	N = PD X DP	N = PD ÷ MDL
Module - MDL		MDL = PD ÷ N
Diametral Pitch - DP	DP = N ÷ PD	
Addendum - ADD	14.5° or 20° ADD = 1 ÷ DP 20° stub ADD = 0.8 ÷ DP	14.5° or 20° ADD = MDL 20° stub ADD = 0.8 X MDL
Dedemdum - DED	14.5° or 20° DED = 1.157 ÷ DP 20° stub DED = 1 ÷ DP	14.5° or 20° DED = 1.157 X MDL 20° stub DED = MDL
Whole Depth - WD	14.5° or 20° WD = 2.157 ÷ DP 20 stub WD = 1.8 ÷ DP	14.5° or 20° WD = 2.157 X MDL 20° stub WD = 1.8 X MDL
Clearance - CL	14.5° or 20° CL = 0.157 ÷ DP 20° stub CL = 0.2 ÷ DP	14.5° or 20° CL = 0.157 X MDL 20° stub CL = 0.2 X MDL
Outside Diameter - OD	14.5° or 20° OD = PD + 2 ADD = (N + 2) ÷ DP	14.5° or 20° OD = PD + 2 ADD = PD + 2 MDL
	20° stub OD = PD + 2 ADD = (N + 1.6) ÷ DP	20° stub OD = PD + 2 ADD = PD + 1.6 MDL
Root Diameter - RD	14.5° or 20° RD = PD - 2 DED = (N - 2.314) ÷ DP	14.5° or 20° RD = PD - 2 DED = PD - 2.314 MDL
	20° stub RD = PD - 2 DED = (N - 2) ÷ DP	20° stub RD = PD - 2 DED = PD - 2 MDL
Base Circle - BC	BC = PD Cos PA	BC = PD Cos PA
Pressure Angle - PA	14.5° or 20°	14.5° or 20°
Circular Pitch - CP	CP = 3.1416 PD ÷ N = 3.1416 ÷ DP	CP = 3.1416 PD ÷ N = 3.1416 MDL
Circular Thickness - T	T = 3.1416 PD ÷ 2 N = 1.5708 ÷ DP	T = 3.1416 PD ÷ 2N = 1.5708 PD ÷ N = 1.5708 MDL
Chordal Thickness - Tc	Tc = PD sin (90° ÷ N)	Tc = PD sin (90° ÷ N)
Chordal Addendum - ADDc	$ADDc = ADD + (T^2 ÷ 4\ PD)$	$ADDc = ADD + (T^2 ÷ 4PD)$
Working Depth - WKG DP	WKG DP = 2 ADD	WKG DP = 2 ADD

FIGURE 38–5 ■ *Spur gear symbols and formulas.*

The *corrected* or *chordal addendum* is the radial distance extending from the addendum circle to the chord.

A gear tooth vernier caliper may be used to measure accurately the thickness of a gear tooth at the pitch line. To use the gear tooth vernier, which measures only a straight line or chordal distance, it is necessary to set the tongue to the computed chordal addendum and then measure the chordal thickness.

Working Drawings of Spur Gears

The working drawings of gears, which are normally cut from blanks are not complicated. A sectional view is sufficient unless a front view is required to show web or arm details. Because the teeth are cut to shape by cutters, they need not be shown in the front view, Assignment A-99.

ANSI recommends the use of phantom lines for the outside and root circles. In the section view, the root and outside circles are shown as solid lines.

The dimensioning for the gear is divided into two groups because the finishing of the gear blank and the cutting of the teeth are separate operations in the shop. The gear blank dimensions are shown on the drawing whereas the gear tooth information is given in a table.

The only differences in terminology between inch-size and metric-size gear drawings are the *terms diametral pitch* and *module.*

For inch-size gears, the term *diametral pitch* is used instead of the term *module.* The diametral pitch is a ratio of the number of teeth to unit length of pitch diameter.

$$\text{Diametral pitch} = DP = \frac{N}{PD}$$

Module is the term used on metric gears. It is the length of pitch diameter per tooth measured in millimeters.

$$\text{Module} = MDL = \frac{PD}{N}$$

From these definitions it can be seen that the module is equal to the reciprocal of the diametral pitch and thus is not its metric dimensional equivalent. If the diametral pitch is known, the module can be obtained.

$$\text{Module} = 25.4 \div \text{diametral pitch}$$

Gears used in North America are designed in the inch system and have a standard diametral pitch instead of a preferred standard module. Therefore, it is recommended that the diametral pitch be referenced beneath the module when gears designed with standard inch pitches are used. For gears designed with standard modules, the diametral pitch need not be referenced on the gear drawing. The standard modules for metric gears are 0.8, 1, 1.25, 2.25, 3, 4, 6, 7, 8, 9, 10, 12, and 16. See Figure 38–6 for gear teeth sizes.

Examples of Spur Gear Calculations

The pitch diameter of a gear can easily be found if the number of teeth and diametral pitch are known.

MODULE FOR METRIC SIZE GEARS	DIAMETRAL PITCH FOR INCH SIZE GEARS	PRESSURE ANGLE 14.5°	PRESSURE ANGLE 20°
6.35	4		
5.08	5		
4.23	6		
3.18	8		
2.54	10		
2.17	12		
1.59	16		
1.27	20		
1.06	24		

NOTE: MODULE SIZES SHOWN ARE CONVERTED INCH SIZES

FIGURE 38–6 ■ *Gear teeth sizes.*

The outside diameter is equal to the pitch diameter plus two addendums. The addendum for a 14.5- or 20-degree spur gear tooth is equal to 1 ÷ DP.

Examples

1. A 14.5-degree spur gear has a DP of 4 and 34 teeth.

$$\text{Pitch diameter} = \frac{N}{DP} = 34 \div 4 = 8.500 \text{ in.}$$

$$OD = PD + 2\ ADD = 8.500 + 2(1/4) = 9.000 \text{ in.}$$

2. The outside diameter of a 14.5-degree spur gear is 6.500 in. The gear has 24 teeth.

$$OD = \frac{N + 2}{DP} = \frac{24 + 2}{DP}$$

$$DP = \frac{26}{6.500} = 4$$

$$\text{Addendum} = \frac{1}{DP} = 1/4 = .250 \text{ in.}$$

$$\text{Pitch diameter} = OD - 2\ ADD$$

$$= 6.500 - 2\ (.250)$$

$$= 6.000 \text{ in.}$$

3. A 14.5-degree spur gear has a module of 6.35 and 38 teeth.

$$\text{Pitch diameter} = N \times MDL$$

$$= 38 \times 6.35$$

$$= 241.3 \text{ mm}$$

$$OD =$$

$$PD + 2\ ADD = 241.3 + 2\ (6.35)$$

$$= 254 \text{ mm}$$

REFERENCE

ASME Y14.7.1-1971 (R2003) Gear Drawing Standards—Part 1

INTERNET RESOURCES

Animated Worksheets. For information on spur gears, see: http://www.animatedworksheets. co.uk (mechanisms)

Boston Gear. For information on all aspects of gears and gear drives, see: http://www. bostongear.com

Flying Pig. For information on mechanisms and movements, see: http://www.flying-pig.co.uk (mechanisms and movements)

Howstuffworks. For information on all types of gears and gear applications, see: http://science.howstuffworks.com/gear3.htm

Machine Design. For information on spur gears and related topics, see *Machine Design, Mechanical Reference* at: http://www.machinedesign.com

TechStudent.Com. For information, including illustrations, on spur gears and simple gear trains, see: http://www.technologystudent.com (Gears and Pulleys)

QUESTIONS:

1. What is the hub diameter?

2. What is the maximum thickness of the spokes?

3. What is the average width of the spokes?

4. How many surfaces indicate that allowance must be added to pattern for finishing?

5. Determine distance Ⓙ for the pattern. Assume .10 is allowed on pattern for each surface to be finished.

6. Determine distance Ⓚ for the pattern.

7. What is the outside diameter of the pattern?

8. Determine distance Ⓛ for the pattern.

9. What is the width of the pattern?

10. What is the diametral pitch of the gear?

11. Calculate the center-to-center distance if this gear were to mesh with a pinion having (A) 24 teeth, (B) 36 teeth, (C) 32 teeth.

12. Calculate distances Ⓔ through Ⓗ.

13. Calculate the following: addendum, dedendum, circular pitch, and root diameter.

14. Complete the missing information in the cutting data table.

15. Calculate the limits of size for the Ø1.500 hole if an LT3 fit between the hole and shaft is required.

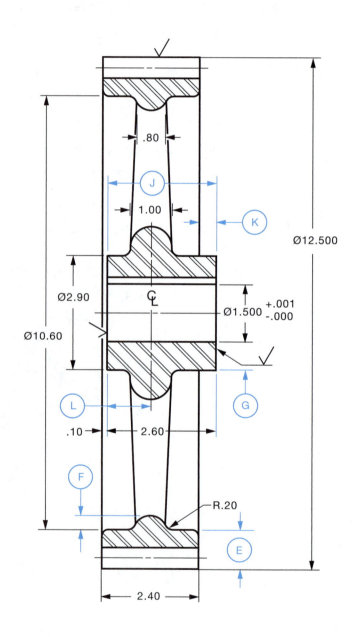

CUTTING DATA	
NUMBER OF TEETH	48
PITCH DIAMETER	
DIAMETRAL PITCH	4
PRESSURE ANGLE	20°
WHOLE DEPTH	
CHORDAL ADDENDUM	
CHORDAL THICKNESS	

NOTE: ROUNDS & FILLETS TO BE R.10
UNLESS OTHERWISE SPECIFIED

SHOWN TO BE .10 $\overset{63}{\bigtriangledown}$

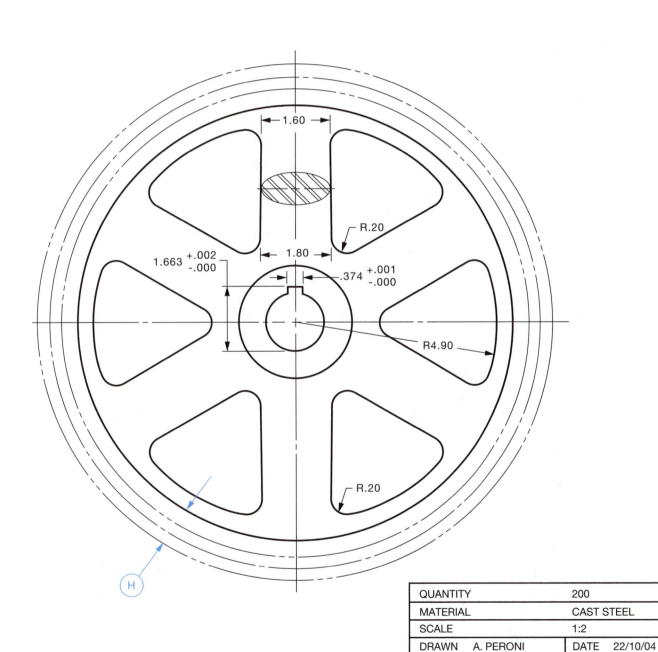

1.60

R.20

1.663 $^{+.002}_{-.000}$

1.80

.374 $^{+.001}_{-.000}$

R4.90

R.20

H

QUANTITY		200	
MATERIAL		CAST STEEL	
SCALE		1:2	
DRAWN A. PERONI		DATE 22/10/04	

SPUR GEAR

A-99

INCH GEARS

TERM AND SYMBOL	GEAR 1	GEAR 2	GEAR 3	GEAR 4
Pitch Diameter - PD		5.000	3.000	2.250
Number of Teeth - N	24	40		36
Diametral Pitch - DP	6		10	
Addendum - ADD				
Dedendum - DED				
Whole Depth - WD				
Clearance - CL				
Outside Diameter - OD				
Root Diameter - RD				
Pressure Angle - PA	20°	14.5°	20°	14.5°
Circular Pitch - CP				
Circular Thickness - T				
Chordal Thickness - Tc				
Chordal Addendum - ADDc				

METRIC GEARS

TERM AND SYMBOL	GEAR 5	GEAR 6	GEAR 7	GEAR 8
Pitch Diameter - PD		89.04		
Number of Teeth - N	40			30
Module - MDL	5.08	3.18	6	4
Addendum - ADD				
Dedendum - DED				
Whole Depth - WD				
Clearance - CL				
Outside Diameter - OD			228	
Root Diameter - RD				
Pressure Angle - PA	14.5°	20°	20°	14.5°
Circular Pitch - CP				
Circular Thickness - T				
Chordal Thickness - Tc				
Chordal Addendum - ADDc				

SPUR GEAR CALCULATIONS

A–100

ASSIGNMENT:

SKETCH CHARTS SIMILAR TO THOSE SHOWN ABOVE AND COMPLETE THE MISSING INFORMATION.

BEVEL GEARS

Drawings of bevel gears may be more easily interpreted and understood as a result of having a working knowledge of the parts, principles, and formulas underlying spur gears.

Spur gears transmit motion by or through shafts that are parallel and in the same plane, whereas bevel gears transmit motion between shafts that are in the same plane but whose axes would meet if extended. Theoretically, the teeth of a spur gear may be said to be built about the original frictional cylindrical surface known as *pitch circle*, whereas the teeth of a bevel gear are formed about the frustum of the original conical surface called *pitch cone*, Figure 39–1.

One type of bevel gear that is commonly used is the miter gear. The term miter gear refers to a pair of bevel gears of the same size that transmit motion at right angles.

Whereas any two spur gears of the same diametral pitch will mesh, this is not true of bevel gears except for miter gears. On each pair of mating bevel gears, the diameters of the gears determine the angles at which the teeth are cut.

Working drawings of bevel gears, like spur gears, give only the dimensions of the bevel gear blank. Cutting data for the teeth are given in a note or table. A single section view is used unless a second view is required to show such details as spokes. Sometimes both the bevel gear and pinion are drawn together. Dimensions and cutting data depend on the method used in cutting the teeth, but the information in Figures 39–2 and 39–3 is commonly used.

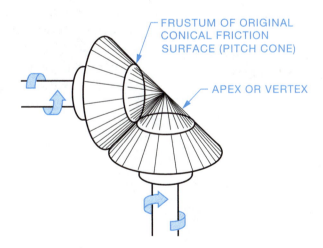

FRUSTUM OF ORIGINAL
CONICAL FRICTION
SURFACE (PITCH CONE)

APEX OR VERTEX

FIGURE 39–1 ■ *Principle as applied to bevel gears.*

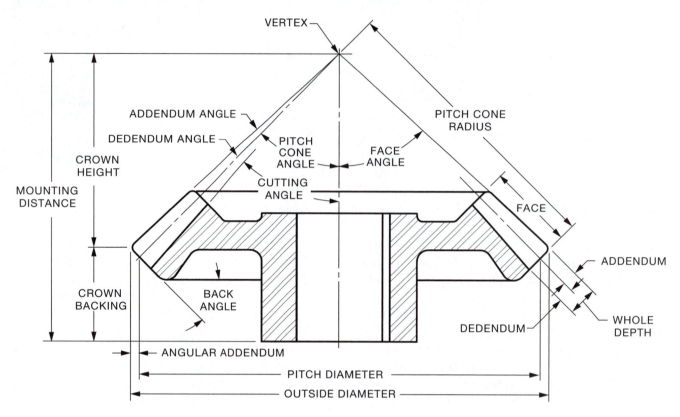

FIGURE 39–2 ■ *Bevel gear nomenclature.*

REFERENCE

ASME Y14.7.2-1978 (2003) Gears and Spline Drawing Standards—Part 2

INTERNET RESOURCES

Animated Worksheets. For information on bevel gears, see: http://www.animatedworksheets. co.uk (mechanisms)

Boston Gear. For information on all aspects of gears and gear drives, see: http://www. bostongear.com

Howstuffworks. For information on all types of gears and gear applications, see: http://science. howstuffworks.com/gear3.htm

Machine Design. For information on miter gears and gear applications, see *Machine Design, Mechanical Reference* at: http://www. machinedesign.com

TechStudent.Com. For information, including illustrations, on bevel gears see: http://www. technologystudent.com (Gears and Pulleys)

TERM	FORMULA
Addendum, dedendum, whole depth, pitch diameter, diametral pitch, number of teeth, circular pitch, chordal thickness, circular thickness	Same as for spur gears. Refer to Figure 38-5
Pitch cone angle (Pitch angle)	$\text{Tan pitch angle} = \dfrac{\text{PD of gear}}{\text{PD of pinion}} = \dfrac{\text{N of gear}}{\text{N of pinion}}$
Pitch cone radius	$\dfrac{\text{PD}}{2 \times \text{sin of pitch angle}}$
Addendum angle	$\text{Tan addendum angle} = \dfrac{\text{Addendum}}{\text{Pitch cone radius}}$
Dedendum angle	$\text{Tan dedendum angle} = \dfrac{\text{Dedendum}}{\text{Pitch cone radius}}$
Face angle	Pitch cone angle plus addendum angle
Cutting angle	Pitch cone angle minus dedendum angle
Back angle	Same as pitch cone angle
Angular addendum	Cosine of pitch cone angle X addendum
Outside diameter	Pitch diameter plus two angular addendums
Crown height	Divide 1/2 the outside diameter by the tangent of the face angle
Face width	$1\frac{1}{2}$ to $2\frac{1}{2}$ times the circular pitch
Chordal addendum	$\text{Addendum} + \dfrac{\text{circular thickness}^2 \times \text{cos pitch cone angle}}{4\ \text{PD}}$

FIGURE 39–3 ■ *Bevel gear formulas.*

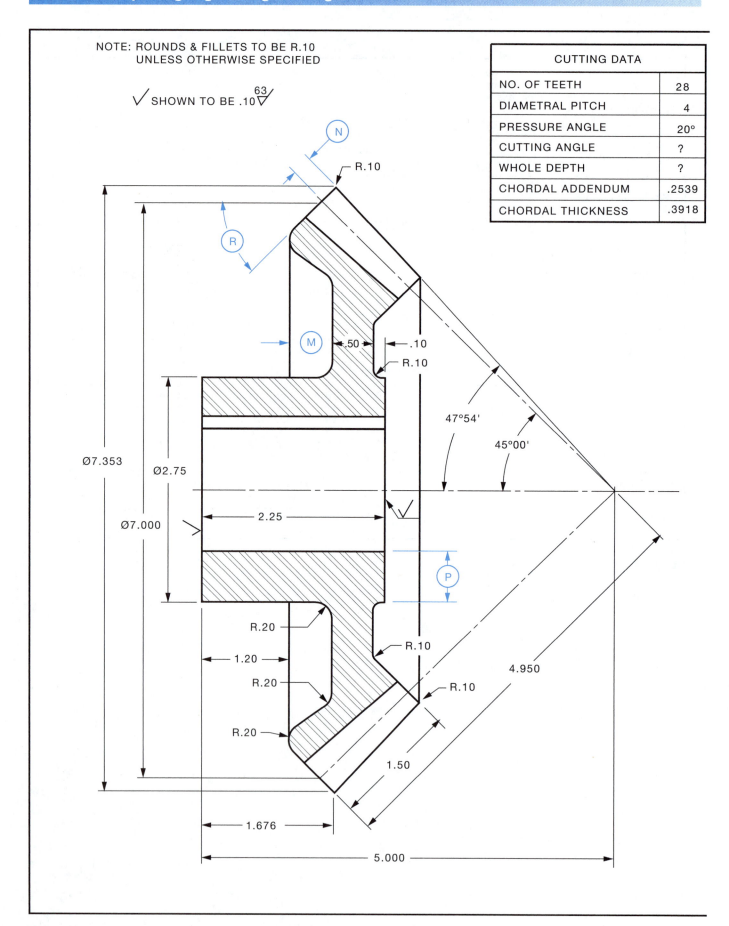

NOTE: ROUNDS & FILLETS TO BE R.10
UNLESS OTHERWISE SPECIFIED

√ SHOWN TO BE .10 ⁶³∇

CUTTING DATA	
NO. OF TEETH	28
DIAMETRAL PITCH	4
PRESSURE ANGLE	20°
CUTTING ANGLE	?
WHOLE DEPTH	?
CHORDAL ADDENDUM	.2539
CHORDAL THICKNESS	.3918

N

R.10

R

M .50 .10

R.10

47°54'

45°00'

Ø7.353

Ø2.75

Ø7.000

2.25

P

R.20

1.20

R.20

R.10

4.950

R.20

R.10

1.50

1.676

5.000

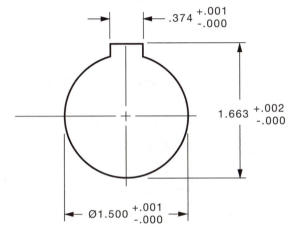

.374 +.001 / -.000

1.663 +.002 / -.000

Ø1.500 +.001 / -.000

QUESTIONS:

1. The drafter neglected to put on angle dimension Ⓡ. What should it be?

2. How many finished surfaces are indicated?

3. List those dimensions shown on the drawing which are not used by the patternmaker. Assume that the hole will not be cored.

4. What is the pitch cone angle?

5. What is the pitch diameter?

6. List those dimensions on the drawing which the machinist would use to machine the blank before the teeth are cut.

7. What is the depth of teeth at the large end?

8. What is the pitch cone radius?

9. What is the face angle?

10. What is the addendum angle?

11. What is the cutting angle?

12. What is the mounting distance?

13. What is the crown height?

14. What is the angular addendum?

15. What is the face width?

16. Determine dimensions Ⓜ, Ⓝ, Ⓟ.

17. Calculate the limits of size for the Ø1.500 hole if an LT2 fit between the hole and shaft is required.

18. What was the width of the cast hub before finishing?

QUANTITY	50
MATERIAL	CAST STEEL
SCALE	1:1
DRAWN L. NICHOLL	DATE 06/09/04
MITER GEAR	A-101

40 UNIT

GEAR TRAINS

Center Distance

The center distance between the two shaft centers is determined by adding the pitch diameter of the two gears together and dividing the sum by 2.

Examples

1. An 8DP, 24-tooth pinion mates with a 96-tooth gear. Find the center distance.

   Pitch diameter of pinion =
   N ÷ DP = 24 ÷ 8 = 3.000 in.

   Pitch diameter of gear =
   N ÷ DP = 96 ÷ 8 = 12.000 in.

   Sum of the two pitch diameters =
   3.000 + 12.000 = 15.000 in.

   Center distance =
   1/2 sum of the two pitch diameters =

   $\dfrac{15.000}{2} = 7.500$ in.

2. A 2.54 module, 28-tooth pinion mates with a 84-tooth gear. Find the center distance.

   Pitch diameter (PD) =
   number of teeth × module =
   28 × 2.54 = 71.12 (pinion) =
   84 × 2.54 = 213.36 (gear)

   Sum of the two pitch diameters =
   71.12 + 213.36 = 284.48 mm

   Center distance =
   1/2 sum of the two pitch diameters =
   284.48 ÷ 2 = 142.24 mm

RATIO

The ratio of gears is a relationship between any of the following:

- The r/min (revolutions per minute) of the gears
- The number of teeth on the gears
- The pitch diameter of the gears

The ratio is obtained by dividing the larger value of any of the three by the corresponding smaller value.

3. A gear rotates at 90 r/min and the pinion at 360 r/min

   $\text{Ratio} = \dfrac{360}{90} = 4$ or ratio = 4:1

4. A gear has 72 teeth; the pinion, 18 teeth.

   $\text{Ratio} = \dfrac{72}{18} = 4$ or ratio = 4:1

5. A gear with a pitch diameter of 8.500 in. meshes with a pinion having a pitch diameter of 2.125 in.

   $\text{Ratio} = \dfrac{\text{PD of gear}}{\text{PD of pinion}} = \dfrac{85.500}{2.125} = 4$

   or ratio = 4:1

Figure 40–1 illustrates how this type of information would be shown on an engineering sketch.

Motor Drive

Assignment A-102 shows a motor drive similar to the type used to operate a load-ratio control switch on a power transformer.

The load-ratio control switch is operated by a small motor with a speed of 1080 r/min. The shaft

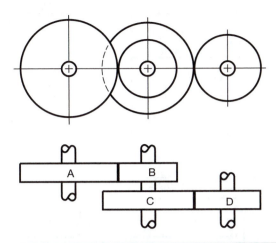

GEAR	PD	N	DP	R/MIN	CENTER DISTANCE
A	6.000	24	4	300	
B	3.000	12	4	600	4.500
C	6.000	48	8	600	
D	2.000	16	8	1800	4.000

FIGURE 40–1 ■ *Gear train data.*

speed at the switch is reduced to 9 r/min by a series of spur and miter gears.

When the operator pushes a button, the motor is activated until the circuit breaker pointer rotates 90 degrees and one of the arms depresses the roller and breaks the circuit. During this time the load-ratio control switch shaft will rotate 360 degrees, moving the contactor in the load-ratio control switch one position. This will be shown on the dial by the position indicator.

To simplify the assembly, only the pitch diameters of the gears are shown and much of the hardware has been omitted.

When referring to the direction in which the gears rotate, the terms clockwise (CWISE) or (CW) and anticlockwise (AWISE) or counterclockwise (CCW) are used.

REFERENCE

BOSTON GEAR INCOM INTERNATIONAL INC.

INTERNET RESOURCES

Boston Gear. For information on all aspects of gears and gear drives, see: http://www.bostongear.com

Howstuffworks. For information on all types of gears and gear applications, see: http://science.howstuffworks.com/gear3.htm

Machine Design. For information on gear trains and gear applications, see *Machine Design, Mechanical Reference* at: http://www.machinedesign.com

TechStudent.Com. For information on gear trains, see: http://www.technologystudent.com/gears/geardex1.htm

GEAR DATA				
GEAR	NO. OF TEETH	PITCH DIAMETER	DP	R/MIN
G_1	24		20	
G_2		4.800		270
G_3	20	1.000		
G_4	100			
G_5		1.000	20	
G_6		6.000		
G_7	18			
G_8		7.200		
G_9	72		20	
G_{10}		2.500	10	
G_{11}	25			

SHAFT DATA			
SHAFT	GEARS ON SHAFT	R/MIN	SHAFT * ROTATION
S_1			
S_2			
S_3			
S_4			
S_5			
S_6			
S_7			C-C WISE

* AS VIEWED FROM FRONT OR BOTTOM OF MOTOR DRIVE ASSEMBLY

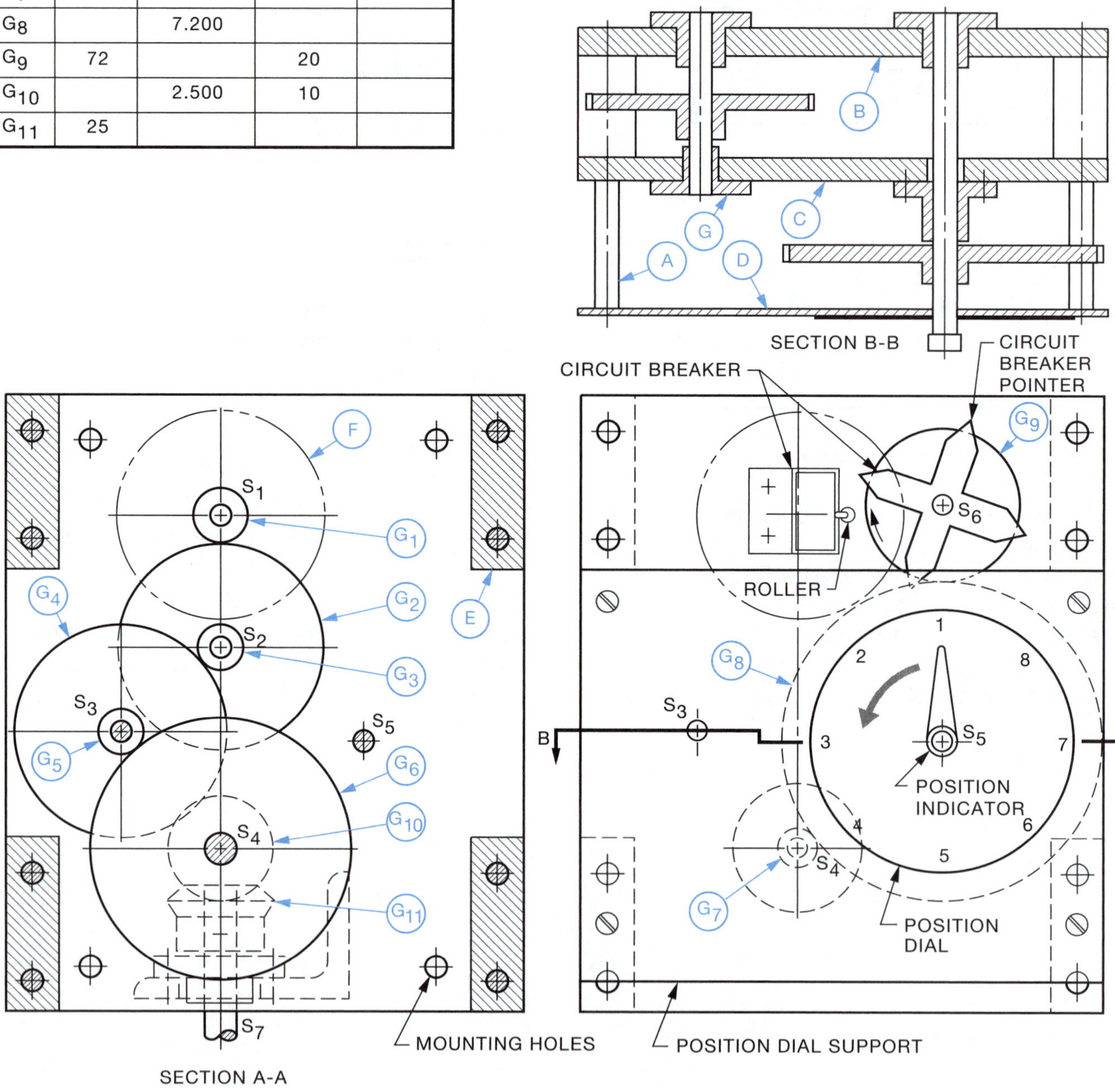

A
B
C
D
E
F
G
SECTION B-B
CIRCUIT BREAKER
CIRCUIT BREAKER POINTER
G1
G2
G3
G4
G5
G6
G7
G8
G9
G10
G11
S1
S2
S3
S4
S5
S6
S7
ROLLER
POSITION INDICATOR
POSITION DIAL
1 2 3 4 5 6 7 8
B
B
MOUNTING HOLES
POSITION DIAL SUPPORT
SECTION A-A

ASSIGNMENT:
ON A GRID SHEET SKETCH THE GEAR AND SHAFT
DATA CHARTS SHOWN AND COMPLETE THE MISSING INFORMATION.

QUESTIONS:

1. What are the names of parts Ⓐ to Ⓖ?

2. How many spur gears are shown?

3. How many miter gears are shown?

4. How many gear shafts are there?

5. What is the ratio between the following gears? (A) G_1 and G_2, (B) G_3 and G_4. (C) G_5 and G_6, (D) G_7 and G_8, (E) G_8 and G_9, (F) G_{10} and G_{11}.

6. What is the center-to-center distance between the following shafts? (A) S_1 and S_2, (B) S_2 and S_3, (C) S_3 and S_4, (D) S_4 and S_5, (E) S_5 and S_6.

7. How many seconds does it take to turn the load ratio control switch one position?

8. How many seconds does it take the position indicator to move continuously from position 4 to position 7?

9. What is the r/min ratio between the motor and the switch?

10. If the shaft S_7 rotates 1800 degrees, how many degrees does the position indicator rotate?

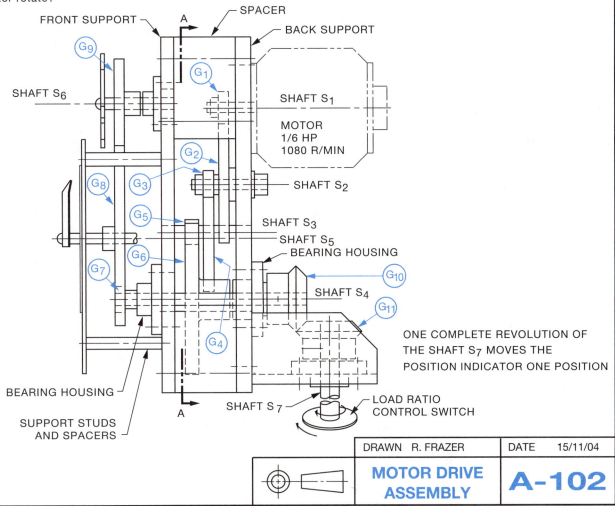

DRAWN R. FRAZER	DATE 15/11/04
MOTOR DRIVE ASSEMBLY	**A-102**

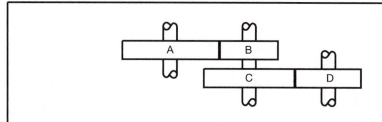

GEAR	PD	N	DP	DIRECTION	R/MIN	CENTER DISTANCE
A	7.000		4	CLOCKWISE	300	
B		12				
C	6.000					
D		12	3			

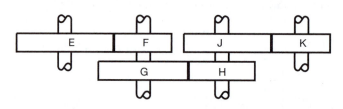

GEAR	PD	N	DP	DIRECTION	R/MIN	CENTER DISTANCE
E	7.500		4	COUNTER-CLOCKWISE	240	
F		18				
G	10.000					
H	3.200	16				
J	8.000		6			
K		40				

ASSIGNMENT:
On a grid sheet sketch the charts shown and complete the missing information.

GEAR TRAIN CALCULATIONS

A-103

CAMS

The cam is invaluable in the design of automatic machinery. Cams make it possible to impart any desired motion to another mechanism.

A *cam* is a rotating, oscillating, or reciprocating machine element that has a surface or groove formed to impart special or irregular motion to a second part called a follower. The follower rides against the curved surface of the cam. The distance that the follower rises and falls in a defined period of time is determined by the shape of the cam profile.

Types of Cams

The type and shape of cam used is dictated by the required relationship of the parts and the motions of both, Figure 41–1. The cams that are generally used are either radial or cylindrical. The follower of a radial or face cam moves in a plane perpendicular to the axis of the cam, while in the cylindrical type of cam the movement of the follower is parallel to the cam axis.

A simple OD (outside diameter) or plate cam is shown in Figure 41–2. The hole in the plate is bored off-center, causing the follower to move up and down as it revolves. The follower can be any type that will roll or slide on the surface of the cam. The follower used with this cam is called a flat face follower.

The cam shown in Figure 41–3 is a drum or barrel-type cam that transmits motion transversely to a lever

connected to a conical follower, which rides in the groove as the cam revolves.

Cam Displacement Diagrams

In preparing cam drawings, a cam displacement diagram is drawn first to plot the motion of the follower. The curve on the drawing represents the path of the follower, not the face of the cam. The diagram may be any convenient length, but often it is drawn equal to the circumference of the base circle of the cam and the height is drawn equal to the follower displacement. The lines drawn on the motion diagram are shown as radial lines on the cam drawing; the sizes are transferred from the motion diagram to the cam drawing.

Figure 41–4 shows a cam displacement diagram having a modified uniform type of motion plus two dwell periods. Most cam displacement diagrams have 360-degree cam displacement angles. For drum or cylindrical grooved cams, the displacement diagram is often replaced by the developed surface of the cam.

The cylindrical feeder cam (Assignment A-104) is a drum or barrel cam. In addition to the working views of the cam, a development of the contour of the grooves is shown. This development aids the machinist in scribing and laying out the contour of the cam action lobes on the surface of the cam blank preparatory to machining grooves.

Regardless of the type of cam or follower, the purpose of all cams is to impart motion to other mechanisms in various directions in order to actuate mechanisms to do specific jobs.

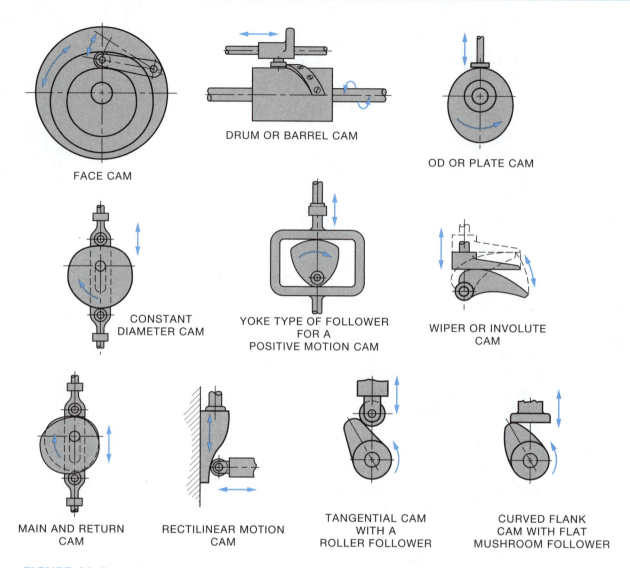

FACE CAM

DRUM OR BARREL CAM

OD OR PLATE CAM

CONSTANT
DIAMETER CAM

YOKE TYPE OF FOLLOWER
FOR A
POSITIVE MOTION CAM

WIPER OR INVOLUTE
CAM

MAIN AND RETURN
CAM

RECTILINEAR MOTION
CAM

TANGENTIAL CAM
WITH A
ROLLER FOLLOWER

CURVED FLANK
CAM WITH FLAT
MUSHROOM FOLLOWER

FIGURE 41–1 ∎ *Common types of cams.*

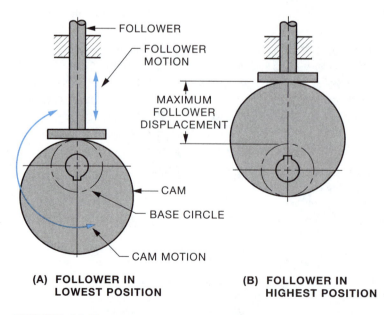

FOLLOWER

FOLLOWER
MOTION

MAXIMUM
FOLLOWER
DISPLACEMENT

CAM

BASE CIRCLE

CAM MOTION

**(A) FOLLOWER IN
LOWEST POSITION**

**(B) FOLLOWER IN
HIGHEST POSITION**

FIGURE 41–2 ∎ *Eccentric plate cam.*

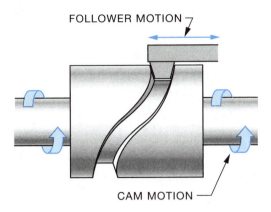

FOLLOWER MOTION

CAM MOTION

FIGURE 41–3 ■ *Drum or barrel-type cam.*

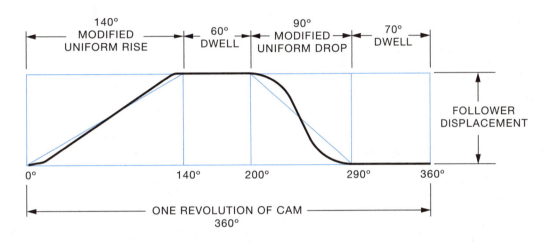

140°
MODIFIED
UNIFORM RISE

60°
DWELL

90°
MODIFIED
UNIFORM DROP

70°
DWELL

FOLLOWER
DISPLACEMENT

0°   140°   200°   290°   360°

ONE REVOLUTION OF CAM
360°

FIGURE 41–4 ■ *Cam displacement diagram.*

INTERNET RESOURCES

Animated Worksheets. For information on cams and cam followers, see: http://www.animatedworksheets.co.uk (mechanisms)

Commercial Cam CO., Inc. For information on cams and cam—followers, see: http://www.camcoindex.com

Design & Technology Online. For information on cams and cam followers, see: http://www.dtonline.org/ (mechanisms)

Flying Pig. For information on mechanisms and movements, see: http://www.flying-pig.co.uk (mechanisms and movements)

Saltire. For information on cams and cam-followers, see: http://www.saltire.com/cams.html

TechStudent.com. For information, including illustrations, on cams, see: http://www.technologystudent.com (mechanisms)

QUESTIONS:

1. Through what thickness of metal is hole ② drilled?

2. Locate line ④ in the right-side view.

3. Locate line ⑪ in the cam displacement diagram.

4. Locate line ⑥ in another view.

5. Locate line ⑨ in the right-side view.

6. What is the maximum permissible diameter of hole ②?

7. Locate line ⑧ in the front view.

8. Locate line ㊸ in the right-side view.

9. What is the total follower displacement of the cam follower for (A) the finishing cut, (B) the roughing cut?

10. Assuming a .06 in. allowance for machining, what would be the outside diameter of the cam before finishing?

11. What is the total number of through holes?

12. Locate lines ㉚ to ㊶ on other view.

13. Determine distances Ⓐ to Ⓜ.

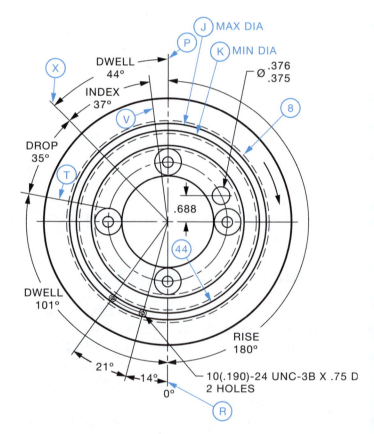

ROUGHING CAM DATA
LEFT-SIDE VIEW

THREAD CONTROLLING ORGANIZATION
AND STANDARD-ASME B1.1-2003

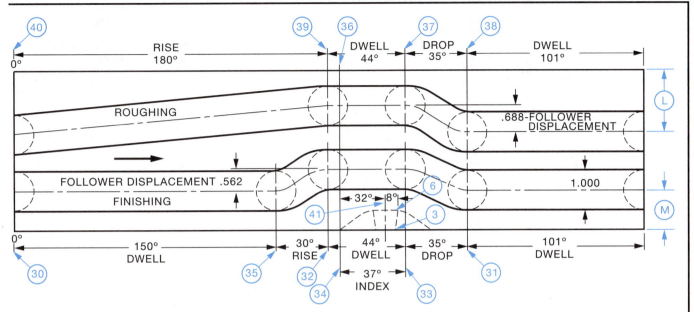

CAM DISPLACEMENT DIAGRAM

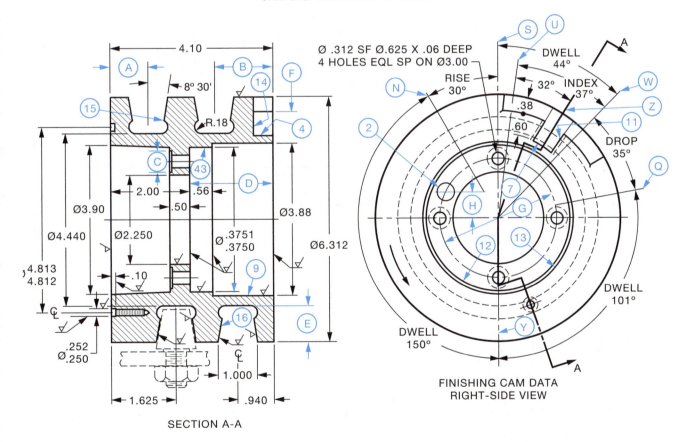

SECTION A-A

FINISHING CAM DATA
RIGHT-SIDE VIEW

NOTE: UNLESS OTHERWISE SHOWN:

- TOLERANCE ON TWO PLACE DECIMAL DIMENSIONS ±.02
- TOLERANCE ON THREE PLACE DECIMAL DIMENSIONS ±.005
- TOLERANCE ON ANGLES ±30'
- SURFACES ∇ TO BE 63∇

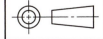

MATERIAL		
SCALE		NOT TO SCALE
DRAWN G. ADAMS		DATE 09/10/04

CYLINDRICAL
FEEDER CAM

A-104

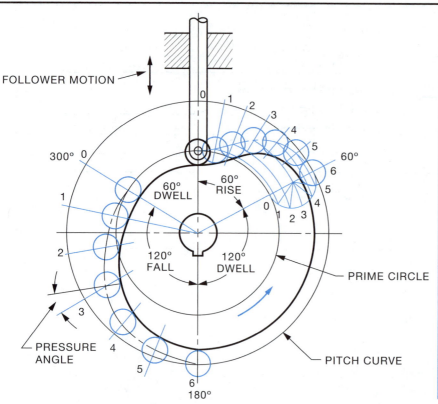

FOLLOWER MOTION

300° 0

60° DWELL   60° RISE

120° FALL   120° DWELL

PRESSURE ANGLE

PRIME CIRCLE

PITCH CURVE

180°

CAM DRAWING

ANSWERS	
POSITION IN DEGREES	DISTANCE FROM THE PRIME CIRCLE
0	
10	
20	
30	
40	
50	
60	
180	
200	
220	
240	
260	
280	
300	
360	

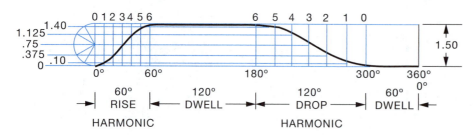

1.125
.75
.375
0

1.40
.10

0 1 2 3 4 5 6   6 5 4 3 2 1 0

1.50

0°   60°   180°   300°   360°
                                                  0°

60° RISE | 120° DWELL | 120° DROP | 60° DWELL

HARMONIC   HARMONIC

DISPLACEMENT DIAGRAM

ASSIGNMENT:

1. ON A GRID SHEET SKETCH THE CHART SHOWN ABOVE. USING 0° AS THE STARTING POSITION AND THE POSITIONS SHOWN, CALCULATE THE DISTANCE THE CENTER OF THE FOLLOWER IS FROM THE PRIME CIRCLE.

2. IF THE CAM SHAFT ROTATES ONE COMPLETE REVOLUTION EVERY THREE MINUTES, HOW MUCH TIME DOES IT TAKE FOR THE CAM FOLLOWER TO (A) RISE 1.50 INCHES, (B) DROP 1.50 INCHES?

3. HOW MUCH TIME DOES THE CAM FOLLOWER REMAIN AT DWELL WHEN (A) IN THE UPPER POSITION, (B) IN THE LOWER POSITION?

PLATE CAM   **A-105**

ANTIFRICTION BEARINGS

Antifriction bearings, also known as *roller-element* bearings, use a type of rolling element between the loaded members. Relative motion is accommodated by rotation of the elements. Roller-element bearings are usually housed in bearing races conforming to the element shapes. In addition, a cage or separator is often used to locate the elements within the bearings. These bearings are usually categorized by the form of the rolling element and in some instances by the load type they carry, Figures 42–1 and 42–2.

Roller-element bearings are generally classified as either ball or roller.

Ball Bearings

Ball bearings may be roughly divided into three categories: radial, angular contact, and thrust. Radial-contact ball bearings are designed for applications in which the load is primarily radial with only low-magnitude thrust loads. Angular-contact bearings are used where loads are combined radial and high

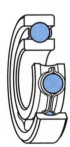

SINGLE ROW, DEEP GROOVE BALL BEARINGS

The *Single Row, Deep Groove Ball Bearing* will sustain, in addition to radial load, a substantial thrust load in either direction... even at very high speeds. This advantage results from the intimate contact existing between the balls and the deep, continuous groove in each ring. When using this type of bearing, careful alignment between the shaft and housing is essential. This bearing is also available with seals, which serve to exclude dirt and retain lubricant.

ANGULAR CONTACT BALL BEARINGS

The *Angular Contact Ball Bearing* supports a heavy thrust load in one direction... sometimes combined with a moderate radial load. A steep contact angle, assuring the highest thrust capacity and axial rigidity, is obtained by a high thrust supporting shoulder on the inner ring and a similar high shoulder on the opposite side of the outer ring. These bearings can be mounted singly or, when the sides are flush ground, in tandem for constant thrust in one direction; mounted in pairs, also when sides are flush ground, for a combined load...either face-to-face or back-to-back.

(continued)

FIGURE 42–1 ■ *Roller-element bearings.*

CYLINDRICAL ROLLER BEARINGS

The *Cylindrical Roller Bearing* has high radial capacity and provides accurate guiding of the rollers, resulting in a close approach to true rolling. Consequent low friction permits operation at high speed. Those types which have flanges on one ring only allow a limited free axial movement of the shaft in relation to the housing. They are easy to dismount even when both rings are mounted with a tight fit. The double row type assures maximum radial rigidity and is particularly suitable for machine tool spindles.

BALL THRUST BEARINGS

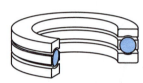

The *Ball Thrust Bearing* is designed for thrust load in one direction only. The load line through the balls is parallel to the axis of the shaft... resulting in high thrust capacity and minimum axial deflection. Flat seats are preferred ... particularly where the load is heavy... or where close axial positioning of the shaft is essential; for example, in machine tool spindles.

SPHERICAL ROLLER THRUST BEARINGS

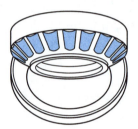

The *Spherical Roller Thrust Bearing* is designed to carry heavy thrust loads, or combined loads which are predominantly thrust. This bearing has a single row of rollers which roll on a spherical outer race with full self-alignment. The cage, centered by an inner ring sleeve, is constructed so that lubricant is pumped directly against the inner ring's unusually high guide flange. This ensures good lubrication between the roller ends and the guide flange.The spherical roller thrust bearing operates best with relatively heavy oil lubrication.

FIGURE 42–1 ■ *Roller-element bearings, continued.*

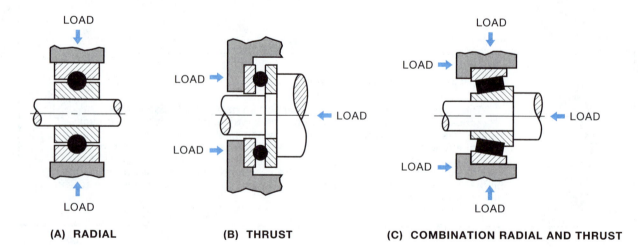

(A) RADIAL **(B) THRUST** **(C) COMBINATION RADIAL AND THRUST**

FIGURE 42–2 ■ *Types of bearing loads.*

thrust, and where precise shaft location is required. Thrust bearings handle loads that are primarily thrust.

Roller Bearings

Roller bearings have higher load capacities than ball bearings for a given envelope size. They are widely used in moderate-speed, heavy-duty applications. The four principal types of roller bearings are cylindrical, needle, tapered, and spherical. Cylindrical roller bearings use cylinders with approximate length-diameter ratios ranging from 1:1 to 1:3 as rolling elements. Needle roller bearings use cylinders or needles of greater length-diameter ratios. Tapered and spherical roller bearings are capable of supporting combined radial and thrust loads.

The rolling elements of tapered roller bearings are truncated cones. Spherical roller bearings are available with both barrel and hourglass roller shapes. The primary advantage of spherical roller bearings is their self-aligning capability. Plain bearings are covered in Unit 25.

RETAINING RINGS

Retaining rings, or snap rings, are designed to provide a removable shoulder to locate, retain, or lock components accurately on shafts and in bored housings, Figures 42–3 and 42–4. They are easily installed and removed. Because they are usually made of spring steel, retaining rings have a high shear strength and impact capacity. In addition to fastening and positioning, a number of rings are designed for taking up end play caused by accumulated tolerances or wear in the parts being retained.

O-RING SEALS

O-rings are used as an axial mechanical seal (a seal that forms a running seal between a moving shaft and a housing) or a static seal (no moving parts). The advantage of using an O-ring as a gasket-type seal, Figure 42–5, over conventional gaskets is that the nuts need not be tightened uniformly and sealing compounds are not required. A rectangular groove is the most common type of groove used for O-rings.

CLUTCHES

Clutches are used to start and stop machines or rotating elements without starting or stopping the prime mover. They are also used for automatic disconnection, quick starts and stops, and to permit shaft rotation in one direction only such as the overrunning clutch shown in Figure 42–6. A full complement of sprags between concentric inner and outer races transmits power from one race to the other by wedging action of the sprags when either race is rotated in the driving direction. Rotation in the opposite direction frees the sprags and the clutch is disengaged or overruns, Figure 42–6. This type of clutch is used in the power drive, Assignment A-106.

BELT DRIVES

A belt drive consists of an endless flexible belt connecting two wheels or pulleys. Belt drives depend on friction between belt and pulley surfaces for transmission of power.

In a V-belt drive, the belt has a trapezoidal cross section, and runs in V-shaped grooves on the pulleys. These belts are made of cords or cables, impregnated and covered with rubber or other organic compound. The covering is formed to produce the required cross section. V-belts are usually manufactured as endless belts, although open-end and link types are available.

In the case of V-belts, the friction for the transmission of the driving force is increased by the wedging of the belt into the grooves on the pulley.

V-Belt Sizes

To facilitate interchangeability and to ensure uniformity V-belt manufacturers have developed industrial standards for the various types of V-belts, Figure 42–7. Industrial V-belts are made in two types: heavy duty (conventional and narrow) and light duty. Conventional belts are available in A, B, C, D, and E sections. Narrow belts are made in 3V, 5V, and 8V sections. Light-duty belts come in 3L, 4L, and 5L sections.

AXIAL ASSEMBLY RINGS

INTERNAL

EXTERNAL

INTERNAL   EXTERNAL

BASIC TYPES: Designed for axial assembly. Internal ring is compressed for insertion into bore or housing, external ring expanded for assembly over shaft. Both rings seat in deep grooves and are secure against heavy thrust loads and high rotational speeds.

INVERTED RINGS: Same tapered construction as basic types, with lugs inverted to abut bottom of groove. Section height increased to provide higher shoulder, uniformly concentric with housing or shaft. Rings provide better clearance, more attractive appearance than basic types.

END PLAY RINGS

INTERNAL

EXTERNAL

EXTERNAL   INTERNAL

BOWED RINGS: For assemblies in which accumulated tolerances cause objectionable end play between ring and retained part. Bowed construction permits rings to provide resilient end-play takeup in axial direction while maintaining tight grip against groove bottom.

RADIAL RINGS: Bowed E-rings are used for providing resilient end-play takeup in an assembly.

SELF-LOCKING RINGS

EXTERNAL

EXTERNAL

INTERNAL

CIRCULAR EXTERNAL RINGS: The push-on type of fastener with inclined prongs which bend from their initial position to grip the shaft. Ring at left has arched rim for increased strength and thrust load capacity; extra-long prongs accommodate wide shaft tolerances. Ring at right has flat rim, shorter locking prongs, smaller OD.

CIRCULAR INTERNAL RINGS: Designed for use in bores and housings. Functions in same manner as external types except that locking prongs are on the outside rim.

RADIAL LOCKING RINGS

EXTERNAL

EXTERNAL

CRESCENT RING: Has a tapered section similar to the basic axial types. Remains circular after installation on a shaft and provides a tight grip against the groove bottom.

E-RINGS: Provide a large bearing shoulder on small-diameter shafts and is often used as a spring retainer. Three heavy prongs, spaced approximately 120 degrees apart, provide contact surface with groove bottom.

FIGURE 42–3 ■ *Stamped retaining rings.*

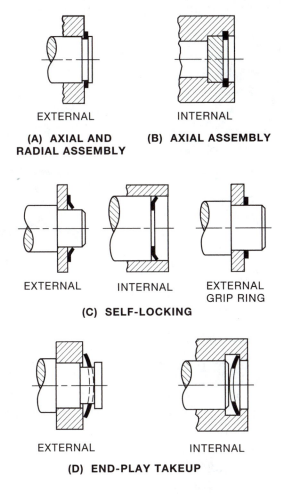

FIGURE 42–4 ■ *Retaining ring application.*

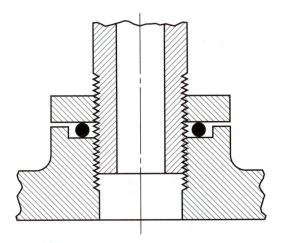

FIGURE 42–5 ■ *O-ring seal.*

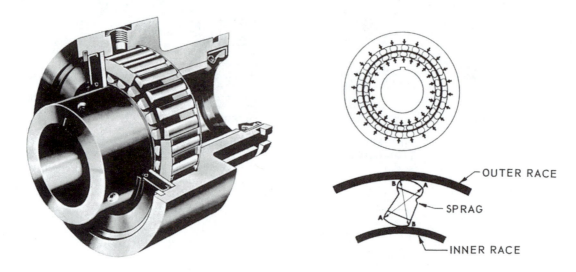

FIGURE 42–6 ■ *Overrunning clutch.*

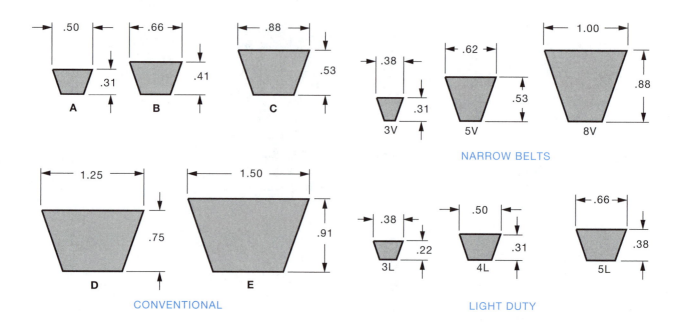

CONVENTIONAL

LIGHT DUTY

NARROW BELTS

FIGURE 42–7 ■ *V-belt sizes.*

FIGURE 42–8 ■ *V-belt sheave and bushing.*

Sheaves and Bushings

Sheaves, the grooved wheels of pulleys, are sometimes equipped with tapered bushings for ease of installation and removal, Figure 42–8. They have extreme holding power, providing the equivalent of a shrink fit. The sheave and bushing used in the power drive, Assignment A-106, have a six-hole drilling arrangement in both the bushing and sheave, making it possible to insert the cap screw from either side. This is especially advantageous for applications where space is at a premium.

REFERENCES

A.O. Dehayt, "Basic Bearing Types," *Machine Design* 40, No. 14

SKF Co. Ltd.

ASME B-18.27-1998 Tapered and Reduced Cross Section Retaining Rings (Inch Series)

ASME B27.7-1977 (R1999) General Purpose Tapered and Reduced Cross Section Retaining Rings (Metric)

INTERNET RESOURCES

Design & Technology Online. For information on pulleys, see: http://www.dtonline.org (mechanisms)

eFunda. For information on O-rings, see www.efunda.com/home.cfm

Gates Rubber Company. For information on automotive and industrial drive belts, see: http://www.gates.com

Machine Design. For information on bearings, retaining rings, clutches, seals, and belt drives, see *Machine Design, Mechanical Reference* at: http://www.machinedesign.com

TechStudent.Com. For information on pulleys and pulley systems, see: http://www.technologystudent. com (Gears and pulleys)

The Torrington Company. For information on precision bearings and motion control components and assemblies, see: http://www.torrington.com/

Truarc Retaining Rings. For information on retaining rings, see: http://www.truarc.com

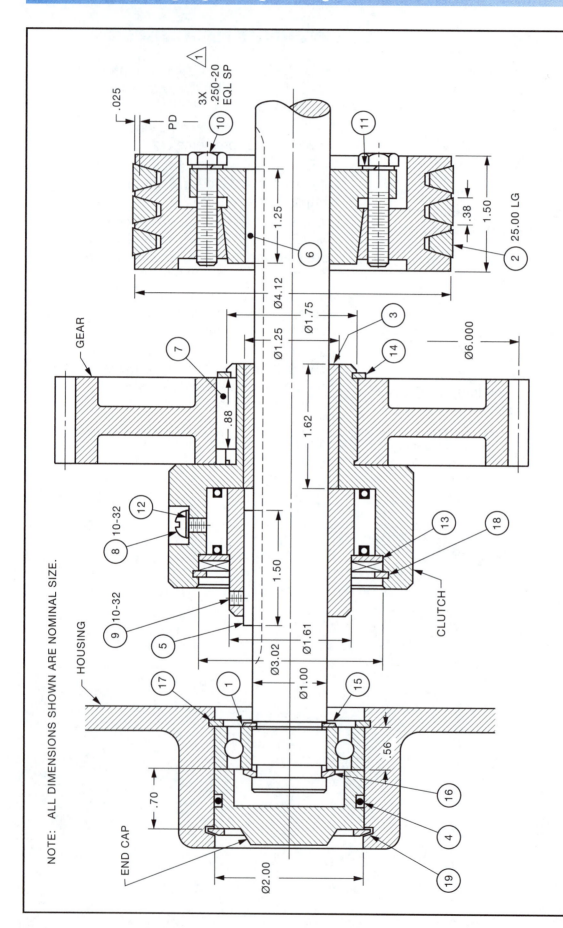

NOTE: ALL DIMENSIONS SHOWN ARE NOMINAL SIZE.

NOTE:
THREAD CONTROLLING ORGANIZATION
AND STANDARD-ASME B.1-2003

QUESTIONS:

1. How many cap screws fasten the sheave to the bushing?

2. List five parts or methods that are used to lock or join parts together on this assembly.

3. How many V-belts are used?

4. How many keys are there?

5. How many retaining rings are used?

6. What type of bearing is part ③ ?

7. What type of bearing is part ① ?

8. What prevents the oil from leaking out between the housing and the end cap?

9. Can the gear be driven in both directions?

10. If the pitch on the gear is 8, what is the number of teeth?

11. What is the pitch diameter of the sheave?

12. What size V-belt is required?

ASSIGNMENT:

1. ON A 1.00 INCH GRID SHEET (.10 IN. SQUARES) MAKE A ONE-VIEW DETAILED SKETCH OF THE END CAP, USING DIMENSIONS TAKEN FROM O-RING CATALOGS. USE AN RC7 FIT BETWEEN THE OUTSIDE DIAMETER OF THE CAP AND THE HOUSING. USE YOUR JUDGMENT FOR DIMENSIONS NOT SHOWN. SCALE 1:1.

2. ON A GRID SHEET SKETCH A BILL OF MATERIAL SIMILAR TO THE ONE SHOWN IN FIGURE 33-2, SHOWING PARTS 1 TO 19. REFER TO MANUFACTURES' CATALOGS AND APPENDIX TABLES.

REVISIONS	⚠	14/11/04	CAMPBELL
		PT 10 WAS 12–24 UNC	

SCALE	NOT TO SCALE	
DRAWN	J. DUGAN	DATE 16/09/04

POWER DRIVE

A–106

43 UNIT

RATCHET WHEELS

Ratchet wheels are used to transform reciprocating or oscillating motion into intermittent motion, to transmit motion in one direction only, or to serve as an indexing device.

Common forms of ratchets and pawls are shown in Figure 43–1. The teeth in the ratchet engage with the teeth in the pawl, permitting rotation in one direction only.

When a ratchet wheel and pawl are designed, points A, B, and C, as shown in Figure 43–1(A) are positioned on the same circle to ensure that the smallest forces are acting on the system.

Mechanical Advantage

Mechanical advantage occurs when a weight in one place lifts a heavier weight in another place, or a force applied at one point on a lever produces a greater force at another point. Examples of mechanical advantage are the teeter-totter, the winch, and the gears shown in Figure 43–2.

The teeter-totter shows how mechanical advantage can be applied. If a 10-pound (lb.) weight is placed on one end of a teeter-totter and a 20 lb. weight is placed on the other end, the 10 lb. weight goes up and the 20 lb. weight goes down when placed equidistant from the fulcrum. If the fulcrum is moved to make the distance from the 10 lb. weight to the fulcrum twice that of the distance between the fulcrum and the 20 lb. weight, the weight becomes balanced. If the fulcrum is moved still closer to the 20 lb. weight, the 10 lb. weight goes down and the 20 lb. weight goes up.

As the distance between the fulcrum and the 10 lb. weight increases, the distance that the 10 lb. weight moves must increase proportionately to maintain the same movement of the 20 lb. weight. Thus, a light mass can move a heavier weight, but in doing so, the smaller weight must travel farther than the larger one.

Mechanical advantage also occurs when a winch or different size gears are used. With the winch design shown in Figure 43–2(B), a mechanical advantage of 10 is obtained because the handle is 10 times the distance from the fulcrum as compared to the center of the rope.

The mechanical advantage of gears can easily be determined by obtaining the ratio between the number of teeth on the gears or the ratio between the pitch diameters.

In the winch, Assignment A-107, the center of the handle bar is 10 in. from the center of the shaft to which the pinion gear is attached. Half the pitch diameter of the pinion is .625 in. This produces a mechanical advantage of 10:.625 or 16:1. Further mechanical advantage is gained through the gear attached to the ∅1.00 shaft on which the rope revolves.

The hand will move a distance of approximately 63 in. when turning the handle one complete revolution. This will turn the rope drum one fifth of a revolution (50:10 teeth ratio), winding a ∅.25 rope up approximately 4 in. A greater force can be exerted using the winch than by simply pulling the rope.

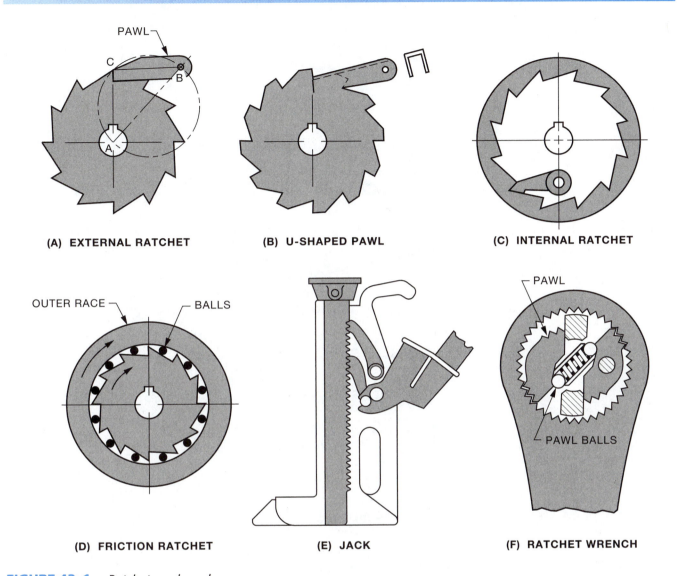

(A) **EXTERNAL RATCHET**

(B) **U-SHAPED PAWL**

(C) **INTERNAL RATCHET**

(D) **FRICTION RATCHET**

(E) **JACK**

(F) **RATCHET WRENCH**

FIGURE 43–1 ■ *Ratchets and pawls.*

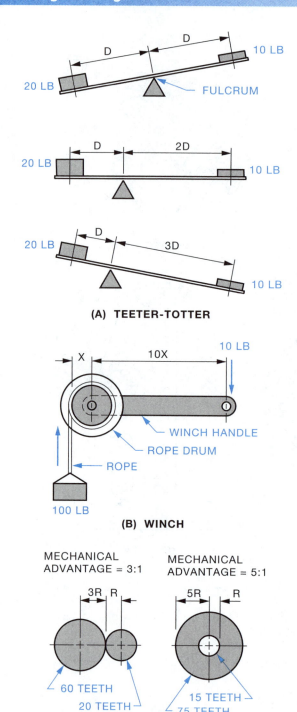

(A) **TEETER-TOTTER**

(B) **WINCH**

(C) **GEARS**

FIGURE 43–2 ■ *Mechanical advantage.*

INTERNET RESOURCES

Animated Worksheets. For information on ratchets and actuators, see: http://www. animatedworksheets.co.uk (mechanisms)

CMP Products Center. For information on ratchets and related mechanisms and applications, see: http://www.cmpmedia.globalspec.com/ SpecSearch/

TechStudent.Com. For information, including illustrations, on ratchets and simple ratchet mechanisms and applications, see: http://www. technologystudent.com (Mechanisms)

ASSIGNMENT: ON ONE-INCH GRID SHEETS (.10 IN. SQUARES), MAKE A DETAILED SKETCH OF THE PARTS OF THE WINCH. DIMENSIONS SHOWN ARE NOMINAL SIZES AND ALLOWANCES AND TOLERANCES ARE TO BE DETERMINED. THE NUMBER OF VIEWS AND THE DRAWING SCALE TO BE SELECTED BY THE STUDENT.

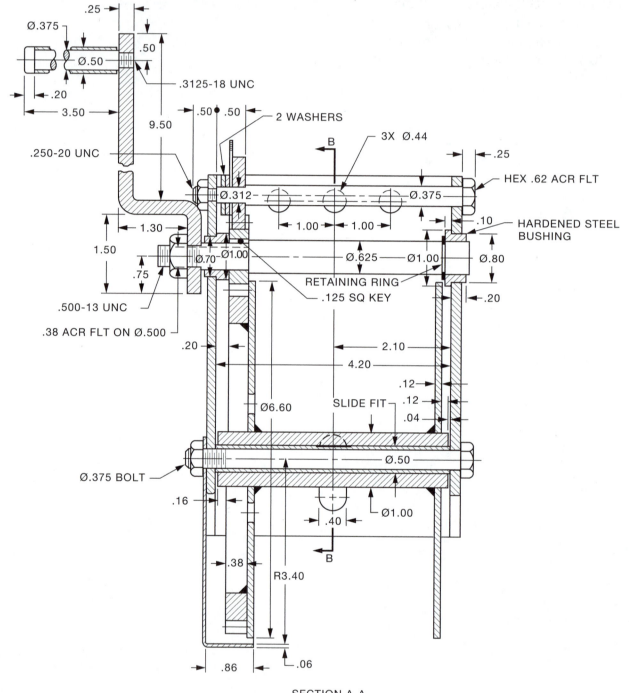

SECTION A-A

THREAD CONTROLLING ORGANIZATION AND STANDARD-ASME B1.1-2003

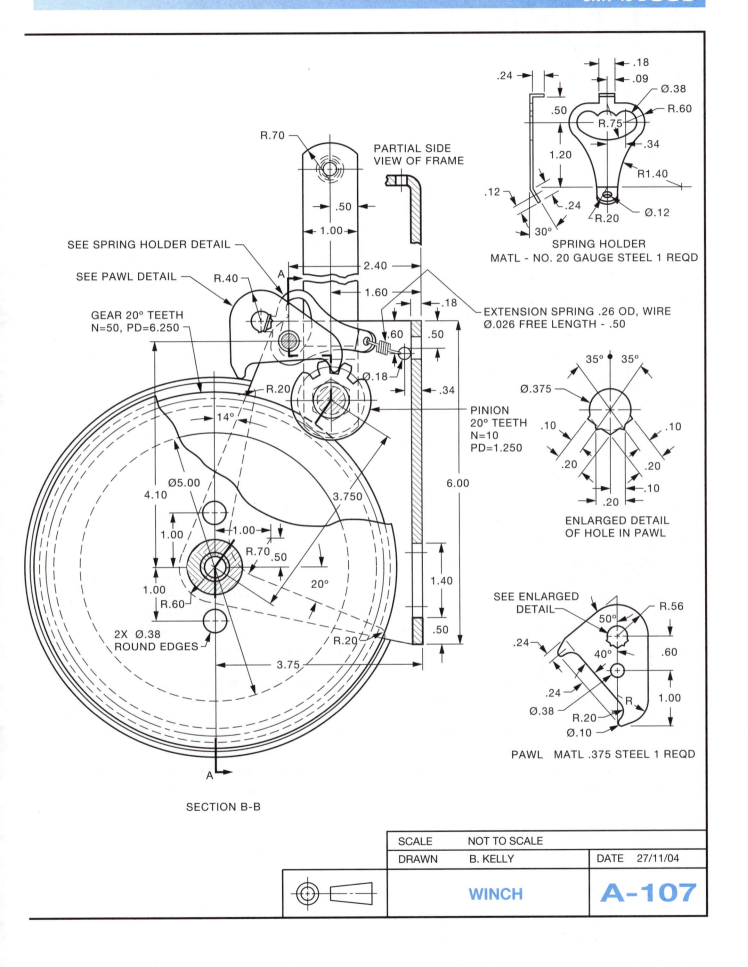

R.70

PARTIAL SIDE
VIEW OF FRAME

.50
1.00

SEE SPRING HOLDER DETAIL

SEE PAWL DETAIL

GEAR 20° TEETH
N=50, PD=6.250

R.40

A

2.40

1.60

.18

EXTENSION SPRING .26 OD, WIRE
Ø.026 FREE LENGTH - .50

.60 .50

Ø.18

.34

PINION
20° TEETH
N=10
PD=1.250

R.20

14°

Ø5.00

4.10

3.750

6.00

1.00

1.00

R.70 .50

20°

1.00

R.60

1.40

2X Ø.38
ROUND EDGES

3.75

R.20

.50

A

SECTION B-B

SPRING HOLDER
MATL - NO. 20 GAUGE STEEL 1 REQD

.24
.18
.09
.50
R.75
Ø.38
R.60
1.20
.34
.12
.24
R1.40
30°
R.20
Ø.12

35° 35°
Ø.375
.10
.10
.20
.20
.10
.20

ENLARGED DETAIL
OF HOLE IN PAWL

SEE ENLARGED
DETAIL
R.56
50°
.24
40°
.60
.24
R
Ø.38
R.20
1.00
Ø.10

PAWL   MATL .375 STEEL 1 REQD

SCALE	NOT TO SCALE	
DRAWN	B. KELLY	
	DATE	27/11/04

WINCH

A-107

44 UNIT

MODERN ENGINEERING TOLERANCING

An engineering drawing of a manufactured part conveys information from the designer to the manufacturer and inspector. It must contain all information necessary for the part to be correctly manufactured. It must also enable the inspector to determine precisely whether the finished parts are acceptable.

Therefore, each drawing must convey three essential items of information: the material to be used, the size or dimensions of the part, and the shape or geometric characteristics of the part. The drawing must also specify the permissible variation of size and form.

The actual size of a feature must be within the size limits specified on the drawing. Each measurement made at any cross section of the feature must not be greater than the maximum limit of size, nor smaller than the minimum limit of size, Figure 44–1. Although each part is within the prescribed tolerance zones, the parts may not be usable because of their deviation from their geometric form.

In order to meet functional requirements, it is often necessary to control errors of form, including squareness, roundness, and flatness, as well as deviation from true size. In the case of mating parts, such as holes and shafts, it is usually necessary to ensure that they do not cross the boundary of perfect form at their maximum material size (the smallest hole or the largest shaft) because of being bent or otherwise deformed. This condition is shown in Figure 44–2, where features are not permitted to cross the boundary of perfect form at the maximum material size (the smallest hole or the largest shaft).

The system of *geometric tolerancing* offers a precise interpretation of drawing requirements. Geometric tolerancing controls geometric characteristics of parts. These characteristics include:

- straightness
- flatness
- circularity
- cylindricity
- angularity
- parallelism
- perpendicularity
- runout
- profile
- position

Other techniques, such as datum systems, datum targets, and projected tolerance zones were developed in order to facilitate this precise interpretation.

Geometric tolerances need not be used for every feature of a part. Generally, if each feature meets all dimensional tolerances, form variations will be adequately controlled by the accuracy of the manufacturing process and equipment used. A geometric tolerance is used when geometric errors must be limited more closely than might ordinarily be expected from the manufacturing process. A geometric tolerance is also used to state functional or interchangeability requirements.

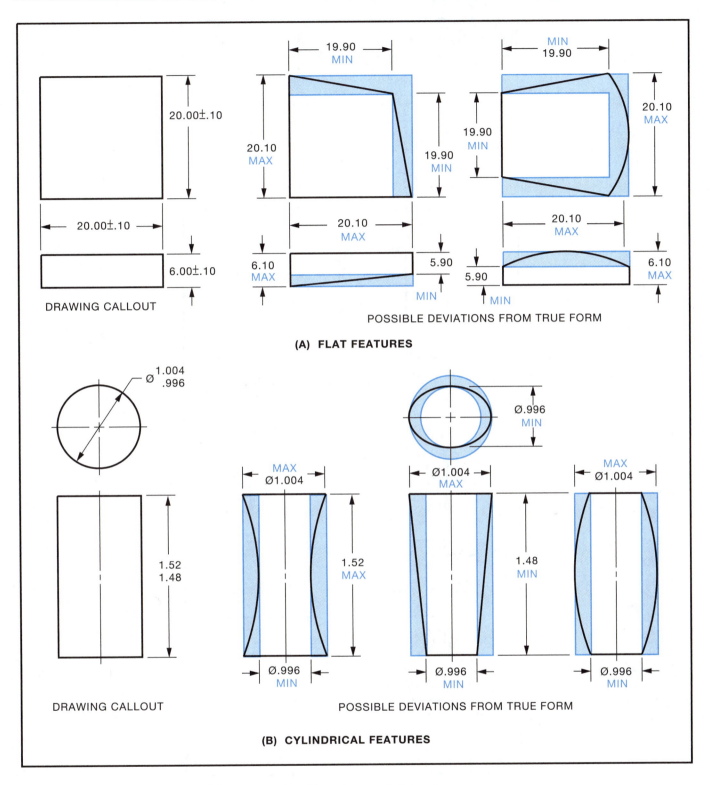

19.90
MIN

20.10
MAX

19.90
MIN

MIN
19.90

20.10
MAX

19.90
MIN

20.10
MAX

6.10
MAX

5.90
MIN

20.10
MAX

5.90
MIN

6.10
MAX

20.00±.10

20.00±.10

6.00±.10

DRAWING CALLOUT

POSSIBLE DEVIATIONS FROM TRUE FORM

(A) FLAT FEATURES

Ø 1.004
.996

Ø.996
MIN

MAX
Ø1.004

Ø1.004
MAX

MAX
Ø1.004

1.52
1.48

1.52
MAX

1.48
MIN

Ø.996
MIN

Ø.996
MIN

Ø.996
MIN

DRAWING CALLOUT

POSSIBLE DEVIATIONS FROM TRUE FORM

(B) CYLINDRICAL FEATURES

FIGURE 44–1 ■ *Deviation of shapes permitted by toleranced dimensions.*

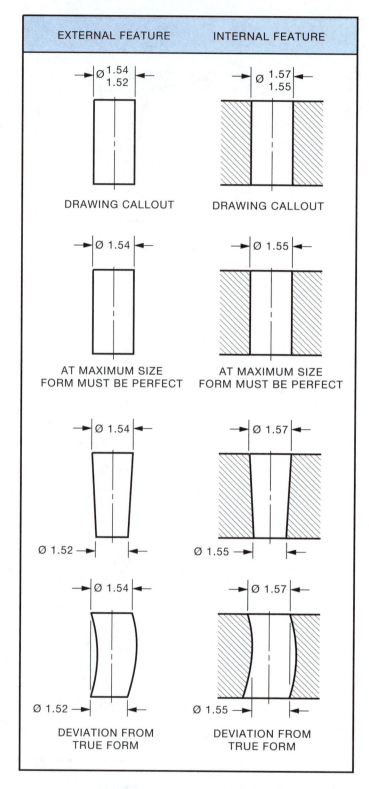

EXTERNAL FEATURE	INTERNAL FEATURE
Ø 1.54 / 1.52	Ø 1.57 / 1.55
DRAWING CALLOUT	DRAWING CALLOUT
Ø 1.54	Ø 1.55
AT MAXIMUM SIZE FORM MUST BE PERFECT	AT MAXIMUM SIZE FORM MUST BE PERFECT
Ø 1.54 — Ø 1.52	Ø 1.57 — Ø 1.55
Ø 1.54 — Ø 1.52	Ø 1.57 — Ø 1.55
DEVIATION FROM TRUE FORM	DEVIATION FROM TRUE FORM

FIGURE 44–2 ■ *Examples of deviation of form when perfect form at the maximum material size is required.*

GEOMETRIC TOLERANCING

A geometric tolerance is the maximum permissible variation of form, orientation, or location of a feature from that indicated or specified on a drawing. The tolerance value represents the width or diameter of the tolerance zone within which the point, line, or surface of the feature should lie.

From this definition, it follows that a feature would be permitted to have any variation of form or take up any position within the specified geometric tolerance zone.

For example, a line controlled in a single plane by a straightness tolerance of .008 in. must be contained within tolerance zone .008 in. wide, Figure 44–3.

Points, Lines, and Surfaces

The production and measurement of engineering parts deal, in most cases, with surfaces of objects. These surfaces may be flat, cylindrical, conical, or spherical, or have some more or less irregular shape or contour.

Measurement, however, usually has to take place at specific points. A line or surface is evaluated dimensionally by making a series of measurements at various points along its length.

Surfaces are considered to be composed of a series of line elements running in two or more directions.

Points have position but no size, therefore, position is the only characteristic that requires control. Lines and surfaces have to be controlled for form, orientation, and location. Therefore, geometric tolerances provide for control of these characteristics, Figure 44–4. (Symbols will be introduced as required, but all are shown in the figure for reference purposes.)

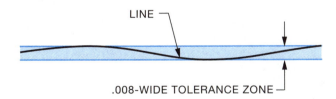

LINE

.008-WIDE TOLERANCE ZONE

FIGURE 44–3 ■ *Tolerance zone for straightness of a line.*

FEATURE CONTROL FRAME

Some geometric tolerances have been used for many years in the form of notes such as PARALLEL WITH SURFACE "A" WITHIN .001" and STRAIGHT WITHIN .12". Although such notes are now obsolete, the reader should be prepared to recognize them on older drawings.

The current method is to specify geometric tolerances by means of the feature control frame, Figure 44–5.

A feature control frame consists of a rectangular frame divided into two or more compartments. The first compartment (starting from the left) contains the geometric characteristic. The second compartment contains the allowable tolerance. Where applicable, the tolerance is preceded by the diameter symbol and followed by a modifying symbol. Other compartments are added when datums must be specified.

The feature control frame is related to the feature by one of the following methods.

Application to Surfaces (Figure 44–6A)

The arrowhead of the leader from the feature control frame should touch the surface of the feature or the extension line of the surface, but not in line with the dimension.

- Attaching a side or end of the frame to an extension line extending from a place on the surface feature.

  The leader from the feature control frame should be directed at the feature in its characteristic profile. Thus, in Figure 44–7, the straightness tolerance is directed to the side view, and the circularity tolerance to the end view. This may not always be possible; a tolerance connected to an alternative view, such as circularity tolerance connected to a side view, is acceptable. When it is more convenient, or when space is limited, the arrowhead may be directed to an extension line, but not in line with the dimension line.

FEATURE	TYPE OF TOLERANCE	CHARACTERISTIC	SYMBOL	SEE UNIT
INDIVIDUAL FEATURES	FORM	STRAIGHTNESS	—	44 & 45
		FLATNESS	▱	46
		CIRCULARITY (ROUNDNESS)	○	46
		CYLINDRICITY	⌭	46
INDIVIDUAL OR RELATED FEATURES	PROFILE	PROFILE OF A LINE	⌒	53
		PROFILE OF A SURFACE	⌓	53
RELATED FEATURES	ORIENTATION	ANGULARITY	∠	48 & 49
		PERPENDICULARITY	⊥	
		PARALLELISM	//	
	LOCATION	POSITION	⊕	51
		CONCENTRICITY	◎	
		SYMMETRY	⩵	
	RUNOUT	CIRCULAR RUNOUT	*↗	54
		TOTAL RUNOUT	*↗↗	
SUPPLEMENTARY SYMBOLS		AT MAXIMUM MATERIAL CONDITION	Ⓜ	45 & 51
		AT LEAST MATERIAL CONDITION	Ⓛ	
		PROJECTED TOLERANCE ZONE	Ⓟ	
		BASIC DIMENSION	XX	47
		DATUM FEATURE	▷─A	47
		DATUM TARGET	Ø.50 / A2	50

* MAY BE FILLED IN

FIGURE 44–4 ■ *Geometric characteristic symbols.*

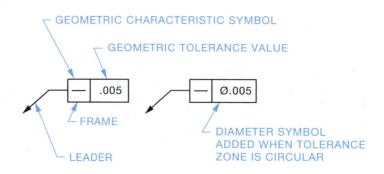

FIGURE 44–5 ■ *Feature control frame.*

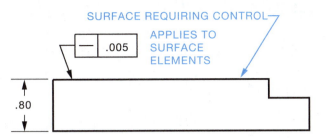

RUNNING A LEADER FROM THE FRAME TO THE FEATURE

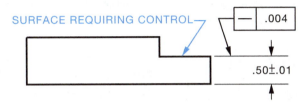

ATTACHED TO AN EXTENSION LINE USING A LEADER

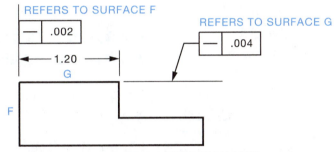

ATTACHED DIRECTLY TO AN EXTENSION LINE

(A) CONTROL OF SURFACE OR SURFACE ELEMENTS

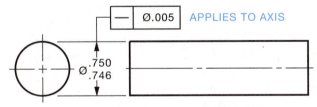

ATTACHED TO THE DIMENSION LINE

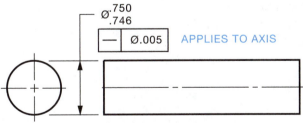

LOCATED BELOW DIMENSION CALLOUT

(B) CONTROL OF FEATURE OF SIZE

FIGURE 44–6 ■ *Application of feature control frame.*

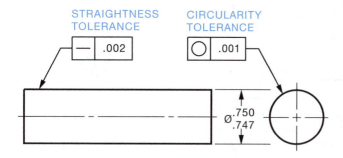

STRAIGHTNESS TOLERANCE

CIRCULARITY TOLERANCE

Ø .750 / .747

FIGURE 44–7 ■ *Preferred location of feature control symbol when referring to a surface.*

Applications to Features of Size (Figure 44–6B)

- Locating the frame below or attached to the leader directed to the callout or dimension pertaining to the feature. (See Unit 45.)

- Locating the frame below the size dimension to control the center line, axis, or center plane of the feature. (See Unit 45.)

When two or more feature control frames apply to the same feature, they are drawn together with a single leader and arrowhead, Figure 44–8.

FORM TOLERANCES

Form tolerances control straightness, flatness, circularity, and cylindricity. Orientation tolerances control angularity, parallelism, and perpendicularity.

Form tolerances are applicable to single (individual) features or elements of single features and, as such, do not require locating dimensions. The form tolerance must be less than the size tolerance.

Form and orientation tolerances critical to function and interchangeability are specified where the tolerances of size and location do not provide sufficient control. A tolerance of form or orientation may be specified where no tolerance of size is given, for example, the control of flatness.

STRAIGHTNESS

Straightness is a condition in which the elements of a surface or its axes are in a straight line. The geometric characteristic symbol for straightness is a horizontal line, Figure 44–9. A straightness tolerance specifies a tolerance zone within which the considered element of the surface or center line must lie. A straightness tolerance is applied to the view where the elements to be controlled are represented by a straight line.

STRAIGHTNESS CONTROLLING SURFACE ELEMENTS

Straightness is fundamentally a characteristic of a line, such as the edge of a part or a line scribed on a surface. A straightness tolerance is specified on a drawing by means of a feature control frame, which is directed by a leader to the line requiring control, Figure 44–10. It states in symbolic form that the line shall be straight within .006 in. This means that the line shall be contained within a tolerance zone consisting of the area between two parallel straight lines in the same plane separated by the specified tolerance.

Theoretically, straightness could be measured by bringing a straightedge into contact with the line and determining that any space between the straightedge and the line does not exceed the specified tolerance.

For cylindrical parts or curved surfaces that are straight in one direction, the feature control frame

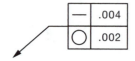

— .004 / ◯ .002

FIGURE 44–8 ■ *Combined feature control frames directed to one surface.*

2H

H = LETTER HEIGHT

FIGURE 44–9 ■ *Straightness symbol.*

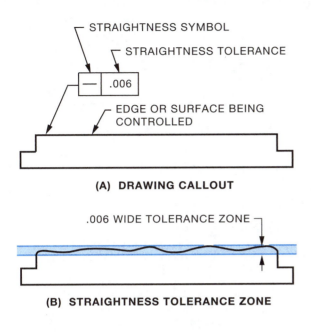

(A) DRAWING CALLOUT

(B) STRAIGHTNESS TOLERANCE ZONE

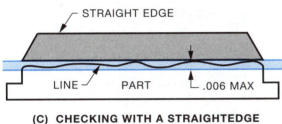

(C) CHECKING WITH A STRAIGHTEDGE

FIGURE 44–10 ■ *Straightness tolerance applied to a flat surface.*

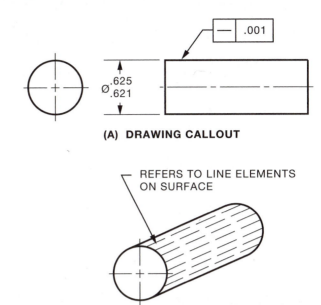

(A) DRAWING CALLOUT

(B) REFERS TO LINE ELEMENTS ON SURFACE

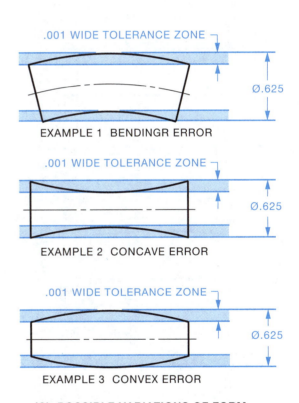

FIGURE 44–11 ■ *Specifying the straightness of line elements of a cylindrical surface.*

should be directed to the side view, where line elements appear as a straight line, Figures 44–11 and 44–12.

A straightness tolerance thus applied to the surface controls surface elements only. Therefore, it would control bending or a wavy condition of the surface or a barrel-shaped part, but it would not necessarily control the straightness of the axis or the conicity of the cylinder.

Straightness of a cylindrical surface is interpreted to mean that each line element of the surface shall be contained within a tolerance zone consisting of the space between two parallel lines, separated by the width of the specified tolerance, when the part is rolled along one of the planes. Theoretically, this could be measured by rolling the part on a flat surface and measuring the space between the part and the plate to ensure that it did not exceed the specified tolerance, Figure 44–13.

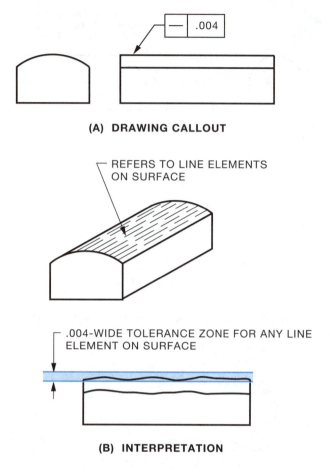

(A) DRAWING CALLOUT

REFERS TO LINE ELEMENTS ON SURFACE

.004-WIDE TOLERANCE ZONE FOR ANY LINE ELEMENT ON SURFACE

(B) INTERPRETATION

FIGURE 44–12 ■ *Straightness of surface line elements.*

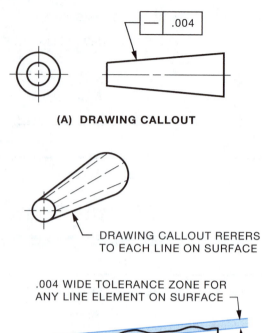

(A) DRAWING CALLOUT

DRAWING CALLOUT RERERS TO EACH LINE ON SURFACE

.004 WIDE TOLERANCE ZONE FOR ANY LINE ELEMENT ON SURFACE

(B) INTERPRETATION

FIGURE 44–14 ■ *Straightness of line elements on a conical surface.*

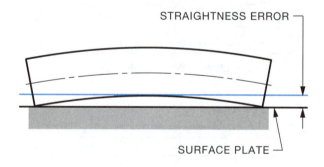

STRAIGHTNESS ERROR

SURFACE PLATE

FIGURE 44–13 ■ *Measuring the straightness of a cylindrical surface.*

A straightness tolerance can be applied to a conical surface in the same manner as for a cylindrical surface, Figure 44–14, and will ensure that the rate of taper is uniform. The actual rate of taper, or the taper angle, must be separately toleranced.

A straightness tolerance applied to a flat surface indicates straightness control in one direction only and must be directed to the line on the drawing representing the surface to be controlled and the direction in which control is required, Figure 44–15. It is then interpreted to mean that each line element on the surface in the indicated direction shall lie within a tolerance zone.

Different straightness tolerances may be specified in two or more directions when required. However, if the same straightness tolerance is required in two coordinate directions on the same surface, a flatness tolerance rather than a straightness tolerance is used.

If it is not otherwise necessary to draw all three views, the straightness tolerances may all be shown on a single view by indicating the direction with short lines terminated by arrowheads, Figure 44–15C.

REFERENCE

ASME Y14.5M-1994 (R1999) Dimensioning and Tolerancing

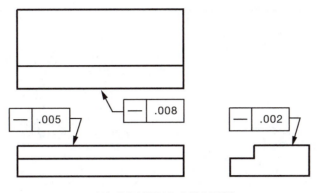

(A) DRAWING CALLOUT

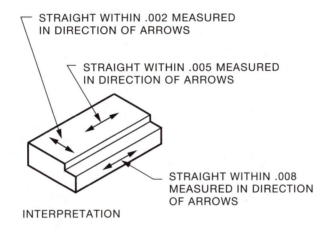

STRAIGHT WITHIN .002 MEASURED IN DIRECTION OF ARROWS

STRAIGHT WITHIN .005 MEASURED IN DIRECTION OF ARROWS

STRAIGHT WITHIN .008 MEASURED IN DIRECTION OF ARROWS

INTERPRETATION

(B) STRAIGHTNESS TOLERANCES IN SEVERAL DIRECTIONS

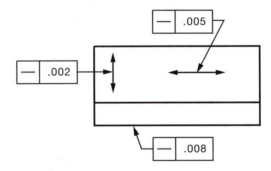

(C) THREE STRAIGHTNESS TOLERANCES ON ONE VIEW

FIGURE 44–15 ■ *Straightness tolerances in several directions.*

INTERNET RESOURCES

Drafting Zone. For information on geometric dimensioning and tolerancing, see: http://www.draftingzone.com

Effective Training Inc. For information on dimensioning and tolerancing, see: htpp://www.etinews.com/eti_solutions.html

eFunda. For information on geometric dimensioning and tolerancing, see http://www.efunda.com/home.cfm

Wikipedia, the Free Encyclopedia. For information on geometric dimensioning and tolerancing, see: http://en.wikipedia.org/wiki/Engineering_drawing

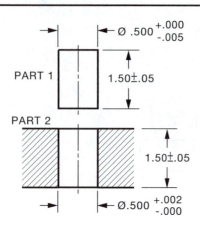

FIGURE 1

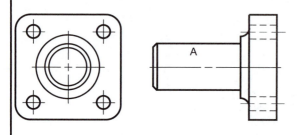

FIGURE 2

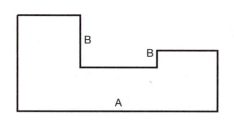

FIGURE 3

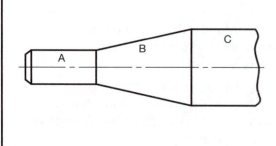

FIGURE 4

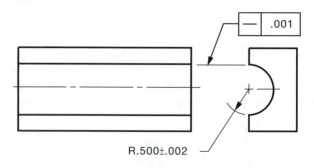

FIGURE 5

ASSIGNMENT:

NOTE: ALL ASSIGNMENTS TO BE SKETCHED ON
ONE INCH GRID SHEETS (.10 IN. SQUARES)

1. SKETCH THREE DIFFERENT PERMISSIBLE
    DEVIATIONS (SIMILAR TO FIGURE 44-2) FOR A
    RECTANGULAR PART .80 X 1.50 X .50 IN. THICK
    HAVING A TOLERANCE OF ±.05 IN. ON THESE
    DIMENSIONS. ADD DIMENSIONS.

2. MAKE A SKETCH, COMPLETE WITH DIMENSIONS,
    OF A RING AND PLUG GAGE TO CHECK THE HOLE
    AND SHAFT IN FIGURE 1.

3. MAKE A SKETCH OF THE PART SHOWN IN FIGURE 2.
    APPLY TWO FEATURE CONTROL FRAMES TO THE PART:
    ONE FRAME TO DENOTE CIRCULARITY (ROUNDNESS),
    THE OTHER FRAME TO CONTROL STRAIGHTNESS OF
    THE SURFACE OF THE CYLINDRICAL FEATURE.

4. MAKE A SKETCH OF THE PART SHOWN IN FIGURE 3.
    ADD THE FOLLOWING GEOMETRIC TOLERANCES TO
    THE SKETCH:
    (A) SURFACE "A" IS TO BE STRAIGHT WITHIN .005 IN.
    (B) SURFACES "B" AND "C" ARE TO BE STRAIGHT
        WITHIN .008 IN. USE A MINIMUM OF TWO DIFFERENT
        METHODS OF CONNECTING THE FEATURE CONTROL
        FRAME TO THE SURFACES REQUIRING STRAIGHTNESS
        CONTROL.

5. MAKE A SKETCH OF THE PART SHOWN IN FIGURE 4.
    ADD THE FOLLOWING STRAIGHTNESS TOLERANCES
    TO THE SURFACES.
    (A) SURFACE "A" - .002 IN.
    (B) SURFACE "B" - .001 IN.
    (C) SURFACE "C" - .005 IN.

6. WHAT IS THE MAXIMUM PERMISSIBLE DEVIATION
    FROM STRAIGHTNESS IF THE RADIUS IN FIGURE 5 IS
    (A) .498 IN.
    (B) .500 IN.
    (C) .502 IN.

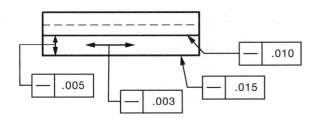

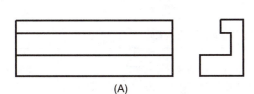

(A)

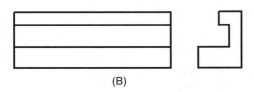

(B)

FIGURE 6

ASSIGNMENT:

7. MAKE A SKETCH OF THE PART SHOWN IN FIGURE 6(B) AND PLACE THE FEATURE CONTROL FRAMES SHOWN IN FIGURE 6(A) ON THIS SKETCH.

8. IS THE PART SHOWN IN FIGURE 7 ACCEPTABLE? STATE YOUR REASON.

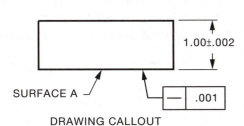

SURFACE A

DRAWING CALLOUT

FIGURE 7

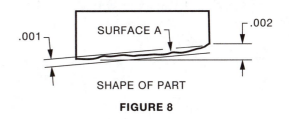

SURFACE A

SHAPE OF PART

FIGURE 8

STRAIGHTNESS TOLERANCE CONTROLLING SURFACE ELEMENTS

A-108

STRAIGHTNESS OF A FEATURE OF SIZE

Features of Size

Geometric tolerances that have so far been considered concern only lines, line elements, and single surfaces. These are features having no diameter or thickness, and the form tolerances applied to them cannot be affected by feature size. In these examples, the feature control frame leader was directed to the surface or extension line of the surface but not to the size dimension. The straightness tolerance had to be less than the size tolerance.

Features of size are features that do have diameter or thickness. These may be cylinders, such as shafts and holes. They may be slots, tabs, or rectangular or flat parts where two parallel flat surfaces are considered to form a single feature. When applying a geometric tolerance to a feature of size, the feature control frame is associated with the size dimension or attached to an extension of the dimension line.

Circular Tolerance Zones

When the resulting tolerance zone is cylindrical, such as when straightness of the axis of a cylindrical feature is specified, a diameter symbol precedes the tolerance value in the feature control frame. The feature control frame is located below the dimension pertaining to the feature, Figure 45–1.

FEATURE OF SIZE DEFINITIONS

Before proceeding with examples of features of size, it is essential to understand certain terms.

Maximum Material Condition (MMC)

When a feature or part is at the limit of size that results in it containing the maximum amount of material, it is said to be at MMC. Thus, it is the maximum limit of size for an external feature, such as a shaft,

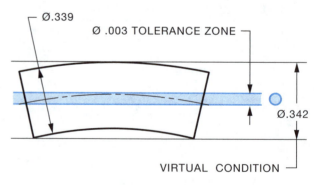

(A) DRAWING CALLOUT

(B) INTERPRETATION

FIGURE 45–1 ■ *Circular tolerance zone.*

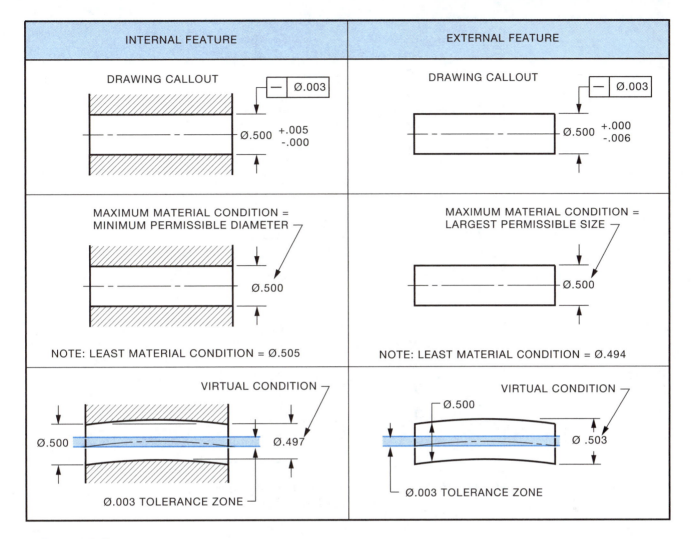

FIGURE 45–2 ■ *Maximum material condition and virtual condition.*

or the minimum limit of size for an internal feature, such as a hole, Figure 45–2.

Virtual Condition

Virtual condition refers to the overall envelope of perfect form within which the feature would just fit. It is the boundary formed by the MMC limit of size of a feature plus or minus the applied geometric tolerances. For an external feature such as a shaft, it is the maximum material size plus the effect of permissible form variations, such as straightness, flatness, roundness, cylindricity, and orientation tolerances. For an internal feature such as a hole, it is the maximum material size minus

the effect of such form variations, Figures 45–2 and 45–3.

Parts are generally toleranced so they will assemble when mating features are at MMC. Additional tolerance on form or location is permitted when features depart from their MMC size.

Least Material Condition (LMC)

This term refers to that size of a feature that results in the part containing the minimum amount of material. Thus it is the minimum limit of size for an external feature, such as a shaft, and the maximum limit of size for an internal feature, such as a hole, Figure 45–3.

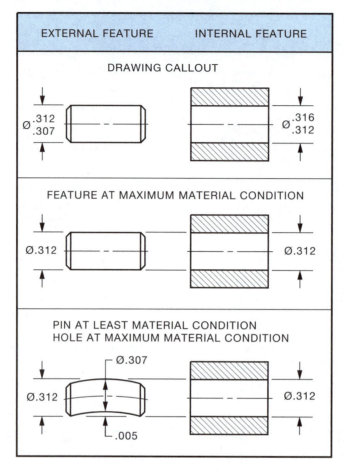

FIGURE 45–3 ■ *Effect of form variation when only feature of size is specified.*

Regardless of Feature Size (RFS)

This term means that the size of the geometric tolerance remains the same for any feature lying within its limits of size.

An example based on the location of features is shown in Figure 45–4. This shows a part with two projecting pins required to assemble into a mating part having two holes at the same center distance.

The worst assembly condition exists when the pins and holes are at their maximum material condition, which is Ø.250 in. Theoretically, these parts would just assemble if their form, orientation, and center distance were perfect. However, if the pins and holes were at their least material condition of Ø.247 and Ø.253, respectively, it would be evident that one center distance could be increased and the other decreased by .003 in. without jeopardizing the assembly condition.

MATERIAL CONDITION SYMBOLS

The modifying symbols used to indicate "at maximum material condition," and "at least material condition" are shown in Figure 45–5. The use of these symbols in local or general notes is prohibited.

Prior to 1994, the ANSI Y14.5 Dimensioning and Tolerancing Standard used a material condition symbol for "regardless of feature size," but this practice is now discontinued.

Applicability of RFS, MMC, and LMC

Applicability of RFS, MMC, and LMC is limited to features subject to variations in size. They may be datum features or other features whose axes or center planes are controlled by geometric tolerances. In

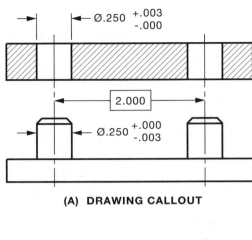

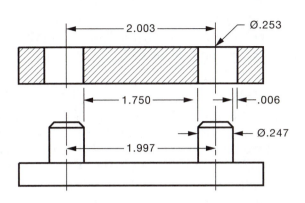

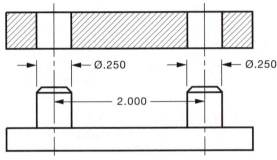

(A) DRAWING CALLOUT

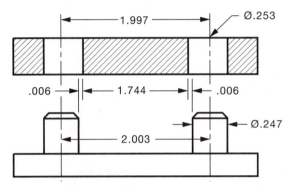

CENTER DISTANCE MUST BE PERFECT IN ORDER TO ASSEMBLE

EACH CENTER DISTANCE MAY BE INCREASED OR DECREASED BY .003

(B) PINS AND HOLES AT MAXIMUM MATERIAL CONDITION

(C) PINS AND HOLES AT LEAST MATERIAL CONDITION

FIGURE 45–4 ■ *Effect of location.*

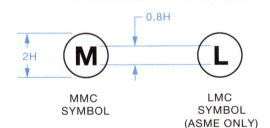

H = LETTER HEIGHT OF DIMENSIONS

0.8H

2H

M MMC SYMBOL

L LMC SYMBOL (ASME ONLY)

FIGURE 45–5 ■ *Modifying symbols.*

such cases, the following practices apply: RFS applies, with respect to the individual tolerance, datum reference, or both, where no modifying symbol is shown. (See Figures 45–2 and 45–11.) MMC or LMC must be specified on the drawing where it is required. (See Figure 45–12.)

EXAMPLES

If freedom of assembly of mating parts is the chief criterion for establishing a geometric tolerance for a feature of size, the least favorable assembly condition exists when the parts are made to the maximum material condition, that is, the largest diameter pin allowed entering a hole produced to the smallest allowable size. Further geometric variations can then be permitted, without jeopardizing assembly, as the features approach their least material condition.

Example 1

The effect of a form tolerance is shown in Figure 45–3, where a cylindrical pin of ∅.307−.312 in. is intended to assemble into a round hole of ∅.312−.316 in. If both parts are at their maximum material condition of ∅.312 in., it is evident that both would have to be perfectly round and straight in order to assemble. However, if the pin was at its least material condition of Ø.307 in., it could be bent up to .005 in. and still assemble in the smallest permissible hole.

Example 2

Another example, based on the location of features, is shown in Figure 45–4. This shows a part with two projecting pins required to assemble into a mating part having two holes at the same center distance.

The worst assembly condition exists when the pins and holes are at their maximum material condition, which is Ø.250 in. Theoretically, these parts would just assemble if their form, orientation (squareness to the surface), and center distances were perfect. However, if the pins and holes were at their least material condition of Ø.247 in. and Ø.253 in., respectively, it would be evident that one center distance could be increased and the other decreased by .003 in. without jeopardizing the assembly condition.

MAXIMUM MATERIAL CONDITION (MMC)

The symbol for maximum material condition is shown in Figure 45–6. The symbol dimensions are based on percentages of the recommended letter height of dimensions.

If a geometric tolerance is required to be modified on an MMC basis, it is specified on the drawing by including the symbol Ⓜ immediately after the tolerance value in the feature control frame, Figure 45–6.

A form tolerance modified in this way can be applied only to a feature of size; it cannot be applied to a single surface. It controls the boundary of the feature, such as a complete cylindrical surface, or two parallel surfaces of a flat feature. This permits the feature surface or surfaces to cross the maximum material boundary by the amount of the form tolerance. If design requirements are such that the virtual condition must be kept within the maximum material boundary, the form tolerance must be specified as zero at MMC, Figure 45–7.

Application of MMC to geometric symbols is shown in Figure 45–8.

Application with Maximum Value

It is sometimes necessary to ensure that the geometric tolerance does not vary over the full range permitted by the size variations. For such applications, a maximum limit may be set to the geometric tolerance and this is shown in addition to that permitted at MMC, Figure 45–9.

REGARDLESS OF FEATURE SIZE (RFS)

When MMC or LMC is not specified with a geometric tolerance for a feature of size, no relationship is intended to exist between the feature size and the geometric tolerance. In other words, the tolerance applies regardless of feature size.

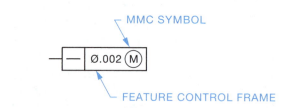

FIGURE 45–6 ■ *Application of MMC symbol.*

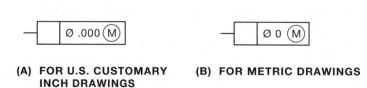

(A) FOR U.S. CUSTOMARY INCH DRAWINGS   **(B) FOR METRIC DRAWINGS**

FIGURE 45–7 ■ *MMC symbol with zero tolerances.*

CHARACTERISTIC TOLERANCE		FEATURE BEING CONTROLLED
STRAIGHTNESS	——	NO FOR A PLANE SURFACE OR A LINE ON A SURFACE
PARALLELISM	//	
PERPENDICULARITY	⊥	YES FOR A FEATURE OF SIZE WHICH IS SPECIFIED BY A TOLERANCED DIMENSION, SUCH AS A HOLE, SHAFT, OR A SLOT
ANGULARITY	∠	
POSITION	⊕	
FLATNESS	▱	NO FOR ALL FEATURES
CIRCULARITY (ROUNDNESS)	○	
CYLINDRICITY	⌭	
CONCENTRICITY	◎	
PROFILE OF A LINE	⌒	
PROFILE OF A SURFACE	⌓	
CIRCULAR RUNOUT	↗	
TOTAL RUNOUT	↗↗	
SYMMETRY	≡	

FIGURE 45–8 ■ *Application of MMC to geometric symbols.*

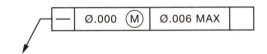

FIGURE 45–9 ■ *Tolerance with a maximum value.*

In this case, the geometric tolerance controls the form, orientation, or location of the axis, or median plane of the feature.

LEAST MATERIAL CONDITION (LMC)

The symbol for LMC is shown in Figure 45–5. It is the condition in which a feature of size contains the least amount of material within the stated limits of size.

Specifying LMC is limited to positional tolerance applications where MMC does not provide the desired

control and RFS is too restrictive. LMC is used to maintain a desired relationship between the surface of a feature and its true position at tolerance extremes. It is used only with a tolerance of position. See Unit 51.

STRAIGHTNESS OF A FEATURE OF SIZE

Figures 45–10 and 45–11 show examples of cylindrical parts where all circular elements of the surface are to be within the specified size tolerance; however, the boundary of perfect form at MMC may be violated. This violation is permissible when the feature control frame is associated with the size dimensions or attached to an extension of the dimension line. In these two figures, a diameter symbol precedes the tolerance value and the tolerance is applied on an RFS and an MMC basis, respectively. Normally, the straightness tolerance is

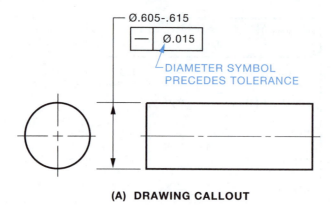

(A) DRAWING CALLOUT

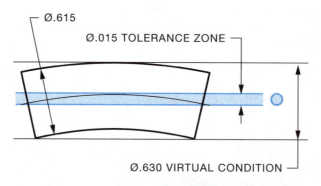

(B) INTERPRETATION

FEATURE SIZE	DIAMETER TOLERANCE ZONE ALLOWED
.615	.015
.614	.015
.613	.015
↓	↓
.606	.015
.605	.015

FIGURE 45–10 ■ *Specifying straightness—RFS.*

smaller than the size tolerance, but a specific design may allow the situation depicted in the figures. The collective effect of size and form variations can produce a virtual condition equal to the MMC size plus the straightness tolerance. (See Figure 45–11.) The derived median plane of the feature must lie within a cylindrical tolerance zone as specified.

Straightness — RFS

When applied on an RFS basis (as in Figure 45–10), the maximum permissible deviation from straightness is .015 in., regardless of the feature size. Note the absence of a modifying symbol indicates that RFS applies.

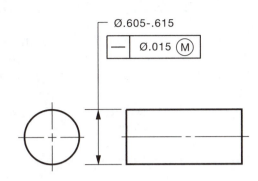

Ø.605-.615

| — | Ø.015 Ⓜ |

NOTE: PERFECT FORM AT MMC NOT REQ'D

(A) DRAWING CALLOUT

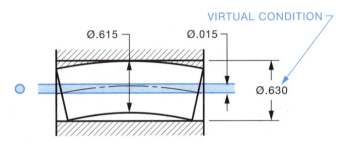

MAXIMUM DIAMETER OF THE PIN WITH PERFECT FORM IN A GAGE

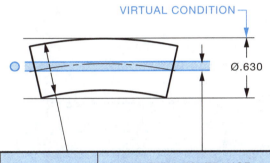

VIRTUAL CONDITION

Ø.630

FEATURE SIZE	DIAMETER TOLERANCE ZONE ALLOWED
.615	.015
.614	.016
.613	.017
↓	↓
.606	.025
.605	.024

(B) INTERPRETATION

PIN AT MAXIMUM DIAMETER (.615 IN.) WITH THE GAGE WILL ACCEPT THE PIN WITH UP TO .015-IN. VARIATION IN STRAIGHTNESS

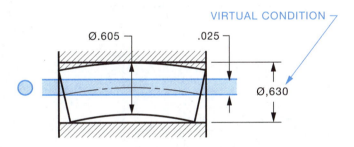

WITH PIN AT MINIMUM DIAMETER (.605 IN.) THE GAGE WILL ACCEPT THE PIN WITH UP TO .025 IN. VARIATION IN STRAIGHTNESS

(C) ACCEPTANCE BOUNDARY

FIGURE 45–11 ■ *Specifying straightness—MMC.*

Straightness — MMC

If the straightness tolerance of .015 in. is required only at MMC, further straightness error can be permitted without jeopardizing assembly, as the feature approaches its least material size (Figure 45–11). The maximum straightness tolerance is the specified tolerance plus the amount the feature departs from its MMC size. The center line of the actual feature must lie within the derived cylindrical tolerance zone such as given in the table of Figure 45–11.

Straightness — Zero MMC

It is quite permissible to specify a geometric tolerance of zero MMC, which means that the virtual condition coincides with the maximum material size, Figure 45–12. Therefore, if a feature is at its maximum material limit everywhere, no errors of straightness are permitted.

Straightness on the MMC basis can be applied to any part or feature having straight-line elements in a plane that includes the diameter or thickness. This also includes parts toleranced on an RFS basis. However, it should not be used for features that do not have a uniform cross section.

Straightness with a Maximum Value

If it is desired to ensure that the straightness error does not become too great when the part approaches the least material condition, a maximum value may be added, Figure 45–13. The maximum overall tolerance follows in a separate compartment in the feature control frame.

Straightness per Unit Length

Straightness may be applied on a unit-length basis as a means of preventing an abrupt surface variation within a relatively short length of the

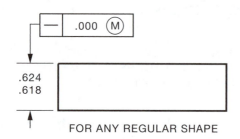

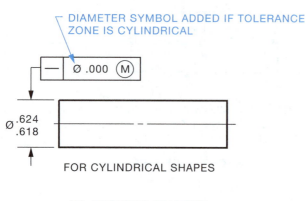

(A) DRAWING CALLOUT

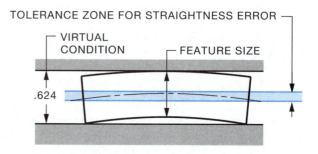

FEATURE SIZE	PERMISSIBLE STRAIGHTNESS ERROR
.624	.000
.623	.001
.622	.002
.621	.003
.620	.004
.619	.005
.618	.006

(B) PERMISSIBLE VARIATIONS

FIGURE 45–12 ■ *Straightness—zero MMC.*

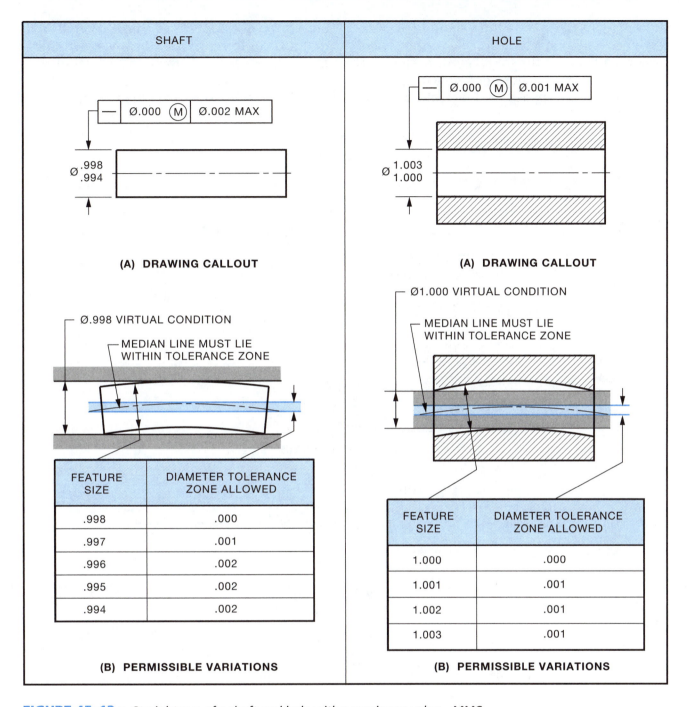

SHAFT	HOLE

SHAFT

─ | Ø.000 (M) | Ø.002 MAX

Ø .998 / .994

(A) DRAWING CALLOUT

─ Ø.998 VIRTUAL CONDITION

─ MEDIAN LINE MUST LIE WITHIN TOLERANCE ZONE

FEATURE SIZE	DIAMETER TOLERANCE ZONE ALLOWED
.998	.000
.997	.001
.996	.002
.995	.002
.994	.002

(B) PERMISSIBLE VARIATIONS

HOLE

─ | Ø.000 (M) | Ø.001 MAX

Ø 1.003 / 1.000

(A) DRAWING CALLOUT

─ Ø1.000 VIRTUAL CONDITION

─ MEDIAN LINE MUST LIE WITHIN TOLERANCE ZONE

FEATURE SIZE	DIAMETER TOLERANCE ZONE ALLOWED
1.000	.000
1.001	.001
1.002	.001
1.003	.001

(B) PERMISSIBLE VARIATIONS

FIGURE 45–13 ■ *Straightness of a shaft and hole with a maximum value—MMC.*

feature, Figure 45–14. Caution should be exercised when using unit control without specifying a maximum limit for the total length because of the relatively large variations that may result if no such restriction is applied. If the feature has a uniformly continuous bow throughout its length that just conforms to the tolerance applicable to the unit length, then the overall tolerance may result in an unsatisfactory part. Figure 43–15 illustrates the possible condition if the straight-

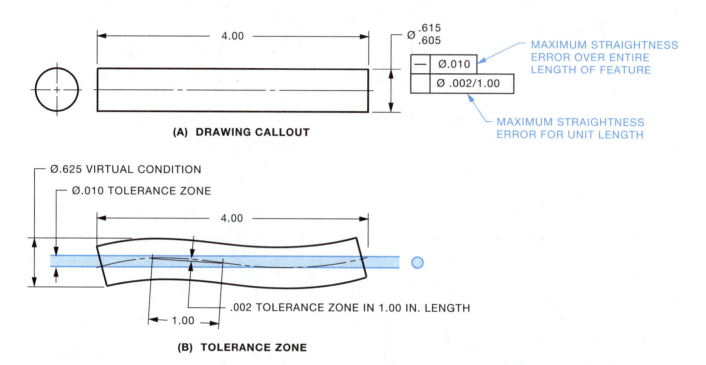

(A) DRAWING CALLOUT

Ø .615
Ø .605

— Ø.010
Ø .002/1.00

MAXIMUM STRAIGHTNESS ERROR OVER ENTIRE LENGTH OF FEATURE

MAXIMUM STRAIGHTNESS ERROR FOR UNIT LENGTH

Ø.625 VIRTUAL CONDITION
Ø.010 TOLERANCE ZONE
4.00
.002 TOLERANCE ZONE IN 1.00 IN. LENGTH
1.00

(B) TOLERANCE ZONE

FIGURE 45–14 ■ *Specifying straightness per unit length with specified total straightness, both RFS.*

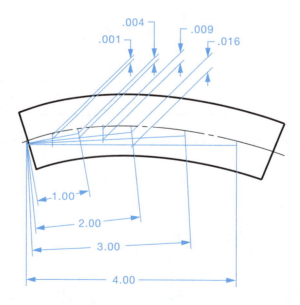

.001 .004 .009 .016
1.00
2.00
3.00
4.00

FIGURE 45–15 ■ *Possible results of specifying straightness per unit length RFS with no maximum value.*

ness per unit length given in Figure 45–14 is used alone, that is, if straightness for the total length is not specified.

REFERENCE

ASME Y14.5M-1994 (R1999) Dimensioning and Tolerancing

INTERNET RESOURCES

Drafting Zone. For information on geometric dimensioning and tolerancing, see: http://www.draftingzone.com

Effective Training Inc. For information on dimensioning and tolerancing, see: http://etinews.com/eti_solutions.html

eFunda. For information on geometric dimensioning and tolerancing, see: http://www.efunda.com/home.cfm

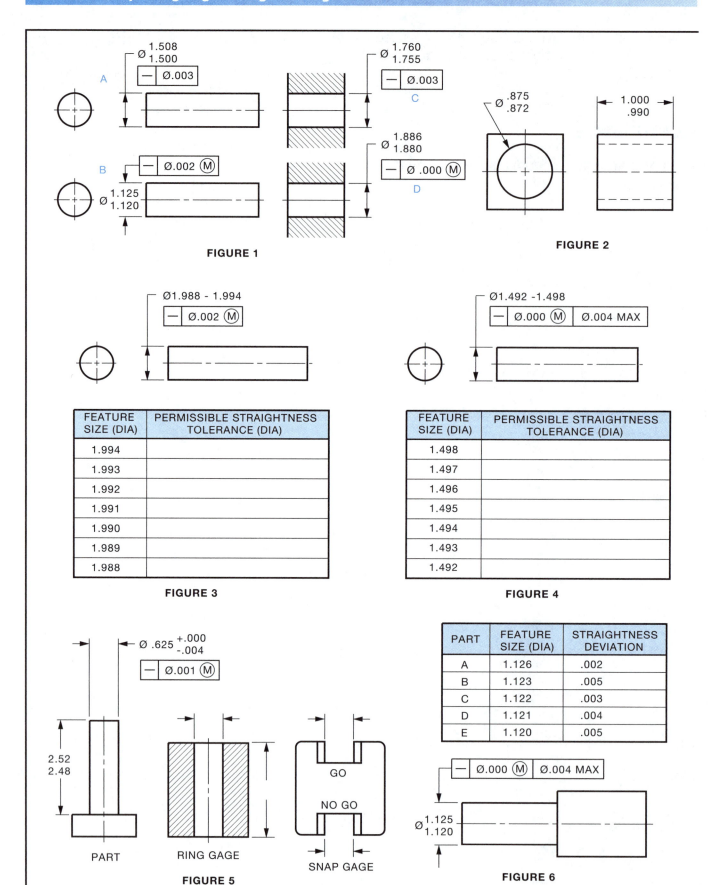

FIGURE 1

FIGURE 2

FIGURE 3

FIGURE 4

FIGURE 5

FIGURE 6

FEATURE SIZE (DIA)	PERMISSIBLE STRAIGHTNESS TOLERANCE (DIA)
1.994	
1.993	
1.992	
1.991	
1.990	
1.989	
1.988	

FEATURE SIZE (DIA)	PERMISSIBLE STRAIGHTNESS TOLERANCE (DIA)
1.498	
1.497	
1.496	
1.495	
1.494	
1.493	
1.492	

PART	FEATURE SIZE (DIA)	STRAIGHTNESS DEVIATION
A	1.126	.002
B	1.123	.005
C	1.122	.003
D	1.121	.004
E	1.120	.005

ASSIGNMENT:

NOTE:
ALL ASSIGNMENTS TO BE SKETCHED ON
ONE INCH GRID SHEETS (.10 IN. SQUARES)

1. WHAT IS THE VIRTUAL CONDITION FOR EACH
   PART SHOWN IN FIGURE 1?

2. THE HOLE IN FIGURE 2 DOES NOT HAVE A
   STRAIGHTNESS TOLERANCE. WHAT IS THE
   MAXIMUM PERMISSIBLE DEVIATION FROM
   STRAIGHTNESS IF PERFECT FORM AT THE
   MAXIMUM MATERIAL SIZE IS REQUIRED?

3. SKETCH CHARTS SIMILAR TO THAT SHOWN
   IN FIGURES 3 AND 4. COMPLETE THE CHARTS
   SHOWING THE LARGEST PERMISSIBLE
   STRAIGHTNESS FOR THE FEATURE SIZES SHOWN.

4. SKETCH AND DIMENSION THE RING AND SNAP
   GAGE SHOWN IN FIGURE 5 TO CHECK THE PIN
   SHOWN. THE RING GAGE SHOULD BE SUCH A SIZE
   AS TO CHECK THE LENGTH OF THE ENTIRE PIN.
   THE TWO OPEN ENDS OF THE SNAP GAGE SHOULD
   MEASURE THE MINIMUM AND MAXIMUM
   ACCEPTABLE PIN DIAMETERS.

5. WITH REFERENCE TO FIGURE 6, ARE PARTS A TO E
   ACCEPTABLE? STATE YOUR REASONS IF THE PART
   IS NOT ACCEPTABLE.

6. SKETCH THE SHAFT SHOWN IN FIGURE 7 AND
   ADD A STRAIGHTNESS TOLERANCE OF .004 IN.
   REGARDLESS OF FEATURE SIZE TO THE SHAFT.

7. WITH REFERENCE TO FIGURE 8, WHAT IS THE
   MAXIMUM DEVIATION PERMITTED FROM
   STRAIGHTNESS FOR THE CYLINDRICAL SURFACE
   IF IT WAS (A) AT MMC? (B) AT LMC? (C) Ø.623?

8. SKETCH THE SHAFT SHOWN IN FIGURE 9 AND ADD
   A MAXIMUM STRAIGHTNESS TOLERANCE OF .002 IN.
   FOR ANY 1.00 INCH OF ITS LENGTH, BUT HAVE A
   MAXIMUM STRAIGHTNESS TOLERANCE OF .005 IN.
   OVER THE ENTIRE LENGTH. APPLY THE APPROPRIATE
   STRAIGHTNESS TOLERANCE TO THE SKETCH.

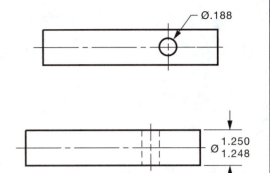

FIGURE 7

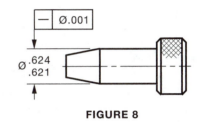

FIGURE 8

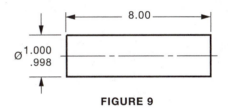

FIGURE 9

| | STRAIGHTNESS OF A FEATURE OF SIZE | A-109 |

46 UNIT

FORM TOLERANCES

Form tolerances are used to control straightness, flatness, circularity, and cylindricity.

FLATNESS

Flatness of a surface is a condition in which all surface elements are in one plane. The symbol for flatness is a parallelogram with angles of 60°.

A flatness tolerance is applied to a line representing the surface of a part by means of a feature control frame, Figure 46–1.

A flatness tolerance means that all points on the surface shall be contained within a tolerance zone consisting of the space between two parallel planes that are separated by the specified tolerance. These two parallel planes must lie within the limits of size. These planes may be oriented in any manner to contain the surface; that is, they are not necessarily parallel to the base.

The flatness tolerance must be less than the size tolerance and be contained within the limits of size.

When flatness tolerances are applied to opposite surfaces of a part and size tolerances are also shown, as in Figure 46–2, the flatness tolerance must be less than the size tolerance and lie within the limits of size.

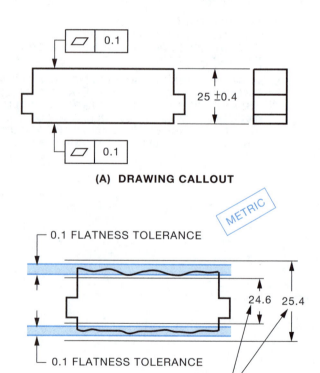

(A) DRAWING CALLOUT

METRIC

0.1 FLATNESS TOLERANCE

0.1 FLATNESS TOLERANCE

LIMITS OF SIZE

(B) TOLERANCE ZONES

FIGURE 46–2 ■ *Location of flatness tolerance within limits of size.*

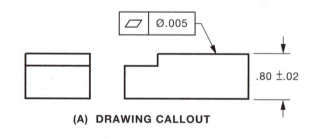

(A) DRAWING CALLOUT

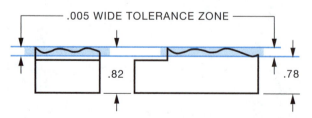

.005 WIDE TOLERANCE ZONE

THE SURFACE MUST LIE BETWEEN TWO PARALLEL PLANES .005 IN. APART. ADDITIONALLY, THE SURFACE MUST BE LOCATED WITHIN THE SPECIFIED LIMITS OF SIZE.

(B) TOLERANCE ZONE

FIGURE 46–1 ■ *Specifying flatness for a surface.*

Flatness per Unit Area

Flatness may be applied, as in the case of straightness, on a unit basis as a means of preventing an abrupt surface variation within a relatively small area of the feature. The unit variation is used either in combination with a specified total variation or alone. Caution should be exercised when using unit control alone for the same reason as was given to straightness.

Because flatness involves surface area, the size of the unit area, for example, 1.00 × 1.00 in., is specified to the right of the flatness tolerance, separated by a slash line, Figure 46–3.

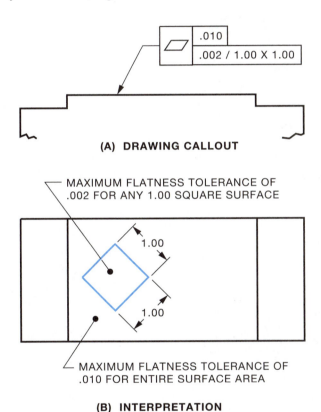

(A) DRAWING CALLOUT

MAXIMUM FLATNESS TOLERANCE OF .002 FOR ANY 1.00 SQUARE SURFACE

1.00

1.00

MAXIMUM FLATNESS TOLERANCE OF .010 FOR ENTIRE SURFACE AREA

(B) INTERPRETATION

FIGURE 46–3 ■ *Overall flatness tolerance combined with a flatness for a unit area.*

CIRCULARITY

Circularity refers to a condition of a circular line or the surface of a circular feature where all points on the line or on the circumference of a plane cross section of the feature are the same distance from a common axis or center point. It is similar to straightness except that it is wrapped around a circular cross section.

Examples of circular features would include disks, spheres, cylinders, and cones.

Errors of circularity (out-of-roundness) of a circular line on the periphery of a cross section of a circular feature may occur (1) as ovality, where differences appear between the major and minor axes; (2) as lobing, where in some instances the diametral values may be constant or nearly so; or (3) as random irregularities from a true circle. All these errors are illustrated in Figure 46–4.

The geometric characteristic symbol for circularity is a circle. A circularity tolerance may be specified by using this symbol in the feature control frame, Figure 46–5.

A circularity tolerance is measured radially and specifies the width between two concentric circular rings for a particular cross section within which the circular line or the circumference of the feature in that plane shall lie, Figure 46–6. Additionally, each circular element of the surface must be within the specified limits of size.

Because circularity is a form of tolerance, it is not related to datums.

A circularity tolerance may be specified by using the circularity symbol in the feature control frame. It is expressed on an RFS basis. The absence of a modifying symbol in the feature control frame means that RFS applies to the circularity tolerance.

A circularity tolerance cannot be modified on an MMC basis because it controls surface elements only. The circularity tolerance must be less than half

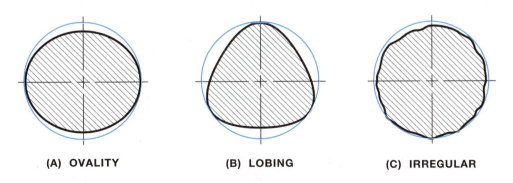

(A) OVALITY   **(B) LOBING**   **(C) IRREGULAR**

FIGURE 46–4 ■ *Common types of circularity errors.*

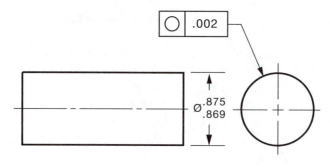

(A) DRAWING CALLOUT

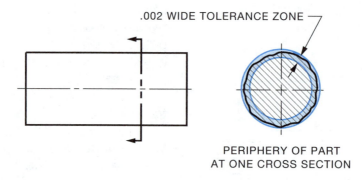

.002 WIDE TOLERANCE ZONE

PERIPHERY OF PART
AT ONE CROSS SECTION

PERIPHERY OF PART MUST LIE WITHIN LIMITS OF SIZE

(B) TOLERANCE ZONE

FIGURE 46–5 ∎ *Circularity tolerance applied to a cylindrical feature.*

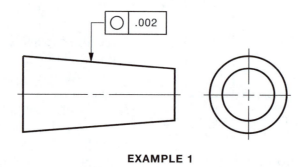

EXAMPLE 1

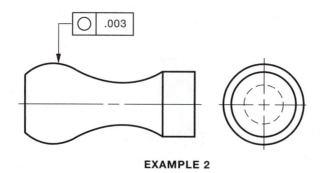

EXAMPLE 2

FIGURE 46–6 ∎ *Circularity tolerance applied to non-cylindrical features.*

the size tolerance because it must lie in a space equal to half the size tolerance.

Circularity of Noncylindrical Parts

Noncylindrical parts refer to conical parts and other features that are circular in cross section but have variable diameters, such as those shown in Figure 46–6. Because many sizes of circles may be involved in the end view, it is usually best to direct the circularity tolerance to the longitudinal surfaces as shown.

CYLINDRICITY

Cylindricity is a condition of a surface in which all points of the surface are the same distance from a common axis. The cylindricity tolerance is a composite control of form that includes circularity, straightness, and parallelism of the surface element of a cylindrical feature. It is like a flatness tolerance wrapped around a cylinder.

The geometric characteristic symbol for cylindricity consists of a circle with two tangent lines at 60°.

A cylindricity tolerance that is measured radially specifies a tolerance zone bounded by two concentric cylinders within which the surface must lie. The cylindricity tolerance must be within the specified limits of size. In the case of cylindricity, unlike that of circularity, the tolerance applies simultaneously to both circular and longitudinal elements of the surface, Figure 46–7. The leader from the feature control symbol

may be directed to either view. The cylindricity tolerance must be less than half of the size tolerance.

Because each part is measured for form deviation, it becomes obvious the total range of the specified cylindricity tolerance will not always be available.

The cylindricity tolerance zone is controlled by the measured size of the actual part. The part size is first determined; then the cylindricity tolerance is added as a refinement to the actual size of the part. If, in the example shown in Figure 46–7, the largest measurement of the produced part is Ø.748 in., which is near the high limit of size (.750 in.), the largest diameter of the two concentric cylinders for the cylindricity tolerance would be Ø.748 in. The smaller of the concentric cylinders would be .748 minus twice the cylindricity tolerance (2 × .002) = Ø.744 in. The cylindricity tolerance zone must also lie between the limits of size and the entire cylindrical surface of the part must lie between these two concentric circles to be acceptable.

If, on the other hand, the largest diameter measured for a part was Ø.743 in., which is near the lower limit of size (.740 in.), the cylindricity deviation of that part cannot be greater than .0015 in. because it would exceed the lower limit of size.

Likewise, if the smallest measured diameter of a part was .748 in., which is near the high limit of size, the largest diameter of the two concentric cylinders for the cylindricity tolerance would be Ø.750 in., which is the maximum permissible diameter of the part. In this case, the cylindricity tolerance could not be greater than (.750−.748)/2 or .001 in.

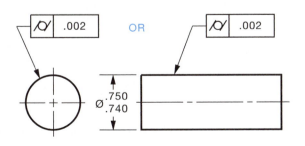

(A) DRAWING CALLOUT

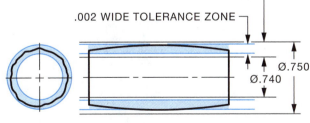

NOTE: CYLINDRICITY TOLERANCE MUST LIE WITHIN LIMITS OF SIZE

(B) TOLERANCE ZONE

FIGURE 46–7 ■ *Cylindrical tolerance directed to either view.*

Figure 46–8 shows some permissible form errors for the part shown in Figure 46–7.

Cylindricity tolerances can be applied only to cylindrical surfaces, such as round holes and shafts. No specific geometric tolerances have been devised for other circular forms, which require the use of several geometric tolerances. A conical surface, for example, must be controlled by a combination of tolerances for circularity, straightness, and angularity.

Because cylindricity is a form tolerance much like that of a flatness tolerance in that it controls surface elements only, it cannot be modified on an MMC basis. The absence of a modifying symbol in the feature control frame indicates that RFS applies.

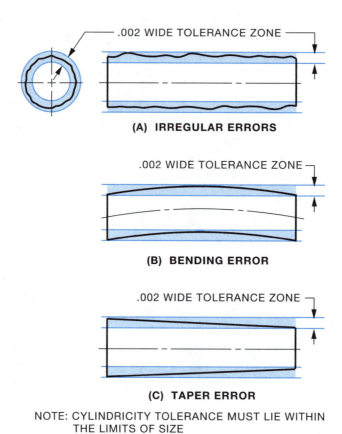

(A) IRREGULAR ERRORS

(B) BENDING ERROR

(C) TAPER ERROR

NOTE: CYLINDRICITY TOLERANCE MUST LIE WITHIN
THE LIMITS OF SIZE

FIGURE 46–8 ■ *Permissible form errors for part shown in Figure 46–7.*

REFERENCE

ASME Y14.5M-1994 (R1999) Dimensioning and
Tolerancing

INTERNET RESOURCES

Drafting Zone. For information on geometric
dimensioning and tolerancing, see:
http://www.draftingzone.com

Effective Training Inc. For information on
dimensioning and tolerancing, see:
http://etinews.cometi_solutions.html

eFunda. For information on geometric
dimensioning and tolerancing, see:
http://www.efunda.com/home.cfm

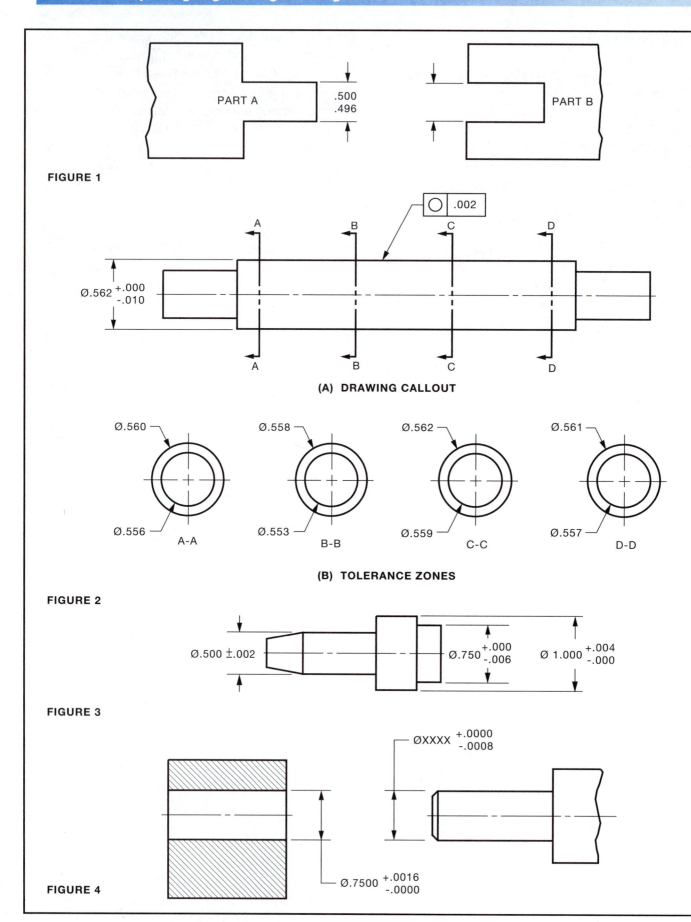

PART A

.500
.496

PART B

FIGURE 1

⌀.562 +.000 / -.010

○ | .002

A   B   C   D

A   B   C   D

(A) DRAWING CALLOUT

Ø.560 / Ø.556 A-A

Ø.558 / Ø.553 B-B

Ø.562 / Ø.559 C-C

Ø.561 / Ø.557 D-D

(B) TOLERANCE ZONES

FIGURE 2

Ø.500 ±.002

Ø.750 +.000 / -.006

Ø 1.000 +.004 / -.000

FIGURE 3

ØXXXX +.0000 / -.0008

Ø.7500 +.0016 / -.0000

FIGURE 4

ASSIGNMENT:

NOTE: ALL ASSIGNMENTS TO BE SKETCHED ON
ONE INCH GRID SHEETS (.10 IN. SQUARES).

1. SKETCH FIGURE 1. PART "A" MUST FIT INTO
PART "B" SO THAT THERE WILL NOT BE ANY
INTERFERENCE AND THE MAXIMUM CLEARANCE
WILL NEVER EXCEED .006 IN. ADD THE MAXIMUM
LIMITS OF SIZE TO PART "B". FLATNESS TOLERANCES
OF .001 IN. ARE TO BE ADDED TO THE TWO SURFACES
OF EACH PART.

2. MEASUREMENTS FOR CIRCULARITY FOR THE PARTS
SHOWN IN FIGURE 2 WERE MADE AT THE CROSS-
SECTIONS A-A TO D-D. ALL POINTS ON THE PERIPHERY
FELL WITHIN THE TWO RINGS. THE OUTER RING WAS
THE SMALLEST THAT COULD BE CIRCUMSCRIBED ABOUT
THE PROFILE AND THE INNER RING THE LARGEST THAT
COULD BE INSCRIBED WITHIN THE PROFILE. STATE
WHICH SECTIONS MEET DRAWING REQUIREMENTS.

3. SKETCH THE PART SHOWN IN FIGURE 3. ADD THE
LARGEST PERMISSIBLE CIRCULARITY TOLERANCE TO
EACH OF THE THREE DIAMETERS.

4. SKETCH THE PARTS SHOWN IN FIGURE 4.
(A) THE PARTS MUST ASSEMBLE WITH A MINIMUM
RADIAL CLEARANCE OF .0010 IN. (PER SIDE).
DIMENSION THE SHAFT ACCORDINGLY.

(B) WOULD ADDING A CYLINDRICITY TOLERANCE
ALTER THE SIZE OF THE SHAFT?

(C) WHAT IS THE LARGEST CYLINDRICITY TOLERANCE
THAT COULD BE REALIZED FOR THE HOLE AND
SHAFT IF THE FOLLOWING MEASUREMENTS WERE
RECORDED: .7510 IN. IN HOLE, .7476 IN. IN SHAFT?

5. SKETCH THE PART SHOWN IN FIGURE 5. APPLY
CYLINDRICITY TOLERANCES TO THE THREE FEATURES.
THE CYLINDRICITY TOLERANCES ARE TO BE 25% OF THE
SIZE OF TOLERANCES SHOWN ON EACH SHAFT.

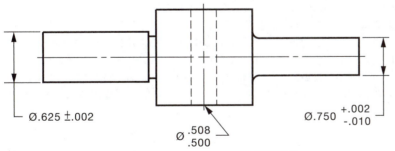

Ø.625 ±.002

Ø .508 / .500

Ø.750 +.002 / -.010

FIGURE 5

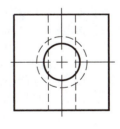

FORM TOLERANCES   **A-110**

47 UNIT

DATUMS AND THE THREE-PLANE CONCEPT

A datum is a point, axis, or plane, from which dimensions are measured, or to which geometric tolerances are referenced. A datum has an exact form and represents an exact or fixed location for purposes of manufacture or measurement.

A datum feature is a feature of a part, such as an edge, surface, or hole, which forms the basis for a datum or is used to establish the location of a datum.

DATUMS FOR GEOMETRIC TOLERANCING

Datums are exact geometric points, axes, or surfaces, each based on one or more datum features of the part. Surfaces are usually either flat or cylindrical, but other shapes are used when necessary. Because the datum features are physical surfaces of the part, they are subject to manufacturing errors and variations. For example, a flat surface of a part, if greatly magnified, will show some irregularity. If brought into contact with a perfect plane, this flat surface will touch only at the highest points, Figure 47–1. The true datums exist only in theory but are considered to be in the form of locating surfaces of machines, fixtures, and gaging equipment on

which the part rests or with which it makes contact during manufacture and measurement.

THREE-PLANE SYSTEM

Geometric tolerances, such as straightness and flatness, refer to unrelated lines and surfaces and do not require the use of datums.

Orientation and locational tolerances refer to related features; that is, they control the relationship of features to one another or to a datum or datum system. Such datum features must be properly identified on the drawing.

Usually only one datum is required for orientation purposes, but positional relationships may require a datum system consisting of two or three datums. These datums are designated as *primary, secondary,* and *tertiary.* When these datums are plane surfaces that are mutually perpendicular, they are commonly referred to as a three-plane datum system or a datum reference frame.

Primary Datum

If the primary datum feature is a flat surface, it could be laid on a suitable plane surface, such as the surface of a gage, which would then become a primary datum, Figure 47–2. Theoretically, there will be a

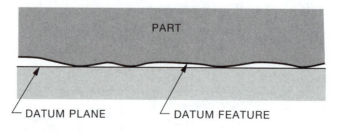

PART

DATUM PLANE   DATUM FEATURE

FIGURE 47–1 ■ *Magnified section of a flat surface.*

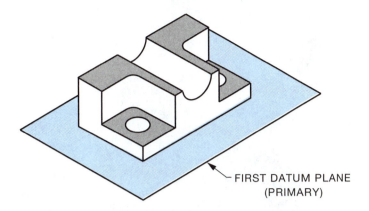

PRIMARY DATUM FEATURE MUST TOUCH PRIMARY
DATUM PLANE AT A MINIMUM OF THREE PLACES

FIGURE 47–2 ■ *Primary datum.*

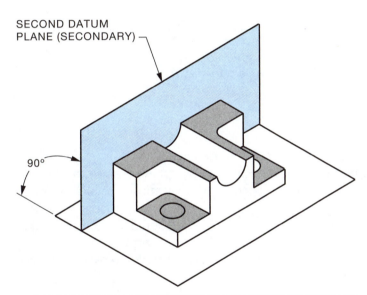

SECONDARY DATUM FEATURE MUST TOUCH SECONDARY
DATUM PLANE AT A MINIMUM OF TWO PLACES WHILE
RESTING ON DATUM PLANE A

FIGURE 47–3 ■ *Secondary datum.*

minimum of three high spots on the flat surface coming in contact with the gage surface.

Secondary Datum

If the part is brought into contact with a secondary plane while lying on the primary plane, it will theoretically touch at a minimum of two points, Figure 47–3.

Tertiary Datum

The part can be slid along while maintaining contact with both the primary and secondary planes until it contacts a third plane, Figure 47–4. This plane then becomes the tertiary datum and the part will, in theory, touch it at only one point.

These three planes constitute a datum system from which measurements can be taken. They will

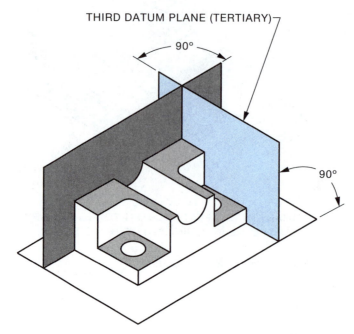

THIRD DATUM PLANE (TERTIARY)

90°

90°

TERTIARY DATUM FEATURE MUST TOUCH TERTIARY
DATUM PLANE AT ONE PLACE

FIGURE 47–4 ■ *Tertiary datum.*

appear on the drawing as shown in Figure 47–5, except that the datum features should be identified in their correct sequence by the methods described later in the unit.

UNEVEN SURFACES

When establishing a datum plane from a datum-feature surface, it is assumed that the surface will be reasonably flat and that the part would normally rest on three high spots on the surface. If the surface has a tendency toward concavity, Figure 47–6, no particular problems would arise.

However, if the surface was somewhat convex, it would have a tendency to rock on one or two high spots. In such cases it is intended that the datum plane should lie in the direction where the rock is equalized as much as possible. This usually results in the least possible deviation of the actual surface from the datum plane.

For example, in Figure 47–7, the datum plane is plane B and not plane A, because this results in deviation Z, which is less than deviation X.

If such conditions are likely to exist, the surface should be controlled by a flatness tolerance (and may have to be machined) or the datum target method should be used as explained in Unit 50.

DATUM FEATURE SYMBOL

Datum symbols have two functions. They indicate the datum surface or feature on the drawing and identify the datum feature so it can be easily referred to in other requirements.

The datum feature symbol is shown in Figure 47–8. The datum is identified by a capital letter placed horizontally in a square frame attached by a leader to a triangular base, which terminates at the datum feature. The only difference between the ASME and ISO datum feature symbols is the shape of the triangular base.

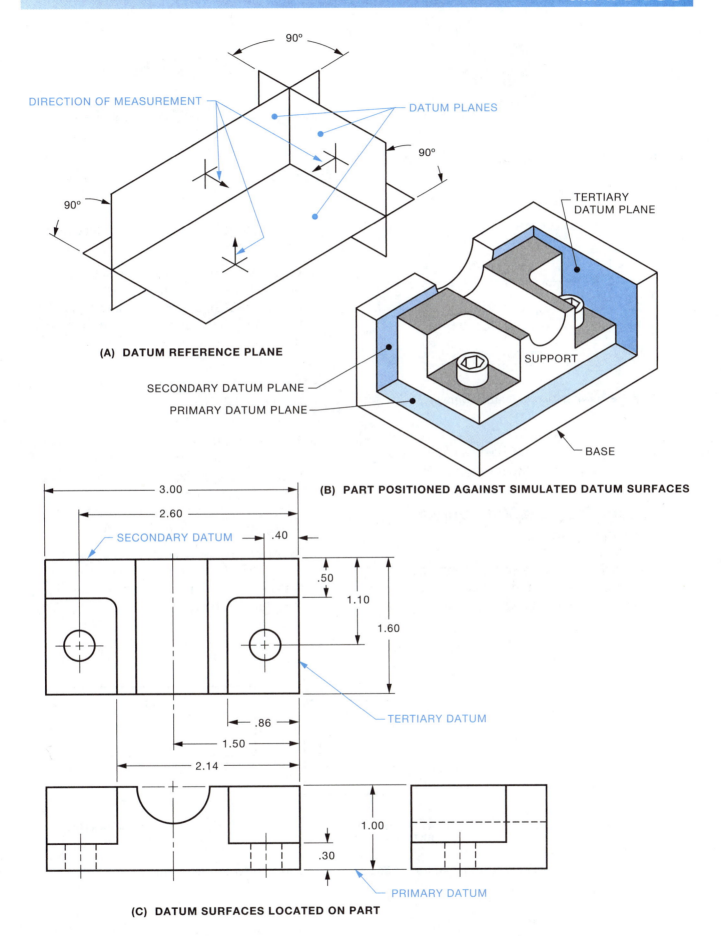

(A) DATUM REFERENCE PLANE

(B) PART POSITIONED AGAINST SIMULATED DATUM SURFACES

(C) DATUM SURFACES LOCATED ON PART

FIGURE 47–5 ■ *Three-plane datum system.*

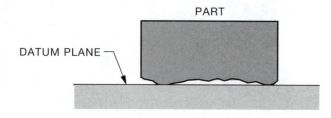

FIGURE 47–6 ■ *Concave surface as a datum feature.*

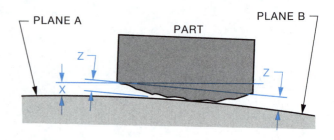

FIGURE 47–7 ■ *Datum plane for convex feature.*

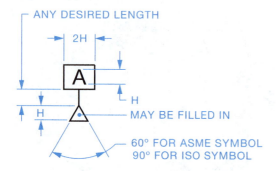

FIGURE 47–8 ■ *Datum feature symbol.*

This identifying symbol may be directed to the datum feature in any one of the following ways.

For Datum Features Not Subject to Size Variation

■ By attaching the base of the triangle to an extension line from the feature, providing it is a plane surface, but clearly separated from the dimension line, or to the surface itself. See Figure 47–9(A) and (B).

■ When only a part of a surface is to be designated as a datum, such as shown in Figure 47–9(C), a chain line drawn parallel to the surface is used to indicate the portion of the surface acting as the datum.

For Datum Features Subject to Size Variation

■ By attaching the base of the triangle to an extension of the dimension line pertaining to the feature of size when the datum is the axis or center plane. The datum feature symbol may replace part of the dimension line as shown in Figure 47–10(A).

■ By attaching the base of the triangle to the leader of a dimension where no feature control frame is used, as shown in Figure 45–10(C).

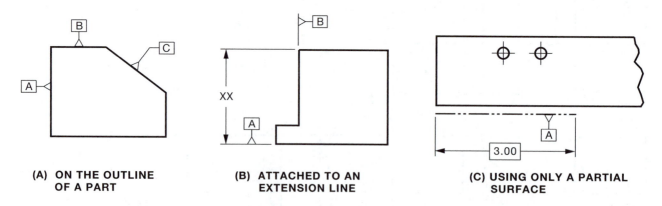

(A) ON THE OUTLINE OF A PART

(B) ATTACHED TO AN EXTENSION LINE

(C) USING ONLY A PARTIAL SURFACE

FIGURE 47–9 ■ *Placement of datum feature symbols for features not subject to size variation.*

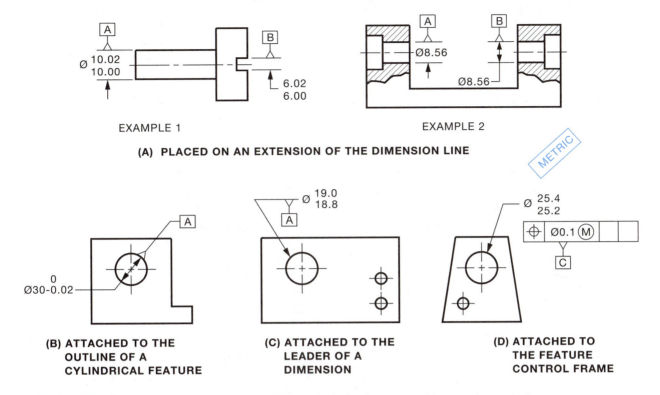

(A) PLACED ON AN EXTENSION OF THE DIMENSION LINE

EXAMPLE 1   EXAMPLE 2

(B) ATTACHED TO THE OUTLINE OF A CYLINDRICAL FEATURE

(C) ATTACHED TO THE LEADER OF A DIMENSION

(D) ATTACHED TO THE FEATURE CONTROL FRAME

FIGURE 47–10 ■ *Placement of datum feature symbols for features subject to size variations.*

■ By attaching the base of the triangle above or below the feature control frame, as shown in Figure 45–10(D).

Former ANSI Datum Feature Symbol

Prior to 1994, the United States (ANSI) used the symbol shown in Figures 47–11 and 47–12 to identify the datum feature. It is shown here as many drawings presently in existence show this symbol.

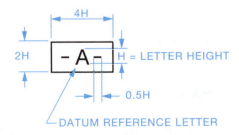

FIGURE 47–11 ■ *Former ANSI datum feature symbol.*

Association with Geometric Tolerances

The datum letter is placed in the feature control frame by adding an extra compartment for the datum reference, Figure 47–13.

If two or more datum references are involved, additional frames are added and the datum references are placed in these frames in the correct order, that is, primary, secondary, and tertiary datums, Figure 47–14.

Multiple Datum Features

If a single datum is established by two datum features, such as two flat or cylindrical surfaces, Figure 47–15, the features are identified by separate letters. Both letters are then placed in the same compartment of the feature control frame, separated by a dash. The datum in this case is the common axis or plane between the two datum features.

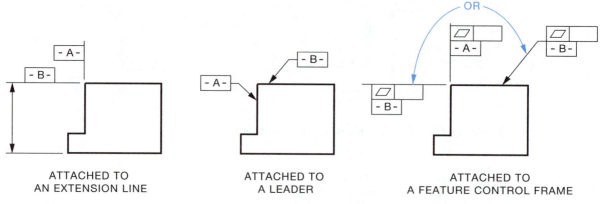

ATTACHED TO
AN EXTENSION LINE

ATTACHED TO
A LEADER

ATTACHED TO
A FEATURE CONTROL FRAME

(A) FEATURES NOT SUBJECT TO SIZE VARIATION

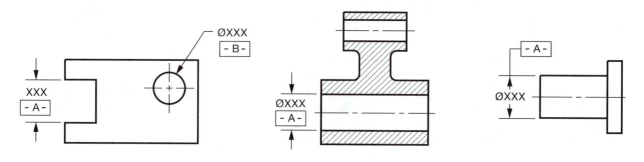

(B) FEATURES SUBJECT TO SIZE VARIATION

FIGURE 47–12 ■ *Placement of former ANSI datum feature symbol.*

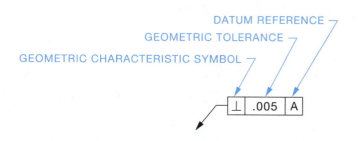

FIGURE 47–13 ■ *Feature control symbol referenced to a datum.*

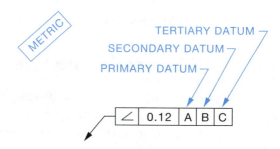

FIGURE 47–14 ■ *Multiple datum references.*

Partial Surfaces as Datums

It is often desirable to specify only part of a surface, instead of the entire surface, to serve as a datum feature. This may be indicated by means of a thick chain line drawn parallel to the surface profile (dimension for length and location), Figure 47–16, or by a datum target area as described in Unit 49. Figure 47–16 illustrates a long part where holes are located only at one end.

REFERENCE

ASME Y14.5M-1994 (R1999) Dimensioning and Tolerancing

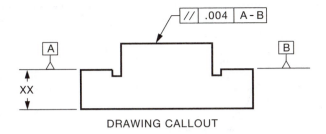

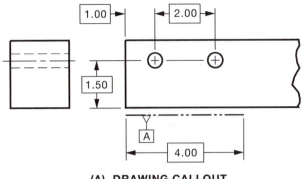

(A) DRAWING CALLOUT

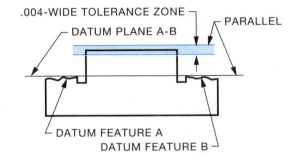

INTERPRETATION

(A) COPLANAR DATUM FEATURES

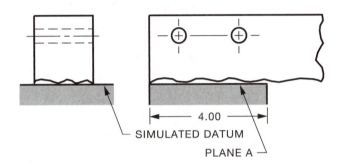

(B) INTERPRETATION

FIGURE 47–16 ■ *Partial datum.*

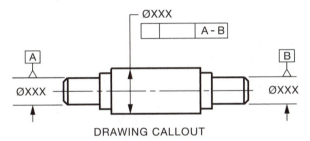

DRAWING CALLOUT

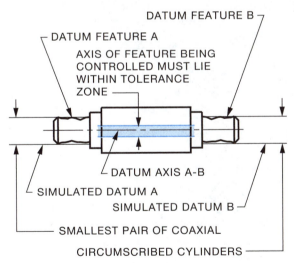

INTERPRETATION

(B) COAXIAL DATUM FEATURES

FIGURE 47–15 ■ *Two datum features for one datum.*

INTERNET RESOURCES

Drafting Zone. For information on geometric dimensioning and tolerancing, see: http://www.draftingzone.com

Effective Training Inc. For information on dimensioning and tolerancing, see: http://etinews.com/eti_solutions.html

eFunda. For information on geometric dimensioning and toleracing, see: http://www.efunda.com/home.cfm

Engineering Edge. For information on geometric tolerancing and dimensioning, see: http://www.engineeringedge.com/gdt/htm

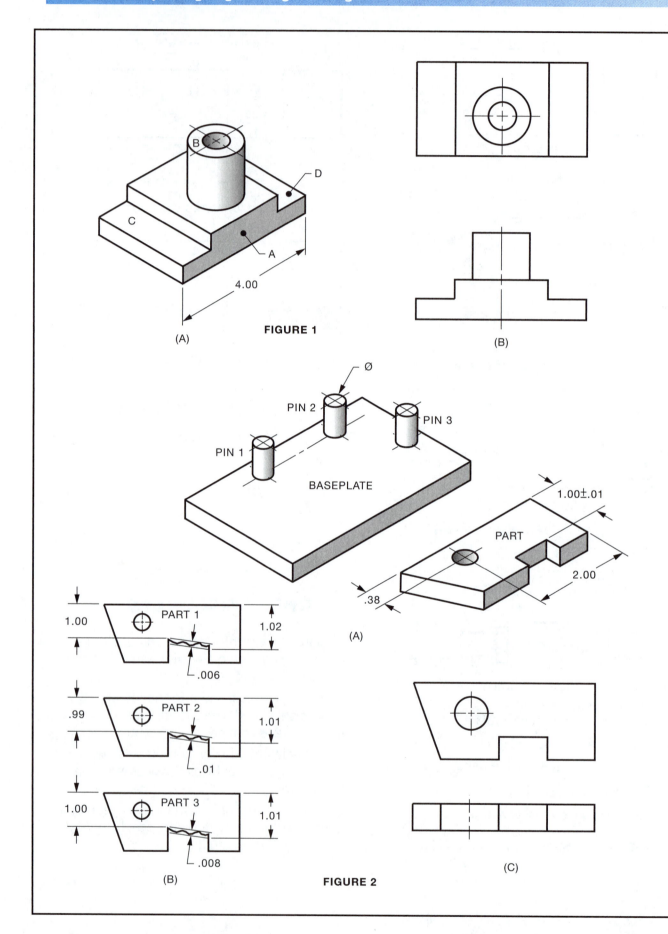

FIGURE 1

(A)

(B)

FIGURE 2

(A)

(B)

(C)

ASSIGNMENT:

Note: Use one-inch grid sheets (.10 in. squares) for the sketching assignments.

1. Sketch the two views shown in Figure 1(B) and add the following information to the sketch:

   a) Surface A is datum A and is to be straight within .008 in. for the 4.00 in. length, but the straightness error should not exceed .002 in. for any 1.00 in. length.
   b) Surface B is datum B and is to be flat within .004 in.
   c) The base is to be flat within .005 in.
   d) Surfaces C and D are datum features C and D respectively which form a single datum.
   e) The surface of the cylinder is to be straight within .003 in.

2. A. Pins 1, 2, and 3 are used to establish the secondary and tertiary datums for the part shown in Figure 2. What is used for the primary datum?
   B. Sketch the two views shown in Figure 2(C) and identify the primary, secondary, and tertiary datum planes as A, B, and C respectively.
   C. How far is the center of the hole from (1) tertiary datum? (2) secondary datum?
   D. The back of the slot is to be flat within .008 in. and the secondary datum is to be flat within .004 in. Place these form tolerances on the sketch.
   E. Are the parts shown in Figure 2B acceptable? If not, state your reasons.

3. A. Sketch the part shown in Figure 3 and add the following to the sketch. The bottom surface is to be flat within .004 in. and is to be identified as datum B. on the drawing.
   B. What is the minimum height of the part?

4. What is the minimum number of contact points in a three-plane datum system for (a) primary datum? (b) secondary datum? (c) tertiary datum?

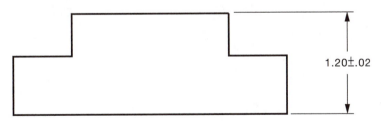

1.20±.02

FIGURE 3

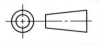 **DATUMS** **A-111**

ASSIGNMENT:

Anyone involved with the use of technical drawings must be capable of interpreting drawings containing current and formerly used symbols and standards. On a centimeter grid sheet (1 mm squares) sketch the axle drawing twice, and add geometric tolerances and datums to these sketches. One drawing is to use current ASME drawing practices and symbols. The other drawing is to use former ANSI drawing practices and symbols.

Show the following information on both drawings:
1. Diameter M to be datum A
2. The end face of diameter N to be used as datum B
3. The width of the slot to be datum C
4. The end face to be flat within 0.25 mm
5. The axis of diameter M must be straight within 0.1 mm regardless of feature size
6. The surface of diameter L to be straight within 0.2 mm

QUESTIONS:

1. What are the names given to the planes of a three-plane datum system?

2. What is the name of the symbol that identifies a datum on a drawing?

3. What is the minimum number of contact points between the secondary datum feature and the datum plane?

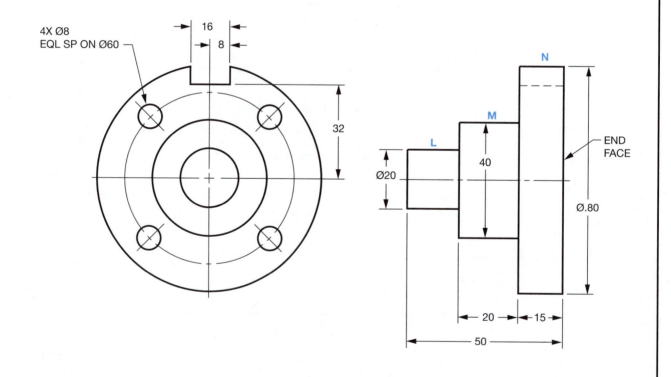

METRIC
DIMENSIONS ARE IN MILLIMETERS

| | AXLE | **A-112M** |

ORIENTATION TOLERANCES

Orientation refers to the angular relationship that exists between two or more lines, surfaces, or other features. Orientation tolerances control angularity, parallelism, and perpendicularity. Because, to a certain degree, the limits of size control form and parallelism, and tolerances of location (see Unit 51) control orientation, the extent of this control should be considered before specifying form or orientation tolerances.

A tolerance of form or orientation may be specified where the tolerances of size and location do not provide sufficient control.

Orientation tolerances, when applied to plane surfaces, control flatness if a flatness tolerance is not specified.

The general geometric characteristic for orientation is termed *angularity*. This term may be used to describe angular relationships, of any angle, between straight lines or surfaces with straight line elements, such as flat or cylindrical surfaces. For two particular types of angularity special terms are used. These are *perpendicularity*, or squareness, for features related to each other by a 90° angle, and *parallelism* for features related to one another by an angle of zero.

An orientation tolerance specifies a zone within which the considered feature, its line elements, its axis, or its center plane must be contained.

Reference to a Datum

An orientation tolerance indicates a relationship between two or more features. Whenever possible, the feature to which the controlled feature is related should be designated as a datum. Sometimes this does not seem possible, for example, where two surfaces are equal and cannot be distinguished from one another. The geometric tolerance could theoretically be applied to both surfaces without a datum, but it is generally preferable to specify two

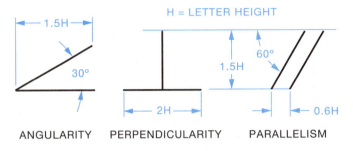

FIGURE 48–1 ■ *Orientation symbols.*

similar requirements, using each surface in turn as the datum.

Angularity, parallelism, and perpendicularity are orientation tolerances applicable to related features. Relation to more than one datum feature should be considered if required to stabilize the tolerance zone in more than one direction.

There are three geometric symbols for orientation tolerances, Figure 48–1. The proportions are based on the height of the lettering used on the drawing.

Angularity Tolerance

Angularity is the condition of a surface or axis at a specified angle (other than 0° and 90°) from a datum plane or axis. An angularity tolerance for a flat surface specifies a tolerance zone, the width of which is defined by two parallel planes at a specified basic angle from a datum plane or axis. The surface of the considered feature must lie within this tolerance zone, Figure 48–2.

For geometric tolerancing of angularity, the angle between the datum and the controlled feature should be stated as a basic angle. Therefore, it should be enclosed in a rectangular frame (basic dimension symbol) as shown in Figure 48–2 to indicate that the general tolerance note does not apply. However, the angle need not be stated for either perpendicularity (90°) or parallelism (0°).

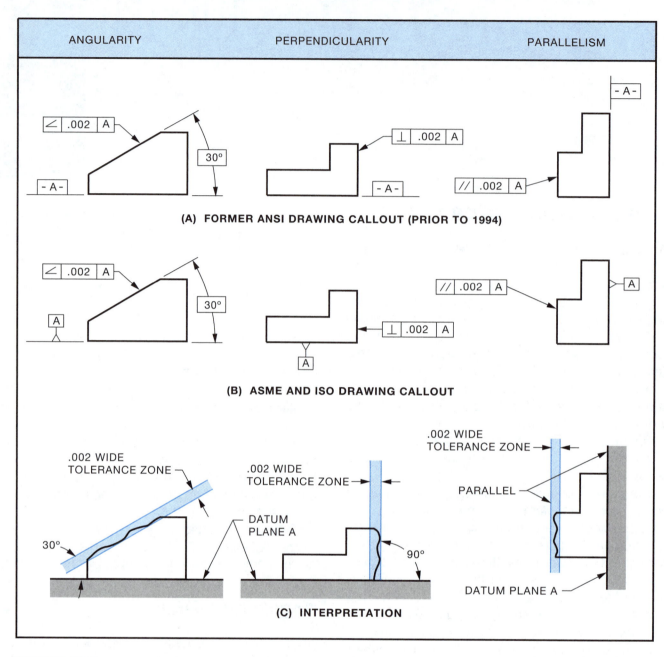

FIGURE 48–2 ■ *Orientation tolerancing of flat surfaces.*

Perpendicularity Tolerance

Perpendicularity is the condition of a surface at 90° to a datum plane or axis. A perpendicularity tolerance for a flat surface specifies a tolerance zone defined by two parallel planes perpendicular to a datum plane or axis. The surface of the considered feature must lie within this, Figure 48–2.

Parallelism Tolerance

Parallelism is the condition of a surface equidistant at all points from a datum plane. A parallelism tolerance for a flat surface specifies a tolerance zone defined by two planes or lines parallel to a datum plane or axis. The line elements of the surface must lie within this tolerance zone, Figure 48–2.

ORIENTATION TOLERANCING FOR FLAT SURFACES

Figure 48–2 shows three simple parts in which one flat surface is designated as a datum feature and another flat surface is related to it by one of the orientation tolerances.

Each of these tolerances is interpreted to mean that the designated surface shall be contained within a tolerance zone consisting of the space between two parallel planes, separated by the specified tolerance (.002 in.) and related to the datum by the basic angle specified (30°, 90°, or 0°).

When orientation tolerances apply to a line or surface, a leader is attached to the feature control frame and is directed to the line or surface requiring control.

An orientation tolerance applied to a feature automatically ensures that the form of the feature is within the same tolerance.

Therefore, when an orientation tolerance is specified, there is no need to also specify a form tolerance for the same feature unless a smaller tolerance is necessary.

Control in Two Directions

The measuring principles for angularity indicate the method of aligning the part prior to making angularity measurements. Proper alignment ensures that line elements of the surface perpendicular to the angular line elements are parallel to the datum.

For example, the part in Figure 48–3 will be aligned so that line elements running horizontally in the right-hand view will be parallel to datum A. However, these line elements will bear a proper relationship with the sides, ends, and top faces only if these surfaces are true and square with datum B.

Applying Form and Orientation Tolerances to a Single Feature

When both form and orientation tolerances are applied to a feature, the form tolerance must be less than the orientation tolerance. An example is shown in Figure 48–4 where the flatness of the surface must be controlled to a greater degree than its orientation. The flatness tolerance must lie within the angularity tolerance zone.

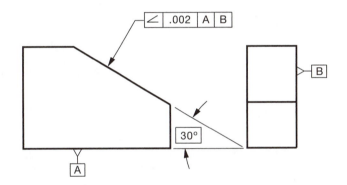

FIGURE 48–3 ■ *Angularity referenced to a datum system.*

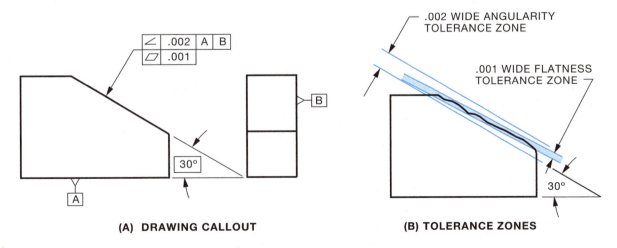

(A) DRAWING CALLOUT　　　　(B) TOLERANCE ZONES

FIGURE 48–4 ■ *Applying both an angularity and flatness tolerance to a feature.*

REFERENCE

ASME Y14.5M-1994 (R1999) Dimensioning and Tolerancing

INTERNET RESOURCES

Drafting Zone. For information on geometric dimensioning and tolerancing, see: http://www. draftingzone.com

Effective Training Inc. For information on dimensioning and tolerancing, see: http://etinews.com/eti_solutions.html

eFunda. For information on geometric dimensioning and tolerancing, see: http://www.efunda.com/home.cfm

Engineering Edge. For information on geometric tolerancing and dimensioning, see: http://www.engineeringedge.com/gdt/htm

ASSIGNMENT:

On a one-inch grid sheet (.10 in. squares) sketch three views of the stand shown below and add the following geometric tolerances to the drawing.

1. Surfaces A, B, and D are to be datums A, B, and D, respectively.

2. The back is to be perpendicular to the bottom within .01 in. and be flat within .006 in.

3. The top is to be parallel with the bottom within .005 in.

4. Surface C is to have an angularity tolerance of .008 in. with the bottom. Surface D is to be the secondary datum for this requirement.

5. The bottom is to be flat within .002 in.

6. The sides of the slot are to be parallel to each other within .002 in. and perpendicular to the back (datum B) within .004 in. One side of the slot is to be datum E.

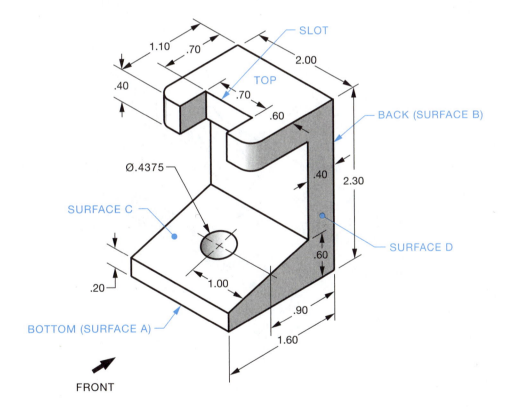

SLOT
1.10
.70
2.00
.40
TOP
.70
.60
BACK (SURFACE B)
Ø.4375
.40
2.30
SURFACE C
SURFACE D
.20
.60
1.00
.90
BOTTOM (SURFACE A)
1.60
FRONT

STAND   **A-113**

ASSIGNMENT:

On a centimeter grid sheet (1 mm squares) sketch the top, front, and left side views of the cut-off stand to the scale of 1 : 2. From the information shown on the drawing below, add the following geometric tolerances and basic dimensions to the surfaces.

1. Surfaces A, B, C, D, and E are to be datums A, B, C, D, and E, respectively.

2. Surface C is to have a flatness tolerance of 0.2 mm.

3. Surfaces F and G of the dovetail are to have an angularity tolerance of 0.05 mm with a single datum established by the two datum features D and E. These surfaces are to be flat within 0.02 mm.

4. Surface H is to be parallel to surface B within 0.05 mm.

5. Surface C is to be perpendicular to surfaces D and E within 0.04 mm.

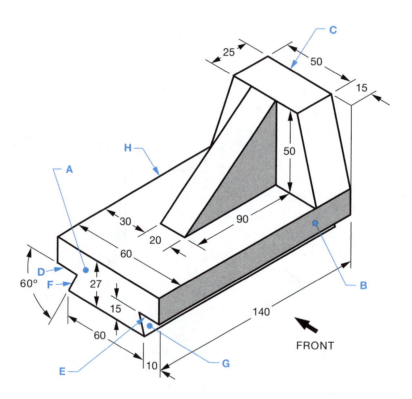

METRIC
DIMENSIONS ARE IN MILLIMETERS

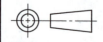 CUT-OFF STAND | A–114M

ORIENTATION TOLERANCING FOR FEATURES OF SIZE

When orientation tolerances apply to the axis of cylindrical features or to the datum planes of two flat surfaces, the feature control frame is associated with the size dimension of the feature requiring control, Figure 49–1.

Tolerances intended to control orientation of the axis of a feature are applied to drawings as shown in Figure 49–2. Although this unit deals mostly with cylindrical features, methods similar to those given here can be applied to noncircular features, such as square and hexagonal shapes.

The axis of the cylindrical feature must be contained within a tolerance zone consisting of the space between two parallel planes separated by the specified tolerance. The parallel planes are related to the datum by the basic angles of 45°, 90°, or 0° in Figure 49–2.

The absence of a modifying symbol in the tolerance compartment of the feature control frame indicates that RFS applies.

Angularity Tolerance

The tolerance zone is defined by two parallel planes at the specified basic angle from a datum plane or axis, within which the axis of the considered feature

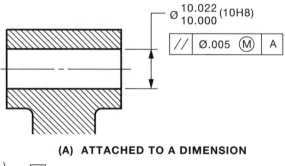

$\varnothing \, ^{10.022}_{10.000}$ (10H8)

| // | Ø.005 Ⓜ | A |

(A) ATTACHED TO A DIMENSION

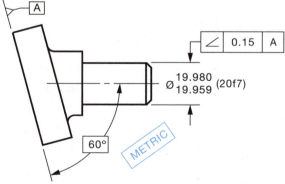

A

| ∠ | 0.15 | A |

$\varnothing \, ^{19.980}_{19.959}$ (20f7)

60°

METRIC

(B) ATTACHED TO THE EXTENSION OF THE DIMENSION LINE

FIGURE 49–1 ■ *Feature control frame associated with size dimension.*

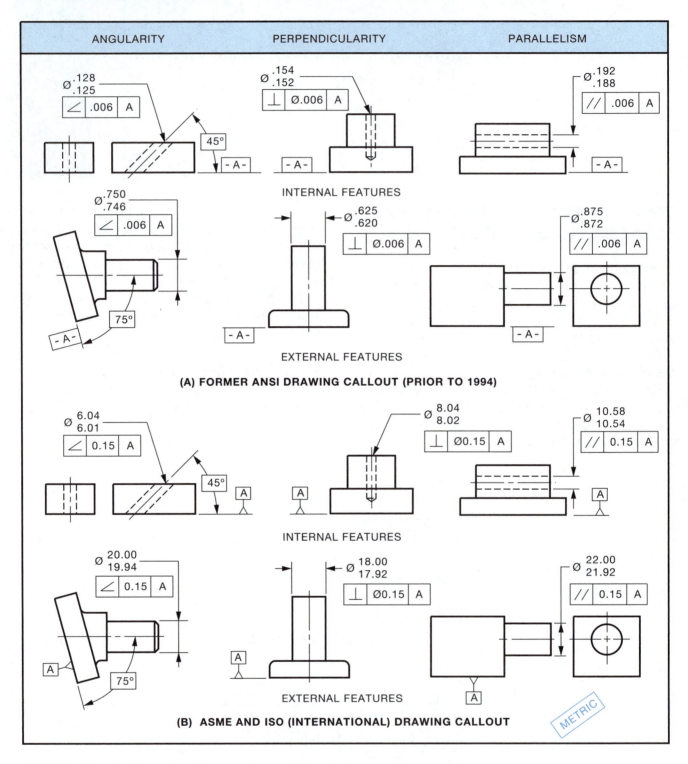

FIGURE 49–2 ■ *Orientation tolerances for cylindrical features—RFS.*

must lie. Figure 49–3 illustrates the tolerance zone for angularity.

Perpendicularity Tolerance

A perpendicularity tolerance specifies one of the following:

1. A cylindrical tolerance zone perpendicular to a datum plane or axis within which the center line of the considered feature must lie. (See Figure 49–2.)

2. A tolerance zone defined by two parallel planes perpendicular to a datum axis within which the axis of the considered feature must lie. (See Figure 49–13.)

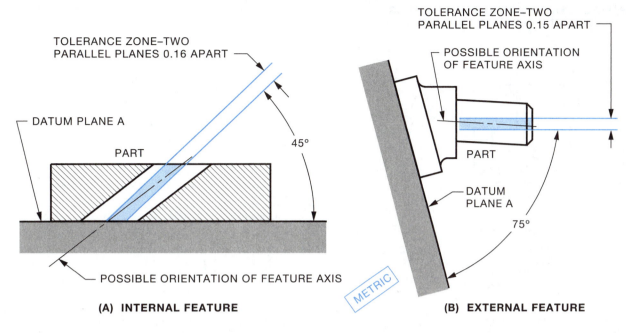

FIGURE 49–3 ■ *Tolerance zone for angularity shown in Figure 49–2.*

When the tolerance is one of perpendicularity, the tolerance zone planes can be revolved around the feature axis without affecting the angle. The tolerance zone therefore becomes a cylinder. This cylindrical zone is perpendicular to the datum and has a diameter equal to the specified tolerance, Figure 49–4. A diameter symbol precedes the perpendicularity tolerance.

Parallelism Tolerance

Parallelism is the condition of a surface equidistant at all points from a datum plane or an axis equidistant along its length from a datum axis or plane. A parallelism tolerance specifies a tolerance zone defined by two planes or lines parallel to a datum plane or axis, within which the axis of the considered feature must lie (see Figure 49–5); or a cylindrical tolerance zone, the axis of which is parallel to the datum axis within which the axis of the considered feature must lie (see Figure 49–10).

Control In Two Directions

The feature control frame for angularity shown in Figure 49–2 controls angularity with the base (datum A) only. If control with a side is also required, the side should be designated as the secondary datum, Figure 49–6. The center line of the hole must lie within the two parallel planes.

Control On an MMC Basis

Example 1

As a hole is a feature of size, any of the tolerances shown in Figure 49–2 can be modified on an MMC basis. This is specified by adding the symbol Ⓜ after the tolerance; Figure 49–7 shows an example.

Examples 2 and 3

Because the cylindrical features represent features of size, orientation tolerances may be applied on an MMC basis. This is indicated by adding the modifying symbol after the tolerance as shown in Figures 49–8 and 49–9.

INTERNAL CYLINDRICAL FEATURES

Figure 49–2 shows some simple parts in which the axis or center line of a hole is related by an orientation tolerance to a flat surface. The flat surface is designated as the datum feature.

The axis of each hole must be contained within a tolerance zone consisting of the space between two parallel planes. These planes are separated by a specified tolerance of .006 in. for the parts shown in Figure 49–2(A), and by a specified tolerance of 0.15 mm for the parts shown in Figure 49–2(B).

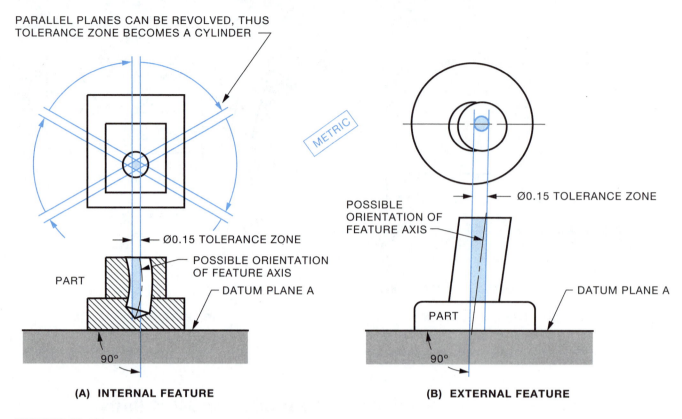

PARALLEL PLANES CAN BE REVOLVED, THUS
TOLERANCE ZONE BECOMES A CYLINDER

Ø0.15 TOLERANCE ZONE

POSSIBLE ORIENTATION
OF FEATURE AXIS

PART

DATUM PLANE A

90°

METRIC

POSSIBLE
ORIENTATION OF
FEATURE AXIS

Ø0.15 TOLERANCE ZONE

PART

DATUM PLANE A

90°

(A) INTERNAL FEATURE

(B) EXTERNAL FEATURE

FIGURE 49–4 ■ *Tolerance zone for perpendicularity shown in Figure 49–2.*

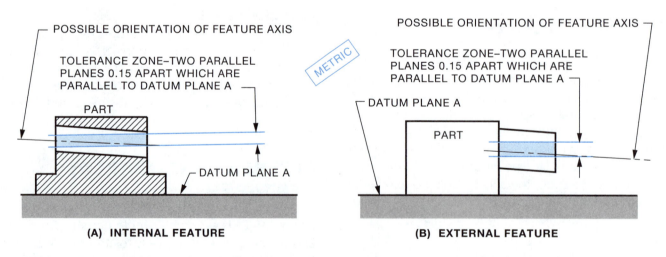

POSSIBLE ORIENTATION OF FEATURE AXIS

TOLERANCE ZONE–TWO PARALLEL
PLANES 0.15 APART WHICH ARE
PARALLEL TO DATUM PLANE A

PART

DATUM PLANE A

POSSIBLE ORIENTATION OF FEATURE AXIS

TOLERANCE ZONE–TWO PARALLEL
PLANES 0.15 APART WHICH ARE
PARALLEL TO DATUM PLANE A

DATUM PLANE A

PART

METRIC

(A) INTERNAL FEATURE

(B) EXTERNAL FEATURE

FIGURE 49–5 ■ *Tolerance zones for parallelism shown in Figure 49–2.*

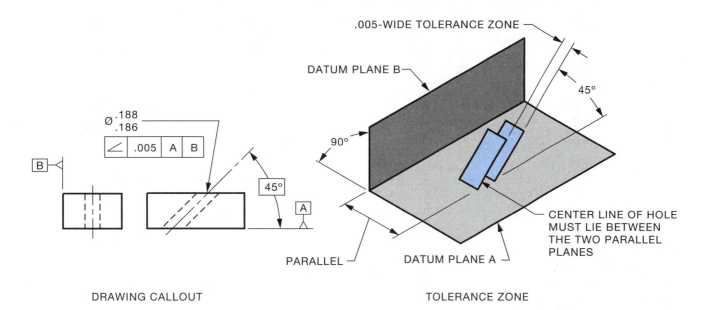

.005-WIDE TOLERANCE ZONE

DATUM PLANE B

90°

45°

PARALLEL

DATUM PLANE A

CENTER LINE OF HOLE MUST LIE BETWEEN THE TWO PARALLEL PLANES

Ø.188 / .186

∠ | .005 | A | B

B

45°

A

DRAWING CALLOUT

TOLERANCE ZONE

FIGURE 49–6 ■ *Angularity tolerances referenced to two datums.*

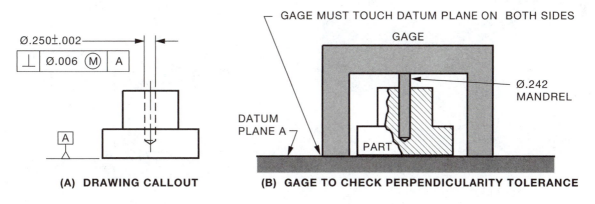

GAGE MUST TOUCH DATUM PLANE ON BOTH SIDES

GAGE

Ø.242 MANDREL

DATUM PLANE A

PART

Ø.250±.002

⊥ | Ø.006 Ⓜ | A

A

(A) DRAWING CALLOUT

(B) GAGE TO CHECK PERPENDICULARITY TOLERANCE

FIGURE 49–7 ■ *Perpendicularity tolerance for a hole on an MMC basis.*

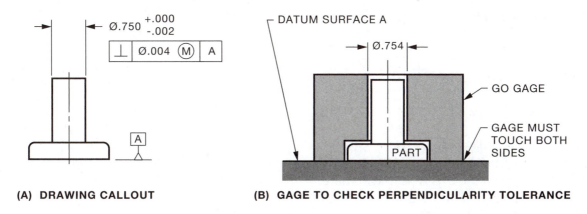

DATUM SURFACE A

Ø.754

GO GAGE

GAGE MUST TOUCH BOTH SIDES

PART

Ø.750 +.000 / -.002

⊥ | Ø.004 Ⓜ | A

A

(A) DRAWING CALLOUT

(B) GAGE TO CHECK PERPENDICULARITY TOLERANCE

FIGURE 49–8 ■ *Perpendicularity tolerance for a shaft on an MMC basis.*

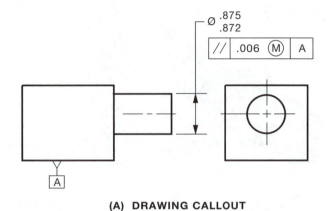

(A) DRAWING CALLOUT

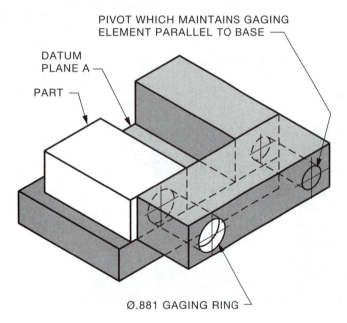

(B) GAGE TO CHECK PARALLELISM TOLERANCE

FIGURE 49–9 ■ *Parallelism tolerance for a shaft on an MMC basis.*

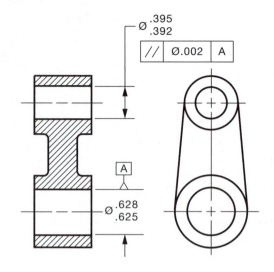

(A) DRAWING CALLOUT

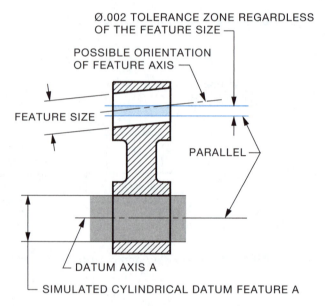

(B) TOLERANCE ZONE

FIGURE 49–10 ■ *Specifying parallelism for an axis (both feature and datum feature RFS).*

Specifying Parallelism for an Axis

Figure 49–10 specifies parallelism for an axis when both the feature and the datum feature are shown on an RFS basis. Regardless of feature size, the feature axis must lie within a cylindrical tolerance zone of .002-in. diameter whose axis is parallel to datum axis A. Additionally, the feature axis must be within any specified tolerance of location.

Figure 49–11 specifies parallelism for an axis when the feature is shown on an MMC basis and the datum feature is shown on an RFS basis. Where the feature is at the maximum material condition (.392 in.), the maximum parallelism tolerance is .002 in. diameter. Where the feature departs from its MMC size, an increase in the

parallelism tolerance is allowed equal to the amount of such departure. Additionally, the feature axis must be within any specified tolerance of location.

Perpendicularity for a Median Plane

Regardless of feature size, the center plane of the feature shown in Figure 49–12 must lie between two parallel planes, .005 in. apart, that are perpendicular to datum plane A. Additionally, the feature center plane must be within any specified tolerance of location.

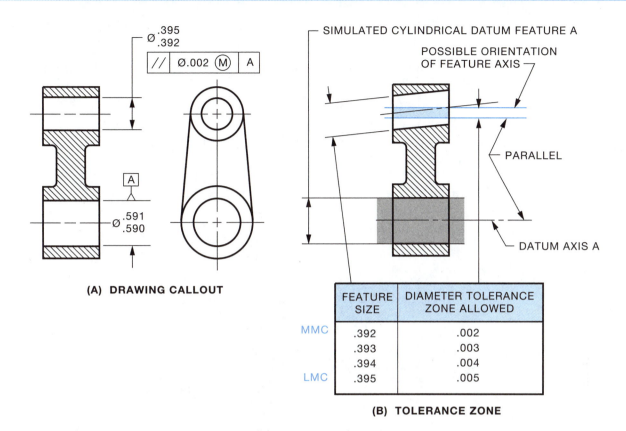

(A) DRAWING CALLOUT

SIMULATED CYLINDRICAL DATUM FEATURE A

POSSIBLE ORIENTATION OF FEATURE AXIS

PARALLEL

DATUM AXIS A

FEATURE SIZE	DIAMETER TOLERANCE ZONE ALLOWED
.392	.002
.393	.003
.394	.004
.395	.005

MMC — .392 .002
LMC — .395 .005

(B) TOLERANCE ZONE

FIGURE 49–11 ■ *Specifying parallelism for an axis (feature at MMC and datum feature RFS).*

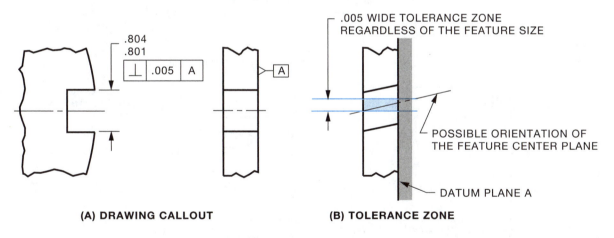

(A) DRAWING CALLOUT

.005 WIDE TOLERANCE ZONE REGARDLESS OF THE FEATURE SIZE

POSSIBLE ORIENTATION OF THE FEATURE CENTER PLANE

DATUM PLANE A

(B) TOLERANCE ZONE

FIGURE 49–12 ■ *Specifying perpendicularity for a median plane (feature RFS).*

Perpendicularity for an Axis (Both Feature and Datum RFS)

Regardless of feature size, the feature axis shown in Figure 49–13 must lie between two parallel planes, .005 in. apart, that are perpendicular to datum axis A. Additionally, the feature axis must be within any specified tolerance of location.

Perpendicularity for an Axis (Tolerance at MMC)

Where the feature shown in Figure 49–14 is at the MMC (Ø2.000), its axis must be perpendicular within .002 in. to the datum plane A. Where the feature departs from MMC, an increase in the perpendicularity tolerance is allowed equal to the amount of such

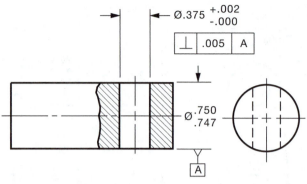

(A) DRAWING CALLOUT

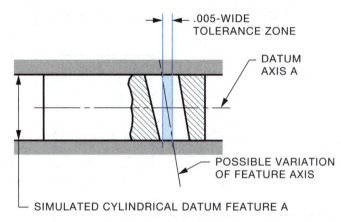

(B) TOLERANCE ZONE

FIGURE 49–13 ■ *Specifying perpendicularity for an axis (both feature and datum feature RFS).*

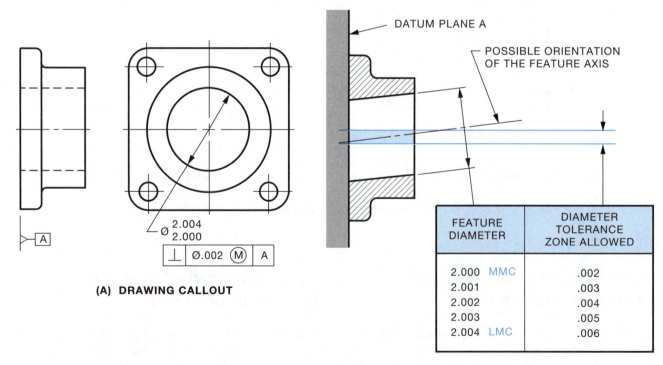

(A) DRAWING CALLOUT

FEATURE DIAMETER		DIAMETER TOLERANCE ZONE ALLOWED
2.000	MMC	.002
2.001		.003
2.002		.004
2.003		.005
2.004	LMC	.006

(B) TOLERANCE ZONE

FIGURE 49–14 ■ *Specifying perpendicularity for an axis (tolerance at MMC).*

departure. Additionally, the feature axis must be within the specified tolerance of location.

Perpendicularity for an Axis (Zero Tolerance at MMC)

Where the feature shown in Figure 49–15 is at the MMC (Ø2.000), its axis must be perpendicular to datum plane A. Where the feature departs from MMC, an increase in the perpendicularity tolerance

is allowed equal to the amount of such departure. Additionally, the feature axis must be within any specified tolerance of location.

Perpendicularity with a Maximum Tolerance Specified

Where the feature shown in Figure 49–16 is at MMC (50.00 mm), its axis must be perpendicular to datum plane A. Where the feature departs from

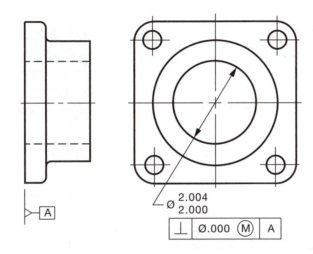

(A) DRAWING CALLOUT

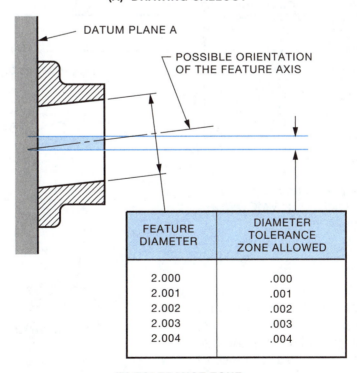

(B) TOLERANCE ZONE

FIGURE 49–15 ■ *Specifying perpendicularity for an axis (zero tolerance at MMC).*

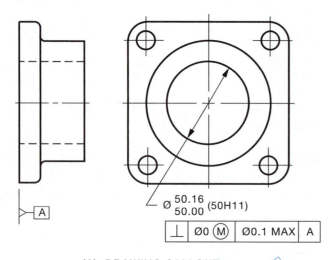

Ø 50.16 / 50.00 (50H11)

⊥ | Ø0 Ⓜ | Ø0.1 MAX | A

(A) DRAWING CALLOUT

METRIC

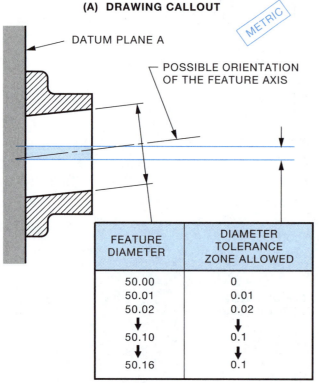

DATUM PLANE A

POSSIBLE ORIENTATION OF THE FEATURE AXIS

FEATURE DIAMETER	DIAMETER TOLERANCE ZONE ALLOWED
50.00	0
50.01	0.01
50.02	0.02
↓	↓
50.10	0.1
↓	↓
50.16	0.1

(B) TOLERANCE ZONE

FIGURE 49–16 ■ *Specifying perpendicularity for an axis (zero tolerance at MMC with a maximum specified).*

MMC, an increase in the perpendicularity tolerance is allowed equal to the amount of such departure, up to the 0.1 mm maximum. Additionally, the feature axis must be within any specified tolerance of location.

EXTERNAL CYLINDRICAL FEATURES

Perpendicularity for an Axis (Pin or Boss RFS)

Regardless of feature size, the feature axis shown in Figure 49–17 must lie within a cylindrical zone (.001-in. diameter) that is perpendicular to and projects from datum plane A for the feature height. Additionally, the feature axis must be within any specified tolerance of location.

Perpendicularity for an Axis (Pin or Boss at MMC)

Where the feature shown in Figure 49–18 is at MMC (Ø.625 in.), the maximum perpendicularity

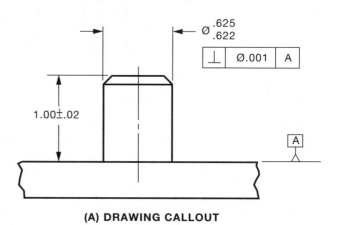

Ø .625 / .622

⊥ | Ø.001 | A

1.00±.02

(A) DRAWING CALLOUT

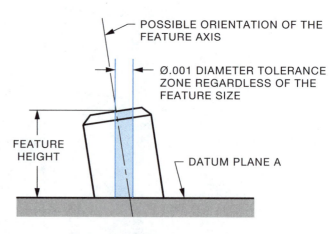

POSSIBLE ORIENTATION OF THE FEATURE AXIS

Ø.001 DIAMETER TOLERANCE ZONE REGARDLESS OF THE FEATURE SIZE

FEATURE HEIGHT

DATUM PLANE A

(B) TOLERANCE ZONE

FIGURE 49–17 ■ *Specifying perpendicularity for an axis (pin or boss RFS).*

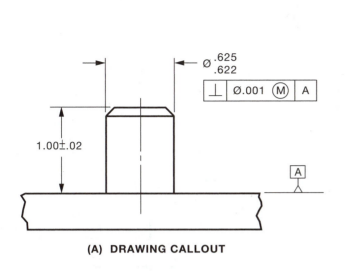

(A) DRAWING CALLOUT

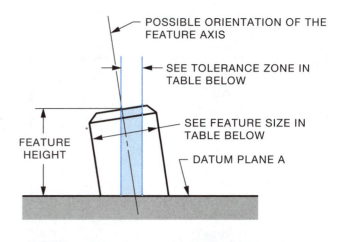

FEATURE SIZE	DIAMETER TOLERANCE ZONE ALLOWED
MMC .625	.001
.624	.002
.623	.003
LMC .622	.004

(B) TOLERANCE ZONE

FIGURE 49–18 ■ *Specifying perpendicularity for an axis (pin or boss at MMC).*

tolerance is .001-in. diameter. Where the feature departs from its MMC size, an increase in the perpendicularity tolerance is allowed equal to the amount of such departure. Additionally, the feature axis must be within any specified tolerance of location.

REFERENCE

ASME Y14.5M-1994 (R1999) Dimensioning and Tolerancing

INTERNET RESOURCES

Drafting Zone. For information on geometric dimensioning and tolerancing, see: http://www.draftingzone.com

Effective Training Inc. For information on dimensioning and tolerancing, see: http://etinews.com/eti_solutions.html

eFunda. For information on geometric dimensioning and tolerancing, see: http://www.efunda.com/home.cfm

Engineering Edge. For information on geometric tolerancing and dimensioning, see: http://www.engineeringedge.com/gdt/htm

⊥	Ø.004	A

FEATURE SIZE Ø	DIAMETER TOLERANCE ZONE ALLOWED
2.000 2.001 ↕ 2.008 2.009	

⊥	Ø.000 Ⓜ	A

FEATURE SIZE Ø	DIAMETER TOLERANCE ZONE ALLOWED
2.000 2.001 ↕ 2.008 2.009	

⊥	Ø.000 Ⓜ	Ø.005 MAX	A

FEATURE SIZE Ø	DIAMETER TOLERANCE ZONE ALLOWED
2.000 2.001 ↕ 2.008 2.009	

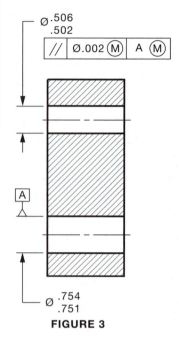

FIGURE 3

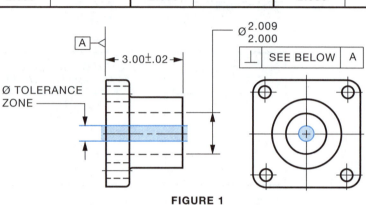

FIGURE 1

ASSIGNMENTS:

1. Sketch the tables shown in Figure 1 on a grid sheet and complete the tables showing the maximum permissible tolerance zones for the three callouts.

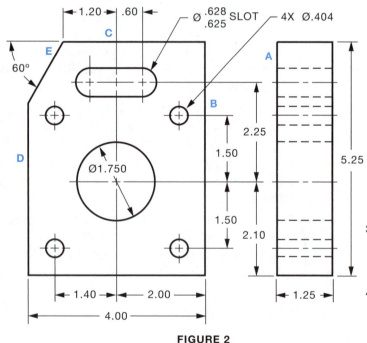

FIGURE 2

2. On an inch grid sheet (.10 in. squares) sketch the views shown in Figure 2, and add the following data to the sketch:

   (A) Surfaces marked A, B, and C are datums A, B, and C, respectively.
   (B) Surface A is perpendicular to surfaces B and C within .01 in.
   (C) Surface D is parallel to surface B within .004 in.
   (D) The slot is parallel to surface C within .002 in. and perpendicular to surface A within .001 in. at MMC.
   (E) The Ø1.750 hole has an RC7 fit (show the size of the hole as limits), and is perpendicular to surface A within .002 in. at MMC.
   (F) Surface E has an angularity tolerance of .010 in. with surface C.
   (G) Surface A is to be flat within .002 in. for any one-inch-square surface with a maximum flatness tolerance of .005 in.
   (H) Indicate which dimensions are basic.

3. With reference to Figure 3, what is the maximum diameter of the tolerance zone allowed when the Ø.506 in. hole is at (a) MMC? (b) LMC? (c) Ø.504 in.?

4. If the symbol M is removed from the tolerance in Figure 3, what is the maximum diameter of the tolerance zone allowed when the hole is at (a) MMC? (b) LMC? (c) Ø.504 in.?

ORIENTATION TOLERANCING FOR FEATURES OF SIZE

A-115

DATUM TARGETS

The full feature surface was used to establish a datum for the features so far designated as datum features. This may not always be practical for the following reasons:

1. The surface of a feature may be so large that a gage designed to make contact with the full surface may be too expensive or too cumbersome to use.
2. Functional requirements of the part may necessitate the use of only a portion of a surface as a datum feature, for example, the portion that contacts a mating part in assembly.
3. A surface selected as a datum feature may not be sufficiently true and a flat datum feature may rock when placed on a datum plane, so that accurate and repeatable measurements from the surface would not be possible. This is particularly so for surfaces of castings, forgings, weldments, and some sheet-metal and formed parts.

A useful technique to overcome such problems is the datum-target method. In this method, certain points, lines, or small areas on the surfaces are selected as the bases for establishment of datums. For flat surfaces, this usually requires three target points or areas for a primary datum, two for a secondary datum, and one for a tertiary datum.

It is not necessary to use targets for all datums. It is quite logical, for example, to use targets for the primary datum and other surfaces or features for secondary and tertiary datums if required; or to use a flat surface of a part as the primary datum and to locate fixed points or lines on the edges as secondary and tertiary datums.

Datum targets should be spaced as far apart from each other as possible to provide maximum stability for making measurements. Dimensions locating target areas are basic and are shown enclosed in a rectangular frame.

DATUM-TARGET SYMBOL

Points, lines, and areas on datum features are designated on the drawing by means of a datum-target symbol, Figure 50–1. The symbol is placed outside the part outline with an arrowless leader directed to the target point (indicated by an "X") target line, or target area, as applicable, Figure 50–2. The use of a solid leader line indicates that the datum target is on the near (visible) surface. The use of a dashed leader

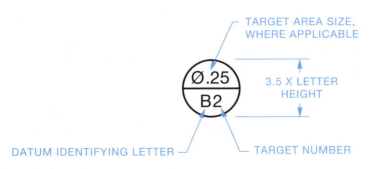

TARGET AREA SIZE, WHERE APPLICABLE

Ø.25

B2

3.5 X LETTER HEIGHT

DATUM IDENTIFYING LETTER

TARGET NUMBER

FIGURE 50–1 ■ *Datum-target symbol.*

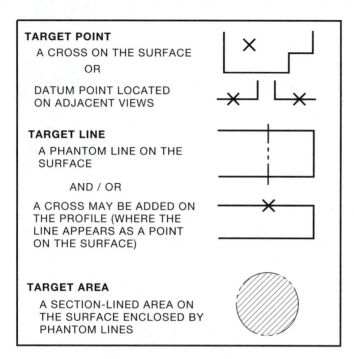

FIGURE 50–2 ■ *Identification of datum targets.*

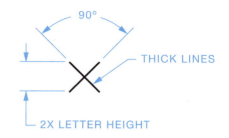

FIGURE 50–3 ■ *Symbol for a datum target point.*

a letter identifying the associated datum, followed by the target number assigned sequentially starting with 1 for each datum. For example, in a three-plane, six-point datum system, if the datums are A, B, and C, the datums would be A1, A2, A3, B1, B2, and C1 (see Figure 50–14). Where the datum target is an area, the area size may be entered in the upper half of the symbol; otherwise, the upper half is left blank.

Identification of Targets

DATUM TARGET POINTS

Each target point is shown on the surface, in its desired location, by means of a cross, drawn at approximately 45° to the coordinate dimensions. The cross is twice the height of the lettering used, Figures 50–3 and 50–4(A). When the view that would show the location of the datum target point is not drawn, its point location is dimensioned on the two adjacent views, Figure 50–4(B).

line (as in Figure 50–10B) indicates that the datum target is on the far (hidden) surface. The leader should not be shown in either a horizontal or vertical position. ASME drawing standards omit showing an arrow at the end of this leader, while ISO and CSA standards show one. The datum feature itself is identified in the usual manner with a datum-feature symbol.

The datum-target symbol is a circle having a diameter approximately 3.5 times the height of the lettering used on the drawing. The circle is divided horizontally into two halves. The lower half contains

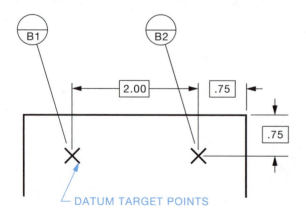

(A) **DATUM POINTS SHOWN ON SURFACE**

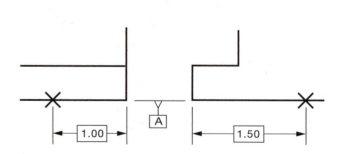

(B) **DATUM POINTS LOCATED BY TWO VIEWS**

FIGURE 50–4 ■ *Datum target points.*

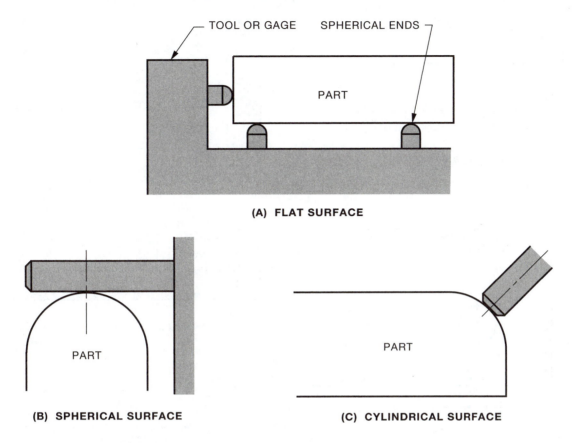

(A) FLAT SURFACE

(B) SPHERICAL SURFACE

(C) CYLINDRICAL SURFACE

FIGURE 50–5 ■ *Location of part on datum target points.*

Target points may be represented on tools, fixtures, and gages by spherically ended pins, Figure 50–5.

DATUM TARGET LINES

A datum target line is indicated by the symbol X on an edge view of a surface, a phantom line on the direct view, or both, Figure 50–6. When the length of the datum target line must be controlled, its length and location are dimensioned.

Datum target lines can be represented in tooling and gaging by the side of a round pin, Figure 50–7.

It should be noted that if a line is designated as a tertiary datum feature, it will touch the gage pin theoretically at only one point. If it is a secondary datum feature, it will touch at two points.

The application and use of a surface and three lines as datum features are shown in Figures 50–8 and 50–9.

DATUM TARGET AREAS

Where it is determined that an area or areas of flat contact are necessary to ensure establishment of the datum (that is where spherical or pointed pins would

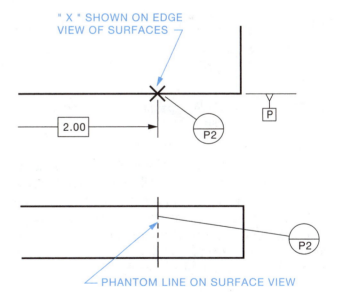

FIGURE 50–6 ■ *Datum target line.*

be inadequate), a target area of the desired shape is specified. The datum target area is indicated by section lines inside a phantom outline of the desired shape, with controlling dimensions added. The diameter of circular areas is given in the upper half of

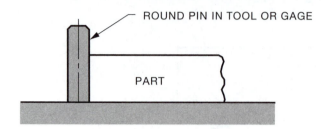

FIGURE 50–7 ▪ *Locating a datum line.*

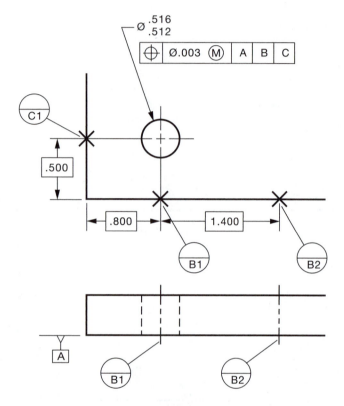

FIGURE 50–8 ▪ *Three target lines used as datum features.*

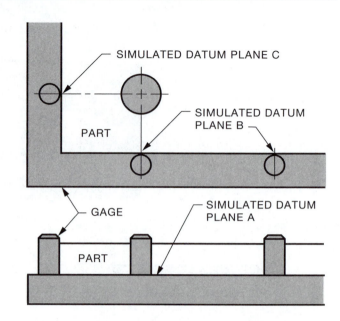

FIGURE 50–9 ▪ *Location of part in Figure 50–8 in a gage.*

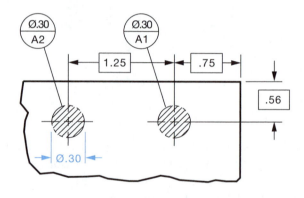

(A) TARGET AREAS ON NEAT SIDE

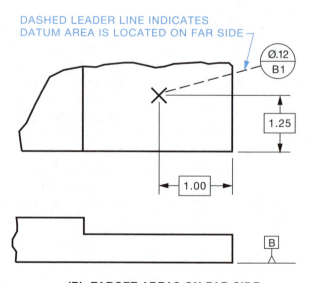

(B) TARGET AREAS ON FAR SIDE

FIGURE 50–10 ▪ *Datum target areas.*

the datum-target symbol, Figure 50–10(A). Where a circular target area is too small to be drawn to scale, the method shown in Figure 50–10(B) may be used.

Datum target areas may have any desired shape, a few of which are shown in Figure 50–11. Target areas should be kept as small as possible but consistent with functional requirements.

Targets Not in the Same Plane

In most applications, datum target points that form a single datum are all located on the same surface (as shown in Figure 50–4(A)). However, this is not essential. They may be located on different surfaces to meet functional requirements, Figure 50–12. In

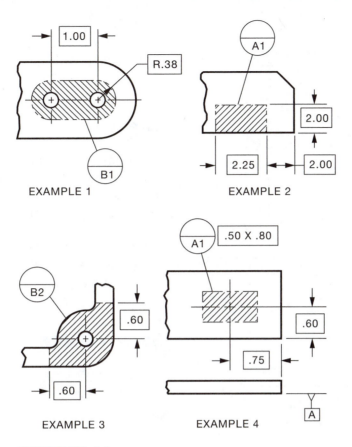

EXAMPLE 1

EXAMPLE 2

EXAMPLE 3

EXAMPLE 4

FIGURE 50–11 ■ *Typical target areas.*

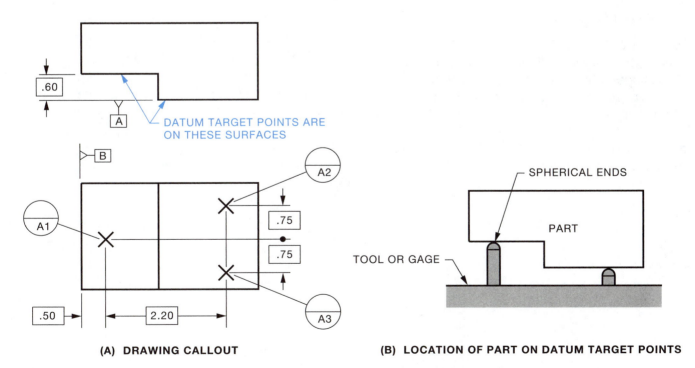

(A) **DRAWING CALLOUT**

(B) **LOCATION OF PART ON DATUM TARGET POINTS**

FIGURE 50–12 ■ *Datum target points on different planes used as the primary datum.*

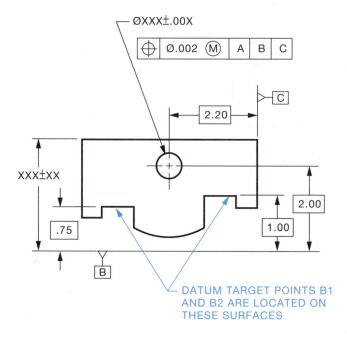

DATUM TARGET POINTS B1 AND B2 ARE LOCATED ON THESE SURFACES

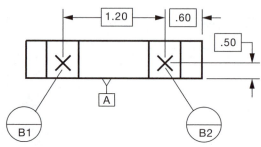

FIGURE 50–13 ■ *Datum outside of part profile.*

some cases, the datum plane may be located in space, not actually touching the part, Figure 50–13. In such applications, the controlled features must be dimensioned from the specified datum, and the position of the datum from the datum targets must be shown by means of exact datum dimensions. For example, in Figure 50–13, datum B is positioned by means of datum dimensions .75 in., 1.00 in., and

2.00 in. The top surface is controlled from this datum by means of a toleranced dimension. The hole is positioned by means of the basic dimension 2.00 in. and a positional tolerance.

Dimensioning for Target Location

The location of datum targets is shown by means of basic dimensions. Each dimension is shown, without tolerances, enclosed in a rectangular frame, indicating that the general tolerance does not apply. Dimensions locating a set of datum targets should be dimensionally related or have a common origin.

Application of datum targets and datum dimensioning is shown in Figure 50–14.

REFERENCE

ASME Y14.5M-1994 (R1999) Dimensioning and Tolerancing

INTERNET RESOURCES

Drafting Zone. For information on geometric dimensioning and tolerancing, see: http://www.draftingzone.com

Effective Training Inc. For information on dimensioning and tolerancing, see: http://etinews.com/eti_solutions.html

eFunda. For information on geometric dimensioning and tolerancing, see: http://www.efunda.com/home.cfm

Engineering Edge. For information on geometric tolerancing and dimensioning, see: http://www.engineeringedge.com/gdt/htm

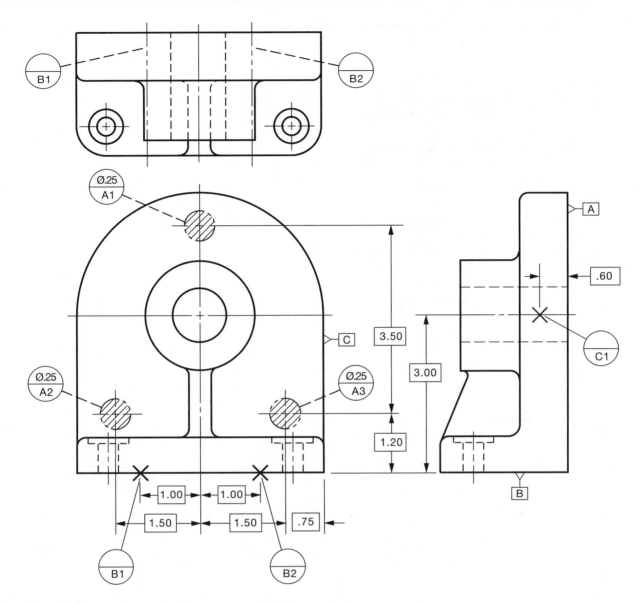

FIGURE 50–14 ■ *Application of datum targets and dimensioning.*

ASSIGNMENT:

ON A ONE INCH GRID SHEET (.10 IN. SQUARES) SKETCH THE BEARING HOUSING SHOWN BELOW AND ADD THE DATUM INFORMATION SHOWN IN THE TABLE AND THE DIMENSIONS RELATED TO THE DATUMS ON THIS SKETCH. SCALE 1:2.

QUESTIONS:

1. WHAT ARE THE THREE TYPES OF DATUM TARGETS?

2. WHAT TYPE OF DIMENSIONS ARE USED TO LOCATE DATUM TARGETS?

3. WHAT IS THE MINIMUM NUMBER OF CONTACT POINTS FOR A PRIMARY DATUM?

4. WHAT IS THE MINIMUM NUMBER OF CONTACT POINTS FOR A TERTIARY DATUM?

5. WHAT TYPE OF LEADER LINE IS USED TO INDICATE THAT THE DATUM TARGET IS ON THE FAR (HIDDEN) SURFACE?

6. WHAT INFORMATION IS CONTAINED IN THE LOWER HALF OF THE DATUM-TARGET SYMBOL?

7. HOW IS A TARGET POINT IDENTIFIED ON A DRAWING?

8. HOW IS A DATUM TARGET LINE IDENTIFIED ON THE EDGE VIEW OF A SURFACE?

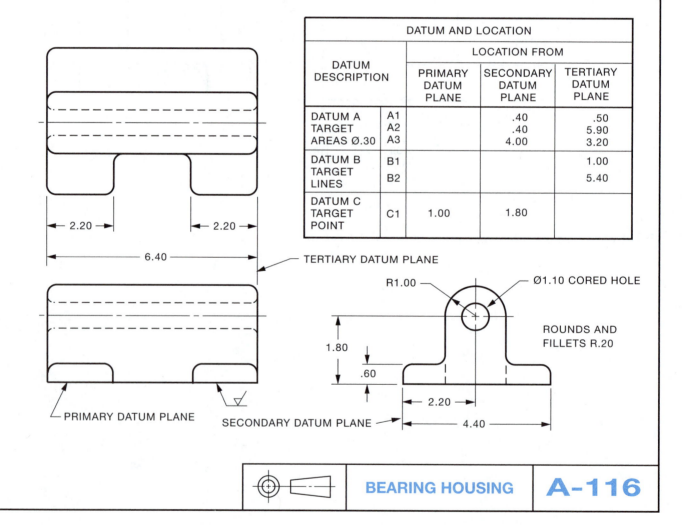

DATUM AND LOCATION				
DATUM DESCRIPTION		LOCATION FROM		
		PRIMARY DATUM PLANE	SECONDARY DATUM PLANE	TERTIARY DATUM PLANE
DATUM A TARGET AREAS Ø.30	A1		.40	.50
	A2		.40	5.90
	A3		4.00	3.20
DATUM B TARGET LINES	B1			1.00
	B2			5.40
DATUM C TARGET POINT	C1	1.00	1.80	

2.20   2.20

6.40

TERTIARY DATUM PLANE

R1.00   Ø1.10 CORED HOLE

ROUNDS AND FILLETS R.20

1.80
.60
2.20
4.40

PRIMARY DATUM PLANE

SECONDARY DATUM PLANE

BEARING HOUSING   **A-116**

TOLERANCING OF FEATURES BY POSITION

The location of features is one of the most frequently used applications of dimensions on technical drawings. Tolerancing may be accomplished either by coordinate tolerances applied to the dimensions or by geometric (Positional) tolerancing.

Positional tolerancing is especially useful when applied on an MMC basis to groups or patterns of holes or other small features in the mass production of parts. This method meets functional requirements in most cases and permits assessment with simple gaging procedures.

Most units in this section are devoted to the principles involved in the location of small round holes, because they represent the most commonly used applications.

TOLERANCING METHODS

The location of a single hole is usually indicated by means of rectangular coordinate dimensions extending from suitable edges or other features of the part to the axis of the hole. Other dimensioning methods, such as polar coordinates, may be used when circumstances warrant.

There are two standard methods of tolerancing the location of holes: coordinate and positional tolerancing.

1. Coordinate tolerancing, Figure 51–1 (A), refers to tolerances applied directly to the coordinate dimensions or to applicable tolerances specified in a general tolerance note.
2. Positional tolerancing, Figures 51–1 (B) to 51–1 (D), refers to a tolerance zone within which the center line of the hole or shaft is

permitted to vary from its true position. In ASME drawing practices, positional tolerancing can be further classified according to the type of modifying associated with the tolerance. These are:

- Positional tolerancing, regardless of feature size (RFS)
- Positional tolerancing, maximum material condition basis (MMC)
- Positional tolerancing, least material condition basis (LMC)

These positional tolerancing methods are part of the system of geometric tolerancing.

When the MMC or LMC modifying symbol is not shown in the feature control frame it is understood that regardless of feature size applies.

Any of these tolerancing methods can be substituted one for the other, although with differing results. It is necessary, however, to first analyze the widely used method of coordinate tolerancing in order to then explain and understand the advantages and disadvantages of the positional tolerancing methods.

COORDINATE TOLERANCING

Coordinate dimensions and tolerances may be applied to the location of a single hole, Figure 51–2. They indicate the location of the hole axis and result in a rectangular or wedge-shaped tolerance zone within which the axis of the hole must lie.

If the two coordinate tolerances are equal, the tolerance zone formed will be a square. Unequal tolerances result in a rectangular tolerance zone. Polar dimensioning, in which one of the locating dimensions is a radius, gives an annular segment (circular ring section) tolerance zone. For simplicity, square tolerance zones are used in the analysis of most of the examples in this section.

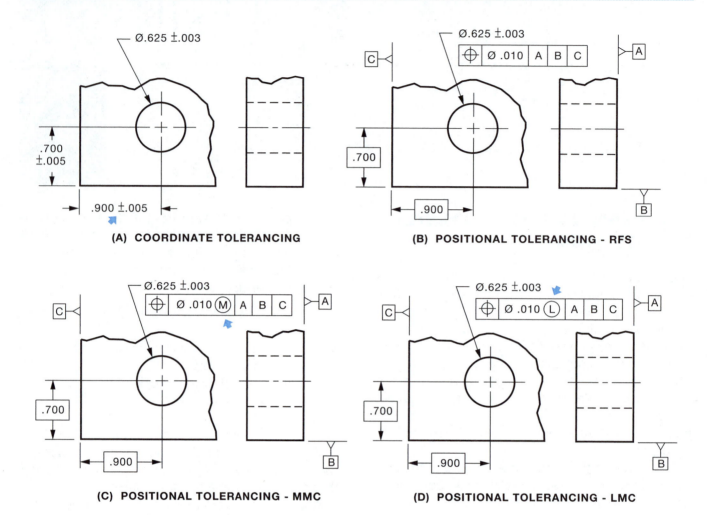

FIGURE 51–1 ■ *Comparison of tolerancing methods.*

It should be noted that the tolerance zone extends for the full depth of the hole, that is, the whole length of the axis. This is illustrated in Figure 51–3. In most of the illustrations, tolerances will be analyzed as they apply at the surface of the part, where the axis is represented by a point.

Maximum Permissible Error

The actual position of the feature axis may be anywhere within the rectangular tolerance zone. For square tolerance zones, the maximum allowable variation from the desired position occurs in a direction of 45° from the direction of the coordinate dimensions, Figure 51–4.

For rectangular tolerance zones, this maximum tolerance is the square root of the sum of the squares of the individual tolerances. This is expressed mathematically as

$$\sqrt{X^2 + Y^2}$$

For the examples shown in Figure 51–2, the tolerance zones are shown in Figure 51–5. The maximum tolerance values are:

Example 1

$$\sqrt{.010^2 + .010^2} = .014 \text{ in.}$$

Example 2

$$\sqrt{.010^2 + .020^2} = .022 \text{ in.}$$

For polar coordinates, the extreme variation is:

$$\sqrt{A^2 + T^2}$$

Where: $A = R \text{ Tan } a$
T = tolerance on radius
R = mean radius
a = angular tolerance

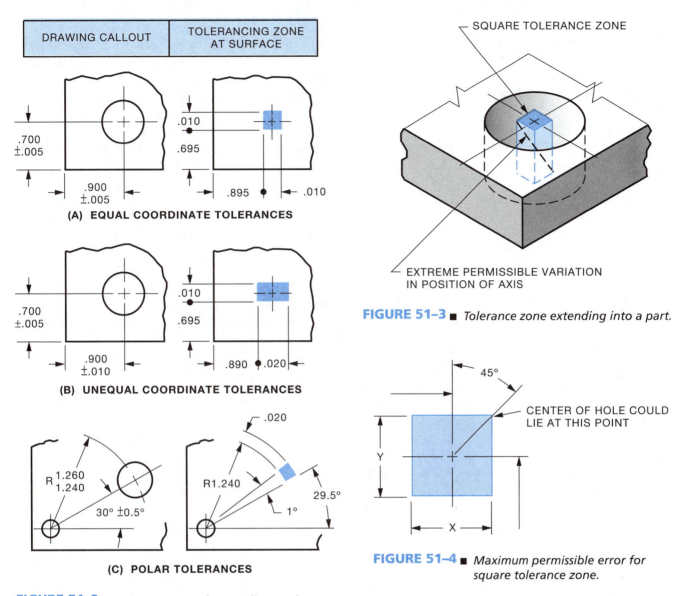

DRAWING CALLOUT	TOLERANCING ZONE AT SURFACE

(A) EQUAL COORDINATE TOLERANCES

(B) UNEQUAL COORDINATE TOLERANCES

(C) POLAR TOLERANCES

FIGURE 51–2 ■ *Tolerance zones for coordinate tolerances.*

SQUARE TOLERANCE ZONE

EXTREME PERMISSIBLE VARIATION IN POSITION OF AXIS

FIGURE 51–3 ■ *Tolerance zone extending into a part.*

CENTER OF HOLE COULD LIE AT THIS POINT

FIGURE 51–4 ■ *Maximum permissible error for square tolerance zone.*

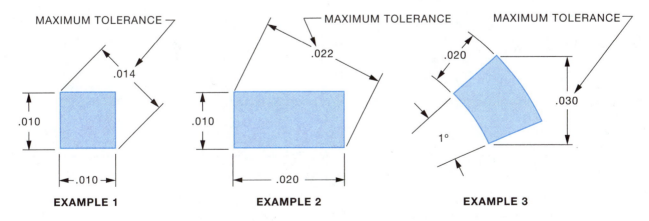

EXAMPLE 1

EXAMPLE 2

EXAMPLE 3

FIGURE 51–5 ■ *Tolerance zones for examples shown in Figure 51–2.*

Example 3

$$\sqrt{(1.25 \times .017)^2 + .020^2} = .030 \text{ in.}$$

Note: Mathematically, the formula for Example 3 is incorrect; but the difference in results using the more complicated correct formula is quite insignificant for the tolerances normally used.

Some values of tan A for commonly used angular tolerances are as follows.

A	TAN A	A	TAN A	A	TAN A
0° 5′	.00145	0° 25′	.00727	0° 45′	.01309
0° 10′	.00291	0° 30′	.00873	0° 50′	.01454
0° 15′	.00436	0° 35′	.01018	0° 55′	01600
0° 20′	.00582	0° 40′	.01164	1° 0′	.01745

USE OF CHART

A quick and easy method of finding the maximum positional error permitted with coordinate tolerancing, without having to calculate squares and square roots, is by use of a chart like that shown in Figure, 51–6.

In the first example shown in Figure 51–2, the tolerance in both directions is .010 in. The extensions of the horizontal and vertical lines of .010 in the chart intersect at point A, which lies between the radii of .014 and .015 in. When rounded to three decimal places, this indicates a maximum permissible variation from true position of .014 in.

In the second example shown in Figure 51–2, the tolerances are .010 in. in one direction and .020 in. in the other. The extensions of the vertical and horizontal lines at .010 and .020 in., respectively, in the chart intersect at point B, which lies between the radii of .022 and .023 in. When rounded off to three decimal places, this indicates a maximum variation of position of .022 in. Figure 51–6 also shows a chart for use with tolerances in millimeters.

ADVANTAGES OF COORDINATE TOLERANCING

The advantages claimed for direct coordinate tolerancing are as follows:

1. It is simple and easily understood and, therefore, is a method commonly used.
2. It permits direct measurements to be made with standard instruments and does not require the use of special purpose functional gages or other calculations.

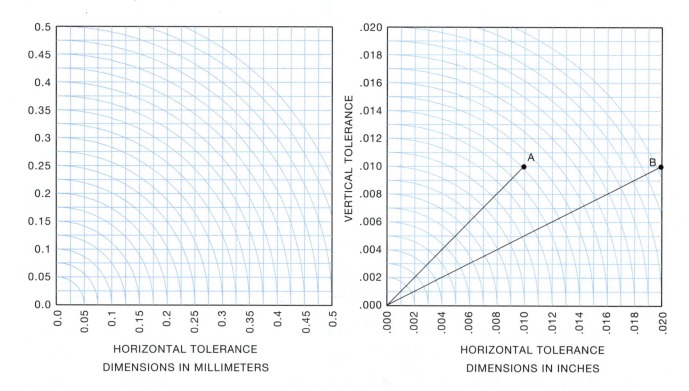

FIGURE 51–6 ■ *Charts for calculating maximum tolerance using coordinate tolerancing.*

DISADVANTAGES OF COORDINATE TOLERANCING

There are a number of disadvantages to the direct tolerancing method. Among these are:

1. It results in a square or rectangular tolerance zone within which the axis must lie. For a square zone, this permits a variation in a 45° direction of approximately 1.4 times the specified tolerance. This amount of variation may necessitate the specification of tolerances that are only 70 percent of those that are functionally acceptable.
2. It may result in an undesirable accumulation of tolerances when several features are involved, especially when chain dimensioning is used.
3. It is more difficult to assess clearances between mating features and components than when positional tolerancing is used, especially when a group or a pattern of features is involved.
4. It does not correspond to the control exercised by fixed functional GO gages often desirable in mass production of parts.

POSITIONAL TOLERANCING

Positional tolerancing is part of the system of geometric tolerancing. It defines a zone within which the center, axis, or center plane of a feature of size is permitted to vary from true (theoretically exact) position. A positional tolerance is indicated by the position symbol, a tolerance, a material condition basis, and appropriate datum references placed in a feature control frame. Basic dimensions represent the exact values to which geometric positional tolerances are applied elsewhere, by symbols or notes on the drawing. They are enclosed in a rectangular frame (basic dimension symbol), Figure 51–7. Where the dimension represents a diameter, the symbol ∅ is included in the rectangular frame. General tolerance notes do not apply to basic dimensions. The frame size need not be any larger than that necessary to enclose the dimension. It is necessary to identify features on the part to establish datums for dimensions locating true positions. The datum features are identified with datum-feature symbols and the applicable datum references are included in the feature control frame.

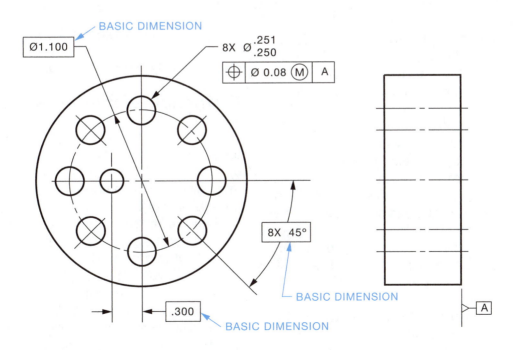

FIGURE 51–7 ■ *Identifying basic dimensions.*

The geometric characteristic symbol for position is a circle with two solid center lines, Figure 51–7. This symbol is used in the feature control frame in the same manner as for other geometric tolerances.

MATERIAL CONDITION BASIS

Positional tolerancing is applied on an MMC, RFS, or LMC basis. When applied on an MMC or LMC basis, the appropriate symbol for the above follows the specified tolerance and where required the applicable datum reference in the feature control frame. When no modifying symbol is shown after the tolerance, the "regardless of feature size" condition applies.

As positional tolerance controls the position of the center, axis, or center plane of a feature of size, the feature control frame is normally attached to the size of the feature, Figure 51–8.

POSITIONAL TOLERANCING FOR CIRCULAR FEATURES

The positional tolerance represents the diameter of a cylindrical tolerance zone, located at true position as determined by the basic dimensions on the drawing. The axis or center line of the feature must lie within this cylindrical tolerance zone.

Except for the fact that the tolerance zone is circular instead of square, a positional tolerance on this basis has exactly the same meaning as direct coordinate tolerancing but with equal tolerances in all directions.

It has already been shown that with rectangular coordinate tolerancing, the maximum permissible error in location is not the value indicated by the horizontal and vertical tolerances, but rather is equivalent to the length of the diagonal between the two tolerances. For square tolerance zones, this is 1.4 times the specified tolerance values. The specified tolerance can therefore be increased to an amount equal to the diagonal of the coordinate tolerance zone without affecting the clearance between the hole and its mating part.

This does not affect the clearance between the hole and its mating part, yet it offers 57 percent more tolerance area, Figure 51–9. Such a change would most likely result in a reduction in the number of parts rejected for positional errors.

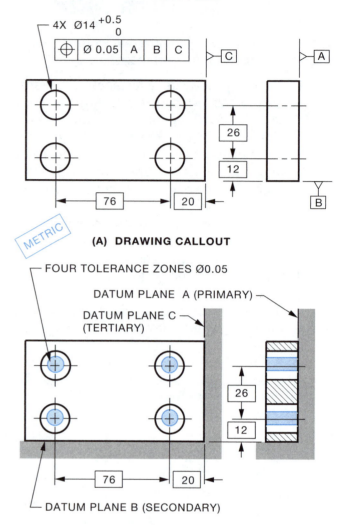

(A) DRAWING CALLOUT

(B) INTERPRETATION

FIGURE 51–8 ■ *Positional tolerancing—RFS.*

Positional Tolerancing—MMC

The positional tolerance and MMC of mating features are considered in relation to each other. MMC by itself means a feature of a finished product contains the maximum amount of material permitted by the toleranced size dimension of that feature. Thus for holes, slots and other internal features, maximum material is the condition in which these factors are at their minimum allowable sizes. For shafts, as well as for bosses, lugs, tabs, and other external features, maximum material is the condition in which these are at their maximum allowable sizes.

A positional tolerance applied on an MMC basis may be explained in either of the following ways:

1. **In terms of the surface of a hole.** While maintaining the specified size limits of the

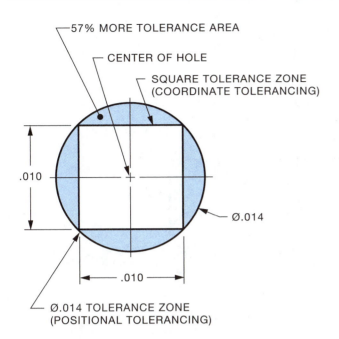

57% MORE TOLERANCE AREA

CENTER OF HOLE

SQUARE TOLERANCE ZONE
(COORDINATE TOLERANCING)

.010

Ø.014

.010

Ø.014 TOLERANCE ZONE
(POSITIONAL TOLERANCING)

FIGURE 51–9 ■ *Relationship of tolerance zones.*

hole, no element of the hole surface shall be inside a theoretical boundary having a diameter equal to the minimum limit of size (MMC) minus the positional tolerance located at true position, Figure 51–10.

2. **In terms of the axis of the hole.** Where a hole is at MMC (minimum diameter), its axis must fall within a cylindrical tolerance zone whose axis is located at true position. The

diameter of this zone is equal to the positional tolerance, Figure 51–11, holes A and B. This tolerance zone also defines the limits of variation in the attitude of the axis of the hole in relation to the datum surface, Figure 51–11, hole C. It is only when the feature is at MMC that the specified positional tolerance applies. Where the actual size of the feature is larger than MMC, additional or bonus positional tolerance is equal to the difference between the specified maximum material limit of size (MMC) and the actual size of the feature.

The problems of tolerancing for the position of holes are simplified when positional tolerancing is applied on an MMC basis. Positional tolerancing simplifies measuring procedures of functional GO gages. It also permits an increase in positional variations as the size departs from the maximum material size without jeopardizing free assembly of mating features.

A positional tolerance on an MMC basis is specified on a drawing, on either the front or the side view, Figure 51–12. The MMC symbol M is added in the feature control frame immediately after the tolerance.

A positional tolerance applied to a hole on an MMC basis means that the boundary of the hole must fall outside a perfect cylinder having a diameter equal to the minimum limit of size minus the

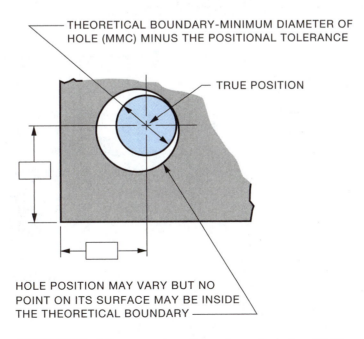

THEORETICAL BOUNDARY-MINIMUM DIAMETER OF
HOLE (MMC) MINUS THE POSITIONAL TOLERANCE

TRUE POSITION

HOLE POSITION MAY VARY BUT NO
POINT ON ITS SURFACE MAY BE INSIDE
THE THEORETICAL BOUNDARY

FIGURE 51–10 ■ *Boundary for surface of a hole—MMC.*

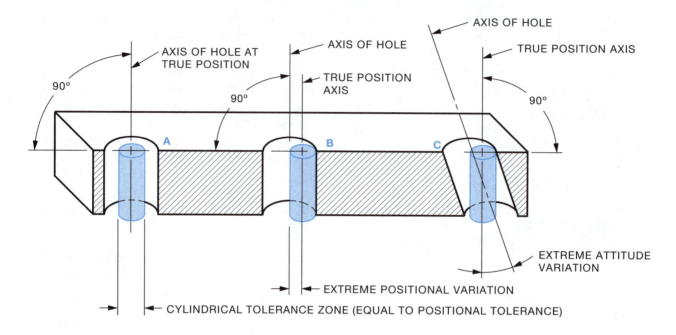

HOLE A - AXIS OF HOLE IS COINCIDENT WITH TRUE POSITION AXIS

HOLE B - AXIS OF HOLE IS LOCATED AT EXTREME POSITION TO THE LEFT OF TRUE POSITION AXIS (BUT WITHIN TOLERANCE ZONE)

HOLE C - AXIS OF HOLE IS INCLINED TO EXTREME ATTITUDE WITHIN TOLERANCE ZONE

NOTE: THE LENGTH OF THE TOLERANCE ZONE IS EQUAL TO THE LENGTH OF THE FEATURE UNLESS OTHERWISE SPECIFIED ON THE DRAWING

FIGURE 51–11 ∎ *Hole axes in relationship to positional tolerance zones.*

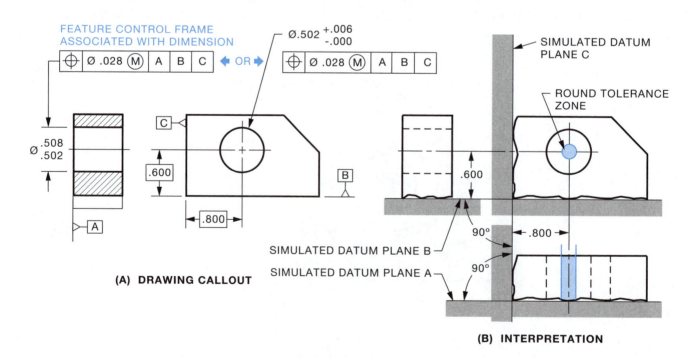

FIGURE 51–12 ∎ *Positional tolerancing—MMC.*

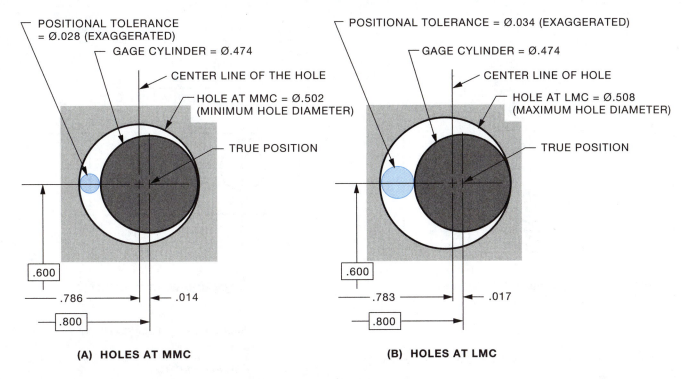

FIGURE 51–13 ■ *Positional variations for tolerancing for Figure 51–12.*

positional tolerance. This cylinder is located with its axis at true position. The hole must, of course, meet its diameter limits.

The effect is illustrated in Figure 51–13, where the gage cylinder is shown at true position and the minimum and maximum diameter holes are drawn to show the extreme permissible variations in position in one direction.

Therefore, if a hole is at its maximum material condition (minimum diameter), the position of its axis must lie within a circular tolerance zone having a diameter equal to the specified tolerance. If the hole is at its maximum diameter (least material condition), the diameter of the tolerance zone for the axis is increased by the amount of the feature tolerance. The greatest deviation of the axis in one direction from true position is therefore

$$\frac{H + P}{2} = \frac{.006 + .028}{2} = .017 \text{ in.}$$

Where: H = hole diameter tolerance
        P = positional tolerance

Positional tolerancing on an MMC basis is preferred when production quantities warrant the provision of functional GO gages, because gaging is then limited to one simple operation, even when a group of holes is involved. This method also facilitates manufacture by permitting larger variations in

position when the diameter departs from the maximum material condition. It cannot be used when it is essential that variations in location of the axis be observed regardless of feature size.

Positional Tolerancing at Zero MMC

The application of MMC permits the tolerance to exceed the value specified, provided features are within size limits and parts are acceptable. This is accomplished by adjusting the minimum size limit of a hole to the absolute minimum required for the insertion of an applicable fastener located precisely at true position, and specifying a zero tolerance at MMC, Figure 51–14. In this case, the positional tolerance allowed is totally dependent on the actual size of the considered feature.

Positional Tolerancing—RFS

In certain cases, the design or function of a part may require the positional tolerance or datum reference, or both, to be maintained regardless of actual feature sizes. RFS, where applied to the positional tolerance of circular features, requires the axis of each feature to be located within the specified positional tolerance regardless of the size of the feature, Figure 51–15. This requirement imposes a closer

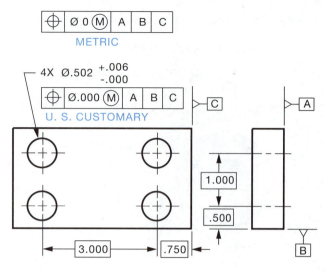

FIGURE 51–14 ■ *Positional tolerancing—zero at MMC.*

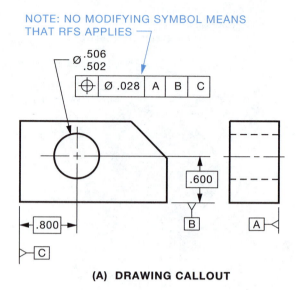

(A) DRAWING CALLOUT

control of the features involved and introduces complexities in verification.

Positional Tolerancing—LMC

Where positional tolerancing at LMC is specified, the stated positional tolerance applies when the feature contains the least amount of material permitted by its toleranced size dimension, Figure 51–16. In this example, LMC is used in order to maintain a maximum wall thickness.

Specifying LMC is limited to applications where MMC does not provide the desired control and RFS is too restrictive.

ADVANTAGES OF POSITIONAL TOLERANCING

It is practical to replace coordinate tolerances with a positional tolerance having a value equal to the diagonal of the coordinate tolerance zone. This provides 57 percent more tolerance area, Figure 51–9, and would probably result in the rejection of fewer parts for positional errors.

A simple method for checking positional tolerance errors is to take coordinate measurements and evaluate them on a chart as shown in Figure 51–17. For example, the four parts shown in Figure 51–18 were rejected when the coordinate tolerances were applied to them.

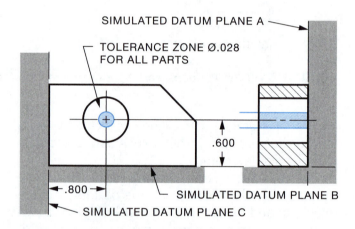

FIGURE 51–15 ■ *Positional tolerancing—RFS.*

If the parts had been toleranced using the positional tolerance—RFS method shown in Figure 51–15 and given a tolerance of ∅.028 in. (equal to the diagonal of the coordinate tolerance zone), three of the parts—A, B, and D—would not have been rejected.

If the parts shown in Figure 51–18 had been toleranced using the positional tolerance (MMC) method and given a tolerance of ∅.028 in. at MMC (Figure 51–12), part C, which was rejected using the RFS tolerancing method (Figure 51–15), would not have been rejected if it had been straight. The positional tolerance can be increased to ∅.034 in. for a part having a diameter of .508 in. (LMC) without jeopardizing the function of the part (Figure 51–13).

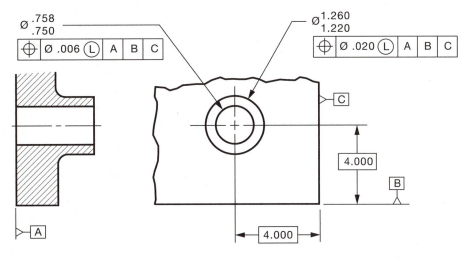

(A) DRAWING CALLOUT

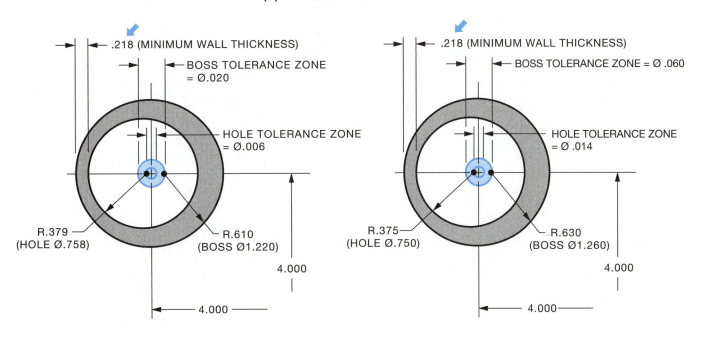

(B) TOLERANCE ZONES WHEN HOLE AT LMC

(C) TOLERANCE ZONES WHEN HOLE AT MMC

FIGURE 51–16 ■ *LMC applied to a boss and a hole.*

REFERENCE

ASME Y14.5M-1994 (R1999) Dimensioning and Tolerancing

INTERNET RESOURCES

Drafting Zone. For information on geometric dimensioning and tolerancing, see: http://www.draftingzone.com

Effective Training Inc. For information on dimensioning and tolerancing, see: http://etinews.com/eti_solutions.html

eFunda. For information on geometric dimensioning and tolerancing, see: http://www.efunda.com/home.cfm

Engineering Edge. For information on geometric tolerancing and dimensioning, see: http://www.engineeringedge.com/gdt/htm

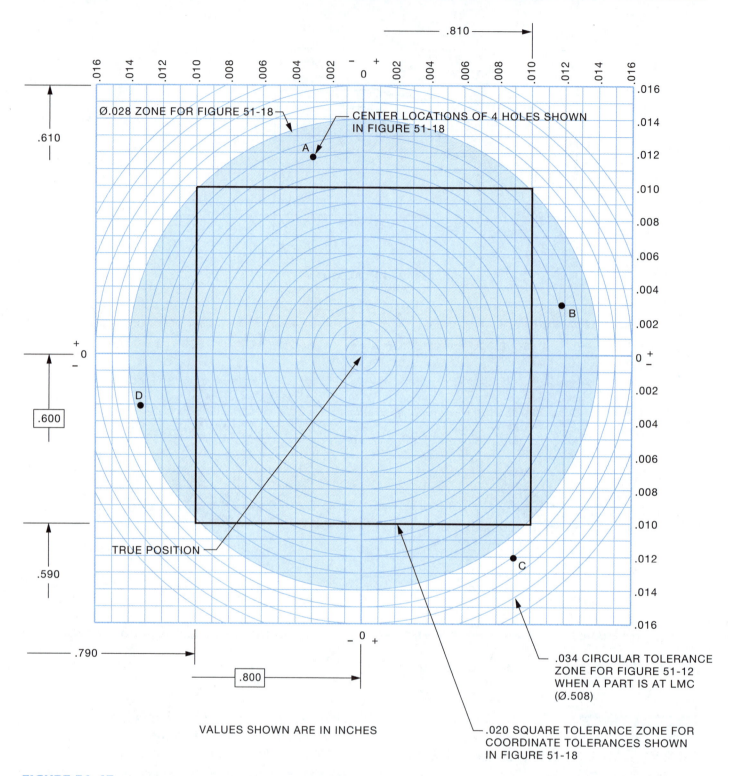

FIGURE 51–17 ■ *Charts for evaluating positional tolerancing.*

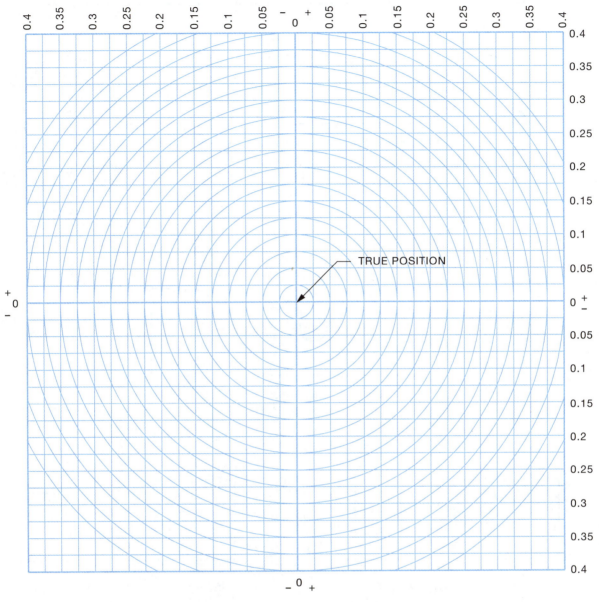

VALUES SHOWN ARE IN MILLIMETERS

FIGURE 51–17 ■ *Continued.*

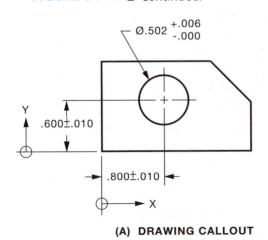

Ø.502 +.006 −.000

.600±.010

.800±.010

(A) DRAWING CALLOUT

PART	HOLE DIA	HOLE LOCATION		COMMENT
		X	Y	
A	.503	.797	.612	REJECTED
B	.504	.812	.603	REJECTED
C	.508	.809	.588	REJECTED
D	.506	.787	.597	REJECTED

REFER TO FIGURE 51-17 FOR LOCATION ON CHART

(B) LOCATION AND SIZE OF REJECTED PARTS

FIGURE 51–18 ■ *Parts A through D rejected because hole centers do not lie within coordinate tolerance zone.*

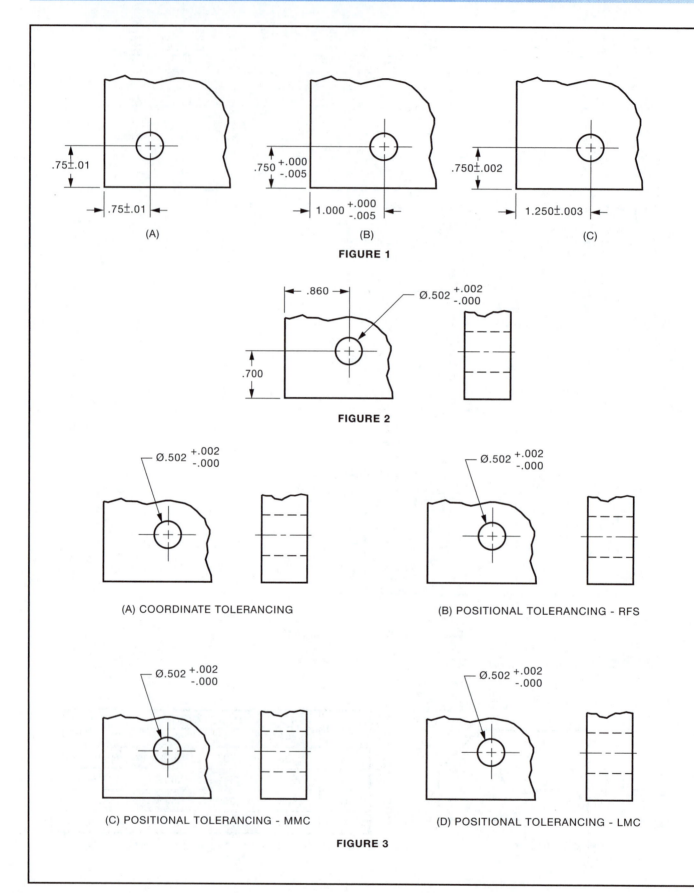

.75±.01

.75±.01

(A)

.750 +.000 -.005

1.000 +.000 -.005

(B)

.750±.002

1.250±.003

(C)

FIGURE 1

.860

Ø.502 +.002 -.000

.700

FIGURE 2

Ø.502 +.002 -.000

(A) COORDINATE TOLERANCING

Ø.502 +.002 -.000

(B) POSITIONAL TOLERANCING - RFS

Ø.502 +.002 -.000

(C) POSITIONAL TOLERANCING - MMC

Ø.502 +.002 -.000

(D) POSITIONAL TOLERANCING - LMC

FIGURE 3

ASSIGNMENT:

Make sketches as needed to answer the following questions.

QUESTIONS:

1. If coordinate tolerances as shown in Figure 1 are given, what is the maximum distance between centers of mating holes for parts made to these drawing callouts?

2. In order to assemble correctly, the holes shown in Figure 2 must not vary more than .0014 in. in any direction from its true position when the hole is at its smallest size. Show suitable tolerancing, dimensioning, and datums where required on the drawings in Figure 2 to achieve this by using:
   a) coordinate tolerancing
   b) positional tolerancing - RFS
   c) positional tolerancing - MMC
   d) positional tolerancing - LMC

3. With reference to question 2 and Figure 3, what would be the maximum permissible deviation from true position when the hole was at its largest size?

4. The part shown in Figure 4A is set on a revolving table, so adjusted that the part revolves about the true position center of the Ø.316 in. hole. If the indicators give identical readings and the results in Figure 4B are obtained which parts are acceptable?

5. What is the positional error for each part in Figure 4B?

6. If MMC instead of RFS had been shown in the feature control frame in Figure 4A, what is the diameter of the mandrel that would be required to check the parts?

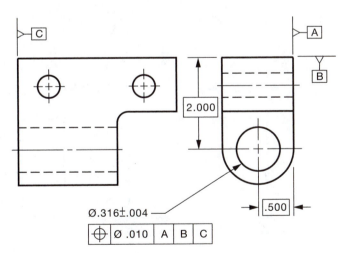

(A) DRAWING CALLOUT

PART NO.	SIZE OF MANDREL	HIGHEST READING	LOWEST READING
1	.316	.014	.008
2	.320	.008	-.004
3	.314	.026	.012
4	.312	.016	.006
5	.316	.015	.009
6	.318	.018	.010

(B) READINGS FOR PARTS

FIGURE 4

POSITIONAL TOLERANCING   **A-117**

SELECTION OF DATUM FEATURES FOR POSITIONAL TOLERANCING

When selecting datums for positional tolerancing, the first consideration is to select the primary datum feature. The usual course of action is to specify as the primary datum the surface into which the hole is produced. This will ensure that the true position of the axis is perpendicular to this surface or at a basic angle if other than 90°. This surface is resting on the gaging plane or surface plate for measuring purposes. Secondary and tertiary datum features are then selected and identified, if required, Figures 52–1 and 52–2.

Positional tolerancing is also useful for parts having holes not perpendicular to the primary surface. This principle is illustrated in Figure 52–3.

LONG HOLES

It is not always essential to have the true position of a hole perpendicular to the face into which the hole is produced. It may be functionally more important, especially with long holes, to have it parallel to one of the sides. Figure 52–4 is a case in point. In this example, the sides are designated as primary and secondary datums. Gaging is facilitated if the positional tolerance is specified on an MMC basis.

CIRCULAR DATUMS

Example 1

Circular features, such as holes or external cylindrical features, can be used as datums just as readily as flat surfaces. In the simple part

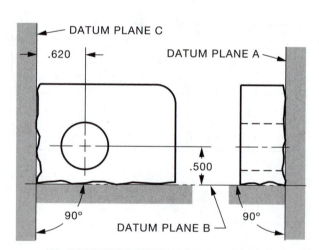

(A) DRAWING CALLOUT

(B) INTERPRETATION OF TRUE POSITION

FIGURE 52–1 ■ *Part with three datum features specified—MMC.*

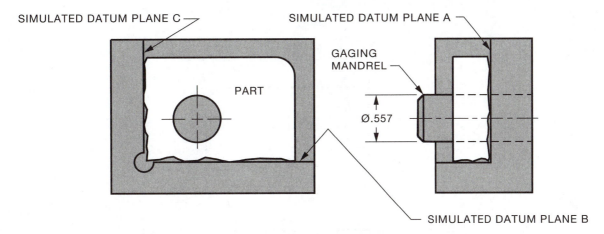

SIMULATED DATUM PLANE C

SIMULATED DATUM PLANE A

GAGING
MANDREL

PART

Ø.557

SIMULATED DATUM PLANE B

PART MUST SLIDE OVER Ø.557 GAGE PIN, MUST LIE FLAT ON BASE OF GAGE
(SIMULATED DATUM PLANE A), AND TOUCH SIMULATED DATUM PLANE B AT
LEAST AT TWO POINTS ALONG ITS LENGTH, WHILE SIMULTANEOUSLY TOUCHING
SIMULATED DATUM PLANE C AT LEAST AT ONE POINT.

FIGURE 52–2 ■ *Gage for the part shown in Figure 52–1.*

shown in Figure 52–5, the true position of the small hole is established from the flat surface, datum A, and the large hole, datum D. Specifying datum hole D on an MMC basis facilitates gaging.

Example 2

In other cases, such as that shown in Figure 52–6, the datum could either be the axis of the hole or the axis of the outside cylindrical surface. In such applications, a determination should be made as to whether the true position should be established perpendicular to the surface as shown or parallel with the datum axis. In the latter case, datum A would not be specified and it would not be necessary to ensure that the gage made full contact with the surface.

In a group of holes, it may be desirable to indicate one of the holes as the datum from which all the other holes are located. This is described in succeeding units. All circular datums of this type may be

specified on an MMC basis when required, and this is preferred for ease of gaging.

MULTIPLE HOLE DATUMS

On an MMC basis, any number of holes or similar features that form a group or pattern, may be specified as a single datum. All features forming such a datum must be related with a positional tolerance on an MMC basis. The actual datum position is based on the virtual condition of all features in the group, that is, the collectible effect of the maximum material sizes of the features and the specified positional tolerance.

Thus, in the example shown in Figure 52–7, the gaging element that locates the datum position would have four Ø.240 pins located at true position with respect to one another. It should be noted that this setup automatically checks the positional tolerance specified for these four holes.

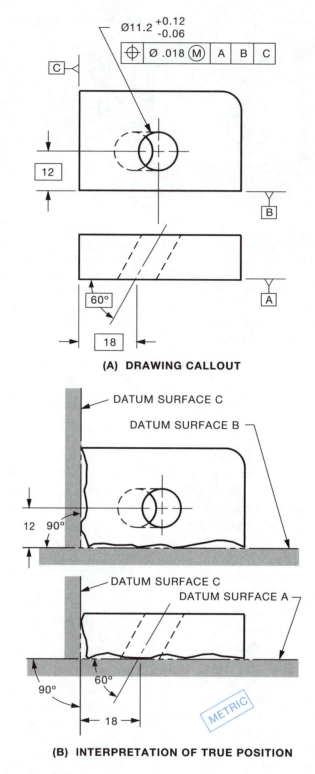

Ø11.2 +0.12 -0.06

⊕ | Ø .018 Ⓜ | A | B | C

(A) DRAWING CALLOUT

DATUM SURFACE C

DATUM SURFACE B

DATUM SURFACE C

DATUM SURFACE A

12   90°

60°

90°

18

METRIC

(B) INTERPRETATION OF TRUE POSITION

FIGURE 52–3 ■ *Part with angular hole referred to a datum system—RFS.*

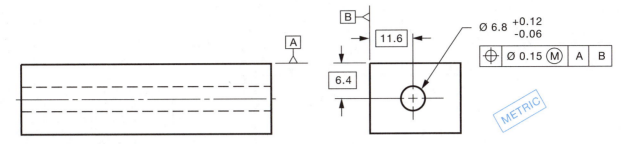

FIGURE 52–4 ■ *Datum system for a long hole—MMC.*

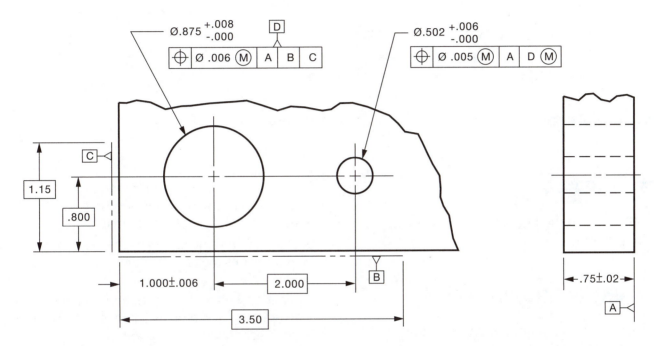

FIGURE 52–5 ■ *Specifying an internal circular feature as a datum—MMC.*

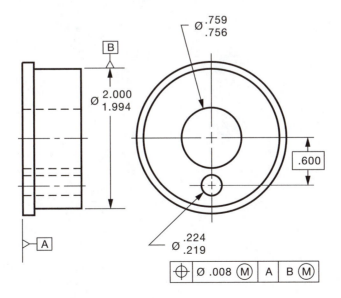

FIGURE 52–6 ■ *Specifying an external circular feature as a datum—MMC.*

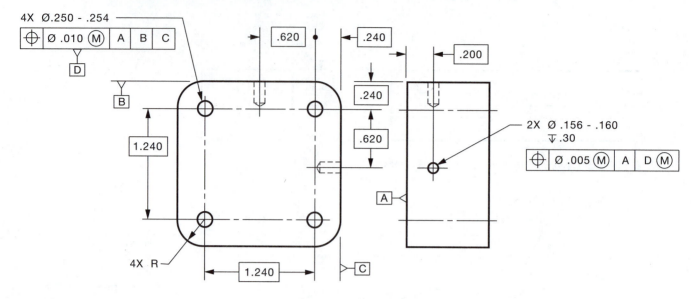

FIGURE 52–7 ■ *Specifying a group of holes as a single datum—MMC.*

REFERENCE

ASME Y14.5M-1994 (R1999) Dimensioning and Tolerancing

INTERNET RESOURCES

Drafting Zone. For information on geometric dimensioning and tolerancing, see: http://www. draftingzone.com

Effective Training Inc. For information on dimensioning and tolerancing, see: http:// etinews.com/eti_solutions.html

eFunda. For information on geometric dimensioning and tolerancing, see: http://www.efunda.com/ home.cfm

Engineering Edge. For information on geometric tolerancing and dimensioning, see: http://www. engineeringedge.com/gdt/htm

ASSIGNMENT:

On a one inch grid sheet (.10 in. squares) sketch a suitable gage to check the positional tolerance for the Ø.750-.755 in. hole.

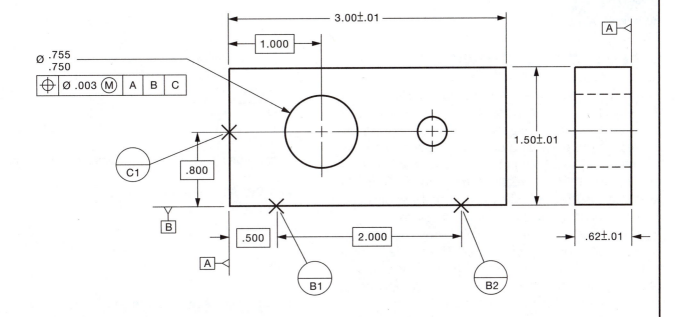

DATUM SELECTION FOR POSITIONAL TOLERANCING

A-118

PROFILE TOLERANCES

A profile is the outline form or shape of a line or surface. A line profile may be the outline of a part or feature as depicted in a view on a drawing. It may represent the edge of a part or it may refer to line elements of a surface in a single direction, such as the outline of cross sections through the part. In contrast, a surface profile outlines the form or shape of a complete surface in three dimensions.

The elements of a line profile may be straight lines, arcs, or other curved lines. The elements of a surface profile may be flat surfaces, spherical surfaces, cylindrical surfaces, or surfaces composed of various line profiles in two or more directions.

A profile tolerance specifies a uniform boundary along the true profile within which the elements of the surface must lie. MMC is not applicable to profile tolerances. Where used as a refinement of size, the profile tolerance must be contained within the size limits.

Profile Symbols

There are two geometric characteristic symbols for profiles: one for lines and one for surfaces. Separate symbols are required because it is often necessary to distinguish between line elements of a surface and the complete surface itself.

The symbol for profile of a line consists of a semicircle with a diameter equal to twice the lettering size used on the drawing. The symbol for profile of a surface is identical except that the semicircle is closed by a straight line at the bottom, Figure 53–1. All other geometric tolerances of form and orientation are merely special cases of profile tolerancing.

Profile tolerances are used to control the position of lines and surfaces that are neither flat nor cylindrical.

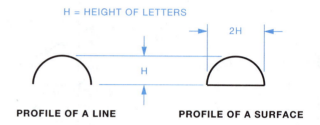

FIGURE 53–1 ■ *Profile symbols.*

PROFILE OF A LINE

A profile-of-a-line tolerance may be directed to a line of any length or shape. With a profile-of-a-line tolerance, datums may be used in some circumstances but would not be used when the only requirement is the profile shape taken cross section by cross section. Profile-of-a-line tolerancing is used where it is not desirable to control the entire surface of the feature as a single entity.

A profile-of-a-line tolerance is specified in the usual manner by including the symbol and tolerance in a feature control frame directed to the line to be controlled, Figure 53–2.

The tolerance zone established by the profile-of-a-line tolerance is two dimensional, extending along the length of the considered feature.

If the line on the drawing to which the tolerance is directed represents a surface, the tolerance applies to all line elements of the surface parallel to the plane of the view on the drawing, unless otherwise specified.

The tolerance indicates a tolerance zone consisting of the area between two parallel lines, separated by the specified tolerance, which are themselves parallel to the basic form of the line being toleranced.

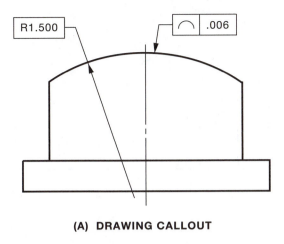

(A) DRAWING CALLOUT

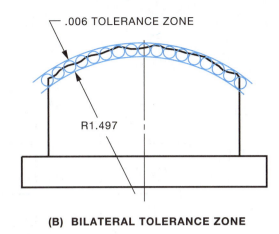

(B) BILATERAL TOLERANCE ZONE

FIGURE 53–2 ■ *Simple profile with a bilateral profile of a line tolerance zone.*

Bilateral and Unilateral Tolerances

The profile tolerance zone, unless otherwise specified, is equally disposed about the basic profile in a form known as a bilateral tolerance zone. The width of this zone is always measured perpendicular to the profile surface. The tolerance zone may be considered to be bounded by two lines enveloping a series of circles, each having a diameter equal to the specified profile tolerance. Their centers are on the theoretical, basic profile as shown in Figure 53–2.

Occasionally, it is desirable to have the tolerance zone wholly on one side of the basic profile instead of equally divided on both sides. Such zones are called unilateral tolerance zones. They are specified by showing a phantom line drawn parallel and close to the profile surface. The tolerance is directed to this line, Figure 53–3. The zone line need extend only a sufficient distance to make its application clear.

Specifying All-Around Profile Tolerancing

Where a profile tolerance applies all around the profile of a part, the symbol used to designate "all around" is placed on the leader from the feature control frame, Figure 53–4.

Method of Dimensioning

The true profile is established by means of basic dimensions, each of which is enclosed in a rectangular frame to indicate that the tolerance in the general tolerance note does not apply.

When the profile tolerance is not intended to control the position of the profile, there must be clear distinction between dimensions that control the position of the profile and those that control the form or shape of the profile.

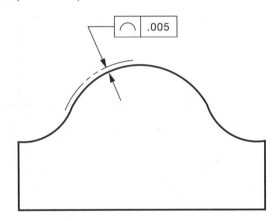

(A) TOLERANCE ZONE ON OUTSIDE OF TRUE PROFILE

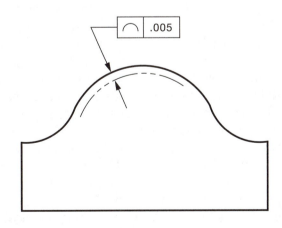

(B) TOLERANCE ZONE ON INSIDE OF TRUE PROFILE

FIGURE 53–3 ■ *Unilateral tolerance zones.*

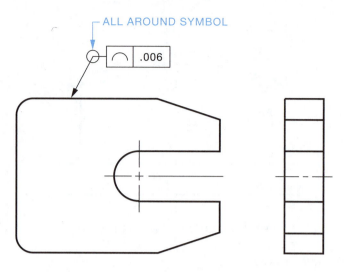

ALL AROUND SYMBOL

FIGURE 53–4 ■ *Profile tolerance required for all around.*

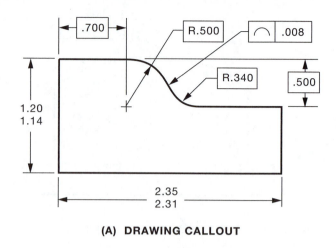

(A) **DRAWING CALLOUT**

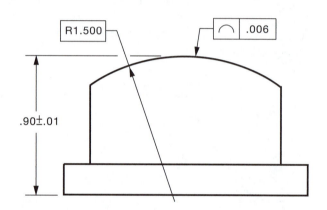

FIGURE 53–5 ■ *Profile and form as separate requirements.*

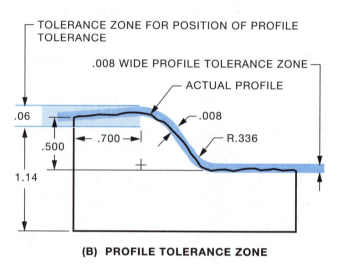

TOLERANCE ZONE FOR POSITION OF PROFILE TOLERANCE

.008 WIDE PROFILE TOLERANCE ZONE

ACTUAL PROFILE

(B) **PROFILE TOLERANCE ZONE**

FIGURE 53–6 ■ *Profile defined by basic dimensions.*

To illustrate, the simple part in Figure 53–5 shows a dimension of .90 ±.01 in. controlling the height of the profile. This dimension must be separately measured. The radius of 1.500 in. is a basic dimension and it becomes part of the profile. Therefore the profile tolerance zone has radii of 1.497 and 1.503 in., but is free to float in any direction within the limits of the positional tolerance zone in order to enclose the curved profile.

Figure 53–6 shows a more complex profile, where the profile is located by a single toleranced dimension. There are, however, four basic dimensions defining the true profile.

In this case, the tolerance on the height indicates a tolerance zone .06 in. wide, extending the full length of the profile. This is because the profile is established by basic dimensions. No other dimension exists to af-

fect the orientation or height. The profile tolerance specifies a .008 in. wide tolerance zone, which may lie anywhere within the .06 in. tolerance zone.

Extent of Controlled Profile

The profile is generally intended to extend to the first abrupt change or sharp corner. For example, in Figure 53–6, it extends from the upper left- to the upper right-hand corners, unless otherwise specified. If the extent of the profile is not clearly identified by sharp corners or by basic profile dimensions, it must be indicated by a note under the feature control symbol, such as A ↔ B, meaning between points A and B, Figure 53–7.

If the controlled profile includes a sharp corner, the tolerance boundary is considered to extend to the intersection of the boundary lines, Figure 53–8.

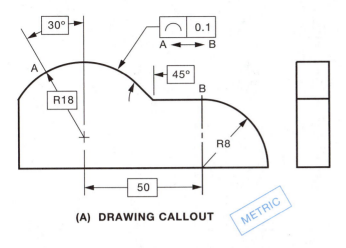

(A) DRAWING CALLOUT METRIC

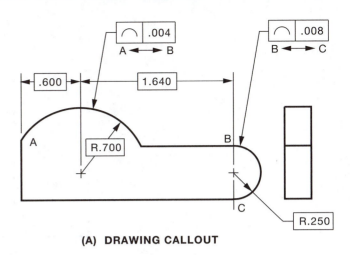

(A) DRAWING CALLOUT

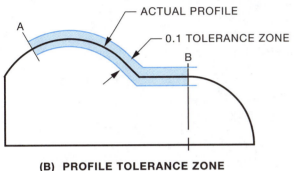

(B) PROFILE TOLERANCE ZONE

FIGURE 53–7 ■ *Specifying extent of profile tolerance.*

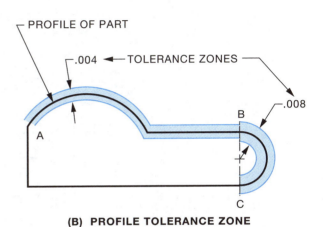

(B) PROFILE TOLERANCE ZONE

FIGURE 53–9 ■ *Dual tolerance zones.*

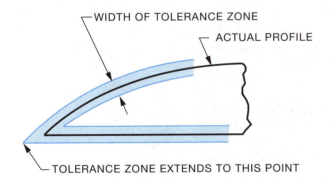

FIGURE 53–8 ■ *Controlling the profile of a sharp corner.*

PROFILE OF A SURFACE

If the same tolerance is intended to apply over the whole surface, instead of to lines or line elements in specific directions, the profile-of-a-surface symbol is used, Figure 53–10. Whereas the profile tolerance may be directed to the surface in either view, it is usually directed to the view showing the shape of the profile.

The profile-of-a-surface tolerance indicates a tolerance zone having the same form as the basic surface, with a uniform width equal to the specified tolerance within which the entire surface must lie. It is used to control form or combinations of size, form, and orientation. Where used as a refinement of size, the profile tolerance must be contained within the size limits.

As previously mentioned MMC is not applicable to profile tolerances.

Because the intersecting surfaces may lie anywhere within the converging zone, the actual part contour could conceivably be round. If this is undesirable, the drawing must indicate the design requirements, such as by specifying the maximum radius.

If different profile tolerances are required on different segments of a surface, the extent of each profile tolerance is indicated by the use of reference letters to identify the extremities, Figure 53–9.

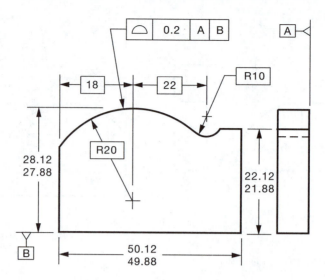

**(A) PROFILE TOLERANCE CONTROLS FORM
AND ORIENTATION OF A PROFILE**

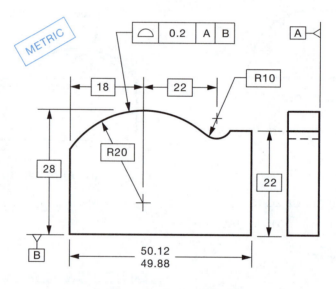

**(B) PROFILE TOLERANCE CONTROLS FORM,
ORIENTATION, AND POSITION OF A PROFILE**

FIGURE 53–10 ■ *Comparison of profile-of-a-surface
tolerance.*

The basic rules for profile-of-a-line tolerancing apply to profile-of-a-surface tolerancing except that in most cases, profile-of-a-surface tolerance requires reference to datums in order to provide proper orientation of the profile. This is specified simply by indicating suitable datums. Figure 53–10 shows a simple part where two datums are designated.

The criterion that distinguishes a profile tolerance as applying to position or to orientation is whether the profile is related to the datum by a basic dimension or by a toleranced dimension.

Profile tolerancing may be used to control the form and orientation of plane surfaces. In Figure 53–11, profile of a surface is used to control a plane surface inclined to a datum feature.

Where a profile-of-a-surface tolerance applies all around the profile of a part, the symbol used to designate "all around" is placed on the leader from the feature control frame, Figure 53–12.

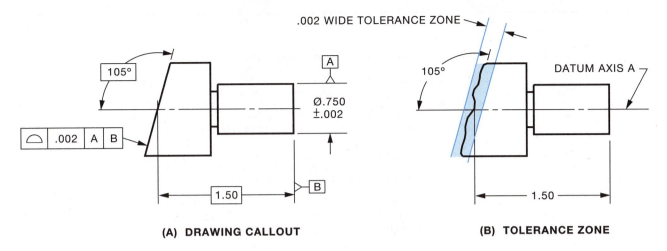

FIGURE 53–11 ■ *Specifying profile-of-a-surface tolerance for a plane surface.*

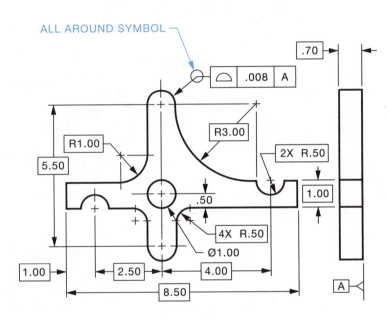

FIGURE 53–12 ■ *Profile-of-a-surface tolerance required for all around the surface*

REFERENCE

ASME Y14.5M-1994 (R1999) Dimensioning and Tolerancing

INTERNET RESOURCES

Drafting Zone. For information on geometric dimensioning and tolerancing, see: http://www.draftingzone.com

Effective Training Inc. For information on dimensioning and tolerancing, see: http://etinews.com/eti_solutions.html

eFunda. For information on geometric dimensioning and tolerancing, see: http://www.efunda.com/home.cfm

Engineering Edge. For information on geometric tolerancing and dimensioning, see: http://www.engineeringedge.com/gdt/htm

ASSIGNMENTS:
Use one inch grid sheets (.10 in. squares) for the sketching assignments below.

1. The profile form B to A (clockwise) as shown in Figure 1 requires a profile-of-a-line tolerance .004 in. It is essential that the point between B and A remains sharp, having a maximum .010-in. radius. The remainder of the profile requires a profile-of-a-line tolerance of .020 in. Sketch Figure 1 showing the geometric tolerance and basic dimensions to meet these requirements.

2. The part shown in Figure 2 requires an all-around profile-of-a-line tolerance of .005 in. located on the outside of the true profile. Sketch Figure 2 showing the geometric tolerance and basic dimensions to meet these requirements.

3. With the information given below and that on Figure 3, make a sketch and add dimensions showing the geometric tolerances, datums, and basic dimensions. Profile-of-a-surface tolerances are to be applied to the part as follows:

   a) Between points A and B - .005 in.
   b) Between points B and C - .004 in.
   c) Between points C and D - .002 in.

   These tolerances are to be referenced to datum surfaces marked E and F, in that order.

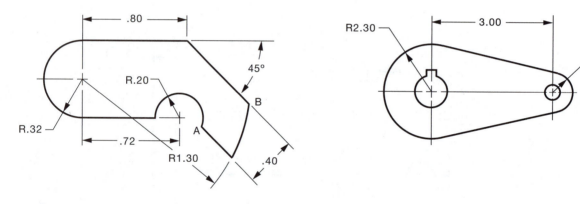

FIGURE 1

FIGURE 2

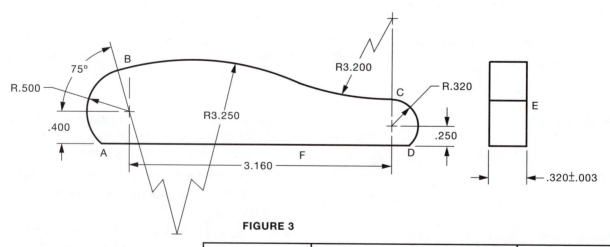

FIGURE 3

PROFILE TOLERANCING | **A-119**

RUNOUT TOLERANCES

Runout is a composite tolerance used to control the functional relationship of one or more features of a part to a datum axis. The types of features controlled by runout tolerances include those surfaces constructed around a datum axis and those constructed at right angles to a datum axis, Figure 54–1.

Each feature must be within its runout tolerance when rotated about the datum axis.

The datum axis is established by a diameter of sufficient length, two diameters having sufficient axial separation, or a diameter and a face at right angles to it. Features used as datums for establishing axes should be functional, such as mounting features that establish an axis of rotation.

The tolerance specified for a controlled surface is the total tolerance or full indicator movement (FIM) in inspection and international terminology. Both the tolerance and the datum feature apply only on an RFS basis.

There are two types of runout control: circular runout and total runout. The type used is dependent on design requirements and manufacturing considerations. The geometric characteristic symbols for runout are shown in Figure 54–2.

CIRCULAR RUNOUT

Circular runout provides control of circular elements of a surface. The tolerance is applied independently at any cross section as the part is rotated 360°. Where applied to surfaces constructed around a datum axis, circular runout controls variations such as circularity and coaxiality. Where applied to surfaces constructed at right angles to the datum axis, circular runout controls wobble at all diametral positions.

Thus in Figure 54–3 the surface is measured at several positions along the surface, as shown by the three indicator positions. At each position the indicator movement during one revolution of the part must not exceed the specified tolerance, in this case .005 inch. For a cylindrical feature such as this, runout error is caused by eccentricity and errors of roundness. It is not affected by taper (conicity) or errors of straightness of the straight line elements such as barrel shaping.

Figure 54–4 shows a part where the tolerance is applied to a surface that is at right angles to the axis. In this case, an error—generally referred to as wobble—will be shown if the surface is flat but not perpendicular to the axis, as shown at B. No error will

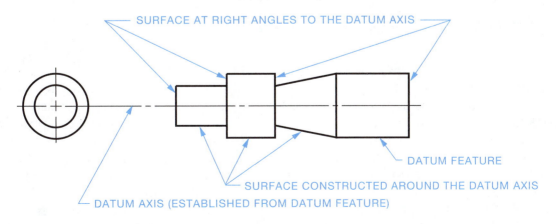

FIGURE 54–1 ■ *Features applicable to runout tolerances.*

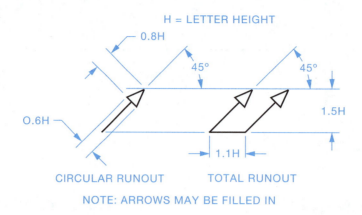

FIGURE 54–2 ■ *Runout symbols.*

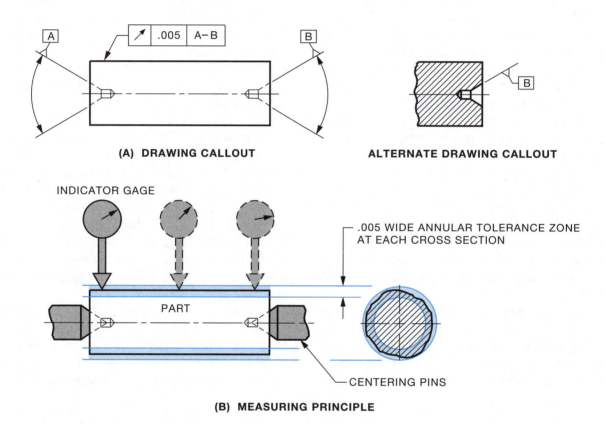

(A) DRAWING CALLOUT

ALTERNATE DRAWING CALLOUT

(B) MEASURING PRINCIPLE

FIGURE 54–3 ■ *Circular runout for cylindrical features.*

be indicated if the surface is convex or concave but otherwise perfect, as shown at C.

Circular runout can also be applied to curved surfaces. Unless otherwise specified, measurement is always made normal to the surface.

A runout tolerance directed to a surface applies to the full length of the surface up to an abrupt change in direction.

If a control is intended to apply to more than one portion of a surface, additional leaders and arrow-

heads may be used where the same tolerance applies. If different tolerance values are required, separate tolerances must be specified.

Where a runout tolerance applies to a specific portion of a surface, a thick chain line is drawn adjacent to the surface profile to show the desired length. Basic dimensions are used to define the extent of the portion so indicated, Figure 54–5.

If only part of a surface or several consecutive portions require the same tolerance, the length to

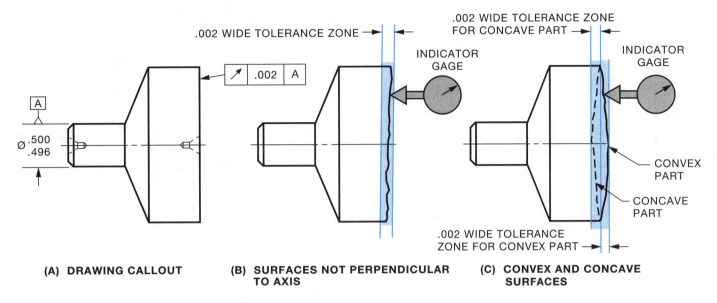

(A) DRAWING CALLOUT

(B) SURFACES NOT PERPENDICULAR TO AXIS

(C) CONVEX AND CONCAVE SURFACES

FIGURE 54–4 ■ *Circular runout perpendicular to datum axis.*

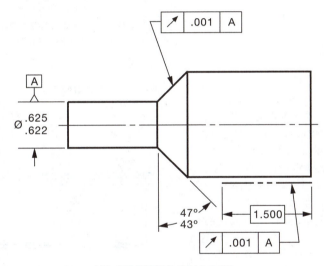

(A) DRAWING CALLOUT

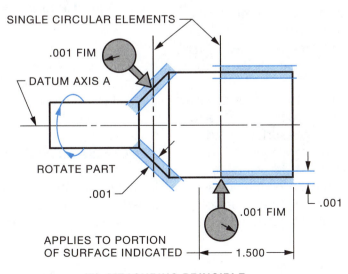

(B) MEASURING PRINCIPLE

FIGURE 54–5 ■ *Specifying circular runout relative to a datum diameter.*

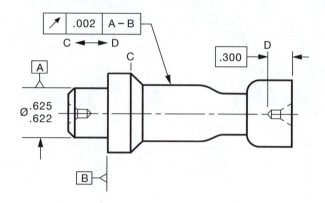

FIGURE 54–6 ■ *Indication of length for a runout tolerance.*

which the tolerance applies may be indicated as shown in Figure 54–6.

Circular runout tolerances can only be applied on an RFS basis.

TOTAL RUNOUT

Total runout concerns the runout of a complete surface, not merely the runout of each circular element. For measurement purposes, the checking indicator must traverse the full length or extent of the surface while the part is revolved about its datum axis. Measurements are made over the whole surface without resetting the indicator. Total runout is the difference between the lowest indicator reading in any position and the highest reading in that or in any other position on the same surface. Thus, in Figure 54–7, the tolerance zone is the space between two concentric cylinders separated by the specified tolerance and coaxial with the datum axis.

Note in this case that the runout is affected not only by eccentricity and errors of roundness, but also by errors of straightness and conicity of the cylindrical surface.

A total runout may be applied to surfaces at various angles, as described for circular runout, and may therefore control profile of the surface in addition to runout. However, for measurement purposes, the indicator gage must be capable of following the true profile direction of the surface. This is comparatively simple for straight surfaces, such as cylindrical surfaces and flat faces. For coni-

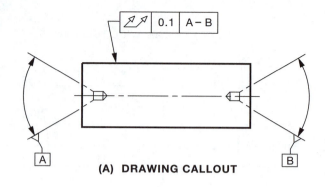

(A) DRAWING CALLOUT

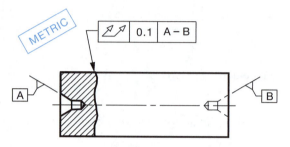

(B) ALTERNATIVE DRAWING CALLOUT FOR DATUM FEATURES

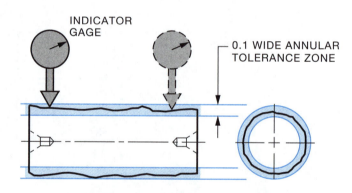

(C) MEASURING PRINCIPLES

FIGURE 54–7 ■ *Tolerance zones for total runout.*

cal surfaces, the datum axis can be tilted to the taper angle so that the measured surface becomes parallel to a surface plate.

ESTABLISHING DATUMS

In many examples the datum axis has been established from centers drilled in the two ends of the part, in which case the part is mounted between

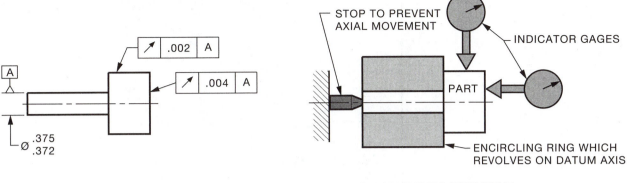

FIGURE 54–8 ■ *External cylindrical datum feature for runout tolerance.*

centers for measurement purposes. This is an ideal method of mounting and revolving the part when such centers have been provided for manufacturing purposes. When centers are not provided, any cylindrical or conical surface may be used to establish the datum axis if chosen on the basis of the functional requirements of the part. In some cases, a runout tolerance may also be applied to the datum feature. Some examples of suitable datum features and methods of establishing datum axes are discussed next.

Measuring Principles

Example 1:

Figure 54–8 shows a simple external cylindrical feature specified as the datum feature.

Measurement would require the datum feature to be held in an encircling ring capable of being revolved about the datum axis. Parts with these types of datum features are sometimes mounted in a vee-block, although this practice permits precise measurements only if there are no significant roundness errors of the datum feature.

Example 2

The L-support method is particularly useful when two datum features are used, Figure 54–9.

Measuring the part by using two L-supports is quite simple. Measurement for this part would be complicated were it necessary to fit the features into concentric encircling rings.

Example 3

Figure 54–10 illustrates the application of runout tolerances where two datum diameters act as a single datum axis to which the features are related. For measurement purposes, the part may be mounted on a mandrel having a diameter equal to the maximum size of the hole.

When required, runout tolerances may be referenced to a datum system, usually consisting of two datum features perpendicular to one another. For measuring purposes, the part is mounted on a flat surface capable of being rotated. Centering on the secondary datum requires some form of centralizing device, such as an expandable arbor.

Example 4

It may be necessary to control individual datum surface variations with respect to flatness, circularity, parallelism, straightness, or cylindricity. Where such control is required, the appropriate tolerances are specified. See Figure 54–11 for applying cylindricity to the datum.

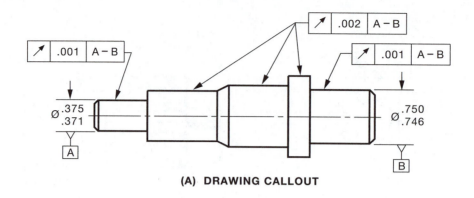

(A) DRAWING CALLOUT

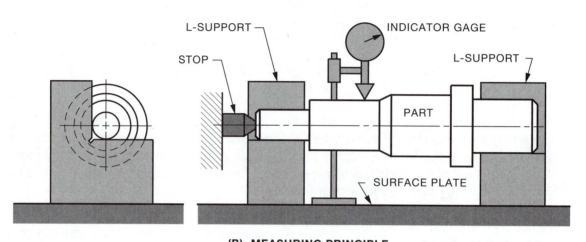

(B) MEASURING PRINCIPLE

FIGURE 54–9 ■ *Runout tolerance with two datum features.*

REFERENCE

ASME Y14.5M-1994 (R1999) Dimensioning and Tolerancing

INTERNET RESOURCES

Drafting Zone. For information on geometric dimensioning and tolerancing, see: http://www.draftingzone.com

Effective Training Inc. For information on dimensioning and tolerancing, see: http://etinews.com/eti_solutions.html

eFunda. For information on geometric dimensioning and tolerancing, see http://www.efunda.com/home.cfm

Engineering Edge. For information on geometric tolerancing and dimensioning, see: http://www.engineeringedge.com/gdt/htm

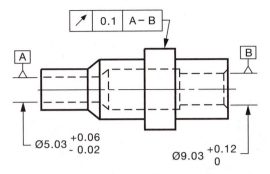

(A) DRAWING CALLOUT

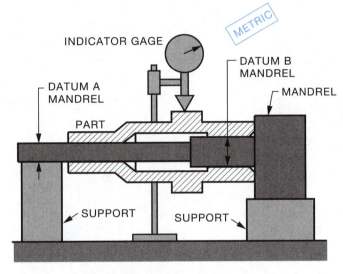

(B) MEASURING PRINCIPLE

FIGURE 54–10 ■ *Specifying runout relative to two datum diameters.*

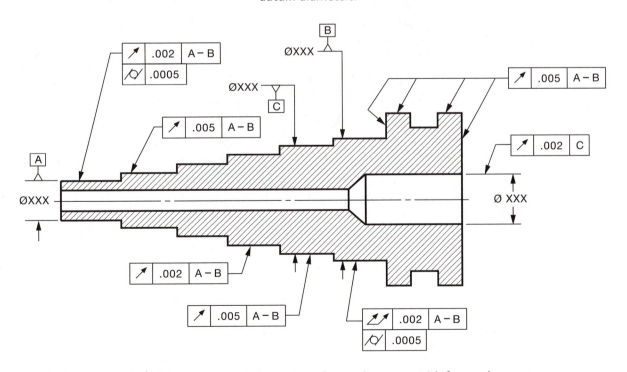

FIGURE 54–11 ■ *Specifying runout relative to two datum diameters with form tolerances.*

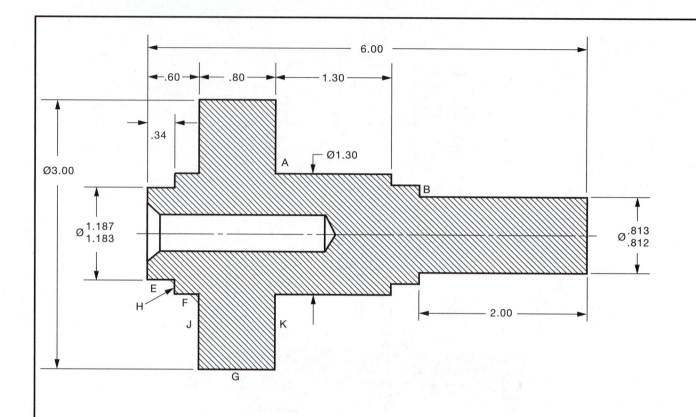

FIGURE 1

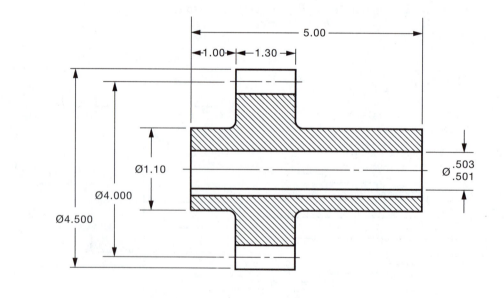

FIGURE 2

ASSIGNMENT:

Use inch grid sheets (.10 in. squares) for the sketching assignments below.

1. Sketch the part shown in Figure 1. Add the following runout tolerances and datums to the sketch:
   a) The Ø1.187 in. is to be datum C.
   b) A 1.20-in. length starting .40 in. from the right end of the part is to be datum D.
   c) Runout tolerances are related to the axis established by datums C and D.
   d) A total runout tolerance of .005 in. between positions A and B
   e) A circular runout tolerance of .002 in. for diameters E and F
   f) A circular runout tolerance of .005 in. for diameter G
   g) A circular runout tolerance of .004 in. for surface H
   h) A circular runout tolerance of .003 in. for surfaces J and K

2. Make a sketch of the gear shown in Figure 2. Add circular runout tolerances referenced to datum A. Both side faces of the gear portion require a tolerance of .015 in. The two hub portions require a tolerance of .010 in. The hole is to be datum A.

3. Make a sketch of the part shown in Figure 3. The part is intended to function by rotating with the two ends (datum diameters A and B) supported in bearings. These two datums collectively act as a coaxial datum for the larger diameters, which are required to have a total runout tolerance of .001 in. Add the above requirements to the drawing.

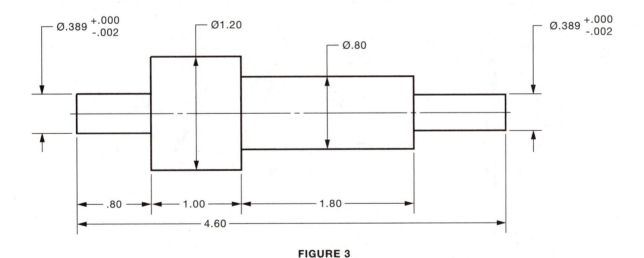

FIGURE 3

RUNOUT TOLERANCES | **A-120**

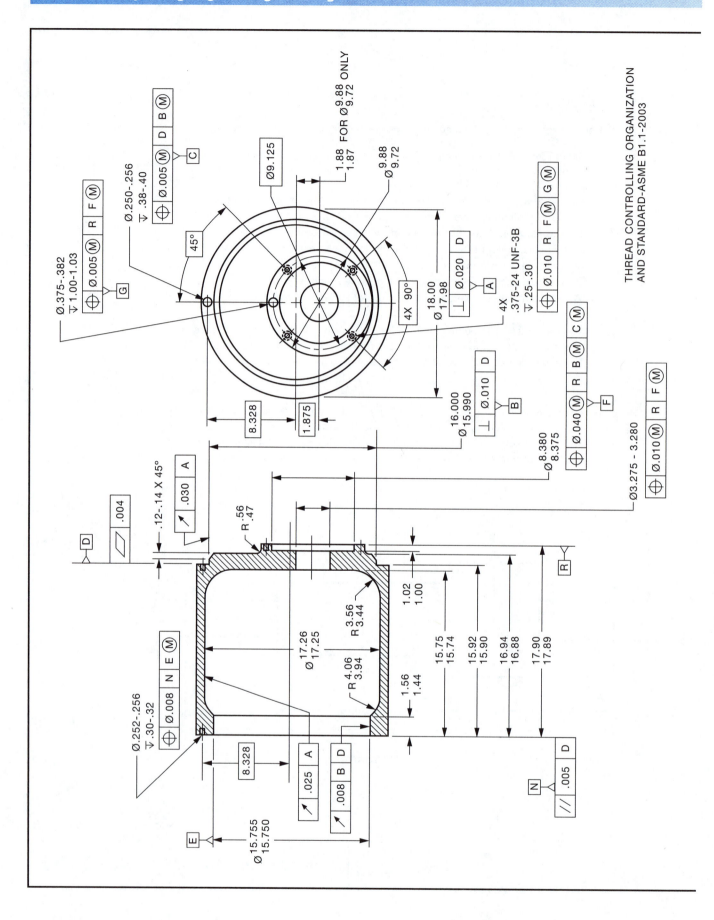

THREAD CONTROLLING ORGANIZATION
AND STANDARD-ASME B1.1-2003

ASSIGNMENT:

Use inch grid sheets (.10 in. squares) for the sketching assignments

1. Sketch a suitable gage to check the Ø3.275-3.280 hole.

2. Make a sketch of datum surface D showing the permissible tolerance zone.

3. Make a sketch of datum surface N showing the permissible tolerance zone.

QUESTIONS:

1. How many datum surfaces or points are indicated?

2. How many basic dimensions are indicated?

3. How many datum surfaces are flat?

4. How many datum surfaces are circular?

5. How many dimensions show positional tolerancing?

6. How many form tolerances are required?

7. How many orientation tolerances are shown?

8. How many features use datum A as a reference?

9. If the diameter of datum F was Ø8.375, what would be the maximum permissible positional tolerance?

10. What is the tertiary datum for the positional tolerance of the diameter shown as datum F?

11. The geometric tolerance placed on datum surface D controls_____.

12. With reference to the Ø3.275-3.280 hole, what is the maximum deviation permitted from true position when the hole is: (A) Ø3.275, (B) Ø3.280?

13. With reference to the Ø.252-.256 hole, what is the maximum deviation permitted from true position if the hole was (A) Ø.252, (B) Ø.256?

14. With reference to the Ø.250-.256 hole, what is the maximum deviation permitted from true position if the hole was: (A) Ø.250, (B) Ø.254, (C) Ø.256?

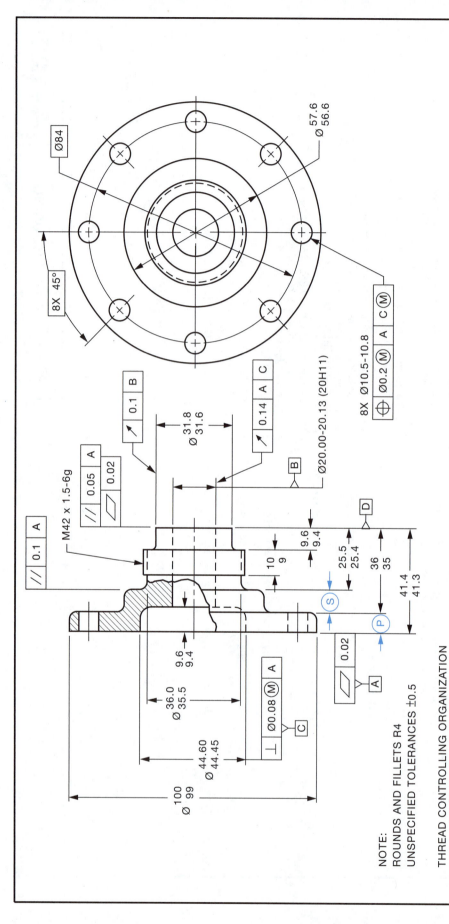

NOTE:
ROUNDS AND FILLETS R4
UNSPECIFIED TOLERANCES ±0.5

THREAD CONTROLLING ORGANIZATION
STANDARD-ASME B1.13M-2001

ASSIGNMENT:

Use centimeter grid sheets (1mm squares) for the sketching assignments.

1. Sketch a suitable gage to check the eight Ø10.5-10.8 holes.
2. Make a sketch of datum surface C showing the permissible tolerances zone.
3. Make a sketch of datum surface D showing the permissible tolerance zones.

QUESTIONS:

1. How many datum surfaces are indicated?
2. How many basic dimensions are shown?
3. How many different geometric tolerancing symbols are shown?
4. How many datum surfaces are circular?
5. How many features use datum A as a reference?
6. How many form tolerances are shown?
7. With reference to the Ø10.5-10.8 holes, what variation from the true position is permissible if the holes are (A) Ø10.5 (B) Ø10.8?
8. Calculate the following dimensions (A) P min., (B) P max., (C) S min., (D) S max.

METRIC
DIMENSIONS ARE IN MILLIMETERS

END PLATE   **A–122M**

8X Ø10.5-10.8
⊕ | Ø0.2 Ⓜ | A | C Ⓜ

Ø 57.6
56.6

Ø84

8X 45°

M42 x 1.5-6g

∥ | 0.1 | A

∥ | 0.05 | A
⟋ | 0.02

↗ | 0.1 | B

Ø 31.8
Ø 31.6

↗ | 0.14 | A | C

B

Ø20.00-20.13 (20H11)

9.6
9.4

10
9

25.5
25.4

36
35

41.4
41.3

D

Ⓢ

Ⓟ

⟋ | 0.02
A

9.6
9.4

Ø 36.0
Ø 35.5

⊥ | Ø0.08 Ⓜ | A
C

Ø 44.60
Ø 44.45

Ø 100
Ø 99

APPENDIX

Fraction	Decimals Inch		Millimeters	Fraction	Decimals Inch		Millimeters
	Two Place	Three Place			Two Place	Three Place	
1/64	.02	.016	.04	33/64	.52	.516	13.1
1/32	.03	.031	.08	17/32	.53	.531	13.5
3/64	.05	.047	1.2	35/64	.55	.547	13.9
1/16	.06	.062	1.6	9/16	.56	.562	14.3
5/64	.08	.078	2	37/64	.58	.578	14.7
3/32	.09	.094	2.4	19/32	.59	.594	15.1
7/64	.11	.109	2.8	39/64	.61	.609	15.5
1/8	.12	.125	3.2	5/8	.62	.625	15.9
9/64	.14	.141	3.6	41/64	.64	.641	16.3
5/32	.16	.156	4	21/32	.66	.656	16.7
11/64	.17	.172	4.4	43/64	.67	.672	17.1
3/16	.19	.188	4.8	11/16	.69	.688	17.5
13/64	.20	.203	5.2	45/64	.70	.703	17.9
7/32	.22	.219	5.6	23/32	.72	.719	18.3
15/64	.23	.234	6	47/64	.73	.734	18.7
1/4	.25	.250	6.4	3/4	.75	.750	19.1
17/64	.27	.266	6.8	49/64	.77	.766	19.5
9/32	.28	.281	7.1	25/32	.78	.781	19.9
19/64	.30	.297	7.5	51/64	.80	.797	20.2
5/16	.31	.312	7.9	13/16	.81	.812	20.6
21/64	.33	.328	8.3	53/64	.83	.828	21
11/32	.34	.344	8.7	27/32	.84	.844	21.4
23/64	.36	.359	9.1	55/64	.86	.859	21.8
3/8	.38	.375	9.5	7/8	.88	.875	22.2
25/64	.39	.391	9.9	57/64	.89	.891	22.6
13/32	.41	.406	10.3	29/32	.91	.906	23
27/64	.42	.422	10.7	59/64	.92	.922	23.4
7/16	.44	.438	11.1	15/16	.94	.938	23.8
29/64	.45	.453	11.5	61/64	.95	.953	24.2
15/32	.47	.469	11.9	31/32	.97	.969	24.6
31/64	.48	.484	12.3	63/64	.98	.984	25
1/2	.50	.500	12.7	1	1.00	1.000	25.4

TABLE 1 ■ *Chart for converting inch dimensions to millimeters.*

And	.....&		Long	.....LG
Across Flats	.....ACR FLT		Machined	.....√
Approximate	.....APPROX		Machine Steel	.....MST
Assembly	.....ASSY		Malleable Iron	.....MI
Bill of Material	.....B/M		Material	.....MATL
Bolt Circle	.....BC		Maximum	.....MAX
Brass	.....BR		Maximum Material Condition	...MMC or Ⓜ
Bronze	.....BRZ		Meter	.....m
Brown and Sharpe Gage	.....B & S GA		Metric Thread	.....M
Carbon Steel	.....CS		Micrometer	.....μm
Cast Iron	CI		Mild Steel	.....MS
Centimeter	.....cm		Millimeter	.....mm
Center Line	.....℄ or CL		Minimum	.....MIN
Center to Center	.....C to C		Minute (Angle)	.....(′)
Chamfer	.....CHAM		Nominal	.....NOM
Circularity	.....CIR		Not to Scale	XXX
Cold Rolled Steel	.....CRS		Number	.....NO
Concentric	.....CONC		Outside Diameter	.....OD
Conical Taper	.....▷		Parallel	.....PAR
Copper	.....COP		Perpendicular	.....PERP
Counterbore	.....CBORE or ⌴		Pitch Circle Diameter	.....PCD
Countersink	.....CSK or ∨		Pitch Diameter	.....PD
Cubic Centimeter	.....cm³		Projected Tolerance Zone	.....Ⓟ
Cubic Meter	.....m³		Radius	.....R
Datum	.....Ⓐ		Referencde or Reference Dimension	.....( )
Deep or Depth	.....▽		Regardless of Feature Size	.....RFS
Degree (Angle)	.....°		Revolutions per Minute	.....R/MIN
Diameter	.....∅ or DIA		Right Hand	.....RH
Diametral Pitch	.....DP		Second (Arc)	.....″
Dimension	.....DIM		Section	.....SECT
Dimension Origin	.....⊕▶		Slope	.....◁
Drawing	.....DWG		Spherical Radius	.....SR
Eccentric	.....ECC		Spotface	.....SF or ⌴
Equally Spaced	.....EQL SP		Square	.....SQ or □
Figure	.....FIG		Square Centimeter	.....cm²
Finish All Over	.....FAO		Square Meter	.....m²
Gage	.....GA		Steel	.....STL
Gray Iron	.....GI		Symmetrical	.....SYM or ⌇
Head	.....HD		Symmetry	.....⌇
Heat Treat	.....HT TR		Taper Pipe Thread	.....NPT
Hexagon	.....HEX		Thick	.....THK
Inside Diameter	.....ID		Through	.....THRU
International Organization for Standardization	.....ISO		Undercut	.....UCUT
International Pipe Standard	.....IPS		United States Gage	.....USG
Kilogram	.....kg		Wrought Iron	.....WI
Kilometer	.....km		Wrought Steel	.....WS
Least Material Condition	.....LMC or Ⓛ			
Left Hand	.....LH			

TABLE 2 ■ *Abbreviations and symbols used on technical drawings.*

Decimal Inch and Millimeter Equivalents of Number Size Drills						Decimal Inch and Millimeter Equivalents of Letter Size Drills		
No.	Decimal Inch	mm	No.	Decimal Inch	mm	Letter	Decimal Inch	mm
1	.2280	5.8	31	.1200	3.0	A	.234	5.9
2	.2210	5.6	32	.1160	2.9	B	.238	6.0
3	.2130	5.4	33	.1130	2.9	C	.242	6.1
4	.2090	5.3	34	.1110	2.8	D	.246	6.2
5	.2055	5.2	35	.1100	2.8	E	.250	6.4
6	.2040	5.2	36	.1065	2.7	F	.257	6.5
7	.2010	5.1	37	.1040	2.6	G	.261	6.6
8	.1990	5.1	38	.1015	2.6	H	.266	6.8
9	.1960	5.0	39	.0095	2.5	I	.272	6.9
10	.1935	4.9	40	.0980	2.5	J	.277	7.0
11	.1910	4.9	41	.0960	2.4	K	.281	7.1
12	.1890	4.8	42	.0935	2.4	L	.290	7.4
13	.1850	4.7	43	.0890	2.3	M	.295	7.5
14	.1820	4.6	44	.0860	2.2	N	.302	7.7
15	.1800	4.6	45	.0820	2.1	O	.316	8.0
16	.1770	4.5	46	.0810	2.1	P	.323	8.2
17	.1730	4.4	47	.0785	2.0	Q	.332	8.4
18	.1695	4.3	48	.0760	1.9	R	.339	8.6
19	.1660	4.2	49	.0730	1.9	S	.348	8.8
20	.1610	4.1	50	.0700	1.8	T	.358	9.1
21	.1590	4.0	51	.0670	1.7	U	.368	9.3
22	.1570	4.0	52	.0635	1.6	V	.377	9.6
23	.1540	3.9	53	.0595	1.5	W	.386	9.8
24	.1520	3.9	54	.0550	1.4	X	.397	10.1
25	.1495	3.8	55	.0520	1.3	Y	.404	10.3
26	.1470	3.7	56	.0465	1.2	Z	.413	10.5
27	.1440	3.7	57	.0430	1.1			
28	.1405	3.6	58	.0420	1.1			
29	.1360	3.5	59	.0410	1.0			
30	.1285	3.3	60	.0400	1.0			

TABLE 3 ■ Number and letter-size drills.

Metric Drill Sizes (mm)		Reference Decimal Equivalent (Inches)	Metric Drill Sizes (mm)		Reference Decimal Equivalent (Inches)	Metric Drill Sizes (mm)		Reference Decimal Equivalent (Inches)
Preferred	Available		Preferred	Available		Preferred	Available	
–	0.40	.0157	2.2	–	.0866	10	–	.3937
–	0.42	.0165	–	2.3	.0906	–	10.3	.4055
–	0.45	.0177	2.4	–	.0945	10.5	–	.4134
–	0.48	.0189	2.5	–	.0984	–	10.8	.4252
0.5	–	.0197	2.6	–	.1024	11	–	.4331
–	0.52	.0205	–	2.7	.1063	–	11.5	.4528
0.55	–	.0217	2.8	–	.1102	12	–	.4724
–	0.58	.0228	–	2.9	.1142	12.5	–	.4921
0.6	–	.0236	3	–	.1181	13	–	.5118
–	0.62	.0244	–	3.1	.1220	–	13.5	.5315
0.65	–	.0256	3.2	–	.1260	14	–	.5512
–	0.68	.0268	–	3.3	.1299	–	14.5	.5709
0.7	–	.0276	3.4	–	.1339	15	–	.5906
–	0.72	.0283	–	3.5	.1378	–	15.5	.6102
0.75	–	.0295	3.6	–	.1417	16	–	.6299
–	0.78	.0307	–	3.7	.1457	–	16.5	.6496
0.8	–	.0315	3.8	–	.1496	17	–	.6693
–	0.82	.0323	–	3.9	.1535	–	17.5	.6890
0.85	–	.0335	4	–	.1575	18	–	.7087
–	0.88	.0346	–	4.1	.1614	–	18.5	.7283
0.9	–	.0354	4.2	–	.1654	19	–	.7480
–	0.92	.0362	–	4.4	.1732	–	19.5	.7677
0.95	–	.0374	4.5	–	.1772	20	–	.7874
–	0.98	.0386	–	4.6	.1811	–	20.5	.8071
1	–	.0394	4.8	–	.1890	21	–	.8268
–	1.03	.0406	5	–	.1969	–	21.5	.8465
1.05	–	.0413	–	5.2	.2047	22	–	.8661
–	1.08	.0425	5.3	–	.2087	–	23	.9055
1.1	–	.0433	–	5.4	.2126	24	–	.9449
–	1.15	.0453	5.6	–	.2205	25	–	.9843
1.2	–	.0472	–	5.8	.2283	26	–	1.0236
1.25	–	.0492	6	–	.2362	–	27	1.0630
1.3	–	.0512	–	6.2	.2441	28	–	1.1024
–	1.35	.0531	6.3	–	.2480	–	29	1.1417
1.4	–	.0551	–	6.5	.2559	30	–	1.1811
–	1.45	.0571	6.7	–	.2638	–	31	1.2205
1.5	–	.0591	–	6.8	.2677	32	–	1.2598
–	1.55	.0610	–	6.9	.2717	–	33	1.2992
1.6	–	.0630	7.1	–	.2795	34	–	1.3386
–	1.65	.0650	–	7.3	.2874	–	35	1.3780
1.7	–	.0669	7.5	–	.2953	36	–	1.4173
–	1.75	.0689	–	7.8	.3071	–	37	1.4567
1.8	–	.0709	8	–	.3150	38	–	1.4961
–	1.85	.0728	–	8.2	.3228	–	39	1.5354
1.9	–	.0748	8.5	–	.3346	40	–	1.5748
–	1.95	.0768	–	8.8	.3465	–	41	1.6142
2	–	.0787	9	–	.3543	42	–	1.6535
–	2.05	.0807	–	9.2	.3622	–	43.5	1.7126
2.1	–	.0827	9.5	–	.3740	45	–	1.7717
–	2.15	.0846	–	9.8	.3858	–	46.5	1.8307

TABLE 4 ▪ *Metric twist drill sizes.*

Size		Coarse Thread Series UNC & NC		Fine Thread Series UNF & NF		Extra Fine Series UNEF & NEF		8-Pitch Thread Series 8 N		12-Pitch Thread Series 12 N		16-Pitch Thread Series 16 N	
Number or Fraction	Decimal	Threads Per Inch	Tap Drill	Threads Per Inch	Tap Drill	Threads Per Inch	Tap Drill	Threads Per Inch	Tap Drill	Threads Per Inch	Tap Drill	Threads Per Inch	Tap Drill
0	.060			80	3/64								
1	.073	64	No. 53	72	No. 53								
2	.086	56	No. 50	64	No. 50								
3	.099	48	No. 47	56	No. 45								
4	.112	40	No. 43	48	No. 42								
5	.125	40	No. 38	44	No. 37								
6	.138	32	No. 36	40	No. 33								
8	.164	32	No. 29	36	No. 29								
10	.190	24	No. 25	32	No. 21								
12	.216	24	No. 16	28	No. 14	32	No. 13						
1/4	.250	20	No. 7	28	No. 3	32	7/32						
5/16	.312	18	F	24	I	32	9/32						
3/8	.375	16	5/16	24	Q	32	11/32						
7/16	.438	14	U	20	25/64	28	13/32						
1/2	.500	13	27/64	20	29/64	28	15/32			12	27/64		
9/16	.562	12	31/64	18	33/64	24	33/64			12	31/64		
5/8	.625	11	17/32	18	37/64	24	37/64			12	35/64		
3/4	.750	10	21/32	16	11/16	20	45/64			12	43/64	16	11/16
7/8	.875	9	49/64	14	13/16	20	53/64			12	51/64	16	13/16
1	1.000	8	7/8	12	59/64	20	61/64	8	7/8	12	59/64	16	15/16
1 1/8	1.125	7	63/64	12	1 3/64	18	1 5/64	8	1	12	1 3/64	16	1 1/16
1 1/4	1.250	7	1 7/64	12	1 11/64	18	1 3/16	8	1 1/8	12	1 11/64	16	1 3/16
1 3/8	1.375	6	1 7/32	12	1 19/64	18	1 5/16	8	1 1/4	12	1 19/64	16	1 5/16
1 1/2	1.500	6	1 11/32	12	1 27/64	18	1 7/16	8	1 3/8	12	1 27/64	16	1 7/16
1 3/4	1.750	5	1 9/16			16	1 11/16	8	1 5/8	12	1 43/64	16	1 11/16
2	2.000	4 1/2	1 25/32			16	1 15/16	8	1 7/8	12	1 59/64	16	1 15/16
2 1/4	2.250	4 1/2	2 1/32					8	2 1/8	12	2 11/64	16	2 3/16
2 1/2	2.500	4	2 1/4					8	2 3/8	12	2 27/64	16	2 7/16
2 3/4	2.750	4	2 1/2					8	2 5/8	12	2 43/64	16	2 11/16
3	3.000	4	2 3/4					8	2 7/8	12	2 59/64	16	2 15/16

Color shows unified thread

TABLE 5 ■ Unified and American (inch) threads.

Series with Graded Pitches and **Series with Constant Pitches**

Nominal Size DIA (mm) Preferred	Coarse		Fine		4		3		2		1.5		1.25		1		0.75		0.5		0.35	
	Thread Pitch	Tap Drill Size	Thread Pitch	Tap Drill Size	Thread Pitch	Tap Drill Size	Thread Pitch	Tap Drill Size	Thread Pitch	Tap Drill Size	Thread Pitch	Tap Drill Size	Thread Pitch	Tap Drill Size	Thread Pitch	Tap Drill Size	Thread Pitch	Tap Drill Size	Thread Pitch	Tap Drill Size	Thread Pitch	Tap Drill Size
1.6	0.35	1.25																				
1.8	0.35	1.45																				
2	0.4	1.6																				
2.2	0.45	1.75																				
2.5	0.45	2.05																			0.35	2.15
3	0.5	2.5																			0.35	2.65
3.5	0.6	2.9																			0.35	3.15
4	0.7	3.3																	0.5	3.5		
4.5	0.75	3.7																	0.5	4		
5	0.8	4.2																	0.5	4.5		
*6	1	5															0.75	5.2				
**6.3	1	5.3																				
8	1.25	6.7	1	7											1	7	0.75	7.2				
10	1.5	8.5	1.25	8.7									1.25	8.7	1	9	0.75	9.2				
12	1.75	10.2	1.25	10.8							1.5	10.5	1.25	10.7	1	11						
14	2	12	1.5	12.5							1.5	12.5	1.25	12.7	1	13						
16	2	14	1.5	14.5							1.5	14.5			1	15						
18	2.5	15.5	1.5	16.5					2	16	1.5	16.5			1	17						
20	2.5	17.5	1.5	18.5					2	18	1.5	18.5			1	19						
22	2.5	19.5	1.5	20.5					2	20	1.5	20.5			1	21						
24	3	21	2	22					2	22	1.5	22.5			1	23						
27	3	24	2	25					2	25	1.5	25.5			1	26						
30	3.5	26.5	2	28					2	28	1.5	28.5			1	29						
33	3.5	29.5	2	31					2	31	1.5	31.5										
36	4	32	3	33					2	34	1.5	34.5										
39	4	35	3	36					2	37	1.5	37.5										
42	4.5	37.5	3	39	4	38	3	39	2	40	1.5	40.5										
45	4.5	39	3	42	4	41	3	42	2	43	1.5	43.5										
48	5	43	3	45	4	44	3	45	2	46	1.5	46.5										

* ISO thread size

** ASME thread size (to be discontinued)

TABLE 6 ■ *Metric threads.*

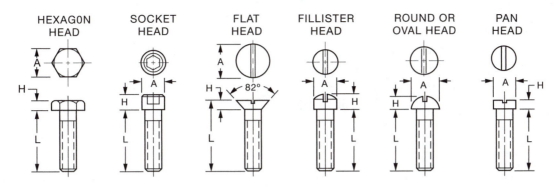

	Nominal Size		Hexagon Head		Socket Head		Flat Head		Fillister Head		Round or Oval Head	
	Fraction	Decimal	A	H	A	H	A	H	A	H	A	H
U.S. Customary (Inches)	1/4	.250	.44	.17	.38	.25	.50	.14	.38	.22	.44	.19
	5/16	.312	.50	.22	.47	.31	.62	.18	.44	.25	.56	.25
	3/8	.375	.56	.25	.56	.38	.75	.21	.56	.31	.62	.27
	7/16	.438	.62	.30	.66	.44	.81	.21	.62	.36	.75	.33
	1/2	.500	.75	.34	.75	.50	.88	.21	.75	.41	.81	.35
	5/8	.625	.94	.42	.94	.62	1.12	.28	.88	.50	1.00	.44
	3/4	.750	1.12	.50	1.12	.75	1.38	.35	1.00	.59	1.25	.55
	7/8	.875	1.31	.58	1.31	.88	1.62	.42	1.12	.69		
	1	1.000	1.50	.67	1.50	1.00	1.88	.49	1.31	.78		
	1 1/8	1.125	1.69	.75	1.69	1.12	2.06	.53				
	1 1/4	1.250	1.88	.84	1.88	1.25	2.31	.60				
	1 1/2	1.500	2.25	1.00	2.25	1.50	2.81	.74				

	Nominal Size	Hexagon Head		Socket Head			Flat Head		Fillister Head		Pan Head	
		A	H	A	H	Key Size	A	H	A	H	A	H
Metric (Millimeters) Sizes	M3	5.5	2	5.5	3	2.5	5.6	1.6	6	2.4	5.6	1.9
	4	7	2.8	7	4	3	7.5	2.2	8	3.1	7.5	2.5
	5	8.5	3.5	9	5	4	9.2	2.5	10	3.8	9.2	3.1
	6	10	4	10	6	5	11	3	12	4.6	11	3.8
	8	13	5.5	13	8	6	14.5	4	16	6	14.5	5
	10	17	7	16	10	8	18	5	20	7.5	18	6.2
	12	19	8	18	12	10						
	14	22	9	22	14	12						
	16	24	10	24	16	14						
	18	27	12	27	18	14						
	20	30	13	30	20	17						
	22	36	15	33	22	17						
	24	36	15	36	24	19						
	27	41	17	40	27	19						
	30	46	19	45	30	22						

NOTE: Length sizes normally available in .25 inch and 10mm increments

TABLE 7 ■ *Common cap screws.*

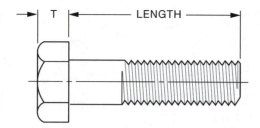

U.S. Customary (Inches)			
Nominal Size		Width Across	
Fraction	Decimal	Flats	Thickness
1/4	.250	.44	.17
5/16	.312	.50	.22
3/8	.375	.56	.25
7/16	.438	.62	.30
1/2	.500	.75	.34
5/8	.625	.94	.42
3/4	.750	1.12	.50
7/8	.875	1.31	.58
1	1.000	1.50	.67
1 1/8	1.125	1.69	.75
1 1/4	1.250	1.88	.84
1 3/8	1.375	2.06	.91
1 1/2	1.500	2.25	1.00

Metric (Millimeters)		
Nominal Size (Millimeters)	Width Across Flats	Thickness
4	7	2.8
5	8	3.5
6	10	4
8	13	5.5
10	17	7
12	19	8
14	22	9
16	24	10
18	27	12
20	30	13
22	32	14
24	36	15
27	41	17
30	46	19
33	50	21
36	55	23

NOTE: For bold and cap screw sizes below 7/16 inch and 8mm length sizes normally available in .25 inch and 10mm increments

TABLE 8 ▪ *Hexagon-head bolts and cap screws.*

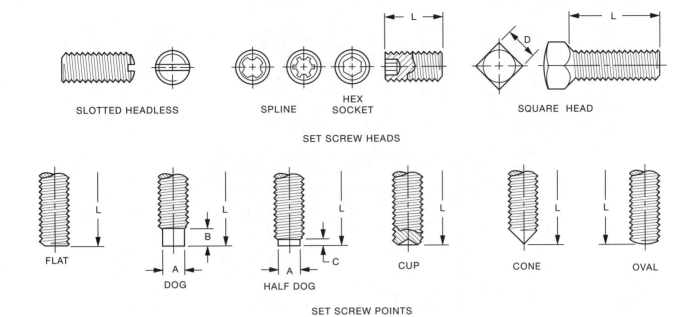

SET SCREW HEADS

SET SCREW POINTS

U.S. Customary (Inches)			Metric (Millimeters)	
Nominal Size		Key Size	Nominal Size	Key Size
Number	Decimal			
4	.112	.050	M 1.4	0.7
5	.125	.062	2	0.9
6	.138	.062	3	1.5
8	.164	.078	4	2
10	.190	.094	5	2.5
12	.216	.109	6	3
1/4	.250	.125	8	4
5/16	.312	.156	10	5
3/8	.375	.188	12	6
1/2	.500	.250	16	8

TABLE 9 ■ *Set screws.*

WASHER FACE

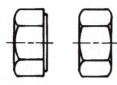

REGULAR

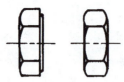

JAM

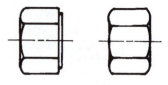

THICK

	Nominal Size		Distance Across Flats	Thickness		
	Fraction	Decimal		Regular	Jam	Thick
U.S. Customary (Inches)	1/4	.250	.44	.22	.16	.28
	5/16	.312	.50	.27	.19	.33
	3/8	.375	.56	.33	.22	.41
	7/16	.438	.69	.38	.25	.45
	1/2	.500	.75	.44	.31	.56
	9/16	.562	.88	.48	.31	.61
	5/8	.625	.94	.55	.38	.72
	3/4	.750	1.12	.64	.42	.81
	7/8	.875	1.31	.75	.48	.91
	1	1.000	1.50	.86	.55	1.00
	1 1/8	1.125	1.69	.97	.61	1.16
	1 1/4	1.250	1.88	1.06	.72	1.25
	1 3/8	1.375	2.06	1.17	.78	1.38
	1 1/2	1.500	2.25	1.28	.84	1.50

	Nominal Size (Millimeters)	Distance Across Flats	Thickness		
			Regular	Jam	Thick
Metric (Millimeters)	4	7	3	2	5
	5	8	4	2.5	5
	6	10	5	3	6
	8	13	6.5	5	8
	10	17	8	6	10
	12	19	10	7	12
	14	22	11	8	14
	16	24	13	8	16
	18	27	15	9	18.5
	20	30	16	9	20
	22	32	18	10	22
	24	36	19	10	24
	27	41	22	12	27
	30	46	24	12	30
	33	50	26		
	36	55	29		
	39	60	31		

TABLE 10 ∎ *Hexagon head nuts.*

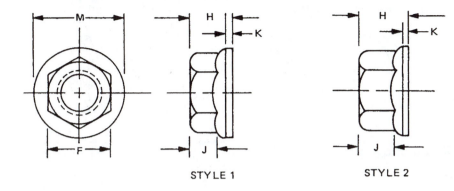

STYLE 1   STYLE 2

Metric (Millimeters)							
Nominal Nut Size and Thread Pitch	Width Across Flats F	Style 1				Style 2	
		H	J	K	M	H	J
M6 X 1	10	5.8	3	1	14.2	6.7	3.7
M8 X 1.25	13	6.8	3.7	1.3	17.6	8	4.5
M10 X 1.5	15	9.6	5.5	1.5	21.5	11.2	6.7
M12 X 1.75	18	11.6	6.7	2	25.6	13.5	8.2
M14 X 2	21	13.4	7.8	2.3	29.6	15.7	9.6
M16 X 2	24	15.9	9.5	2.5	34.2	18.4	11.7
M20 X 2.5	30	19.2	11.1	2.8	42.3	22	12.6

TABLE 11 ■ *Hex flanged nuts.*

FLAT WASHER   LOCKWASHER

Nominal Screw Size		Flat Washer			Lockwasher		
Number or Fraction	Decimal	Inside Dia A	Outside Dia B	Thickness C	Inside Dia A	Outside Dia B	Thickness C
6	.138	.16	.38	.05	.14	.25	.03
8	.164	.19	.44	.05	.17	.29	.04
10	.190	.22	.50	.05	.19	.33	.05
12	.216	.25	.56	.07	.22	.38	.06
1/4	.250 N	.28	.63	.07	.26	.49	.06
1/4	.250 W	.31	.73	.07			
5/16	.312 N	.34	.69	.07	.32	.59	.08
5/16	.312 W	.38	.88	.08			
3/8	.375 N	.41	.81	.07	.38	.68	.09
3/8	.375 W	.44	1.00	.08			
7/16	.438 N	.47	.92	.07	.45	.78	.11
7/16	.438 W	.50	1.25	.08			
1/2	.500 N	.53	1.06	.10	.51	.87	.12
1/2	.500 W	.56	1.38	.11			
5/8	.625 N	.66	1.31	.10	.64	1.08	.16
5/8	.625 W	.69	1.75	.13			
3/4	.750 N	.81	1.47	.13	.76	1.27	.19
3/4	.750 W	.81	2.00	.15			
7/8	.875 N	.94	1.75	.13	.89	1.46	.22
7/8	.875 W	.94	2.25	.17			
1	1.000 N	1.06	2.00	.13	1.02	1.66	.25
1	1.000 W	1.06	2.50	.17			
1 1/8	1.125 N	1.25	2.25	.13	1.14	1.85	.28
1 1/8	1.125 W	1.25	2.75	.17			
1 1/4	1.250 N	1.38	2.50	.17	1.27	2.05	.31
1 1/4	1.250 W	1.38	3.00	.17			
1 3/8	1.375 N	1.50	2.75	.17	1.40	2.24	.34
1 3/8	1.375 W	1.50	3.25	.18			
1 1/2	1.500 N	1.62	3.00	.17	1.53	2.43	.38
1 1/2	1.500 W	1.62	3.50	.18			

N–SAE Sizes (Narrow)
W–Standard Plate (Wide)   INCH SIZES

TABLE 12 ■ *Common washer sizes.*

FLAT WASHER  LOCKWASHER   SPRING LOCKWASHER

Bolt Size	Flat Washers			Lockwashers			Spring Lockwashers		
	ID	OD	Thickness	ID	OD	Thickness	ID	OD	Thickness
2	2.2	5.5	0.5	2.1	3.3	0.5			
3	3.2	7	0.5	3.1	5.7	0.8			
4	4.3	9	0.8	4.1	7.1	0.9	4.2	8	0.3 / 0.4
5	5.3	11	1	5.1	8.7	1.2	5.2	10	0.4 / 0.5
6	6.4	12	1.5	6.1	11.1	1.6	6.2	12.5	0.5 / 0.7
7	7.4	14	1.5	7.1	12.1	1.6	7.2	14	0.5 / 0.8
8	8.4	17	2	8.2	14.2	2	8.2	16	0.6 / 0.9
10	10.5	21	2.5	10.2	17.2	2.2	10.2	20	0.8 / 1.1
12	13	24	2.5	12.3	20.2	2.5	12.2	25	0.9 / 1.5
14	15	28	2.5	14.2	23.2	3	14.2	28	1 / 1.5
16	17	30	3	16.2	26.2	3.5	16.3	31.5	1.2 / 1.7
18	19	34	3	18.2	28.2	3.5	18.3	35.5	1.2 / 2
20	21	36	3	20.2	32.2	4	20.4	40	1.5 / 2.25
22	23	39	4	22.5	34.5	4	22.4	45	1.75 / 2.5
24	25	44	4	24.5	38.5	5			
27	28	50	4	27.5	41.5	5			
30	31	56	4	30.5	46.5	6			

MILLIMETER SIZES

TABLE 12 (CONT'D) ■ *Common washer sizes.*

U.S. Customary (Inches)					Metric (Millimeters)					
Diameter of Shaft	Square Key		Flat Key		Diameter of Shaft (mm)		Square Key		Flat Key	
	Nominal Size		Nominal Size				Nominal Size		Nominal Size	
Inclusive	W	H	W	H	Over	Up To	W	H	W	H
.500– .562	.125	.125	.125	.094	12	17	5	5		
.625– .875	.188	.188	.188	.125	17	22	6	6		
.938–1.250	.250	.250	.250	.188	22	30	7	7	8	7
1.312–1.375	.312	.312	.312	.250	30	38	8	8	10	8
1.438–1.750	.375	.375	.375	.250	38	44	9	9	12	8
1.812–2.250	.500	.500	.500	.375	44	50	10	10	14	9
2.312–2.750	.625	.625	.625	.438	50	58	12	12	16	10

TABLE 13 ■ *Square and flat stock keys.*

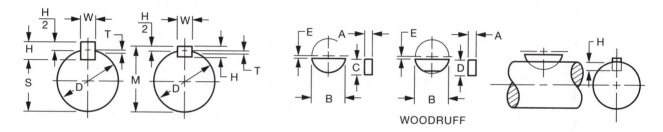

WOODRUFF

C = ALLOWANCE FOR PARALLEL KEYS = .005 IN. OR 0.12 MM

$$S = D - \frac{H}{2} - T = \frac{D - H + \sqrt{D^2 - W^2}}{2} \qquad T = \frac{D - \sqrt{D^2 - W^2}}{2} \qquad \text{W = NORMAL KEY WIDTH (INCH OR MILLIMETERS)}$$

$$M = D - T + \frac{H}{2} + C = \frac{D + H + \sqrt{D^2 - W^2} + C}{2}$$

Key No.	Nominal (A x B)		U.S. Customary (Inches)				Metric (Millimeters)			
			Key			Keyseat	Key			Key Seat
	Millimeters	Inches	E	C	D	H	E	C	D	H
204	1.6 x 6.4	0.062 x 0.250	.05	.20	.19	.10	0.5	2.8	2.8	4.3
304	2.4 x 12.7	0.094 x 0.500	.05	.20	.19	.15	1.3	5.1	4.8	3.8
305	2.4 x 15.9	0.094 x 0.625	.06	.25	.24	.20	1.5	6.4	6.1	5.1
404	3.2 x 12.7	0.125 x 0.500	.05	.20	.19	.14	1.3	5.1	4.8	3.6
405	3.2 x 15.9	0.125 x 0.625	.06	.25	.24	.18	1.5	6.4	6.1	4.6
406	3.2 x 19.1	0.125 x 0.750	.06	.31	.30	.25	1.5	7.9	7.6	6.4
505	4.0 x 15.9	0.156 x 0.625	.06	.25	.24	.17	1.5	6.4	6.1	4.3
506	4.0 x 19.1	0.156 x 0.750	.06	.31	.30	.23	1.5	7.9	7.6	5.8
507	4.0 x 22.2	0.156 x 0.875	.06	.38	.36	.29	1.5	9.7	9.1	7.4
606	4.8 x 19.1	0.188 x 0.750	.06	.31	.30	.21	1.5	7.9	7.6	5.3
607	4.8 x 22.2	0.188 x 0.875	.06	.38	.36	.28	1.5	9.7	9.1	7.1
608	4.8 x 25.4	0.188 x 1.000	.06	.44	.43	.34	1.5	11.2	10.9	8.6
609	4.8 x 28.6	0.188 x 1.250	.08	.48	.47	.39	2.0	12.2	11.9	9.9
807	6.4 x 22.2	0.250 x 0.875	.06	.38	.36	.25	1.5	9.7	9.1	6.4
808	6.4 x 25.4	0.250 x 1.000	.06	.44	.43	.31	1.5	11.2	10.9	7.9

TABLE 14 ■ *Woodruff keys.*

| Nominal Pipe Size | U.S. Customary (Inches) | | | | | Metric | | | |
| | Outside Diameter | Wall Thickness | | | Outside Diameter (Millimeters) | Wall Thickness (Millimeters) | | |
		Schedule 40 Pipe *	Schedule 80 Pipe **	Schedule 160 Pipe		Standard	Extra Strong	Double Extra Strong
.125 (1/8)	.405	.068	.095	—	10.29	1.75	2.44	—
.250 (1/4)	.540	.088	.119	—	13.72	2.29	3.15	—
.375 (3/8)	.675	.091	.126	—	17.15	2.36	3.28	—
.500 (1/2)	.840	.109	.147	.188	21.34	2.82	3.84	7.80
.750 (3/4)	1.050	.113	.154	.219	26.67	2.92	3.99	8.08
1.00	1.315	.133	.179	.250	33.4	3.45	4.65	9.37
1.25	1.660	.140	.191	.250	42.4	3.63	4.98	9.98
1.50	1.900	.145	.200	.281	48.3	3.76	5.18	10.44
2.00	2.375	.154	.218	.344	60.3	3.99	5.66	11.35
2.50	2.875	.203	.276	.375	73.0	5.26	7.16	14.40
3.00	3.500	.216	.300	.438	88.9	5.61	7.77	15.62
3.50	4.000	.226	.318	—	101.6	5.87	8.25	16.54
4.00	4.500	.237	.337	.531	114.3	6.15	8.74	17.53
5.00	5.563	.258	.375	.625	141.3	6.68	9.73	19.51
6.00	6.625	.280	.432	.719	168.3	7.26	11.20	22.45
8.00	8.625	.322	.500	.096	219.1	8.36	12.98	22.73
10.00	10.750	.365	.594	1.125	273.1	9.45	12.95	—
12.00	12.750	.406	.688	1.312	323.9	—	12.95	—
14.00	14.000	.438	.750	1.406	355.6	9.73	12.95	—
16.00	16.000	.500	.844	1.594	406.4	9.73	12.95	—

*Standard Pipe
**Extra Strong Pipe

Nominal pipe sizes are specified in inches.
Outside diameter and wall thicknesses are specified in millimeters.

TABLE 15 ■ American standard wrought iron pipe.

North American Gages								European Gages			
Ferrous Metals, Such as Galvanized Steel, Tin Plate				Galvanized Steel, Tin Plate, Copper, Strip Steel and Steel, Copper, and Aluminum Tubes				Nonferrous Metals, Such as Copper, Brass, Aluminum		Nonferrous	
U.S. Standard (USS)		U.S. Standard (Revised)		Birmingham (BWG)		New Birmingham (BG)		Browne And Sharpe (B & S)		Imperial Standard (SWG)	
Gage	In.	Gage	In.	Gage	In.	Gage	In.	Gage	In.	Gage	In.
		3	.239					3	.229		
4	.234	4	.224	4	.238	4	.250	4	.204	4	.232
5	.219	5	.209	5	.220	5	.223	5	.182	5	.212
6	.203	6	.194	6	.203	6	.198	6	.162	6	.192
7	.188	7	.179	7	.180	7	.176	7	.144	7	.176
8	.172	8	.164	8	.165	8	.157	8	.129	8	.160
9	.156	9	.149	9	.148	9	.140	9	.114	9	.144
10	.141	10	.135	10	.134	10	.125	10	.102	10	.128
11	.125	11	.120	11	.120	11	.111	11	.091	11	.116
12	.109	12	.105	12	.109	12	.099	12	.081	12	.104
13	.094	13	.090	13	.095	13	.088	13	.072	13	.092
14	.078	14	.075	14	.083	14	.079	14	.064	14	.080
15	.070	15	.067	15	.072	15	.070	15	.057	15	.072
16	.063	16	.060	16	.065	16	.063	16	.051	16	.064
17	.056	17	.054	17	.058	17	.056	17	.045	17	.056
18	.050	18	.048	18	.049	18	.050	18	.040	18	.048
19	.044	19	.042	19	.042	19	.044	19	.036	19	.040
20	.038	20	.036	20	.035	20	.039	20	.032	20	.036
21	.034	21	.033	21	.032	21	.035	21	.029	21	.032
22	.031	22	.030	22	.028	22	.031	22	.025	22	.028
23	.028	23	.027	23	.025	23	.028	23	.023	23	.024
24	.025	24	.024	24	.022	24	.025	24	.020	24	.022
25	.022	25	.021	25	.020	25	.022	25	.018	25	.020
26	.019	26	.018	26	.018	26	.020	26	.016	26	.018
27	.017	27	.016	27	.016	27	.017	27	.014	27	.016
28	.016	28	.015	28	.014	28	.016	28	.013	28	.015
29	.014	29	.014	29	.013	29	.014	29	.011	29	.014
30	.012	30	.012	30	.012	30	.012	30	.010	30	.012
31	.011	31	.011	31	.010	31	.011	31	.009		
32	.010	32	.010	32	.009			32	.008	32	.011
33	.009	33	.009	33	.008	33	.009	33	.007	33	.010
34	.008	34	.008	34	.007	34	.008	34	.006	34	.009
				35	.005	35	.007			35	.008
36	.007	36	.007	36	.004	36	.006	36	.005		

Table 16 Sheet Metal Gages and Thicknesses

TABLE 16 ■ *Sheet metal gages and thicknesses.*

**EXAMPLE: RC2 SLIDING FIT FOR A
Ø1.50 NOMINAL HOLE DIAMETER**

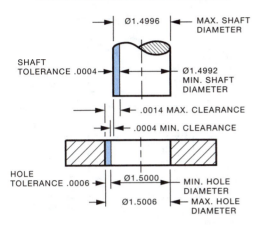

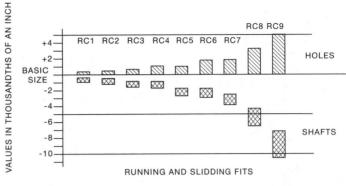

RUNNING AND SLIDING FITS

Nominal Size Range Inches		Class RC1 Precision Sliding			Class RC2 Sliding Fit			Class RC3 Precision Running			Class RC4 Close Running			Class RC5 Medium Running		
		Hole Tol. GR5	Minimum Clearance	Shaft Tol. GR4	Hole Tol. GR6	Minimum Clearance	Shaft Tol. GR5	Hole Tol. GR7	Minimum Clearance	Shaft Tol. GR6	Hole Tol. GR8	Minimum Clearance	Shaft Tol. GR7	Hole Tol. GR8	Minimum Clearance	Shaft Tol. GR7
Over	To	-0		+0	-0		+0	-0		+0	-0		+0	-0		+0
0	.12	+0.15	0.10	-0.12	+0.25	0.10	-0.15	+0.40	0.30	-0.25	+0.60	0.30	-0.40	+0.60	0.60	-0.40
.12	.24	+0.20	0.15	-0.15	+0.30	0.15	-0.20	+0.50	0.40	-0.30	+0.70	0.40	-0.50	+0.70	0.80	-0.50
.24	.40	+0.25	0.20	-0.15	+0.40	0.20	-0.25	+0.60	0.50	-0.40	+0.90	0.50	-0.60	+0.90	1.00	-0.60
.40	.71	+0.30	0.25	-0.20	+0.40	0.25	-0.30	+0.70	0.60	-0.40	+1.00	0.60	-0.70	+1.00	1.20	-0.70
.71	1.19	+0.40	0.30	-0.25	+0.50	0.30	-0.40	+0.80	0.80	-0.50	+1.20	0.80	-0.80	+1.20	1.60	-0.50
1.19	1.97	+0.40	0.40	-0.30	+0.60	0.40	-0.40	+1.00	1.00	-0.60	+1.60	1.00	-1.00	+1.60	2.00	-1.00
1.97	3.15	+0.50	0.40	-0.30	+0.70	0.40	-0.50	+1.20	1.20	-0.70	+1.80	1.20	-1.20	+1.80	2.50	-1.20
3.15	4.73	+0.60	0.50	-0.40	+0.90	0.50	-0.60	+1.40	1.40	-0.90	+2.20	1.40	-1.40	+2.20	3.00	-1.40
4.73	7.09	+0.70	0.60	-0.50	+1.00	0.60	-0.70	+1.60	1.60	-1.00	+2.50	1.60	-1.60	+2.50	3.50	-1.60
7.09	9.85	+0.80	0.60	-0.60	+1.20	0.60	-0.80	+1.80	2.00	-1.20	+2.80	2.00	-1.80	+2.80	4.50	-1.80
9.85	12.41	+0.90	0.80	-0.60	+1.20	0.80	-0.90	+2.00	2.50	-1.20	+3.00	2.50	-2.00	+3.00	5.00	-2.00
12.41	15.75	+1.00	1.00	-0.70	+1.40	1.00	-1.00	+2.20	3.00	-1.40	+3.50	3.00	-2.20	+3.50	6.00	-2.20

Nominal Size Range Inches		Class RC6 Medium Running			Class RC7 Free Running			Class RC8 Loose Running			Class RC9 Loose Running		
		Hole Tol. GR9	Minimum Clearance	Shaft Tol. GR8	Hole Tol. GR9	Minimum Clearance	Shaft Tol. GR8	Hole Tol. GR10	Minimum Clearance	Shaft Tol. GR9	Hole Tol. GR11	Minimum Clearance	Shaft Tol. GR10
Over	To	-0		+0	-0		+0	-0		+0	-0		+0
0	.12	+1.00	0.60	-0.60	+1.00	1.00	-0.60	+1.60	2.50	-1.00	+2.50	4.00	-1.60
.12	.24	+1.20	0.80	-0.70	+1.20	1.20	-0.70	+1.80	2.80	-1.20	+3.00	4.50	-1.80
.24	.40	+1.40	1.00	-0.90	+1.40	1.60	-0.90	+2.20	3.00	-1.40	+3.50	6.00	-2.20
.40	.71	+1.60	1.20	-1.00	+1.60	2.00	-1.00	+2.80	3.50	-1.60	+4.00	6.00	-2.80
.71	1.19	+2.00	1.60	-1.20	+2.00	2.50	-1.20	+3.50	4.50	-2.00	+5.00	7.00	-3.50
1.19	1.97	+2.50	2.00	-1.60	+2.50	3.00	-1.60	+4.00	5.00	-2.50	+6.00	8.00	-4.00
1.97	3.15	+3.00	2.50	-1.80	+3.00	4.00	-1.80	+4.50	6.00	-3.00	+7.00	9.00	-4.50
3.15	4.73	+3.50	3.00	-2.20	+3.50	5.00	-2.20	+5.00	7.00	-3.50	+9.00	10.00	-5.00
4.73	7.09	+4.00	3.50	-2.50	+4.00	6.00	-2.50	+6.00	8.00	-4.00	+10.00	12.00	-6.00
7.09	9.85	+4.50	4.00	-2.80	+4.50	7.00	-2.80	+7.00	10.00	-4.50	+12.00	15.00	-7.00
9.85	12.41	+5.00	5.00	-3.00	+5.00	8.00	-3.00	+8.00	12.00	-5.00	+12.00	18.00	-8.00
12.41	15.75	+6.00	6.00	-3.50	+6.00	10.00	-3.50	+9.00	14.00	-6.00	+14.00	22.00	-9.00

TABLE 17 ■ *Running and sliding fits (Values in thousandths of an inch).*

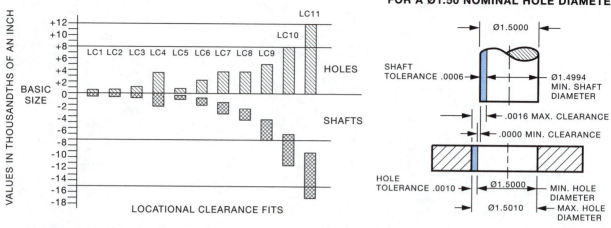

EXAMPLE: LC2 LOCATIONAL FIT
FOR A Ø1.50 NOMINAL HOLE DIAMETER

Nominal Size Range Inches		Class LC1			Class LC2			Class LC3			Class LC4			Class LC5			Class LC6		
		Hole Tol. GR6	Minimum Clearance	Shaft Tol. GR5	Hole Tol. GR8	Minimum Clearance	Shaft Tol. GR7	Hole Tol. GR10	Minimum Clearance	Shaft Tol. GR9	Hole Tol. GR7	Minimum Clearance	Shaft Tol. GR6	Hole Tol. GR9	Minimum Clearance	Shaft Tol. GR8	Hole Tol. GR9	Minimum Clearance	Shaft Tol. GR8
Over	To	-0		+0	-0		+0	-0		+0	-0		+0	-0		+0	-0		+0
0	.12	+0.25	0	-0.15	+0.4	0	-0.25	+0.6	0	-0.4	+1.6	0	-1.0	+0.4	0.10	-0.25	+1.0	0.3	-0.6
.12	.24	+0.30	0	-0.20	+0.5	0	-0.30	+0.7	0	-0.5	+1.8	0	-1.2	+0.5	0.15	-0.30	+1.2	0.4	-0.7
.24	.40	+0.40	0	-0.25	+0.6	0	-0.40	+0.9	0	-0.6	+2.2	0	-1.4	+0.6	0.20	-0.40	+1.4	0.5	-0.9
.40	.71	+0.40	0	-0.30	+0.7	0	-0.40	+1.0	0	-0.7	+2.8	0	-1.6	+0.7	0.25	-0.40	+1.6	0.6	-1.0
.71	1.19	+0.50	0	-0.40	+0.8	0	-0.50	+1.2	0	-0.8	+3.5	0	-2.0	+0.8	0.30	-0.50	+2.0	0.8	-1.2
1.19	1.97	+0.60	0	-0.40	+1.0	0	-0.60	+1.6	0	-1.0	+4.0	0	-2.5	+1.0	0.40	-0.60	+2.5	1.0	-1.6
1.97	3.15	+0.70	0	-0.50	+1.2	0	-0.70	+1.8	0	-1.2	+4.5	0	-3.0	+1.2	0.40	-0.70	+3.0	1.2	-1.8
3.15	4.73	+0.90	0	-0.60	+1.4	0	-0.90	+2.7	0	-1.4	+5.0	0	-3.5	+1.4	0.50	-0.90	+3.5	1.4	-2.2
4.73	7.09	+1.00	0	-0.70	+1.6	0	-1.00	+2.5	0	-1.6	+6.0	0	-4.0	+1.6	0.60	-1.00	+4.0	1.6	-2.5
7.09	9.85	+1.20	0	-0.80	+1.8	0	-1.20	+2.8	0	-1.8	+7.0	0	-4.5	+1.8	0.60	-1.20	+4.5	2.0	-2.8
9.85	12.41	+1.20	0	-0.90	+2.0	0	-1.20	+3.0	0	-2.0	+8.0	0	-5.0	+2.0	0.70	-1.20	+5.0	2.2	-3.0
12.41	15.75	+1.40	0	-1.00	+2.2	0	-1.40	+3.5	0	-2.2	+9.0	0	-6.0	+2.2	0.70	-1.40	+6.0	2.5	-3.5

Nominal Size Range Inches		Class LC7			Class LC8			Class LC9			Class LC10			Class LC11		
		Hole Tol. GR10	Minimum Clearance	Shaft Tol. GR9	Hole Tol. GR10	Minimum Clearance	Shaft Tol. GR9	Hole Tol. GR11	Minimum Clearance	Shaft Tol. GR10	Hole Tol. GR12	Minimum Clearance	Shaft Tol. GR11	Hole Tol. GR13	Minimum Clearance	Shaft Tol. GR12
Over	To	-0		+0	-0		+0	-0		+0	-0		+0	-0		+0
0	.12	+1.6	0.6	-1.0	+1.6	1.0	-1.0	+2.5	2.5	-1.6	+1.0	4.0	-2.5	+6.0	5.0	-4.0
.12	.24	+1.8	0.8	-1.2	+1.8	1.2	-1.2	+3.0	2.8	-1.8	+5.0	4.5	-3.0	+7.0	6.0	-5.0
.24	.40	+2.2	1.0	-1.4	+2.2	1.6	-1.4	+3.5	3.0	-2.2	+6.0	5.0	-3.5	+9.0	7.0	-6.0
.40	.71	+2.8	1.2	-1.6	+2.8	2.0	-1.6	+4.0	3.5	-2.8	+7.0	6.0	-4.0	+10.0	8.0	-7.0
.71	1.19	+3.5	1.6	-2.0	+3.5	2.5	-2.0	+5.0	4.5	-3.5	+8.0	7.0	-5.0	+12.0	10.0	-8.0
1.19	1.97	+4.0	2.0	-2.5	+4.0	3.6	-2.5	+6.0	5.0	-4.0	+10.0	8.0	-6.0	+16.0	12.0	-10.0
1.97	3.15	+4.5	2.5	-3.0	+4.5	4.0	-3.0	+7.0	6.0	-4.5	+12.0	10.0	-7.0	+18.0	14.0	-12.0
3.15	4.73	+5.0	3.0	-3.5	+5.0	5.0	-3.5	+9.0	7.0	-5.0	+14.0	11.0	-9.0	+22.0	16.0	-14.0
4.73	7.09	+6.0	3.5	-4.0	+6.0	6.0	-4.0	+10.0	8.0	-6.0	+16.0	12.0	-10.0	+25.0	18.0	-16.0
7.09	9.85	+7.0	4.0	-4.5	+7.0	7.0	-4.5	+12.0	10.0	-7.0	+18.0	16.0	-12.0	+28.0	22.0	-18.0
9.85	12.41	+8.0	4.5	-5.0	+8.0	7.0	-5.0	+12.0	12.0	-8.0	+20.0	20.0	-12.0	+30.0	28.0	-20.0
12.41	15.75	+9.0	5.0	-6.0	+9.0	8.0	-6.0	+14.0	14.0	-9.0	+22.0	22.0	-14.0	+35.0	30.0	-22.0

TABLE 18 ■ *Locational clearance fits (Values in thousandths of an inch).*

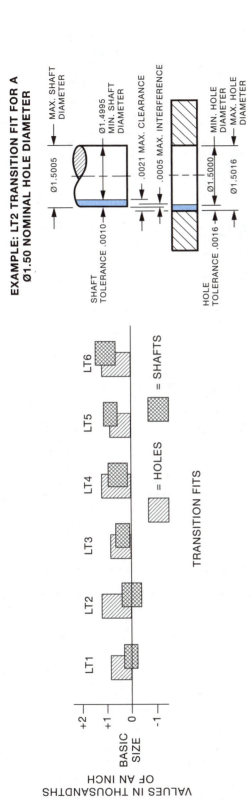

EXAMPLE: LT2 TRANSITION FIT FOR A Ø1.50 NOMINAL HOLE DIAMETER

TRANSITION FITS

= HOLES   = SHAFTS

VALUES IN THOUSANDTHS OF AN INCH

Nominal Size Range Inches		Class LT1			Class LT2			Class LT3			Class LT4			Class LT5			Class LT6		
		Hole Tol. GR7	Maximum Interference	Shaft Tol. GR6	Hole Tol. GR8	Maximum Interference	Shaft Tol. GR7	Hole Tol. GR7	Maximum Interference	Shaft Tol. GR6	Hole Tol. GR8	Maximum Interference	Shaft Tol. GR7	Hole Tol. GR7	Maximum Interference	Shaft Tol. GR6	Hole Tol. GR8	Maximum Interference	Shaft Tol. GR7
Over	To	−0		+0	−0		+0	−0		+0	−0		+0	−0		+0	−0		+0
0	.12	+0.4	0.10	−0.25	+0.6	0.20	−0.4	+0.4	0.25	−0.25	+0.6	0.4	−0.4	+0.4	0.5	−0.25	+0.6	0.65	−0.4
.12	.24	+0.5	0.15	−0.30	+0.7	0.25	−0.5	+0.5	0.40	−0.30	+0.7	0.6	−0.5	+0.5	0.6	−0.30	+0.7	0.80	−0.5
.24	.40	+0.6	0.20	−0.40	+0.9	0.30	−0.6	+0.6	0.50	−0.40	+0.9	0.7	−0.6	+0.6	0.8	−0.40	+0.9	1.00	−0.6
.40	.71	+0.7	0.20	−0.40	+1.0	0.30	−0.7	+0.7	0.50	−0.40	+1.0	0.8	−0.7	+0.7	0.9	−0.40	+1.0	1.20	−0.7
.71	1.19	+0.8	0.25	−0.50	+1.2	0.40	−0.8	+0.8	0.60	−0.50	+1.2	0.9	−0.8	+0.8	1.1	−0.50	+1.2	1.40	−0.8
1.19	1.97	+1.0	0.30	−0.60	+1.6	0.50	−1.0	+1.0	0.70	−0.60	+1.6	1.1	−1.0	+1.0	1.3	−0.60	+1.6	1.70	−1.0
1.97	3.15	+1.2	0.30	−0.70	+1.8	0.60	−1.2	+1.2	0.80	−0.70	+1.8	1.3	−1.2	+1.2	1.5	−0.70	+1.8	2.00	−1.2
3.15	4.73	+1.4	0.40	−0.90	+2.2	0.70	−1.4	+1.4	1.00	−0.90	+2.2	1.5	−1.4	+1.4	1.9	−0.90	+2.2	2.40	−1.4
4.73	7.09	+1.6	0.50	−1.00	+2.5	0.80	−1.6	+1.6	1.10	−1.00	+2.5	1.7	−1.6	+1.6	2.2	−1.00	+2.5	2.80	−1.6
7.09	9.85	+1.8	0.60	−1.20	+2.8	0.90	−1.8	+1.8	1.40	−1.20	+2.8	2.0	−1.8	+1.8	2.6	−1.20	+2.8	3.20	−1.8
9.85	12.41	+2.0	0.60	−1.20	+3.0	1.00	−2.0	+2.0	1.40	−1.20	+3.0	2.2	−2.0	+2.0	2.6	−1.20	+3.0	3.40	−2.0
12.41	15.75	+2.2	0.70	−1.40	+3.5	1.00	−2.2	+2.2	1.60	−1.40	+3.5	2.4	−2.2	+2.2	3.0	−1.40	+3.5	3.80	−2.2

TABLE 19 ■ *Locational transition fits (Values in thousandths of an inch).*

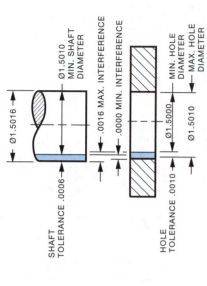

EXAMPLE: LN2 LOCATIONAL INTERFERENCE FIT FOR A Ø1.50 NOMINAL HOLE DIAMETER

Ø1.5016
Ø1.5010 MIN. SHAFT DIAMETER
SHAFT TOLERANCE .0006
.0016 MAX. INTERFERENCE
.0000 MIN. INTERFERENCE
Ø1.5000
Ø1.5010
MIN. HOLE DIAMETER
MAX. HOLE DIAMETER
HOLE TOLERANCE .0010

VALUES IN THOUSANDTHS OF AN INCH
BASIC SIZE
LOCATIONAL INTERFERENCE FITS
SHAFTS — HOLES
LN1 LN2 LN3 LN4 LN5 LN6

Nominal Size Range Inches		Class LN1 Light Press Fit			Class LN2 Medium Press Fit			Class LN3 Heavy Press Fit			Class LN4			Class LN5			Class LN6		
Over	To	Hole Tol. GR6 −0	Maximum Interference	Shaft Tol. GR5 +0	Hole Tol. GR7 −0	Maximum Interference	Shaft Tol. GR6 +0	Hole Tol. GR7 −0	Maximum Interference	Shaft Tol. GR6 +0	Hole Tol. GR8 −0	Maximum Interference	Shaft Tol. GR7 +0	Hole Tol. GR9 −0	Maximum Interference	Shaft Tol. GR8 +0	Hole Tol. GR10 −0	Maximum Interference	Shaft Tol. GR9 +0
0	.12	+0.25	0.40	−0.15	+0.4	0.65	−0.25	+0.4	0.75	−0.25	+0.6	1.2	−0.4	+1.0	1.8	−0.6	+1.6	3.0	−1.0
.12	.24	+0.30	0.50	−0.20	+0.5	0.80	−0.30	+0.5	0.90	−0.30	+0.7	1.5	−0.5	+1.2	2.3	−0.7	+1.8	3.6	−1.2
.24	.40	+0.40	0.65	−0.25	+0.6	1.00	−0.40	+0.6	1.20	−0.40	+0.9	1.8	−0.6	+1.4	2.8	−0.9	+2.2	4.4	−1.4
.40	.71	+0.40	0.70	−0.30	+0.7	1.10	−0.40	+0.7	1.40	−0.40	+1.0	2.2	−0.7	+1.6	3.4	−1.0	+2.8	5.6	−1.6
.71	1.19	+0.50	0.90	−0.40	+0.8	1.30	−0.50	+0.8	1.70	−0.50	+1.2	2.6	−0.8	+2.0	4.2	−1.2	+3.5	7.0	−2.0
1.19	1.97	+0.60	1.00	−0.40	+1.0	1.60	−0.60	+1.0	2.00	−0.60	+1.6	3.4	−1.0	+2.5	5.3	−1.6	+4.0	8.5	−2.5
1.97	3.15	+0.70	1.30	−0.50	+1.2	2.10	−0.70	+1.2	2.30	−0.70	+1.8	4.0	−1.2	+3.0	6.3	−1.8	+4.5	10.0	−3.0
3.15	4.73	+0.90	1.60	−0.60	+1.4	2.50	−0.90	+1.4	2.90	−0.90	+2.2	4.8	−1.4	+4.0	7.7	−2.2	+5.0	11.5	−3.5
4.73	7.09	+1.00	1.90	−0.70	+1.6	2.80	−1.00	+1.6	3.50	−1.00	+2.5	5.6	−1.6	+4.5	8.7	−2.5	+6.0	13.5	−4.0
7.09	9.85	+1.20	2.20	−0.80	+1.8	3.20	−1.20	+1.8	4.20	−1.20	+2.8	6.6	−1.8	+5.0	10.3	−2.8	+7.0	16.5	−4.5
9.85	12.41	+1.20	2.30	−0.90	+2.0	3.40	−1.20	+2.0	4.70	−1.20	+3.0	7.5	−2.0	+6.0	12.0	−3.0	+8.0	19.0	−5.0
12.41	15.75	+1.40	2.60	−1.00	+2.2	3.90	−1.40	+2.2	5.90	−1.40	+3.5	8.7	−2.2	+6.0	14.5	−3.5	+9.0	23.0	−6.0

TABLE 20 ■ *Locational interference fits (Values in thousandths of an inch).*

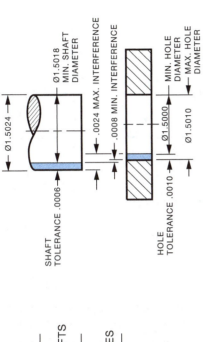

EXAMPLE: FN2 MEDIUM DRIVE FIT FOR A
Ø1.50 NOMINAL HOLE DIAMETER

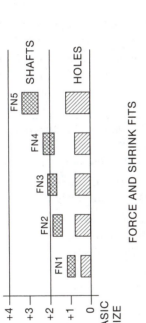

VALUES IN THOUSANDTHS
OF AN INCH

FORCE AND SHRINK FITS

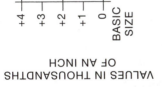

Nominal Size Range Inches		Class FN1 Light Drive Fit			Class FN2 Medium Drive Fit			Class FN3 Heavy Drive Fit			Class FN4 Shrink Fit			Class FN5 Heavy Shrink Fit		
Over	To	Hole Tol. GR6 (-0)	Maximum Interference	Shaft Tol. GR5 (+0)	Hole Tol. GR7 (-0)	Maximum Interference	Shaft Tol. GR6 (+0)	Hole Tol. GR7 (-0)	Maximum Interference	Shaft Tol. GR6 (+0)	Hole Tol. GR7 (-0)	Maximum Interference	Shaft Tol. GR6 (+0)	Hole Tol. GR8 (-0)	Maximum Interference	Shaft Tol. GR7 (+0)
0	.12	+0.25	0.50	-0.15	+0.40	0.85	-0.25				+0.40	0.95	-0.25	+0.60	1.30	-0.40
.12	.24	+0.30	0.60	-0.20	+0.50	1.00	-0.30				+0.50	1.20	-0.30	+0.70	1.70	-0.50
.24	.40	+0.40	0.75	-0.25	+0.60	1.40	-0.40				+0.60	1.60	-0.40	+0.90	2.00	-0.60
.40	.56	+0.40	0.80	-0.30	+0.70	1.60	-0.40				+0.70	1.80	-0.40	+1.00	2.30	-0.70
.56	.71	+0.40	0.90	-0.30	+0.70	1.60	-0.40				+0.70	1.80	-0.40	+1.00	2.50	-0.70
.71	.95	+0.50	1.10	-0.40	+0.80	1.90	-0.50				+0.80	2.10	-0.50	+1.20	3.00	-0.80
.95	1.19	+0.50	1.20	-0.40	+0.80	1.90	-0.50	+0.80	2.10	-0.50	+0.80	2.30	-0.50	+1.20	3.30	-0.80
1.19	1.58	+0.60	1.30	-0.40	+1.00	2.40	-0.60	+1.00	2.60	-0.60	+1.00	3.10	-0.60	+1.60	4.00	-1.00
1.58	1.97	+0.60	1.40	-0.40	+1.00	2.40	-0.60	+1.00	2.80	-0.60	+1.00	3.40	-0.60	+1.60	5.00	-1.00
1.97	2.56	+0.70	1.80	-0.50	+1.20	2.70	-0.70	+1.20	3.20	-0.70	+1.20	4.20	-0.70	+1.80	6.20	-1.20
2.56	3.15	+0.70	1.90	-0.50	+1.20	2.90	-0.70	+1.20	3.70	-0.70	+1.20	4.70	-0.70	+1.80	7.20	-1.20
3.15	3.94	+0.90	2.40	-0.60	+1.40	3.70	-0.90	+1.40	4.40	-0.70	+1.40	5.90	-0.90	+2.20	8.40	-1.40

TABLE 21 ■ *Force and shrink fits (Values in thousandths of an inch).*

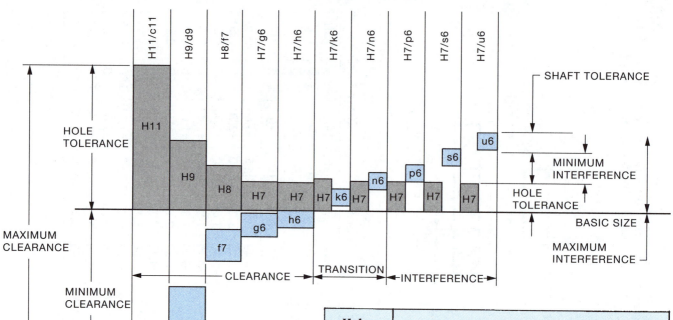

Hole Basis Symbol	Description
H11/c11 (RC9)	*Loose running* fit for wide commercial tolerances or allowances on external members.
H9/d9 (RC7)	*Free running* fit not for use where accuracy is essential, but good for large temperature variations, high running speeds, or heavy journal pressures.
H8/f7 (RC4)	*Close running* fit for running on accurate machines and for accurate location at moderate speeds and journal pressures.
H7/g6 (LC5)	*Sliding* fit not intended to run freely, but to move and turn freely and locate accurately.
H7/h6 (LC2)	*Locational clearance* fit provides snug fit for locating stationary parts; but can be freely assembled and disassembled.
H7/k6 (LT3)	*Locational transition* fit for accurate location, a compromise between clearance and interference.
H7/n6 (LT5)	*Locational transition* fit for more accurate location where greater interference is permissible.
H7/p6 (LN2)	*Locational interference* fit for parts requiring rigidity and alignment with prime accuracy of location but without special bore pressure requirements.
H7/s6 (FN2)	*Medium drive* fit for ordinary steel parts or shrink fits on light sections, the tightest fit usable with cast iron.
H7/u6 (FN4)	*Force* fit suitable for parts which can be highly stressed or for shrink fits where the heavy pressing forces required are impractical.

TABLE 22 ■ *Preferred hole basis metric fits description.*

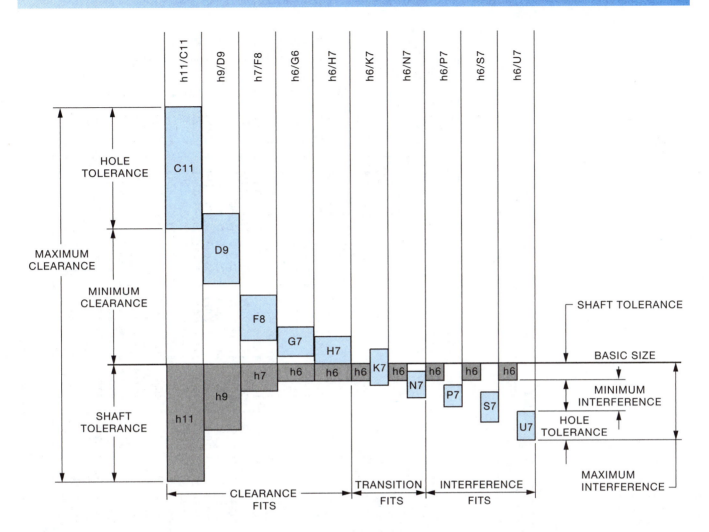

TABLE 23 ■ *Preferred shaft basis metric fits description.*

Shaft Basis Symbol	Description	Shaft Basis Symbol	Description
C11/h11	*Loose running* fit for wide commercial tolerances or allowances on external members.	K7/h6	*Locational transition* fit for accurate location, a compromise between clearance and interference.
D9/h9	*Free running* fit not for use where accuracy is essential, but good for large temperature variations, high running speeds, or heavy journal pressures.	N7/h6	*Locational transition* fit for more accurate location where greater interference is permissible.
F8/h7	*Close running* fit for running on accurate machines and for accurate location at moderate speeds and journal pressures.	P7/h6	*Locational interference* fit for parts requiring rigidity and alignment with prime accuracy of location but without special bore pressure requirements.
G7/h6	*Sliding* fit not intended to run freely, but to move and turn freely and locate accurately.	S7/h6	*Medium drive* fit for ordinary steel parts or shrink fits on light sections, the tightest fit usable with cast iron.
H7/h6	*Locational clearance* fit provides snug fit for locating stationary parts; but can be freely assembled and disassembled.	U7/h6	*Force* fit suitable for parts which can be highly stressed or for shrink fits where the heavy pressing forces required are impractical.

EXAMPLE: RC9 LOOSE RUNNING FIT FOR A Ø20 NOMINAL HOLE DIAMETER

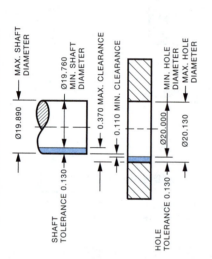

- Ø19.890 MAX. SHAFT DIAMETER
- Ø19.760 MIN. SHAFT DIAMETER
- 0.370 MAX. CLEARANCE
- 0.110 MIN. CLEARANCE
- Ø20.000 MIN. HOLE DIAMETER
- Ø20.130 MAX. HOLE DIAMETER
- SHAFT TOLERANCE 0.130
- HOLE TOLERANCE 0.130

EXAMPLE: LC2 LOCATIONAL CLEARANCE FIT FOR A Ø40 NOMINAL HOLE DIAMETER

- Ø40.000 MAX. SHAFT DIAMETER
- Ø39.984 MIN. SHAFT DIAMETER
- 0.041 MAX. CLEARANCE
- 0.000 MIN. CLEARANCE
- Ø40.000 MIN. HOLE DIAMETER
- Ø40.025 MAX. HOLE DIAMETER
- SHAFT TOLERANCE 0.016
- HOLE TOLERANCE 0.025

Preferred Hole Basis Clearance Fits

Basic Size		Loose Running			Free Running			Close Running			Sliding			Locational Clearance		
		Hole H11	Shaft c11	Fit RC9	Hole H9	Shaft d9	Fit RC7	Hole H8	Shaft f7	Fit RC4	Hole H7	Shaft g6	Fit LC5	Hole H7	Shaft h6	Fit LC2
5	MAX	5.075	4.930	0.220	5.030	4.970	0.090	5.018	4.990	0.040	5.012	4.996	0.024	5.012	5.000	0.020
	MIN	5.000	4.855	0.070	5.000	4.940	0.030	5.000	4.978	0.010	5.000	4.988	0.004	5.000	4.992	0.000
6	MAX	6.075	5.930	0.220	6.030	5.970	0.090	6.018	5.990	0.040	6.012	5.996	0.024	6.012	6.000	0.020
	MIN	6.000	5.855	0.070	6.000	5.940	0.030	6.000	5.978	0.010	6.000	5.988	0.004	6.000	5.992	0.000
8	MAX	8.090	7.920	0.260	8.036	7.960	0.112	8.022	7.987	0.050	8.015	7.995	0.029	8.015	8.000	0.024
	MIN	8.000	7.830	0.080	8.000	7.924	0.040	8.000	7.972	0.013	8.000	7.986	0.006	8.000	7.991	0.000
10	MAX	10.090	9.920	0.260	10.036	9.960	0.112	10.022	9.987	0.050	10.015	9.995	0.029	10.015	10.000	0.024
	MIN	10.000	9.830	0.080	10.000	9.924	0.040	10.000	9.972	0.013	10.000	9.986	0.005	10.000	9.991	0.000
12	MAX	12.110	11.905	0.315	12.043	11.950	0.136	12.027	11.984	0.061	12.018	11.994	0.035	12.018	12.000	0.029
	MIN	12.000	11.795	0.095	12.000	11.907	0.050	12.000	11.966	0.016	12.000	11.983	0.006	12.000	11.989	0.000
16	MAX	16.110	15.905	0.315	16.043	15.950	0.136	16.027	15.984	0.061	16.018	15.994	0.035	16.018	16.000	0.029
	MIN	16.000	15.795	0.095	16.000	15.907	0.050	16.000	15.966	0.016	16.000	15.983	0.006	16.000	15.989	0.000
20	MAX	20.130	19.890	0.370	20.052	19.935	0.169	20.033	19.980	0.074	20.021	19.993	0.041	20.021	20.000	0.034
	MIN	20.000	19.760	0.110	20.000	19.883	0.065	20.000	19.959	0.020	20.000	19.980	0.007	20.000	19.987	0.000
25	MAX	25.130	24.890	0.370	25.052	24.935	0.169	25.033	24.980	0.074	25.021	24.993	0.041	25.021	25.000	0.034
	MIN	25.000	24.760	0.110	25.000	24.883	0.065	25.000	24.959	0.020	25.000	24.980	0.007	25.000	24.987	0.000
30	MAX	30.130	29.890	0.370	30.052	29.935	0.169	30.033	29.980	0.074	30.021	29.993	0.041	30.021	30.000	0.034
	MIN	30.000	29.760	0.110	30.000	29.883	0.065	30.000	29.959	0.020	30.000	29.980	0.007	30.000	29.987	0.000
40	MAX	40.160	39.880	0.440	40.062	39.920	0.204	40.039	39.975	0.089	40.025	39.991	0.050	40.025	40.000	0.041
	MIN	40.000	39.720	0.120	40.000	39.858	0.080	40.000	39.950	0.025	40.000	39.975	0.009	40.000	39.984	0.000
50	MAX	50.160	49.870	0.450	50.062	49.920	0.204	50.039	49.975	0.089	50.025	49.991	0.050	50.025	50.000	0.041
	MIN	50.000	49.710	0.130	50.000	49.858	0.080	50.000	49.950	0.025	50.000	49.975	0.009	50.000	49.984	0.000
60	MAX	60.190	59.860	0.520	60.074	59.900	0.248	60.046	59.970	0.106	60.030	59.990	0.059	60.030	60.000	0.049
	MIN	60.000	59.670	0.140	60.000	59.826	0.100	60.000	59.940	0.030	60.000	59.971	0.010	60.000	59.981	0.000
80	MAX	80.190	79.850	0.530	80.074	79.900	0.248	80.046	79.970	0.106	80.030	79.990	0.059	80.030	80.000	0.049
	MIN	80.000	79.660	0.150	80.000	79.826	0.100	80.000	79.940	0.030	80.000	79.971	0.010	80.000	79.981	0.000
100	MAX	100.220	99.830	0.610	100.087	99.880	0.294	100.054	99.964	0.125	100.035	99.988	0.069	100.035	100.000	0.057
	MIN	100.000	99.610	0.170	100.000	99.793	0.120	100.000	99.929	0.036	100.000	99.966	0.012	100.000	99.978	0.000

TABLE 24 ■ *Preferred hole basis metric fits (Dimensions in millimeters).*

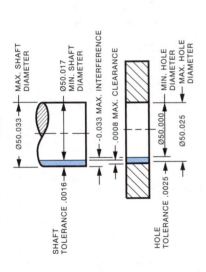

EXAMPLE: FN4 FORCE FIT FOR A Ø30 NOMINAL HOLE DIAMETER

EXAMPLE: LT5 LOCATIONAL TRANSITION FIT FOR A Ø50 NOMINAL HOLE DIAMETER

Preferred Hole Basis Transition and Interference Fits

Basic Size	Locational Transn. Hole H7	Locational Transn. Shaft k6	Locational Transn. Fit LT3	Locational Transn. Hole H7	Locational Transn. Shaft n6	Locational Transn. Fit LT5	Locational Interf. Hole H7	Locational Interf. Shaft p6	Locational Interf. Fit LN2	Medium Drive Hole H7	Medium Drive Shaft s6	Medium Drive Fit FN2	Force Hole H7	Force Shaft u6	Force Fit FN4
5 MAX	5.012	5.009	0.011	5.012	5.016	0.004	5.012	5.020	0.000	5.012	5.027	−0.007	5.012	5.031	−0.011
5 MIN	5.000	5.001	−0.009	5.000	5.008	−0.016	5.000	5.012	−0.020	5.000	5.019	−0.027	5.000	5.023	−0.031
6 MAX	6.012	6.009	0.011	6.012	6.016	0.004	6.012	6.020	0.000	6.012	6.027	−0.007	6.012	6.031	−0.011
6 MIN	6.000	6.001	−0.009	6.000	6.008	−0.016	6.000	6.012	−0.020	6.000	6.019	−0.027	6.000	6.023	−0.031
8 MAX	8.015	8.010	0.014	8.015	8.019	0.005	8.015	8.024	0.000	8.015	8.032	−0.008	8.015	8.037	−0.013
8 MIN	8.000	8.001	−0.010	8.000	8.010	−0.019	8.000	8.015	−0.024	8.000	8.023	−0.032	8.000	8.028	−0.037
10 MAX	10.015	10.010	0.014	10.015	10.019	0.005	10.015	10.024	0.000	10.015	10.032	−0.008	10.015	10.037	−0.013
10 MIN	10.000	10.001	−0.010	10.000	10.010	−0.019	10.000	10.015	−0.024	10.000	10.023	−0.032	10.000	10.028	−0.037
12 MAX	12.018	12.012	0.017	12.018	12.023	0.006	12.018	12.029	0.000	12.018	12.039	−0.010	12.018	12.044	−0.015
12 MIN	12.000	12.001	−0.012	12.000	12.012	−0.023	12.000	12.018	−0.029	12.000	12.028	−0.039	12.000	12.033	−0.044
16 MAX	16.018	16.012	0.017	16.018	16.023	0.006	16.018	16.029	0.000	16.018	16.039	−0.010	16.018	16.044	−0.015
16 MIN	16.000	16.001	−0.012	16.000	16.012	−0.023	16.000	16.018	−0.029	16.000	16.028	−0.039	16.000	16.033	−0.044
20 MAX	20.021	20.015	0.019	20.021	20.028	0.006	20.021	20.035	−0.001	20.021	20.048	−0.014	20.021	20.054	−0.020
20 MIN	20.000	20.002	−0.015	20.000	20.015	−0.028	20.000	20.022	−0.035	20.000	20.035	−0.048	20.000	20.041	−0.054
25 MAX	25.021	25.014	0.019	25.021	25.028	0.006	25.021	25.035	−0.001	25.021	25.048	−0.014	25.021	25.061	−0.027
25 MIN	25.000	25.002	−0.015	25.000	25.015	−0.028	25.000	25.022	−0.035	25.000	25.035	−0.048	25.000	25.048	−0.061
30 MAX	30.021	30.015	0.019	30.021	30.028	0.006	30.021	30.035	−0.001	30.021	30.048	−0.014	30.021	30.061	−0.027
30 MIN	30.000	30.002	−0.015	30.000	30.015	−0.028	30.000	30.022	−0.035	30.000	30.035	−0.048	30.000	30.048	−0.061
40 MAX	40.025	40.018	0.023	40.025	40.033	0.008	40.025	40.042	−0.001	40.025	40.059	−0.018	40.025	40.076	−0.035
40 MIN	40.000	40.002	−0.018	40.000	40.017	−0.033	40.000	40.026	−0.042	40.000	40.043	−0.059	40.000	40.060	−0.076
50 MAX	50.025	50.018	0.023	50.025	50.033	0.008	50.025	50.042	−0.001	50.025	50.059	−0.018	50.025	50.086	−0.045
50 MIN	50.000	50.002	−0.018	50.000	50.017	−0.033	50.000	50.026	−0.042	50.000	50.043	−0.059	50.000	50.070	−0.086
60 MAX	60.030	60.021	0.028	60.030	60.039	0.010	60.030	60.051	−0.002	60.030	60.072	−0.023	60.030	60.106	−0.057
60 MIN	60.000	60.002	−0.021	60.000	60.020	−0.039	60.000	60.032	−0.051	60.000	60.053	−0.072	60.000	60.087	−0.106
80 MAX	80.030	80.021	0.028	80.030	80.039	0.010	80.030	80.051	−0.002	80.030	80.078	−0.029	80.030	80.121	−0.072
80 MIN	80.000	80.002	−0.021	80.000	80.020	−0.039	80.000	80.032	−0.051	80.000	80.059	−0.078	80.000	80.102	−0.121
100 MAX	100.035	100.025	0.032	100.035	100.045	0.012	100.035	100.059	−0.002	100.035	100.093	−0.036	100.035	100.146	−0.089
100 MIN	100.000	100.003	−0.025	100.000	100.023	−0.045	100.000	100.037	−0.059	100.000	100.071	−0.093	100.000	100.124	−0.146

TABLE 24 (CONT'D) ■ Preferred hole basis metric fits (Dimensions in millimeters).

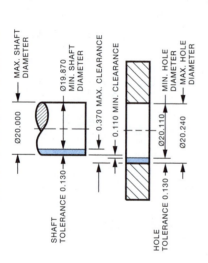

EXAMPLE: RC9 LOOSE RUNNING FIT FOR A Ø20 NOMINAL SHAFT DIAMETER

Ø20.000 — MAX. SHAFT DIAMETER
Ø19.870 MIN. SHAFT DIAMETER
SHAFT TOLERANCE 0.130

0.370 MAX. CLEARANCE
0.110 MIN. CLEARANCE

Ø20.110 MIN. HOLE DIAMETER
Ø20.240 MAX. HOLE DIAMETER
HOLE TOLERANCE 0.130

EXAMPLE: LC2 LOCATIONAL CLEARANCE FIT FOR A Ø40 NOMINAL SHAFT DIAMETER

Ø40.000 — MAX. SHAFT DIAMETER
Ø39.984 MIN. SHAFT DIAMETER
SHAFT TOLERANCE 0.016

0.041 MAX. CLEARANCE
0.000 MIN. CLEARANCE

Ø40.000 MIN. HOLE DIAMETER
Ø40.025 MAX. HOLE DIAMETER
HOLE TOLERANCE 0.025

Preferred Shaft Basis Transition and Interference Fits

Basic Size		Locational Transn. Hole K7	Shaft h6	Fit LT3	Locational Transn. Hole N7	Shaft h6	Fit LT5	Locational Interf. Hole P7	Shaft h6	Fit LN2	Medium Drive Hole S7	Shaft h6	Fit FN2	Force Hole U7	Shaft h6	Fit FN4
5	MAX	5.003	5.000	0.011	4.996	5.000	0.004	4.992	5.000	0.000	4.985	5.000	-0.007	4.981	5.000	-0.011
5	MIN	4.991	4.992	-0.009	4.984	4.992	-0.016	4.980	4.992	-0.020	4.973	4.992	-0.027	4.969	4.992	-0.031
6	MAX	6.003	6.000	0.011	5.996	6.000	0.004	5.992	6.000	0.000	5.985	6.000	-0.007	5.981	6.000	-0.011
6	MIN	5.991	5.992	-0.009	5.984	5.992	-0.016	5.980	5.992	-0.020	5.973	5.992	-0.027	5.969	5.992	-0.031
8	MAX	8.005	8.000	0.014	7.996	8.000	0.005	7.991	8.000	0.000	7.983	8.000	-0.008	7.978	8.000	-0.013
8	MIN	7.990	7.991	-0.010	7.981	7.991	-0.019	7.976	7.991	-0.024	7.968	7.991	-0.032	7.963	7.991	-0.037
10	MAX	10.005	10.000	0.014	9.996	10.000	0.005	9.991	10.000	0.000	9.983	10.000	-0.008	9.978	10.000	-0.013
10	MIN	9.990	9.991	-0.010	9.981	9.991	-0.019	9.976	9.991	-0.024	9.968	9.991	-0.032	9.963	9.991	-0.037
12	MAX	12.006	12.000	0.017	11.995	12.000	0.006	11.989	12.000	0.000	11.979	12.000	-0.010	11.974	12.000	-0.015
12	MIN	11.988	11.989	-0.012	11.977	11.989	-0.023	11.971	11.989	-0.029	11.961	11.989	-0.039	11.956	11.989	-0.044
16	MAX	16.006	16.000	0.017	15.995	16.000	0.006	15.989	16.000	0.000	15.979	16.000	-0.010	15.974	16.000	-0.015
16	MIN	15.988	15.989	-0.012	15.977	15.989	-0.023	15.971	15.989	-0.029	15.961	15.989	-0.039	15.956	15.989	-0.044
20	MAX	20.006	20.000	0.019	19.993	20.000	0.006	19.986	20.000	-0.001	19.973	20.000	-0.014	19.967	20.000	-0.020
20	MIN	19.985	19.987	-0.015	19.972	19.987	-0.028	19.965	19.987	-0.035	19.952	19.987	-0.048	19.946	19.987	-0.054
25	MAX	25.006	25.000	0.019	24.993	25.000	0.006	24.986	25.000	-0.001	24.973	25.000	-0.014	24.960	25.000	-0.027
25	MIN	24.985	24.987	-0.015	24.972	24.987	-0.028	24.965	24.987	-0.035	24.952	24.987	-0.048	24.939	24.987	-0.061
30	MAX	30.006	30.000	0.019	29.993	30.000	0.006	29.986	30.000	-0.001	29.973	30.000	-0.014	29.960	30.000	-0.027
30	MIN	29.985	29.987	-0.015	29.972	29.987	-0.028	29.965	29.987	-0.035	29.952	29.987	-0.048	29.939	29.987	-0.061
40	MAX	40.007	40.000	0.023	39.992	40.000	0.008	39.983	40.000	-0.001	39.966	40.000	-0.018	39.949	40.000	-0.035
40	MIN	39.982	39.984	-0.018	39.967	39.984	-0.033	39.958	39.984	-0.042	39.941	39.984	-0.059	39.924	39.984	-0.076
50	MAX	50.007	50.000	0.023	49.992	50.000	0.008	49.983	50.000	-0.001	49.966	50.000	-0.018	49.939	50.000	-0.045
50	MIN	49.982	49.984	-0.018	49.967	49.984	-0.033	49.958	49.984	-0.042	49.941	49.984	-0.059	49.914	49.984	-0.086
60	MAX	60.009	60.000	0.028	59.991	60.000	0.010	59.979	60.000	-0.002	59.958	60.000	-0.023	59.924	60.000	-0.057
60	MIN	59.979	59.981	-0.021	59.961	59.981	-0.039	59.949	59.981	-0.051	59.928	59.981	-0.072	59.894	59.981	-0.106
80	MAX	80.009	80.000	0.028	79.991	80.000	0.010	79.979	80.000	-0.002	79.952	80.000	-0.029	79.909	80.000	-0.072
80	MIN	79.979	79.981	-0.021	79.961	79.981	-0.039	79.949	79.981	-0.051	79.922	79.981	-0.078	79.879	79.981	-0.121
100	MAX	100.010	100.000	0.032	99.990	100.000	0.012	99.976	100.000	-0.002	99.942	100.000	-0.036	99.889	100.000	-0.089
100	MIN	99.975	99.978	-0.025	99.955	99.978	-0.045	99.941	99.978	-0.059	99.907	99.978	-0.093	99.854	99.978	-0.146

TABLE 25 ■ *Preferred shaft basis metric fits (Values in millimeters).*

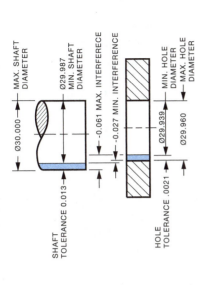

EXAMPLE: FN4 FORCE FIT FOR A Ø30 NOMINAL SHAFT DIAMETER

MAX. SHAFT DIAMETER — Ø30.000
Ø29.987 MIN. SHAFT DIAMETER
SHAFT TOLERANCE 0.013
-0.061 MAX. INTERFERECE
-0.027 MIN. INTERFERENCE
Ø29.939 MIN. HOLE DIAMETER
Ø29.960 MAX. HOLE DIAMETER
HOLE TOLERANCE .0021

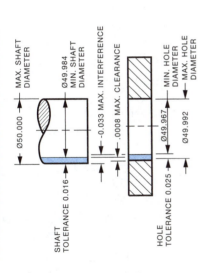

EXAMPLE: LT5 LOCATIONAL TRANSITION FIT FOR A Ø50 NOMINAL SHAFT DIAMETER

MAX. SHAFT DIAMETER — Ø50.000
Ø49.984 MIN. SHAFT DIAMETER
SHAFT TOLERANCE 0.016
-0.033 MAX. INTERFERENCE
.0008 MAX. CLEARANCE
Ø49.967 MIN. HOLE DIAMETER
Ø49.992 MAX. HOLE DIAMETER
HOLE TOLERANCE 0.025

Preferred Shaft Basis Clearance Fits

Basic Size		Loose Running			Free Running			Close Running			Sliding			Locational Clearance		
		Hole C11	Shaft h11	Fit RC9	Hole D9	Shaft h9	Fit RC7	Hole F8	Shaft h7	Fit RC4	Hole G7	Shaft h6	Fit LC5	Hole H7	Shaft h6	Fit LC2
5	MAX	5.145	5.000	0.220	5.060	5.000	0.090	5.028	5.000	0.040	5.016	5.000	0.024	5.012	5.000	0.020
	MIN	5.070	4.925	0.070	5.030	4.970	0.030	5.010	4.988	0.010	5.004	4.992	0.004	5.000	4.992	0.000
6	MAX	6.145	6.000	0.220	6.060	6.000	0.090	6.028	6.000	0.040	6.016	6.000	0.024	6.012	6.000	0.020
	MIN	6.070	5.925	0.070	6.030	5.970	0.030	6.010	5.988	0.010	6.004	5.992	0.004	6.000	5.992	0.000
8	MAX	8.170	8.000	0.260	8.076	8.000	0.112	8.035	8.000	0.050	8.020	8.000	0.029	8.015	8.000	0.024
	MIN	8.080	7.910	0.080	8.040	7.964	0.040	8.013	7.985	0.013	8.005	7.991	0.005	8.000	7.991	0.000
10	MAX	10.170	10.000	0.260	10.076	10.000	0.112	10.035	10.000	0.050	10.020	10.000	0.029	10.015	10.000	0.024
	MIN	10.080	9.910	0.080	10.040	9.964	0.040	10.013	9.985	0.013	10.005	9.991	0.005	10.000	9.991	0.000
12	MAX	12.205	12.000	0.315	12.093	12.000	0.136	12.043	12.000	0.061	12.024	12.000	0.035	12.018	12.000	0.029
	MIN	12.095	11.890	0.095	12.050	11.957	0.050	12.016	11.982	0.016	12.006	11.989	0.006	12.000	11.989	0.000
16	MAX	16.205	16.000	0.315	16.093	16.000	0.136	16.043	16.000	0.061	16.024	16.000	0.035	16.018	16.000	0.029
	MIN	16.095	15.890	0.095	16.050	15.957	0.050	16.016	15.982	0.016	16.006	15.989	0.006	16.000	15.989	0.000
20	MAX	20.240	20.000	0.370	20.117	20.000	0.169	20.053	20.000	0.074	20.028	20.000	0.041	20.021	20.000	0.034
	MIN	20.110	19.870	0.110	20.065	19.948	0.065	20.020	19.979	0.020	20.007	19.987	0.007	20.000	19.987	0.000
25	MAX	25.240	25.000	0.370	25.117	25.000	0.169	25.053	25.000	0.074	25.028	25.000	0.041	25.021	25.000	0.034
	MIN	25.110	24.870	0.110	25.065	24.948	0.065	25.020	24.979	0.020	25.007	24.987	0.007	25.000	24.987	0.000
30	MAX	30.240	30.000	0.370	30.117	30.000	0.169	30.053	30.000	0.074	30.028	30.000	0.041	30.021	30.000	0.034
	MIN	30.110	29.870	0.110	30.065	29.948	0.065	30.020	29.979	0.020	30.007	29.987	0.007	30.000	29.987	0.000
40	MAX	40.280	40.000	0.440	40.142	40.000	0.204	40.064	40.000	0.089	40.034	40.000	0.050	40.025	40.000	0.041
	MIN	40.120	39.840	0.120	40.080	39.938	0.080	40.025	39.975	0.025	40.009	39.984	0.009	40.000	39.984	0.000
50	MAX	50.290	50.000	0.450	50.142	50.000	0.204	50.064	50.000	0.089	50.034	50.000	0.050	50.025	50.000	0.041
	MIN	50.130	49.840	0.130	50.080	49.938	0.080	50.025	49.975	0.025	50.009	49.984	0.009	50.000	49.984	0.000
60	MAX	60.330	60.000	0.510	60.174	60.000	0.248	60.076	60.000	0.106	60.040	60.000	0.059	60.030	60.000	0.049
	MIN	60.140	59.810	0.140	60.100	59.926	0.100	60.030	59.970	0.030	60.010	59.981	0.010	60.000	59.981	0.000
80	MAX	80.340	80.000	0.530	80.174	80.000	0.248	80.076	80.000	0.106	80.040	80.000	0.059	80.030	80.000	0.049
	MIN	80.150	79.810	0.150	80.100	79.926	0.100	80.030	79.970	0.030	80.010	79.981	0.010	80.000	79.981	0.000
100	MAX	100.390	100.000	0.610	100.207	100.000	0.294	100.090	100.000	0.125	100.047	100.000	0.069	100.035	100.000	0.057
	MIN	100.170	99.780	0.170	100.120	99.913	0.120	100.036	99.965	0.036	100.012	99.978	0.012	100.000	99.987	0.000

TABLE 25 (CONT'D) ■ *Preferred shaft basis metric fits (Values in millimeters).*

Quantity	Metric Unit	Symbol	Metric to Inch-Pound Unit	Inch-Pound to Metric Unit
Length	millimeter	mm	1 mm = 0.0394 in.	1 in. = 25.4 mm
	centimeter	cm	1 cm = 0.394 in.	1 ft. = 30.5 cm
	meter	m	1 m = 39.37 in. = 3.28 ft	1 yd. = 0.914 m = 914 mm
	kilometer	km	1 km = 0.62 mile	1 mile = 1.61 km
Area	square millimeter	mm²	1 mm² = 0.001 55 sq. in.	1 sq. in. = 6 452 mm²
	square centimeter	cm²	1 cm² = 0.155 sq. in.	1 sq. ft. = 0.093 m²
	square meter	m²	1 m² = 10.8 sq. ft.	1 sq. yd. = 0.836 m²
			= 1.2 sq. yd.	
Mass	milligram	mg	1 g = 0.035 oz.	1 oz. = 28.3 g
	gram	g	1 kg = 2.205 lb.	1 lb. = 0.454 kg
	kilogram	kg	1 tonne = 1.102 tons	1 ton = 907.2 kg
	tonne	t		= 0.907 tonnes
Volume	cubic centimeter	cm³	1 mm³ = 0.000 061 cu. in.	1 fl. oz. = 28.4 cm³
	cubic meter	m³	1 cm³ = 0.061 cu. in.	1 cu. in. = 16.387 cm³
	milliliter	m	1 m³ = 35.3 cu. ft.	1 cu. ft. = 0.028 m³
			= 1.308 cu. yd.	1 cu. yd. = 0.756 m³
			1 mℓ = 0.035 fl. oz.	
Capacity	liter	L	U.S. Measure	U.S. Measure
			1 pt. = 0.473 L	1 L = 2.113 pt.
			1 pt. = 0.946 L	= 1.057 qt.
			1 gal. = 3.785 L	= 0.264 gal.
			Imperial Measure	Imperial Measure
			1 pt. = 0.568 L	1 L = 1.76 pt.
			1 qt. = 1.137 L	= 0.88 qt.
			1 gal. = 4.546 L	= 0.22 gal.
Temperature	Celsius degree	°C	$°C = \frac{5}{9}(°F\text{-}32)$	$°F = \frac{9}{5} \times °C + 32$
Force	newton	N	1 N = 0.225 lb (f)	1 lb (f) = 4.45N
	kilonewton	kN	1 kN = 0.225 kip (f)	= 0.004 448 kN
			= 0.112 ton (f)	
Energy/Work	joule	J	1 J = 0.737 ft ∘ lb	1 ft ∘ lb = 1.355 J
	kilojoule	kJ	1 J = 0.948 Btu	1 Btu = 1.055 J
	megajoule	MJ	1 MJ = 0.278 kWh	1 kWh = 3.6 MJ
Power	kilowatt	kW	1 kW = 1.34 hp	1 hp (550 ft ∘ lb/s) = 0.746 kW
			1 W = 0.0226 ft ∘ lb/min.	1 ft ∘ lb/min = 44.2537 W
Pressure	kilopascal	kPa	1 kPa = 0.145 psi	1 psi = 6.895 kPa
			= 20.885 psf	1 lb-force/sq. ft. = 47.88 Pa
			= 0.01 ton-force	1 ton-force/sq. ft. = 95.76 kPa
			per sq. ft.	
	*kilogram per			
	square centimeter	kg/cm²	1 kg/cm² = 13.780 psi	
Torque	newton meter	N ∘ m	1 N ∘ m = 0.74 lb ∘ ft	1 lb ∘ ft = 1.36 N ∘ m
	*kilogram meter	kg/m	1 kg/m = 7.24 lb ∘ ft	1 lb ∘ ft = 0.14 kg/m
	*kilogram per centimeter	kg/cm	1 kg/cm = 0.86 lb ∘ in	1 lb ∘ in = 1.2 kg/cm
Speed/Velocity	meters per second	m/s	1 m/s = 3.28 ft/s	1 ft/s = 0.305 m/s
	kilometers per hour	km/h	1 km/h = 0.62 mph	1 mph = 1.61 km/h

*Not SI units, but included here because they are employed on some of the gages and indicators currently in use in industry.

TABLE 26 ∎ *Metric conversion tables.*

INDEX